Springer Proceedings in Mathematics & Statistics

Volume 250

Springer Proceedings in Mathematics & Statistics

This book series features volumes composed of selected contributions from workshops and conferences in all areas of current research in mathematics and statistics, including operation research and optimization. In addition to an overall evaluation of the interest, scientific quality, and timeliness of each proposal at the hands of the publisher, individual contributions are all refereed to the high quality standards of leading journals in the field. Thus, this series provides the research community with well-edited, authoritative reports on developments in the most exciting areas of mathematical and statistical research today.

More information about this series at http://www.springer.com/series/10533

Patrice Bertail • Delphine Blanke •
Pierre-André Cornillon • Eric Matzner-Løber
Editors

Nonparametric Statistics

3rd ISNPS, Avignon, France, June 2016

Springer

Editors
Patrice Bertail
MODAL'X
Paris West University Nanterre La Défense
Nanterre, France

Delphine Blanke
LMA
Avignon University
Avignon, France

Pierre-André Cornillon
MIASHS
University of Rennes 2
Rennes, France

Eric Matzner-Løber
Formation Continue CEPE
Ecole Nationale de la Statistique et de l'Administration
Malakoff, France

ISSN 2194-1009 ISSN 2194-1017 (electronic)
Springer Proceedings in Mathematics & Statistics
ISBN 978-3-319-96940-4 ISBN 978-3-319-96941-1 (eBook)
https://doi.org/10.1007/978-3-319-96941-1

Library of Congress Control Number: 2018964410

Mathematics Subject Classification (2010): 62-00, 62-06, 62G05, 62G15, 62G20, 62G32

This Springer imprint is published by the registered company Springer Nature Switzerland AG
The registered company address is: Gewerbestrasse 11, 6330 Cham, Switzerland

Contents

Contributors

S. Achard CNRS, University of Grenoble Alpes, GIPSA-Lab, Grenoble, France

P. Bertail Modal'X, UPL, Univ. Paris Nanterre, Nanterre, France

J. Bógalo Universidad de Alcalá, Madrid, Spain

T. Burr Los Alamos National Laboratory, Los Alamos, NM, USA

A. Carpentier IMST, Otto von Guericke University Magdeburg, Magdeburg, Germany

G. Ciołek LTCI, Télécom ParisTech, Université Paris-Saclay, Paris, France

G. Ciuperca Université Lyon 1, CNRS, UMR 5208, Institut Camille Jordan, Villeurbanne Cedex, France

P.-A. Cornillon Univ. Rennes, CNRS, IRMAR, UMR 6625, Rennes, France

A. Dobrovidov V.A. Trapeznikov Institute of Control Sciences, Russian Academy of Sciences, Moscow, Russia

A.-K. Fermin Modal'X, Université Paris Nanterre, Nanterre, France

I. Gannaz Univ Lyon, INSA de Lyon, CNRS UMR 5208, Institut Camille Jordan, Villeurbanne, France

S. Ghosh Statistics Lab, Swiss Federal Research Institute WSL, Birmensdorf, Switzerland

A. Gribinski Department of Mathematics, Princeton University, Princeton, NJ, USA

N. Hengartner Los Alamos National Laboratory, Los Alamos, NM, USA

O. Jelassi Telecom ParisTech, Paris, France

V. Jothiprakash Indian Institute of Technology, Bombay, India

J. Kalina Institute of Computer Science of the Czech Academy of Sciences, Prague, Czech Republic

National Institute of Mental Health, Klecany, Czech Republic

T. Kerdreux D.I. UMR 8548, Ecole Normale Supérieure, Paris, France

O. Klopp IDS, ESSEC Business School, Cergy, France

K. Knight University of Toronto, Toronto, ON, Canada

E. Le Pennec CMAP/XPOP, École Polytechnique, Palaiseau, France

S. Lee Department of Statistics, Seoul National University, Seoul, South Korea

M. Löffler StatsLab, University of Cambridge, Cambridge, UK

C. Ludeña Universidad Jorge Tadeo Lozano, Dpto. de Ciencias Básicas y Modelado, Carrera, Bogotá, Colombia

Matúš Maciak Department of Probability and Mathematical Statistics, Charles University, Prague, Czech Republic

L. Markovich Moscow Institute of Physics and Technology, Dolgoprudny, Moscow Region, Russia

Institute for Information Transmission Problems, Moscow, Russia

V. A. Trapeznikov Institute of Control Sciences, Moscow, Russia

N. Markovich V.A. Trapeznikov Institute of Control Sciences of Russian Academy of Sciences, Moscow, Russia

E. Matzner-Løber CREST UMR 9194, Cepe-Ensae, France

S. G. Meintanis National and Kapodistrian University of Athens, Athens, Greece

Unit for Business Mathematics and Informatics, North-West University, Potchefstroom, South Africa

L. Montuelle RTE, La Défense, France

F. Navarro CREST-ENSAI-UBL, Bruz, France

S. Y. Novak Middlesex University, London, UK

Michal Pešta Faculty of Mathematics and Physics, Department of Probability and Mathematical Statistics, Charles University, Prague, Czech Republic

Barbora Peštová Department of Medical Informatics and Biostatistics, Institute of Computer Science, The Czech Academy of Sciences, Prague, Czech Republic

D. N. Politis Department of Mathematics, University of California, San Diego, La Jolla, CA, USA
P. Poncela European Comission, Joint Research Centre (JRC), Ispra, Italy

Universidad Autónoma de Madrid, Madrid, Spain

C. Pretorius Unit for Business Mathematics and Informatics, North-West University, Potchefstroom, South Africa

A. Rebecq Modal'X, UPL, University Paris Nanterre, France

L. Rouvière Univ. Rennes, IRMAR - UMR 6625, Rennes, France

A. Saumard CREST-ENSAI-UBL, Bruz, France

E. Senra Universidad de Alcalá, Madrid, Spain

C. Tillier University of Hamburg, SPST, Hamburg, Germany

J. Tressou MORSE, INRA-MIA, Paris, France

P. Unnikrishnan Indian Institute of Technology, Bombay, India

M. Vaičiulis Institute of Mathematics and Informatics, Vilnius University, Vilnius, Lithuania

V. A. Vasiliev Department of Applied Mathematics and Cybernetics, Tomsk State University, Lenin, Tomsk, Russia

V. Vasilyev Moscow Institute of Physics and Technology (State University), Moscow, Russia

S. E. Vorobeychikov Department of Applied Mathematics and Cybernetics, Tomsk State University, Lenin, Tomsk, Russia

M. Zetlaoui Modal'X, UPL, Univ. Paris Nanterre, Nanterre, France

D. N. Politis Department of Mathematics, University of California, San Diego, La Jolla, CA, USA

Symmetrizing k-nn and Mutual k-nn Smoothers

P.-A. Cornillon, A. Gribinski, N. Hengartner, T. Kerdreux, and E. Matzner-Løber

Abstract In light of Cohen (Ann Math Stat 37:458–463, 1966) and Rao (Ann Stat 4:1023–1037, 1976), who provide necessary and sufficient conditions for admissibility of linear smoothers, one realizes that many of the well-known linear nonparametric regression smoothers are inadmissible because either the smoothing matrix is asymmetric or the spectrum of the smoothing matrix lies outside the unit interval $[0, 1]$. The question answered in this chapter is how can an inadmissible smoother transformed into an admissible one? Specifically, this contribution investigates the spectrum of various matrix symmetrization schemes for k-nearest neighbor-type smoothers. This is not an easy task, as the spectrum of many traditional symmetrization schemes fails to lie in the unit interval. The contribution of this study is to present a symmetrization scheme for smoothing matrices that make the associated estimator admissible. For k-nearest neighbor smoothers, the result of the transformation has a natural interpretation in terms of graph theory.

P.-A. Cornillon
University of Rennes, IRMAR UMR 6625, Rennes, France
e-mail: pac@univ-rennes2.fr

A. Gribinski
Department of Mathematics, Princeton University, Princeton, NJ, USA
e-mail: aurelien.gribinski@princeton.edu

N. Hengartner
Los Alamos National Laboratory, Los Alamos, NM, USA
e-mail: nickh@lanl.gov

T. Kerdreux
UMR 8548, Ecole Normale Supérieure, Paris, France
e-mail: thomas.kerdreux@inria.fr

E. Matzner-Løber (✉)
CREST, UMR 9194, Cepe-Ensae, Palaiseau, France
e-mail: eml@ensae.fr

P. Bertail et al. (eds.), *Nonparametric Statistics*, Springer Proceedings in Mathematics & Statistics 250, https://doi.org/10.1007/978-3-319-96941-1_1

1 Introduction

1.1 The Statistical Background

Consider the nonparametric regression model

$$Y_i = m(X_i) + \varepsilon_i \tag{1}$$

that relates the response $Y_i \in \mathbb{R}$ to predictors $X_i \in \mathbb{R}^d$ through the regression function $m(x) = \mathbb{E}(Y|X = x)$. The disturbances ε_i are mean zero and constant finite variance σ^2 random variables that are independent of the explanatory variables $X_1, \ldots, X_n$. The vector of predicted values $\widehat{Y} = (\widehat{Y}_1, \ldots, \widehat{Y}_n)^\intercal$ for $m = (m(X_1), \ldots, m(X_n))^\intercal$ is called a regression smoother, or simply a *smoother* as these are less variable than the original observations. Linear smoothers, which are linear in the response variable, have been extensively studied in the literature. See [9, 10] for recent expositions on common linear smoothers. These smoothers can be written as:

$$\widehat{m} = S_\lambda Y, \tag{2}$$

where S_λ is the $n \times n$ *smoothing matrix*. That matrix depends on the observed response variables, and typically on a tuning parameter, which we will denote by λ, that governs the trade-off between the variance and the bias of the smoother. For simplicity reason from now on, we will write the smoothing matrix S.

Classical smoothers include smoothing splines where S is symmetric and positive define where λ is the coefficient associated to the penalty term, kernel smoothers which could be written as $S = D^{-1}\mathbb{K}$ where λ is the bandwidth. Usually, $\mathbb{K}$ is symmetric but not always positive definite (see, for example, [6]) and D is diagonal with D_{ii} equal to the row sum of $\mathbb{K}$ so S is row-stochastic. A similar representation holds for k-nearest neighbor (k-nn)-type smoother such as the classical k-nn and the mutual k-nn.

1.2 k-nn and Mutual k-nn Smoothers

K-nn smoother [11] estimates the regression function $m(X_i)$ by averaging the responses Y_j associated to the k nearest observations X_j to X_i.

Let $(X_1, Y_1), \ldots, (X_n, Y_n)$ be independent identically distributed copies of $(X, Y) \in \mathbb{R}^d \times \mathbb{R}$. Given $x \in \mathbb{R}^d$, and a distance $dist()$ (the choice of the distance is beyond the scope of this chapter), reorder the data in a manner such that the distances $d_i(x) = dist(X_{(1,i)}(x), x)$ are nondecreasing and denote the reordering as:

$$(X_{(1,n)}(x), Y_{(1,n)}(x)), \cdots, (X_{(n,n)}(x), Y_{(n,n)}(x)).$$

The k-nearest neighbor (k-nn) smoother is defined as:

$$\hat{m}_n^{Knn}(x) = \frac{1}{k}\sum_{i=1}^{k} Y_{(i,n)}(x). \tag{3}$$

Conditional on the observed covariates $X_1, \ldots, X_n$, the k-nn smoother is linear in vector of responses $Y = (Y_1, \ldots, Y_n)$. Let us denote $\mathcal{N}_k(X_i)$ the set of the k nearest neighbors of X_i in the design points $\{X_1, \ldots, X_n\}$, the (i, j)th entry of the smoothing matrix S_{knn} is

$$(S_{knn})_{ij} = \begin{cases} \frac{1}{k} \text{ if } X_j \in \mathcal{N}_k(X_i) \\ 0 \text{ otherwise} \end{cases}.$$

We refer to [1, 7, 15] for in-depth treatment of consistency and statistical properties of k-nn smoothers in the context of regression or classification.

It is instructive to interpret the matrix $A = kS_{knn}$ as the adjacency matrix of a directed graph on $\{X_1, \ldots, X_n\}$, in which a directed edge from X_i to X_j exists if X_j belongs to $\mathcal{N}_k(X_i)$. While $\sum_{j=1}^{n} A_{ij} = k$ for all i, $\sum_{i=1}^{n} A_{ij}$ depends on the configuration of the covariates $X_1, \ldots, X_n$ and counts the number of k-nn neighborhoods that X_j belongs too. When that number is larger than k, we say that X_j is a hub. While generally speaking, there are always hubs, the emergence of highly connected hub in higher dimensions is associated to the curse of dimensionality (see [24]).

Mutual k-nearest neighbors (mk-nn) have been introduced by Gowda and Krishna [13] in an attempt to build a symmetric adjacency matrix. X_i and X_j are mutual k-nearest neighbors if both X_j belongs to the k-nn of X_i **and** X_i belongs to the k-nn of X_j. The adjacency matrix A^m, defined as:

$$A_{ij}^m = \min(A_{ij}, A_{ji}) = A_{ij}A_{ji}.$$

The number of mutual k-nn of each covariate X_i, $K_i = \sum_{j=1}^{n} A_{ij}^m$ is a random variable bounded from above by k. In principle, it is possible that $K_i = 0$. Guyader and Hengartner [14] provide conditions on the distribution of the covariates to ensure that as the sample size n and the size of the neighborhood k both tend to infinity, we have that $K_i = O(k)$. Define the set $\mathcal{M}_k(x) = \{X_i \in \mathcal{N}_k(x), x \in \mathcal{N}_k(X_i)\}$ the closed form is

$$\hat{m}_n^{mKnn}(x) = \frac{1}{|\mathcal{M}_k(x)|}\sum_{i, X_i \in \mathcal{M}_k(x)} Y_i. \tag{4}$$

The (i, j)th entry of the smoothing matrix S_{mknn} is

$$(S_{mknn})_{ij} = \begin{cases} \frac{1}{K_i} \text{ if } X_j \in \mathcal{M}_k(X_i) \text{ and } K_i > 0 \\ 0 \text{ otherwise} \end{cases},$$

or simply $S_{mknn} = D^{-1}A_m$ where D is a diagonal matrix with entry K_i so that the smoother is row-stochastic. The matrix A_m is symmetric and could be viewed as an adjacency matrix of an undirected graph.

Many papers were devoted to k-nearest neighbor graph see, for example, [8, 21] among others or mutual k-nearest neighbor graph see, for example, [2].

Going back to Eq. (2), we could mostly write the smoothing matrix as a matrix product $S = D^{-1}A$. Generally, the smoothing matrix is row-stochastic (so the eigen values are in $[-1, 1]$) but not always symmetric and as pointed out by Cohen [5] this leads to non-admissible estimator.

1.3 Admissibility

In this contribution, we are concerned with the mean-squared error admissibility of k-nn smoothers within the class of all linear smoothers. Recall that a linear smoother $\widetilde{m}$ is *inadmissible* in mean-square error if there exists another linear smoother $m^\star$ such that

$$\mathbb{E}[\|m^\star - m\|^2] \leq \mathbb{E}[\|\widetilde{m} - m\|^2],$$

for all regression functions m, and with strict inequality for at least one regression function m. That is, there exists a better smoother in terms of mean-squared error, and thus being inadmissible is an undesirable property. If a smoother is not inadmissible, it is admissible. Unless stated otherwise in this work, we will overload the term admissible to mean admissible in the class of all linear smoothers.

Rao [25] showed that the following two conditions were sufficient and necessary for a linear smoother to be admissible within the class of all linear smoothers: (1) The smoothing matrix S_λ is symmetric and (2) the spectrum of S_λ lies in the unit interval $[0, 1]$. We note that Rao's was preceded by Cohen [5] who showed that admissible linear smoothers in the class of all smoothers need to have symmetric smoothing matrix and spectrum in the unit interval $[0, 1]$, with at most two eigenvalues being equal to one.

From that characterization, smoothing splines of any order are admissible, whereas all local polynomial smoothers, see [10] including the Nadaraya-Watson smoother [23], and k-nn-type smoothers are inadmissible because they have asymmetric smoothing matrices.

We may point out here that admissibility does not affect minimax rate of convergence, for example. In general, the difference between an estimator inadmissible with optimal rate and it admissible pendant one is in the analysis of the constant of the second-order term development. Such developments are beyond the scope of this chapter.

1.4 Symmetrization

The proof in [5] is particularly interesting, as it shows constructively how, starting with a smoother with asymmetric smoothing matrix, one can symmetrize the smoothing matrix to produce a smoother with smaller mean-square error. The resulting smoother has a closed form:

$$S_{cohen} = I - [(I - S)^{\mathsf{T}}(I - S)]^{1/2}. \tag{5}$$

Other symmetrization schemes have been proposed by modifying Cohen's estimator [28]:

$$S_{zhao} = I - \rho[(I - S)^{\mathsf{T}}(I - S)]^{1/2} \tag{6}$$

or [19] (who are averaging the smoothing matrix and its transpose) in the context of nonparametric regression.

Symmetrization is of real importance in image analysis and in particular in image denoising. Milanfar [22] advocates for symmetrization which is not only a "mathematical nicety but can have interesting practical advantages." Among them are:

1. Performance improvement,
2. Stability of iterative filtering,
3. Eigen decomposition.

Milanfar [22] studied smoothers of the form $S = D^{-1}\mathbb{K}$ where $\mathbb{K}$ is symmetric and positive define and D is a diagonal matrix that makes S row-stochastic. Applying the Sinkhorn algorithm (see [26]), he constructs a doubly stochastic estimator $\hat{S}$ and controls the behavior of $\hat{S}$ because the change of the eigenvalues due to Sinkhorn normalization is upper bounded by the Frobenius norm $\|S - \hat{S}\|$.

More recently, Haque et al. [16] again starting with a smoother of the form $S = D^{-1}\mathbb{K}$ and working with the Laplacian $L = D - \mathbb{K}$ proposed the following estimator $C = (I + \lambda L'L)^{-1}$ where the value of λ is chosen by optimization. This estimator could be applied to any type of smoother, it does not need a symmetric $\mathbb{K}$ or positive eigen values for S but the interpretation in terms of S is almost impossible to understand. Moreover, it is impossible to apply that smoother at a new observation (the same will be true for [28]). Furthermore, the eigen values of C are bounded away from zero and the resulting variance is much bigger than the initial smoother.

More recently, [3, 4] interpreted Sinkhorn algorithm (in order to obtain a doubly stochastic smoother) as an expectation-maximization algorithm learning a Gaussian mixture model of the image patches.

Stability of the iterative filtering as pointed out by Milanfar [22] was already advocated by Cornillon et al. [6] in the context of L_2 boosting. Friedman et al. [12] showed that L_2 boosted smoother at iteration j is given by:

$$\tilde{m}_j = [I - (I - \nu S)^j]Y.$$

The $0 < \nu \leq 1$ could be seen as the step factor since boosting could be viewed as a functional gradient descent algorithm. For simplicity, let us consider the case $\nu = 1$. In the context of kernel boosting, Cornillon et al. [6] proposed the following:

$$\begin{aligned}\tilde{m}_j &= [I - (D^{-1/2}D^{1/2} - D^{-1/2}D^{-1/2}SD^{-1/2}D^{1/2})^j]Y\\ &= [I - D^{-1/2}(I - D^{-1/2}SD^{-1/2})^j D^{1/2}]Y.\end{aligned}$$

While symmetrization of the smoothing matrix is necessary, it is not sufficient for a linear smoother to be admissible. Specifically, the spectrum of the symmetrized smoothing matrix needs to belong to the unit interval. As indicated above, most of the literature dedicated to symmetrization of smoother assume that S is of the form $D^{-1}\mathbb{K}$ where $\mathbb{K}$ is symmetric and positive definite which is not true for k-nn-based smoothers.

In the next section, we prove mainly negative results for row-stochastic matrices (so the results are directly applicable to k-nn-type smoothers). We show that the arithmetic and the geometric averages of a row-stochastic matrix and its transpose, and even the symmetrization scheme proposed by Cohen, have eigenvalues outside the unit interval. In Sect. 3, we propose an alternative approach to symmetrize k-nn-type smoother that results in a smoothing matrix whose spectrum lies in the unit interval. The new estimators can be evaluated at any arbitrary points and have their own interpretation.

2 Symmetrization Procedures for Row-Stochastic Smoothers

In this section, we relate the spectrum of various symmetrized smoothing matrices to the spectrum of their original smoothing matrix. We assume that the smoothing matrix S is row-stochastic, that is, all of its elements are nonzero and $S\mathbf{1} = \mathbf{1}$, where $\mathbf{1} = (1, 1, \ldots, 1)^\top \in \mathbb{R}^n$ is the vector of ones.

2.1 *Geometric and Arithmetic Mean*

Geometric Mean Given a smoothing matrix S, define the symmetric matrix $\widetilde{S} = (S^\top S)^{1/2}$. The square-root is well defined as $S^\top S$ is symmetric and positive definite. The variance of the resulting smoother is the same as the variance of the initial smoother, but the biases are different. No comparison can be made between the biases of the original smoother and the symmetrized one. While $\widetilde{S}$ is symmetric and nonnegative definite, it is possible for the largest eigenvalue λ_{max} to be strictly larger than one. In those cases, the symmetrized smoother $\tilde{S}$ remains inadmissible.

Lemma 1 *Let S be a row-stochastic matrix. Then, $\lambda_{max}(S^\top S) \geq 1$, with equality if and only if S is a doubly stochastic matrix.*

Proof Consider the Rayleigh quotient $Q(x) = x^\top S^\top Sx/(x^\top x)$ and denote by $\mathbf{1}_n = (\frac{1}{\sqrt{n}}, \frac{1}{\sqrt{n}}, \ldots, \frac{1}{\sqrt{n}})$ the vector of length one. Note that $Q(\mathbf{1}_n) = \mathbf{1}_n^\top S^\top S\mathbf{1}_n = (S\mathbf{1}_n)^\top(S\mathbf{1}_n) = 1$ since S is row-stochastic. Hence, $\lambda_{max}(S^\top S) \geq 1$. Furthermore, if $\lambda_{max}(S^\top S) = 1$, then $\mathbf{1}_n$ belongs to the eigenspace associated to the largest eigenvalue of $S^\top S$. This implies that $S^\top S\mathbf{1}_n = \mathbf{1}_n$. Since S is row-stochastic, we also have that $S^\top S\mathbf{1}_n = S^\top\mathbf{1}_n$. Combining these identities, we conclude that $\lambda_{max}(S^\top S) = 1$ is and only if $S^\top\mathbf{1}_n = \mathbf{1}_n$, which is equivalent to S being doubly stochastic (since S is assumed to be row-stochastic).

The conclusion of the lemma also holds for $SS^\top$ which has the same spectrum as $S^\top S$. It follows from the above lemma that the geometric mean smoother $\tilde{S} = (S^\top S)^{1/2}$ is inadmissible whenever S is not a doubly stochastic matrix.

Arithmetic Mean Given a smoothing matrix S, define the symmetric smoothing matrix $\widetilde{S} = (S + S^\top)/2$ considered by Linton and Jacho-Chavez [19]. For the arithmetic mean smoother $\tilde{S}$, we can show

$$V(SY) - V(\widetilde{S}Y) = \sigma^2\left(\text{trace}(S^\top S) - \frac{1}{2}\text{trace}(S^2) - \frac{1}{2}\text{trace}(S^\top S)\right)$$

$$= \frac{\sigma^2}{2}\left(\text{trace}(S^\top S) - \text{trace}(S^2)\right) \geq 0.$$

The last inequality is justified by Lemma 6 in the Appendix. This shows that the arithmetic average smoother has a smaller variance than the original smoother, a result that first was proven for kernel smoothers by Linton and Jacho-Chavez [19]. As for the geometric mean smoother, nothing can be said of the biases.

Even though the variance of the average smoother is smaller than that of the original smoother, the following theorem proves that the largest eigenvalue of $\widetilde{S}$ is larger than one, unless S is doubly stochastic. As a result, the average smoother is not admissible.

Lemma 2 *Let S be a row-stochastic matrix, then $\lambda_{max}((S^\top + S)/2) \geq 1$, with equality if and only if S is a doubly stochastic matrix.*

Proof Consider the Rayleigh quotient $Q(x) = x^\top(S^\top + S)x/(2(x^\top x))$ and denote by $\mathbf{1}_n = (\frac{1}{\sqrt{n}}, \frac{1}{\sqrt{n}}, \ldots, \frac{1}{\sqrt{n}})$ the vector of length one. It is easy to verify that $Q(\mathbf{1}_n) = 1$ since S is row-stochastic. Hence, $\lambda_{max}((S^\top + S)/2) \geq 1$. If $\lambda_{max}((S^\top + S)/2) = 1$, then $\mathbf{1}_n$ is in the eigenspace associated to the largest eigenvalue, and

$$S^\top\mathbf{1}_n + S\mathbf{1}_n = 2\mathbf{1}_n.$$

Since S is row-stochastic, we have that $S\mathbf{1}_n = \mathbf{1}_n$, and thus $S^\top\mathbf{1}_n = \mathbf{1}_n$, which occurs if and only if S is doubly stochastic (since S is assumed to be row-stochastic).

2.2 Cohen's and Zhao's Symmetrization

Given a smoothing matrix S, Cohen [5] proposed the symmetrized smoothing matrix which has the same bias and smaller variance than the original smoother S. That is, S_{cohen} dominates S. While this shows that S is not admissible, it does not imply that S_{cohen} is itself admissible. The following Lemma gives conditions for S_{cohen} to have an eigenvalue larger than one.

Lemma 3 *If the smoother S admits negative eigenvalues, then Cohen symmetrized smoother S_{cohen} has an eigenvalue larger than one and hence is inadmissible.*

Proof of Lemma 3 Consider the right eigenvector x associated to the eigenvalue $\lambda < 0$. With that vector x, develop the quadratic form:

$$
\begin{aligned}
x^\top (I-S)^\top (I-S)x &= x^\top x - x^\top S^\top x - x^\top S x + x^\top S^\top S x \\
&= x^\top x - (Sx)^\top x - x^\top (Sx) + (Sx)^\top (Sx) \\
&= x^\top x - 2\lambda x^\top x + \lambda^2 x^\top x \\
&= (1-\lambda)^2 x^\top x \\
&> x^\top x.
\end{aligned}
$$

This shows that $\lambda_{max}((I-S)^\top (I-S)) > 1$, which completes the proof.

Zhao [28] recognizes, without proof, that the Cohen estimator may have eigenvalues larger than 1 and proposes a stepped version of the Cohen smoother. When $\rho^{-2} > \lambda_{max}((I-S)^\top (I-S))$, the resulting smoother has spectrum in $[0, 1)$, and hence is admissible. But, the construction of this smoother requires the knowledge of the largest eigenvalue of $(I-S)^\top (I-S)$, or at least an upper bound for that eigenvalue. Also, this smoother cannot be extended out of the initial design.

In order to close that section, we note that Sinkhorn algorithm was derived to obtain from a positive matrix a doubly stochastic one. Here, the authors need a doubly stochastic matrix, positive definite and symmetric. In order to reach their goal, they need to start with a positive symmetric definite matrix [18] which is not the case for k-nn-type smoother as we will see in the next section.

3 Symmetrization of *k*-nn-Type Smoothers

Let us recall that $S_{knn} = k^{-1}A$ and $S_{mknn} = D^{-1}A_m$ where A (resp., A_m) is the adjacency matrix associated to the direct graph (resp. of the indirect graph), and A_m is symmetric. We first state some results concerning the eigenvalues of A and A_m.

Lemma 4 *Let S be the smoothing matrix of the k-nearest neighbor smoother with $k \geq 3$. Denote for each i, the set of indices $\mathcal{N}_i$ of k-nearest neighbors of i. Assume that the set of covariates $\{X_1, \dots, X_n\}$ contain three points E, F, G such that*

$$E \in \mathcal{N}_F \quad \text{and} \quad F \in \mathcal{N}_E \qquad F \in \mathcal{N}_G \quad \text{and} \quad G \in \mathcal{N}_F \qquad E \notin \mathcal{N}_G \quad \text{or} \quad G \notin \mathcal{N}_E.$$

Then, at least one eigen value of $(I - S)'(I - S)$ is bigger than 1.

And, similarly for the mutual k-nn smoother:

Lemma 5 *If the graph related to the adjacency matrix A_m admits a path of length bigger than one, then the mutual k-nn initial smoother has a negative eigenvalue.*

The proofs are given in the Appendix.

Remark 1 If the graph is not connected, the conclusions of Lemma 5 remain true, provided that for at least one of the connected components, the conditions of the lemma hold. Brito et al. [2] showed under technical condition that with probability one, when k is of order of $\log(n)$ the graph is almost surely connected.

3.1 Construction of the Symmetrized Estimator

The previous section demonstrates that most attempts to symmetrize k-nn or mutual k-nn smoothers do not result in an admissible smoother because the considered symmetrizations do not control the spectrum. In this section, we propose a construction specialized to k-nn smoothers and mutual k-nn. These two smoothers could be written as $S = D^{-1}A$ where A is the adjacency matrix and D is a diagonal matrix. We propose the following symmetrization of k-nn-type smoothers:

$$S_{new} = W^{-1/2} A A^{\top} W^{-1/2}, \tag{7}$$

where W is the diagonal matrix of the row sum of $AA^{\top}$. It is easy to see why the spectrum of this symmetric smoother lies in the unit interval: The matrix S_{new} is similar to the row-stochastic matrix $W^{-1}AA^{\top}$, which has eigenvalue in $[-1, 1]$, and is positive definite which implies that all the eigenvalues are nonnegative. Hence, all the eigenvalues are in $[0, 1]$.

Another strategy could have been to propose $V^{-1/2}A^{\top}AV^{-1/2}$. These new estimators (for k-nn and k-mnn) have the advantage of producing an admissible estimator that can be evaluated at any point. Furthermore, the resulting smoother has a compelling interpretation in terms of neighbors. However, such an interpretation is not new. Dealing with $AA^{\top}$ was named *bibliographic coupling* by Kessler [17] and $A^{\top}A$ was named *co-citation* by Small [27].

3.2 Interpretation, Estimation at Any Point x

Our new estimator (7) can be interpreted as a k-nn-type estimator. The general term of the adjacency matrix A_{ij} says if X_j belongs to the k-nn (or mutual) of X_i. The general term (i, j) of $AA^\top$ is the scalar product of lines i and j of A so it counts the number of points in common from the k-nn (or mutual) of X_i and X_j. Let us write $(AA^\top) = n_{ij}$ (for neighbor) and $(A^m A^{m\top}) = m_{ij}$.

In the case $AA^\top$, n_{ij} is the number of points in the intersection of $\mathcal{N}_k(X_i)$ and $\mathcal{N}_k(X_j)$ and on the diagonal obviously there is k.

$$AA^\top = \begin{pmatrix} k \cdots\cdots n_{1,j} \cdots\cdots \\ \ddots \cdots \cdots \cdots\cdots \\ k \;\; n_{i,j} \cdots\cdots \\ \ddots \cdots\cdots \\ \ddots \cdots \\ n_{n,j} \cdots k \end{pmatrix} \quad A_m A_m{}^\top = \begin{pmatrix} K_1 \cdots\cdots m_{1,j} \cdots\cdots \\ \ddots \cdots \cdots \cdots\cdots \\ K_i \;\; m_{i,j} \cdots\cdots \\ \ddots \cdots\cdots \\ \ddots \cdots \\ m_{n,j} \cdots K_n \end{pmatrix}$$

Obviously, this quantity n_{ij} (respectively, m_{ij}) is large when X_i has a lot of common k-nn (or mutual) with X_j but will always be smaller than k. We further note that the quantity m_{ij} arises as a similarity measure in graph theory, see [20].

The quantity $W_{ii}^{1/2}$ is equal $(\sum_{l=1}^n n_{il})^{-1/2}$ (or $(\sum_{l=1}^n m_{il})^{-1/2}$), that is, one divided by the square-root of the sum of the numbers of common k-nn (or mutual) of X_i and the sample. This transformation looks like a weighted k-nearest neighbor, though each weight depends on the considered point of the design.

The transformed k-nn smoother can be evaluated at the design point X_i as:

$$\hat{m}(X_l) = \sum_{i=1}^n \frac{1}{\sqrt{\sum_j n_{lj}}} n_{li} \frac{Y_i}{\sqrt{\sum_j n_{ij}}} = \sum_{i=1}^n W_{ni}(X_l) Y_i.$$

The quantities n_{li} are of course dependent on k. That expression can be extended to be evaluated at arbitrary points x as follows:

$$\hat{m}(x) = \sum_{i=1}^n \frac{1}{\sqrt{\sum_j n_{xj}}} n_{xi} \frac{Y_i}{\sqrt{\sum_j n_{ij}}} = \sum_{i=1}^n W_{ni}(x) Y_i.$$

where n_{xj} is the number of points in common from the k-nn of x and X_j. The new estimator is a weighted nearest neighbor estimator with random weights. The proof of the consistency (which is not immediate) is beyond the scope of this contribution. Being able to predict gives us the possibility to estimate for which k the strategy should be carried on (using data-splitting, for instance).

4 Conclusion

In summary, this chapter makes several contributions to the theory of k-nearest neighbor-type smoothers (mutual and symmetric). First, we show that the symmetrization strategies proposed by Cohen [5] and Linton and Jacho-Chavez [19] can produce smoothing matrices whose spectrum lies outside the unit interval for general row-stochastic smoothers.

Second, we show that the spectrum of k-nearest neighbor smoothers has negative eigenvalues.

Third, we propose an alternative construction of a symmetric smoothing matrix whose eigenvalues are provably in the unit interval. Applying that construction to k-nearest neighbor smoothers results in a novel k-nn smoother. This estimator could be applied by itself as a weighted k-nn one by selecting the parameter k or, by extension, it could be used in L_2 boosting procedures.

Appendix

Lemma 6 *Let A be an $n \times n$ matrix. Then,*

$$trace(A^2) \leq trace(A^\top A).$$

Proof Given A, the matrix $(A - A^\top)^\top(A - A^\top)$ is positive definite. Thus,

$$\begin{aligned} 0 &\leq \text{trace}((A - A^\top)^\top(A - A^\top)) = \text{trace}((A^\top - A)(A - A^\top)) \\ &= \text{trace}(A^\top A - A^\top A^\top + AA^\top - AA) = 2\,\text{trace}(A^\top A) - 2\,\text{trace}(A^2). \end{aligned}$$

The conclusion follows.

Proof of Lemma 5 Recall that $S_{mknn} = DA^m$, where D is a diagonal matrix with nonzero diagonal. The matrices S_{mknn} and $D^{1/2}A^mD^{1/2}$ have the same spectrum. We need only to show that the matrix A^m has a negative eigenvalue. Having a path of length two, then there exist two vertices i_1 and i_2 such that the shortest path between these two vertices is two. As a result, there exists a 3×3 sub-matrix of the form:

$$B = \begin{pmatrix} 1 & 1 & 0 \\ 1 & 1 & 1 \\ 0 & 1 & 1 \end{pmatrix}$$

The smallest eigenvalue of that matrix is negative and thus, by the interweaving theorem

$$\lambda_{min}(A^m) \leq \lambda_{min}(B) < 0.$$

As a result, there exists a vector $u \in \mathbb{R}^n$, $\|u\|^2 = 1$, such that the Rayleigh quotient $u^\top A u < 0$. If we set $w = D^{-1/2}u$, we have $w^\top D^{1/2} A D^{1/2} w < 0$. This strict inequality remains for the normalized vector $w^* = \frac{w}{\|w\|}$. Thus, the symmetric matrix $D^{1/2} A D^{1/2}$ has also a strictly negative eigenvalue. Finally, since $D^{1/2} A D^{1/2}$ and $S = DA$ have the same spectrum, the conclusion of the lemma follows.

Proof of Lemma 4 Let us consider the k-nn smoother the matrix S is of general term

$$S_{ij} = \frac{1}{k} \quad \text{if} \quad X_j \in \mathcal{N}_k(X_i).$$

Consider the eigen values of $(I-S)(I-S)'$, since $A = (I-S)(I-S)'$ is symmetric, we have for any vector u that

$$\lambda_n \leq \frac{u' A u}{u' u} \leq \lambda_1. \tag{8}$$

Let us find a vector u such that $u' A u > u' u$. First notice that $A = I - S - S' + SS'$. Thus, we have that

$$A_{ii} = 1 - \frac{1}{k}.$$

Second, to bound A_{ij}, we need to consider three cases:

1. If $X_i \in \mathcal{N}_k(X_j)$ and $X_j \in \mathcal{N}_k(X_i)$, then $S_{ij} = S_{ji} = 1/k$. This does not mean that all the k-nn neighbors of X_i are the same as those of X_j, but if it is the case, then $(SS')_{ij} \leq k/k^2$ and otherwise in the pessimistic case, we bound $(SS')_{ij} \geq 2/k^2$. It therefore follows that

$$2/k^2 - \frac{2}{k} \leq A_{i,j} \leq \frac{k}{k^2} - \frac{2}{k} = -\frac{1}{k}.$$

2. If $X_i \in \mathcal{N}_k(X_j)$ and $X_j \notin \mathcal{N}_k(X_i)$, then $S_{ij} = 1/k$ and $S_{ji} = 0$. There is at a maximum of $k-1$ points that are in the k-nn of X_i and in the k-nn of X_j so $(SS')_{ij} \leq (k-1)/k^2$. In the pessimistic case, there is only one point, which leads to the bound

$$\frac{1}{k^2} - \frac{1}{k} \leq A_{i,j} \leq \frac{k-1}{k^2} - \frac{1}{k} \leq -\frac{1}{k^2}.$$

3. If $X_i \notin \mathcal{N}_k(X_j)$ and $X_j \notin \mathcal{N}_k(X_i)$, then $S_{ij} = 0$ and $S_{ji} = 0$. However, there are potentially as many as $k-2$ points that are in the k-nn of X_i and in the k-nn of X_j. In that case

$$0 \leq A_{ij} \leq \frac{k-2}{k^2}.$$

Choose three points E, F, and G in the sample X such that

$$E \in \mathcal{N}_k(F) \quad \text{and} \quad F \in \mathcal{N}_k(E)$$
$$F \in \mathcal{N}_k(G) \quad \text{and} \quad G \in \mathcal{N}_k(F)$$
$$E \notin \mathcal{N}_k(G) \quad \text{or} \quad G \notin \mathcal{N}_k(E).$$

Next, consider the vector u of $\mathbb{R}^n$ that is zero everywhere except at position e corresponding at point E (respectively, f and g) where its value is -1 (respectively, 2 and -1). For this choice, we expand $u'Au$ to get

$$\begin{aligned} u'Au &= A_{e,e} + 4A_{f,f} + A_{g,g} - 4A_{e,f} - 4A_{f,g} + 2A_{e,g} \\ &= 6 - \frac{6}{k} - 4A_{e,f} - 4A_{f,g} + 2A_{e,g}. \end{aligned}$$

With the choice of E, F, and G, we have

$$u'Au \geq 6 + \frac{2}{k} + 2A_{e,g}.$$

The latter shows that $u'Au > u'u$ whenever

$$A_{e,g} > -\frac{1}{k},$$

which is always true with the choice of points E and G.

References

1. Biau, G., & Devroye, L. (2015). *Lectures on the nearest neighbor method*. Cham: Springer.
2. Brito, M. R., Chavez, E. L., Quiroz, A. J., & Yukisk, J. E. (1997). Connectivity of the mutual k-nearest-neighbor graph in clustering and outlier detection. *Statistics and Probability Letters, 35*(1), 33–42.
3. Chan, S., Zickler, T., & Lu, Y. (2015). Understanding symmetric smoothing filters via Gaussian mixtures. In *2015 IEEE International Conference on Image Processing (ICIP), Quebec City, QC* (pp. 2500–2504).
4. Chan, S., Zickler, T., & Lu, Y. (2017). Understanding symmetric smoothing filters via Gaussian mixtures. *IEEE Transactions on Image Processing, 26*(11), 5107–5121.
5. Cohen, A. (1966). All admissible linear estimates of the mean vector. *Annals of Mathematical Statistics, 37*, 458–463.
6. Cornillon, P. A., Hengartner, N., & Matzner-Løber, E. (2013). Recursive bias estimation for multivariate regression smoothers. *ESAIM: Probability and Statistics, 18*, 483–502.
7. Devroye, L., Györfi, L., & Lugosi, G. (1996). *A probabilistic theory of pattern recognition*. New York: Springer.

8. Dong, W., Moses, C., & Li, K. (2011). Efficient k-nearest neighbor graph construction for generic similarity measures. In *Proceedings of the 20th International Conference on World Wide Web* (pp. 577–586).
9. Eubank, R. (1999). *Nonparametric regression and spline smoothing* (2nd ed.). New York: Dekker.
10. Fan, J., & Gijbels, I. (1996). *Local polynomial modeling and its application, theory and methodologies*. New York, NY: Chapman et Hall.
11. Fix, E., & Hodges, J. L. (1951). *Discriminatory Analysis, Nonparametric Discrimination: Consistency Properties*. Technical report, USAF School of Aviation Medicine, Randolph Field, TX.
12. Friedman, J., Hastie, T., & Tibshirani, R. (2000). Additive logistic regression: A statistical view of boosting. *The Annals of Statistics, 28*, 337–407.
13. Gowda, K., & Krishna, G. (1978). Agglomerative clustering using the concept of mutual nearest neighbourhood. *Pattern Recognition, 10*(2), 105–112.
14. Guyader, A., & Hengartner, N. (2013). On the mutual nearest neighbors estimate in regression. *The Journal of Machine Learning Research (JMLR), 13*, 2287–2302.
15. Gyorfi, L., Kohler, M., Krzyzak, A., & Walk, H. (2002). *A distribution-free theory of nonparametric regression*. New York: Springer.
16. Haque, S., Pai, G., & Govindu, V. (2014). Symmetric smoothing filters from global consistency constraints. *IEEE Transactions on Image Processing, 24*, 1536–1548.
17. Kessler, K. (1963). Bibliographic coupling between scientific papers. *American Documentation, 14*, 10–25.
18. Knight, P. (2008). The Sinkhorn-Knopp algorithm: Convergence and applications. *SIAM Journal on Matrix Analysis and Applications, 30*, 261–275.
19. Linton, O., & Jacho-Chavez, D. (2010). On internal corrected and symmetrized kernel estimators for nonparametric regression. *Test, 19*, 166–186.
20. Lovasz, L. (2010). *Large networks and graph limits*. London: CRC.
21. Maier, M., von Luxburg, U., & Hein, M. (2013). How the result of graph clustering methods depends on the construction of the graph. *ESAIM Probability and Statistics, 17*, 370–418.
22. Milanfar, P. (2013). Symmetrizing smoothing filters. *SIAM Journal of Imaging Sciences, 6*(1), 263–284.
23. Nadaraya, E. A. (1964). On estimating regression. *Theory Probability and Their Applications, 9*, 134–137.
24. Radovanović, M., Nanopoulos, A., & Ivanović, M. (2010). Hubs in space: Popular nearest neighbors in high-dimensional data. *Journal of Machine Learning Research, 11*, 2487–2531.
25. Rao, C. R. (1976). Estimation of parameters in linear models. *The Annals of Statistics, 4*, 1023–1037.
26. Sinkhorn, R. (1964). A relationship between arbitrary positive matrices and doubly stochastic matrices. *The Annals of Mathematical Statistics, 35*, 876–879.
27. Small, H. (1973). Co-citation in the scientific literature: A new measure of relationship between documents. *Journal of the American Society for Information Science, 24*, 265–269.
28. Zhao, L. (1999). Improved estimators in nonparametric regression problems. *Journal of the American Statistical Association, 94*(445), 164–173.

Nonparametric PU Learning of State Estimation in Markov Switching Model

A. Dobrovidov and V. Vasilyev

Abstract In this contribution, we develop methods of nonlinear filtering and prediction of an unobservable Markov chain which controls the states of observable stochastic process. This process is a mixture of two subsidiary stochastic processes, the switching of which is controlled by the Markov chain. Each of this subsidiary processes is described by conditional distribution density (cdd). The feature of the problem is that cdd's and transition probability matrix of the Markov chain are unknown, but a training sample (positive labeled) from one of the two subsidiary processes and training sample (unlabeled) from the mixture process are available. Construction of process binary classifier using positive and unlabeled samples in machine learning is called PU learning. To solve this problem for stochastic processes, nonparametric kernel estimators based on weakly dependent observations are applied. We examine the novel method performance on simulated data and compare it with the same performance of the optimal Bayesian solution with known cdd's and the transition matrix of the Markov chain. The modeling shows close results for the optimal task and the PU learning problem even in the case of a strong overlapping of the conditional densities of subsidiary processes.

1 Introduction

The hidden Markov chain (HMC) model is widely used in different problems, including signal and image processing, economical filtering and prediction, biological and medical sciences, and so on. In this model, the unobservable or "hidden" signal s_n is assumed to be a realization of a Markov chain $S_n, n \in \{1, 2, \ldots, N\}$ with a finite number M of states. The observed signal x_n is assumed to be a realization of a stochastic process $X_n, n \in \{1, 2, \ldots, N\}$. The links between

A. Dobrovidov (✉)
V.A. Trapeznikov Institute of Control Sciences, Russian Academy of Sciences, Moscow, Russia

V. Vasilyev
Moscow Institute of Physics and Technology (State University), Moscow, Russia

P. Bertail et al. (eds.), *Nonparametric Statistics*, Springer Proceedings
in Mathematics & Statistics 250, https://doi.org/10.1007/978-3-319-96941-1_2

them are described by the conditional density $f(x_n \mid S_n = s_n, X_1^{n-1} = x_1^{n-1})$. One way of generating processes with statistically dependent values is to use the conditional distribution of observations if the multidimensional distribution of the process is known. This case is considered in an example 2 with the multidimensional Rayleigh distribution. Another more simple way to obtain conditional density is the recurrent equation for representation of random sequences. Such equations are convenient for generating processes with dependent observations. These models could be autoregressive models, GARCH models, and others (see [6, 11, 12, 16, 17]). Especially often there are works devoted to the construction of volatility models [2, 8] and many of them use Markov switching to describe the well-known phenomenon of the volatility clusterization. Similar models are obtained as mixed types like a Markov-switching ARMA-GARCH model [10] and MSGARH [7].

In the vast majority of articles, conditional distribution densities are specified parametrically up to unknown parameters. In this case, the parameters of the mixed model can be estimated using the well-known EM-method based on the observation of the mixed sample.

In this chapter, we propose an algorithm for estimating the Markov chain state under conditions when

1. the number of Markov chain states $M = 2$;
2. the conditional distribution densities of the subsidiary processes, corresponding to each mixed process states, are completely unknown;
3. the a priori probabilities and transition probability matrix of Markov chain are unknown;
4. for the restoration of unknown densities, a realization of mixed process (unlabeled) and a training realization from one of the subsidiary processes (labeled) are available.

The availability of sampling from only one class (or process) is very common in applications. It is enough to cite such well-known problems as useful signal detection in sonar or estimating volatility in econometrics. A main feature of these problems is that the useful signal or volatility is never observed in pure form and there is no information to restore signal distribution or generate corresponding sample. On the other hand, the noise process is always observed while the devices that measure the signal against the background of noise are operating. It is this noise that is the source of the positive sample.

For the reconstruction of unknown densities, nonparametric kernel estimation procedures generalized on weakly dependent random variables are used [5].

Two examples demonstrate the algorithm quality. The first one is dedicated to the distinction between two autoregressive processes with different coefficients and the conditional distributions of these processes. Herewith these conditional distributions and the transition probability matrix of the Markov chain are unknown to the experimenter. In the second example, the model of one of the processes does not exist in the form of an equation, and the second process is defined only by its observations (in the language of modern learning theory—positive learning).

2 Problem Statement

Let (S_n, X_n) be a two-component stationary process with strong mixing, where (S_n) is unobservable component and (X_n) is observable one, $n \in \{1, 2, \dots, N\}$, $\mathbb{N} \in \mathbb{N}$. Let (S_n) be a stationary Markov chain with 2 states $\{0, 1\}$ and transition probability matrix $\|p_{i,j}\|$, $p_{i,j} = \mathbf{P}\{S_n = j \mid S_{n-1} = i\}$. Process (S_n) "controls" coefficients of equations, which describe the observable process (X_n):
if $S_n = 0$, then

$$X_n \sim f(x_n \mid S_n = 0, x_1^{n-1}) = f_0(x_n \mid x_1^{n-1}), \tag{1}$$

if $S_n = 1$, then

$$X_n \sim f(x_n \mid S_n = 1, x_1^{n-1}) = f_1(x_n \mid x_1^{n-1}). \tag{2}$$

For example, in case of $S_n = 0$, the process (X_n) may be described by the autoregressive model (AR) of order p:

$$X_n = \mu + \sum_{i=1}^{p} a_i (X_{n-i} - \mu) + b\xi_n, \tag{3}$$

where $\{\xi_n\}$ are i.i.d. random variables with the standard normal distribution, parameters $\mu, a_i \in \mathbb{R}$, $b \in \mathbb{R}^+$, $p \in \mathbb{N}$. Therefore, the conditional pdf (1) equals

$$f_0(x_n \mid x_1^{n-1}) = f_0(x_n \mid x_{n-p}^{n-1}) = \frac{1}{\sqrt{2\pi}b} \exp\left(- \frac{\left(x_n - \mu - \sum\limits_{i=1}^{p} a_i (x_{n-i} - \mu) \right)^2}{2b^2} \right).$$

As a performance of the proposed methods we use mean risk $R = \mathsf{E}L(S_n, \hat{S}_n)$ with a simple loss function

$$L(S_n, \hat{S}_n) = \begin{cases} 1 & S_n \neq \hat{S}_n \\ 0 & S_n = \hat{S}_n. \end{cases} \tag{4}$$

An optimal estimator of S_n (the Bayes decision function) is

$$S_n^* = \begin{cases} 0 & \text{if } \mathbf{P}(S_n = 0 \mid X_1^n = x_1^n) \geq 1/2 \\ 1 & \text{otherwise,} \end{cases} \tag{5}$$

where $\mathbf{P}(S_n = 0 \mid X_1^n = x_1^n)$ is a realization of the posterior probability $\mathbf{P}\{S_n = 0 \mid X_1^n\}$ with respect to a σ-algebra generated by r.v. X_1^n. For brevity, we write x_1^n and x_n instead of events $\{X_1^n = x_1^n\}$ and $\{X_n = x_n\}$. For instance,

$$\mathbf{P}(S_n = m \mid X_1^n = x_1^n) = \mathbf{P}(S_n = m \mid x_1^n). \tag{6}$$

In this work we solve a problem of estimation S_n using testing sample x_1^n (generated from the marginal distribution of the process (X_n)) and two training samples: positive x_p (drawn from the conditional distribution of the process (X_n) given $S_n = 0$) and mixed unlabeled x_u (drawn from the distribution of the mixed process (X_n))

$$x_p = (x_{p,i})_{i=1}^{n_p}, \tag{7}$$

$$x_u = (x_{u,i})_{i=1}^{n_u}, \tag{8}$$

where their sizes $n_p, n_u \in \mathbb{N}$. The last two samples are used to learn a nonlinear filter and the first for testing it.

3 Optimal Filtering

The optimal filtering may be applied, when the conditional densities $f_0(x_n \mid x_1^{n-1})$ and $f_1(x_n \mid x_1^{n-1})$ and transition probability matrix $\|p_{i,j}\|$ are known. In this case the posterior probability of the Markov state (6) is related to the predictive posterior probability of the state by the formula

$$\mathbf{P}(S_n = m \mid x_1^n) = \frac{f_m(x_n \mid x_1^{n-1})}{f(x_n \mid x_1^{n-1})} \mathbf{P}(S_n = m \mid x_1^{n-1}), \quad m \in \{0, 1\} \tag{9}$$

$$f(x_n \mid x_1^{n-1}) = \sum_{m=0}^{1} f_m(x_n \mid x_1^{n-1}) \mathbf{P}(S_n = m \mid x_1^{n-1}). \tag{10}$$

Since transition probability matrix $\|p_{i,j}\|$ is known, then for predictive posterior probability $\mathbf{P}(S_n = m \mid x_1^{n-1})$ an equation

$$\mathbf{P}(S_n = m \mid x_1^{n-1}) = \sum_{i=0}^{1} p_{i,m} \mathbf{P}(S_{n-1} = i \mid x_1^{n-1}) \tag{11}$$

is correct. Then Eq. (9) can be transformed to the well-known evaluation equation [5]

$$\mathbf{P}(S_n = m \mid x_1^n) = \frac{f_m(x_n \mid x_1^{n-1}) \sum_{i=0}^{1} p_{i,m}\mathbf{P}(S_{n-1} = i \mid x_1^{n-1})}{\sum_{j=0}^{1} f_j(x_n \mid x_1^{n-1}) \sum_{i=0}^{1} p_{i,j}\mathbf{P}(S_{n-1} = i \mid x_1^{n-1})}.$$

Substituting this posterior distribution in (5), we obtain the optimal Bayes estimator S_n^* for the nonlinear filtering. This optimal method will be considered as a standard and compared with the proposed method, where $f_0(x_n)$, $f_1(x_n)$, and $\|p_{i,j}\|$ are not available.

4 Nonparametric Filtering

4.1 Main Idea

Let us consider the following estimator of S_n

$$\tilde{S}_n = \begin{cases} 0 & \text{if } \mathbf{P}(S_n = 0 \mid X_{n-\tau}^n = x_{n-\tau}^n) \geq 1/2 \\ 1 & \text{otherwise,} \end{cases} \tag{12}$$

where $\tau \in \{0, 1, \dots, n-1\}$. Here, only last $\tau + 1$ observations are used in the condition of the posterior probability. If $\tau = n - 1$, then $\tilde{S}_n = S_n^*$. We assume that the process (S_n, X_n) is α-mixing. Then $\forall \epsilon > 0, \exists \tau(\alpha) : |\mathbf{P}(S_n = 0 \mid X_{n-\tau}^n = x_{n-\tau}^n) - \mathbf{P}(S_n = 0 \mid X_1^n = x_1^n)| < \epsilon$. It means that $\tilde{S}_n \approx S_n^*$ for some τ. Using simple relation

$$\mathbf{P}(S_n = 0 \mid X_{n-\tau}^n = x_{n-\tau}^n) = \frac{f_0(x_n \mid x_{n-\tau}^{n-1})}{f(x_n \mid x_{n-\tau}^{n-1})}\mathbf{P}(S_n = 0 \mid x_{n-\tau}^{n-1}),$$

where denominator is equal to

$$f(x_n \mid x_{n-\tau}^{n-1}) = f_0(x_n \mid x_{n-\tau}^{n-1})\mathbf{P}(S_n = 0 \mid x_{n-\tau}^{n-1}) + f_1(x_n \mid x_{n-\tau}^{n-1})\mathbf{P}(S_n = 1 \mid x_{n-\tau}^{n-1}), \tag{13}$$

and normalization condition

$$\mathbf{P}(S_n = 0 \mid x_{n-\tau}^{n-1}) + \mathbf{P}(S_n = 1 \mid x_{n-\tau}^{n-1}) = 1,$$

one can rewrite (12) as

$$\tilde{S}_n = \begin{cases} 0 & \text{if } 2 f_0(x_n \mid x_{n-\tau}^{n-1}) \mathbf{P}(S_n = 0 \mid x_{n-\tau}^{n-1}) - f(x_n \mid x_{n-\tau}^{n-1}) > 0 \\ 1 & \text{otherwise.} \end{cases} \tag{14}$$

This estimator will be the base of proposed method. In the next sections nonparametric kernel estimation of the conditional densities $f_0(x_n \mid x_{n-\tau}^{n-1})$, $f(x_n \mid x_{n-\tau}^{n-1})$ and probability $\mathbf{P}(S_n = 0 \mid x_{n-\tau}^{n-1})$ will be considered.

4.2 *Estimators of* $f_0(x_n \mid x_{n-\tau}^{n-1})$, $f(x_n \mid x_{n-\tau}^{n-1})$

In this section estimators of $f_0(x_n \mid x_{n-\tau}^{n-1})$, $f(x_n \mid x_{n-\tau}^{n-1})$ are proposed. Firstly, let us transform positive sample x_p of univariate elements to sample $\mathbf{x}_p = (\mathbf{x}_{p,i})_{i=1}^{n_p-\tau}$ with $(\tau + 1)$-dimensional elements

$$\mathbf{x}_{p,i} = (x_{p,i}, x_{p,i+1}, \dots, x_{p,i+\tau}).$$

Analogous to x_p, construct new sample $\mathbf{x}_u = (\mathbf{x}_{u,i})_{i=1}^{n_u-\tau}$ from unlabeled sample x_u:

$$\mathbf{x}_{u,i} = (x_{u,i}, x_{u,i+1}, \dots, x_{u,i+\tau}).$$

Secondly, let us rewrite conditional densities $f_0(x_n \mid x_{n-\tau}^{n-1})$, $f(x_n \mid x_{n-\tau}^{n-1})$ in the following form

$$f_0(x_n \mid x_{n-\tau}^{n-1}) = \frac{f_0(x_{n-\tau}^{n})}{f_0(x_{n-\tau}^{n-1})} = \frac{f_0(x_{n-\tau}^{n})}{\int_{-\infty}^{\infty} f_0(x_{n-\tau}^{n}) dx_n},$$

$$f(x_n \mid x_{n-\tau}^{n-1}) = \frac{f(x_{n-\tau}^{n})}{f(x_{n-\tau}^{n-1})} = \frac{f(x_{n-\tau}^{n})}{\int_{-\infty}^{\infty} f(x_{n-\tau}^{n}) dx_n}.$$

Finally, for unknown densities $f_0(x_{n-\tau}^{n})$ and $f(x_{n-\tau}^{n})$ estimators

$$f_0(x_{n-\tau}^{n}) \approx f((x_{n-\tau}, \dots, x_n) \mid (\mathbf{x}_{p,i})_{i=1}^{n_p-\tau}),$$
$$f(x_{n-\tau}^{n}) \approx f((x_{n-\tau}, \dots, x_n) \mid (\mathbf{x}_{u,i})_{i=1}^{n_u-\tau})$$

are proposed, where notation $f(\mathbf{x} \mid (\mathbf{x}_i)_{i=1}^{n})$ is the multivariate kernel density estimator (MKDE) in the point vector $\mathbf{x}$ constructed by training set $(\mathbf{x}_i)_{i=1}^{n}$. The next section is devoted to MKDE.

4.3 Estimator of $f(\mathbf{x} \mid (\mathbf{x}_i)_{i=1}^n)$

There are a lot of kernel density estimators and approaches to configure them. In this section is a one of the possible combinations of them, which includes two steps: pilot and subtle estimators. Firstly, for density $f(\mathbf{x} \mid (\mathbf{x}_i)_{i=1}^n)$ next fixed kernel estimator

$$\tilde{f}(\mathbf{x}) = \tilde{f}(\mathbf{x} \mid (\mathbf{x}_i)_{i=1}^n) = \frac{1}{n}\sum_{i=1}^{n} \frac{1}{h^d} K\left(\frac{\mathbf{x}-\mathbf{x}_i}{h}\right) \tag{15}$$

is applied, where $K(\cdot)$ is the kernel function, usually some probability density function with zero mean; h is the bandwidth (tuning parameter), $h > 0$; $\mathbf{x}, \mathbf{x}_i \in \mathbb{R}^{1\times d}$, $d \in \mathbb{N}$. Probability density function of multivariate normal distribution with zero mean and identity covariance matrix

$$\phi(\mathbf{x}) = \frac{1}{(2\pi)^{d/2}} \exp\left(-\frac{\mathbf{x}\mathbf{x}^\top}{2}\right)$$

is used as the kernel function $K(\cdot)$. One of the methods for calculating the bandwidth h is the unbiased cross-validation (UCV)(see [3, 14]). Procedure UCV leads to an estimator

$$\hat{h} = \underset{h>0}{\operatorname{argmin}}\, \mathrm{UCV(h)},$$

with minimization function

$$\mathrm{UCV}(h) = \frac{1}{n(n-1)h^d} \sum_{i=1}^{n} \sum_{\substack{j=1,\\ j\neq i}}^{n} \frac{1}{2^{d/2}}\phi\left(\frac{\mathbf{x}_i-\mathbf{x}_j}{\sqrt{2}h}\right) - 2\phi\left(\frac{\mathbf{x}_i-\mathbf{x}_j}{h}\right) + \frac{1}{nh^d}.$$

Computing minima analytically is a challenge, so a numerical calculation is popular. The function $\mathrm{UCV}(h)$ often has multiple local minima, therefore more correct way is to use brute-force search to find $\hat{h}$, however it is a very slow algorithm. In [9] it was shown that spurious local minima are more likely at too small values of h, so we propose to use golden section search between 0 and h^+, where

$$h^+ = \left(\frac{4}{n(d+2)}\right)^{\frac{1}{d+4}} \max_{i,j\in\{1,\dots,d\}} \sqrt{|\hat{\mathbf{S}}_{i,j}|},$$

and $\hat{\mathbf{S}}$ is the sample covariance matrix of vector sequence $(\mathbf{x}_i)_{i=1}^n$. To improve accuracy of estimator (15) in the second step using more flexible approach is considered. Constructing estimator with fixed kernel is the first step in the methods with "adaptive" kernel like *balloon* estimators (see [13]) and *sample point* estimators (see [1, 4]). Silverman in [15] explored Abramson's implementation and proposed the following estimator

$$f(\mathbf{x} \mid (\mathbf{x}_i)_{i=1}^n) = \frac{1}{n} \sum_{i=1}^{n} \frac{1}{(h\lambda_i)^d} K\left(\frac{\mathbf{x} - \mathbf{x}_i}{h\lambda_i}\right) \tag{16}$$

with local bandwidth factors

$$\lambda_i = \left\{ \frac{\tilde{f}(\mathbf{x}_i)}{g} \right\}^{-1/2},$$

where g is the geometric mean of the $\tilde{f}(\mathbf{x}_i)$

$$g = \left\{ \prod_{i=1}^{n} \tilde{f}(\mathbf{x}_i) \right\}^{1/n}.$$

Silverman noted that using similar bandwidth h for (15) and (16) gives good results.

4.4 *Estimator of* $\mathbf{P}(S_n = 0 \mid x_{n-\tau}^{n-1})$

In this section an estimator of probability $\mathbf{P}(S_n = 0 \mid x_{n-\tau}^{n-1})$ from (14) is explored. Rewrite (13) as

$$\frac{f(x_n \mid x_{n-\tau}^{n-1})}{f_0(x_n \mid x_{n-\tau}^{n-1})} = \mathbf{P}(S_n = 0 \mid x_{n-\tau}^{n-1}) + \frac{f_1(x_n \mid x_{n-\tau}^{n-1})(1 - \mathbf{P}(S_n = 0 \mid x_{n-\tau}^{n-1}))}{f_0(x_n \mid x_{n-\tau}^{n-1})}. \tag{17}$$

Note that $\mathbf{P}(S_n = 0 \mid x_{n-\tau}^{n-1})$ does not depend on x_n. It means that $\forall x_n \in \mathbb{R}$ the last equation is true. If densities $f(x_n \mid x_{n-\tau}^{n-1})$ and $f_0(x_n \mid x_{n-\tau}^{n-1})$ were known, then simple estimator

$$\tilde{\mathbf{P}}(S_n = 0 \mid x_{n-\tau}^{n-1}) = \min_{x \in \mathbb{1} R} \frac{f(x \mid x_{n-\tau}^{n-1})}{f_0(x \mid x_{n-\tau}^{n-1})} \tag{18}$$

would give good results, if there is some $x \in \mathbb{R}$, such that

$$\frac{f_1(x_n \,|\, x_{n-\tau}^{n-1})(1 - \mathbf{P}(S_n = 0 \,|\, x_{n-\tau}^{n-1}))}{f_0(x_n \,|\, x_{n-\tau}^{n-1})} \approx 0.$$

Then the estimator $\tilde{\mathbf{P}}(S_n = 0 \,|\, x_{n-\tau}^{n-1})$ will be close to $\mathbf{P}(S_n = 0 \,|\, x_{n-\tau}^{n-1})$. Since in fact densities $f(x_n \,|\, x_{n-\tau}^{n-1})$ and $f_0(x_n \,|\, x_{n-\tau}^{n-1})$ are unknown, their estimators are substituted in (18). However, in this case estimator $\tilde{\mathbf{P}}(S_n = 0 \,|\, x_{n-\tau}^{n-1})$ is unreasonable, because if only for one point $x_0 \in \mathbb{R}$ the value of $\frac{f(x_0 \,|\, x_{n-\tau}^{n-1})}{f_0(x_0 \,|\, x_{n-\tau}^{n-1})} \approx 0$, then the estimator $\tilde{\mathbf{P}}(S_n = 0 \,|\, x_{n-\tau}^{n-1})$ will be too undervalued in comparison with the true value of $\mathbf{P}(S_n = 0 \,|\, x_{n-\tau}^{n-1})$. Therefore, it is necessary to introduce some cumulative characteristics which will have less influence of particular x. As such characteristics, we propose

$$Q(p) = \int (f_0(x \,|\, x_{n-\tau}^{n-1})p - f(x \,|\, x_{n-\tau}^{n-1}))^+ dx,$$

$$\hat{Q}(p) = \int (\hat{f}_0(x \,|\, x_{n-\tau}^{n-1})p - \hat{f}(x \,|\, x_{n-\tau}^{n-1}))^+ dx,$$

where $(a)^+ = \max(0, a)$ and $p \in [0, 1]$ is a variable parameter. The meaning of the function $Q(p)$ is an area between two functions $pf_0(x \,|\, x_{n-\tau}^{n-1})$ and $f(x \,|\, x_{n-\tau}^{n-1})$, where the first one exceeds the second one. The meaning of the estimator $\hat{Q}(p)$ is the same. It is easy to show that $Q(p) = 0$ for $p \in [0, \tilde{\mathbf{P}}(S_n = 0 \,|\, x_{n-\tau}^{n-1})]$ and $0 < Q(p) < 1$ for $p \in (\tilde{\mathbf{P}}(S_n = 0 \,|\, x_{n-\tau}^{n-1}), 1]$. So $Q(p)$ changes its first and second derivatives in the point $\tilde{\mathbf{P}}(S_n = 0 \,|\, x_{n-\tau}^{n-1})$. We expect the similar changes in derivatives of the estimator $\hat{Q}(p)$. Therefore, we propose to use point p, where the curvature of the function $\hat{Q}(p)$

$$\kappa(p) = \frac{|\hat{Q}''(p)|}{(1 + (\hat{Q}'(p))^2)^{3/2}}$$

reaches maximum, i.e. a final estimator of $\mathbf{P}(S_n = 0 \,|\, x_{n-\tau}^{n-1})$ is

$$\hat{\mathbf{P}}(S_n = 0 \,|\, x_{n-\tau}^{n-1}) = \operatorname*{argmax}_{0 \le p \le 1} \kappa(p). \tag{19}$$

Due to the analytical complexity of the estimator $\hat{\mathbf{P}}(S_n = 0 \,|\, x_{n-\tau}^{n-1})$, its properties are investigated by computer simulation.

5 One-Step Ahead Prediction

Let us consider one-step ahead prediction. Like for filtering we minimize mean risk $\mathsf{E}L(S_n, \hat{S}_n)$ with simple loss function (4). Therefore an optimal predictive estimator of S_n is

$$S_n^+ = \begin{cases} 0 & \text{if } \mathbf{P}(S_n = 0 \,|\, X_1^{n-1} = x_1^{n-1}) \geq 1/2 \\ 1 & \text{otherwise.} \end{cases}$$

Note that probability $\mathbf{P}(S_n = 0 \,|\, X_1^{n-1} = x_1^{n-1})$ is already obtained in the considered approaches to filtering: for optimal method it is written in (11) and for nonparametric method accordingly in (19). It means that we primarily solve problem of one-step ahead prediction and then filtering problem.

6 Examples

6.1 Example 1 (AR(1) + AR(2))

Let the process X_n for each state of S_n be given by the autoregressive model:

$$\begin{aligned} &\text{if } S_n = 0 : X_n = 1 + 0.2X_{n-1} + \xi_n, \\ &\text{if } S_n = 1 : X_n = 4 + 0.3X_{n-1} + 0.2X_{n-2} + 0.8\xi_n \end{aligned}$$

and transition probability matrix equals

$$\|p_{i,j}\| = \begin{pmatrix} 0.92 & 0.08 \\ 0.05 & 0.95 \end{pmatrix}.$$

Parameters $\tau = 1$, $n \in \{0, 1, \ldots, 201\}$, $n_p = 2000$, $n_u = 2000$. One may see illustration of densities and their estimators for some point x_n in Fig. 1. It follows that the estimator $\hat{\mathbf{P}}(S_n = 0 \,|\, x_{n-\tau}^{n-1}) = 0.04$ is very close to real value of the probability $\mathbf{P}(S_n = 0 \,|\, x_{n-\tau}^{n-1}) = 0.05$ from Fig. 2. This experiment shows good quality of the proposed method (see Table 1 and Fig. 3).

6.2 Example 2 (AR(2) + Rayleigh)

Let the process X_n be described as

$$\text{if } S_n = 0: \; X_n = 4 + 0.1X_{n-1} + 0.5X_{n-2} + \xi_n;$$

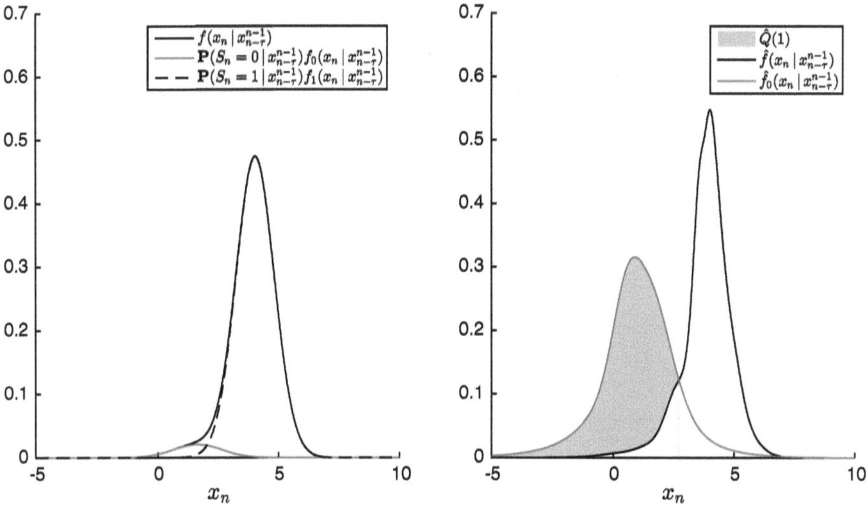

Fig. 1 Left plot: black line is $f(x_n \mid x_{n-\tau}^{n-1})$; gray line is $\mathbf{P}(S_n = 0 \mid x_{n-\tau}^{n-1}) f_0(x_n \mid x_{n-\tau}^{n-1})$; dashed line is $\mathbf{P}(S_n = 1 \mid x_{n-\tau}^{n-1}) f_1(x_n \mid x_{n-\tau}^{n-1})$. Right plot: black line is estimator $\hat{f}(x_n \mid x_{n-\tau}^{n-1})$; gray line is estimator $\hat{f}_0(x_n \mid x_{n-\tau}^{n-1})$; gray area between two densities $\hat{f}_0(x_n \mid x_{n-\tau}^{n-1})$ and $\hat{f}(x_n \mid x_{n-\tau}^{n-1})$, where $\hat{f}_0(x_n \mid x_{n-\tau}^{n-1})$ exceeds $\hat{f}(x_n \mid x_{n-\tau}^{n-1})$

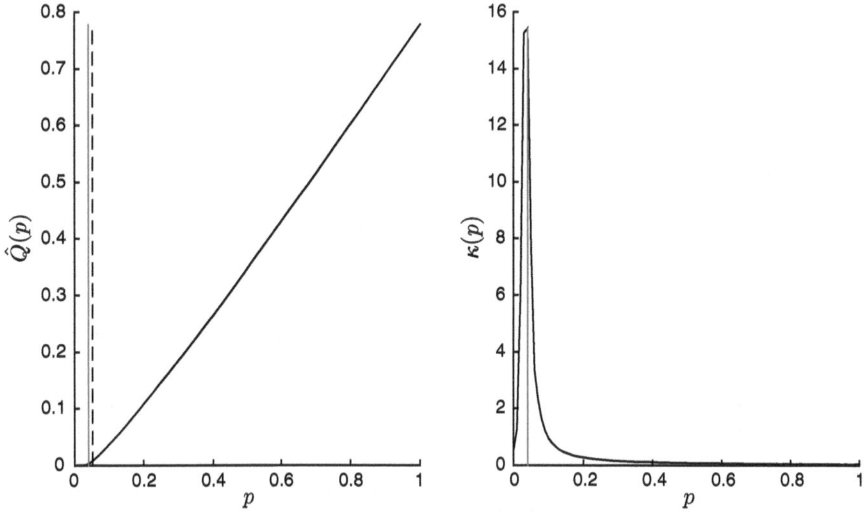

Fig. 2 Left plot: black line is $\hat{Q}(p)$; grey line shows that estimator $\hat{\mathbf{P}}(S_n = 0 \mid x_{n-\tau}^{n-1}) = 0.04$; dashed line represents true $\mathbf{P}(S_n = 0 \mid x_{n-\tau}^{n-1}) = 0.05$. Right plot: black line is function $\kappa(p)$; grey line shows that estimator $\hat{\mathbf{P}}(S_n = 0 \mid x_{n-\tau}^{n-1}) = 0.04$

Table 1 Simulation results in example 1 after 50 launches

n_p	n_u	Optimal error (%)	Nonparametric error (%)	Difference (%)
2000	2000	7.96	13.93	5.97

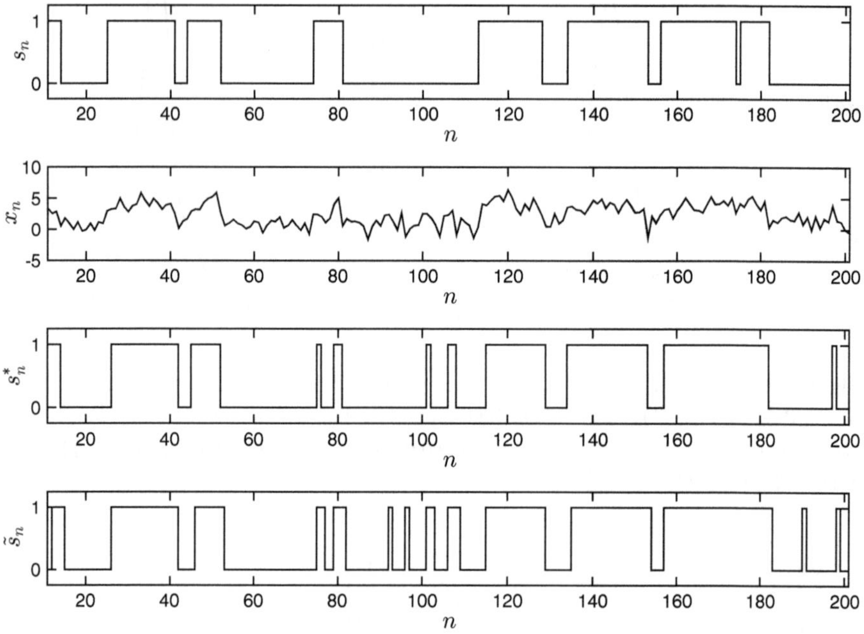

Fig. 3 First plot: unobservable s_n. Second plot: observable x_n. Third plot: optimal estimator s_n^*. Fourth plot: nonparametric $\tilde{s}_n$

if $S_n = 1$, the conditional density is the Rayleigh density

$$f_1(x_n|x_{n-1}) = \frac{x_n}{\sigma^2(1-\rho)} \exp\left\{-\frac{\rho x_{n-1}^2 + x_n^2}{2\sigma^2(1-\rho)}\right\} I_0\left(\frac{\sqrt{\rho}x_{n-1}x_n}{\sigma^2(1-\rho)}\right)$$

with $\rho = 0.2$, $\sigma = 1$, $I_0(x)$ is modified Bessel function, and transition probability matrix equals

$$\|p_{i,j}\| = \begin{pmatrix} 0.87 & 0.13 \\ 0.10 & 0.90 \end{pmatrix}.$$

Parameters $\tau = 2$, $n \in \{0, 1, \ldots, 201\}$, $n_p = 2000$, $n_u = 2000$. In Fig. 4 one may see illustration of densities and their estimators for some point x_n. From Fig. 5 it follows that estimator $\hat{\mathbf{P}}(S_n = 0 \,|\, x_{n-\tau}^{n-1}) = 0.95$ is close to real value of a probability

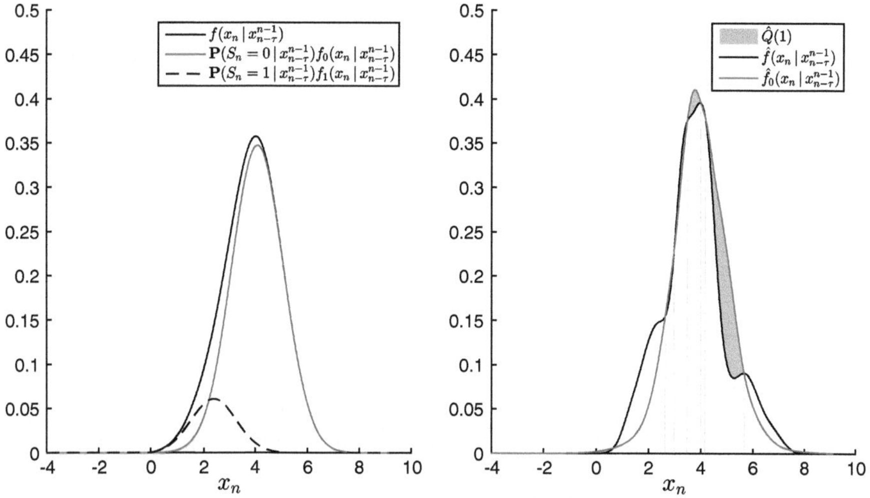

Fig. 4 Left plot: black line is $f(x_n \mid x_{n-\tau}^{n-1})$; gray line is $\mathbf{P}(S_n = 0 \mid x_{n-\tau}^{n-1}) f_0(x_n \mid x_{n-\tau}^{n-1})$; dashed line is $\mathbf{P}(S_n = 1 \mid x_{n-\tau}^{n-1}) f_1(x_n \mid x_{n-\tau}^{n-1})$. Right plot: black line is estimator $\hat{f}(x_n \mid x_{n-\tau}^{n-1})$; gray line is estimator $\hat{f}_0(x_n \mid x_{n-\tau}^{n-1})$; gray area between two densities $\hat{f}_0(x_n \mid x_{n-\tau}^{n-1})$ and $\hat{f}(x_n \mid x_{n-\tau}^{n-1})$, where $\hat{f}_0(x_n \mid x_{n-\tau}^{n-1})$ exceeds $\hat{f}(x_n \mid x_{n-\tau}^{n-1})$

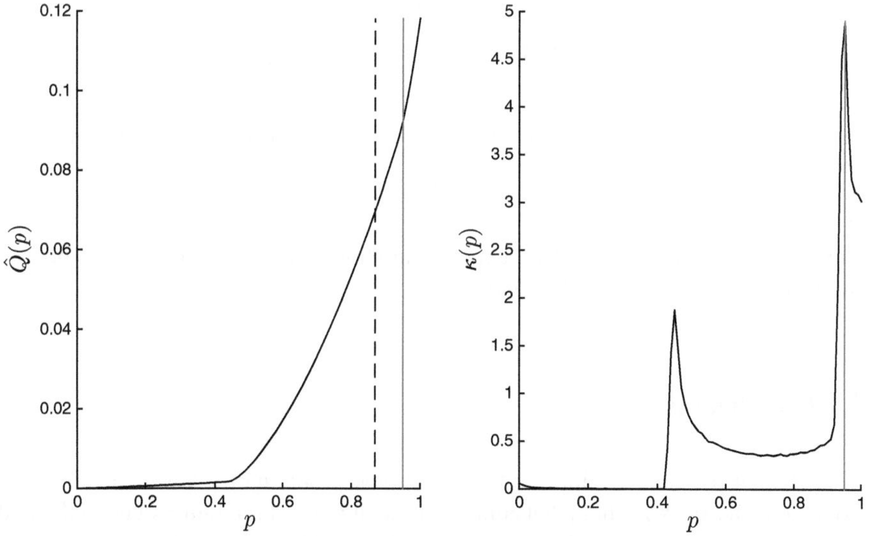

Fig. 5 Left plot: black line is $\hat{Q}(p)$; grey line shows that estimator $\hat{\mathbf{P}}(S_n = 0 \mid x_{n-\tau}^{n-1}) = 0.95$; dashed line represents true $\mathbf{P}(S_n = 0 \mid x_{n-\tau}^{n-1}) = 0.87$. Right plot: black line is function $\kappa(p)$; grey line shows that estimator $\hat{\mathbf{P}}(S_n = 0 \mid x_{n-\tau}^{n-1}) = 0.95$

Table 2 Simulation results in example 2 after 50 launches

n_p	n_u	Optimal error (%)	Nonparametric error (%)	Difference (%)
2000	2000	9.87	19.24	9.37

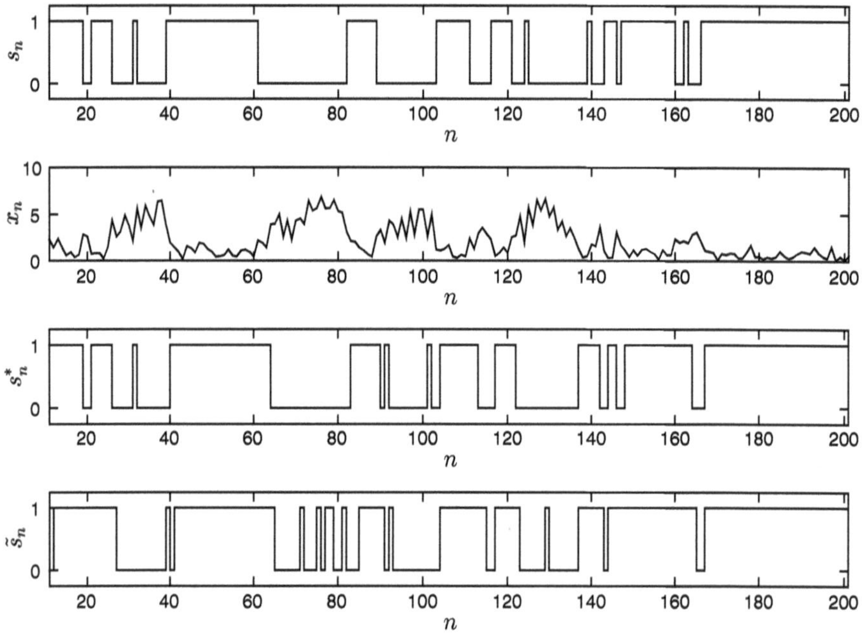

Fig. 6 First plot: unobservable s_n. Second plot: observable x_n. Third plot: optimal estimator s_n^*. Fourth plot: nonparametric $\tilde{s}_n$

$\mathbf{P}(S_n = 0 \,|\, x_{n-\tau}^{n-1}) = 0.87$. This experiment shows good quality of the proposed method in this case (see Table 2 and Fig. 6).

7 Conclusion

This chapter presents a solution of the nonlinear problem of states estimating of a homogeneous Markov chain that controls the switching of random processes defined by their conditional distribution densities under conditions when these densities are completely unknown to the operator. In addition, the transition probability matrix of the Markov chain is also unknown. Only a sample of one of the processes and a mixed sample are available and used to evaluate the state. A novelty of this work is the nonparametric algorithm for estimating the probability of forecasting the state of Markov chain by one step ahead, which makes it possible to construct an estimator of nonlinear filtration. At the same time, in well-known works on PU learning it was repeatedly noted that it is not possible to construct an estimator of the probability of a forecast without additional information. Two examples given in this chapter show the sufficiently high accuracy of the proposed nonparametric estimator, even in the case of a strong overlapping of the conditional densities of the two subsidiary processes. To our knowledge, the a priori conditions adopted in this work are minimal for solving the problem of estimating the states of the mixing process. In what follows we intend to find conditions for the convergence of the proposed state estimates.

References

1. Abramson, I. S. (1982). On bandwidth variation in kernel estimates-a square root law. *Annals of Statistics, 10*(4), 1217–1223.
2. Bollerslev, T. (1986). Generalized autoregressive conditional heteroskedasticity. *Journal of Econometrics, 31*(3), 307–327.
3. Bowman, A. (1984). An alternative method of cross-validation for the smoothing of density estimates. *Biometrika, 7*, 353–360.
4. Breiman, L., Meisel, W., & Purcell, E. (1977). Variable kernel estimates of multivariate densities. *Technometrics, 19*(2), 135–144. ISSN 0040(1706)
5. Dobrovidov, A. V., Koshkin, G. M., & Vasiliev V. A. (2012). *Non-parametric models and statistical inference from dependent observations*. Heber: Kendrick Press.
6. Douc, R., Moulines, E., & Rydén, T. (2004). Asymptotic properties of the maximum likelihood estimator in autoregressive models with Markov regime. *The Annals of Statistics, 32*(5), 2254–2304.
7. Francq, C., & Zakoian, J.-M. (2005). The l2-structures of standard and switching-regime Garch models. *Stochastic Processes and Their Applications, 115*(9), 1557–1582.
8. Giraitis, L., Leipus, R., & Surgailis, D. (2007). Recent advances in arch modelling. In *Long memory in economics* (pp. 3–38). Berlin: Springer.
9. Hall, P., & Marron, J. (1991). Local minima in cross-validation functions. *Journal of the Royal Statistical Society, Series B (Methodological), 53*, 245–252.
10. Hamilton, J. D., & Susmel, R. (1994). Autoregressive conditional heteroskedasticity and changes in regime. *Journal of Econometrics, 64*(1), 307–333.
11. Klaassen, F. (2002). Improving Garch volatility forecasts with regime-switching Garch. *Empirical Economics, 27*(2), 363–394.
12. Lanchantin, P., & Pieczynski, W. (2005). Unsupervised restoration of hidden non stationary Markov chain using evidential priors. *IEEE Transactions on Signal Processing, 53*(8), 3091–3098.
13. Loftsgaarden, D. O., & Quesenberry, C. P. (1965). A nonparametric estimate of a multivariate density function. *Annals of Mathematical Statistics, 36*(3), 1049–1051.
14. Rudemo, M. (1982). Empirical choice of histograms and kernel density estimators. *Scandinavian Journal of Statistics, 9*, 65–78.
15. Silverman, B. (1986). *Density estimation for statistics and data analysis. Monographs on statistics and applied probability*. Boca Raton, London: Chapman and Hall.
16. Yao, J.-F., & Attali, J.-G. (2000). On stability of nonlinear ar processes with Markov switching. *Advances in Applied Probability, 32*(2), 394–407.
17. Zhang, M. Y., Russell, J. R., & Tsay, R. S. (2001). A nonlinear autoregressive conditional duration model with applications to financial transaction data. *Journal of Econometrics, 104*(1), 179–207.

Multiplicative Bias Corrected Nonparametric Smoothers

N. Hengartner, E. Matzner-Løber, L. Rouvière, and T. Burr

Abstract This contribution presents a general multiplicative bias reduction strategy for nonparametric regression. The approach is most effective when applied to an oversmooth pilot estimator, for which the bias dominates the standard error. The practical usefulness of the method was demonstrated in Burr et al. (IEEE Trans Nucl Sci 57:2831–2840, 2010) in the context of estimating energy spectra. For such data sets, it was observed that the method could decrease significantly the bias with only negligible increase in variance. This chapter presents the theoretical analysis of that estimator. In particular, we study the asymptotic properties of the bias corrected local linear regression smoother, and prove that it has zero asymptotic bias and the same asymptotic variance as the local linear smoother with a suitably adjusted bandwidth. Simulations show that our asymptotic results are available for modest sample sizes.

1 Introduction

In nonparametric regression, the bias-variance tradeoff of linear smoothers such as kernel-based regression smoothers, wavelet based smoother, or spline smoothers, is generally governed by a user-supplied parameter. This parameter is often called the bandwidth, which we will denote by h. As an example, assuming that the regression function m is locally twice continuously differentiable at a point x, the local linear

N. Hengartner · T. Burr
Los Alamos National Laboratory, Los Alamos, NM, USA
e-mail: nickh@lanl.gov; tburr@lanl.gov

E. Matzner-Løber (✉)
CREST, UMR 9194, Cepe-Ensae, Palaiseau, France
e-mail: eml@ensae.fr

L. Rouvière
Univ. Rennes, IRMAR, UMR 6625, Rennes, France
e-mail: laurent.rouviere@univ-rennes2.fr

P. Bertail et al. (eds.), *Nonparametric Statistics*, Springer Proceedings in Mathematics & Statistics 250, https://doi.org/10.1007/978-3-319-96941-1_3

smoother with bandwidth h and kernel K has conditional bias at that point

$$\frac{h^2}{2}m''(x)\int u^2K(u)\,du + o_p(h^2)$$

and conditional variance

$$\frac{1}{nh}\frac{\sigma^2(x)}{f(x)}\int K^2(u)\,du + o_p\left(\frac{1}{nh}\right)$$

where f stands for the density of the (one-dimensional) explanatory variable X and $\sigma^2(x)$ is the conditional variance of the response variable given $X = x$. See, for example, the book of [7]. Since the bias increases with the second order derivative of the regression function, the local linear smoother tends to under-estimate in the peaks and over-estimate in the valleys of the regression function. See, for example, [25–27].

The resulting bias in the estimated peaks and valleys is troublesome in some applications, such as the estimation of energy spectrum from nuclear decay. That example motivates the development of our multiplicative bias correction methodology. The interested reader is referred to [2] for a more detailed description and analysis.

All nonparametric smoothing methods are generally biased. There are a large number of methods to reduce the bias, but most of them do so at the cost of an increase in the variance of the estimator. For example, one may choose to undersmooth the energy spectrum. Undersmoothing will reduce the bias but will have a tendency of generating spurious peaks. One can also use higher order smoothers, such as local polynomial smoother with a polynomial of order larger than one. While again this will lead to a smaller bias, the smoother will have a larger variance. Another approach is to start with a pilot smoother and to estimate its bias by smoothing the residuals [3, 4, 6]. Subtracting the estimated bias from the smoother produces a regression smoother with smaller bias and larger variance. For the estimation of an energy spectrum, the additive bias correction and the higher order smoothers have the unfortunate side effect of possibly generating a non-positive estimate.

An attractive alternative to the linear bias correction is the multiplicative bias correction pioneered by [19]. Because the multiplicative correction does not alter the sign of the regression function, this type of correction is particularly well suited for adjusting non-negative regression functions. [20] showed that if the true regression function has four continuous derivatives, then the multiplicative bias reduction is operationally equivalent to using an order four kernel. And while this does remove the bias, it also increases the variance because of the roughness of such a kernel.

Many authors have extended the work of [20]. Glad [9, 10] propose to use a parametrically guided local linear smoother and Nadaraya-Watson smoother by starting with a parametric pilot. This approach is extended to a more general framework which includes both multiplicative and additive bias correction by [21] (see also [16, 22, 28] for an extension to time series conditional variance estimation and spectral estimation). For multiplicative bias correction in density estimation and hazard estimation, we refer the reader to the works of [11, 12, 17, 23, 24].

Although the bias-variance tradeoff for nonparametric smoothers is always present in finite samples, it is possible to construct smoothers whose *asymptotic bias* converges to zero while keeping the same asymptotic variance. Hengartner and Matzner-Løber [13] has exhibited a nonparametric density estimator based on multiplicative bias correction with that property, and has shown in simulations that his estimator also enjoys good finite sample properties. Burr et al. [2] adapts the estimator from [13] to nonparametric regression with aim to estimate energy spectra. They illustrate the benefits of their approach on real and simulated spectra. The goal of this chapter is to study the asymptotic properties of that estimator. It is worth pointing out that these properties have already been studied by [19] for fixed design and further by [20]. We emphasize that there are two major differences between our work and that of [20].

- First, we do not add regularity assumptions on the target regression function. In particular, we do not assume that the regression function has four continuous derivatives as in [20].
- Second, we show that the multiplicative bias reduction procedure performs a bias reduction with no cost to the asymptotic variance. It is exactly the same as the asymptotic variance of the local linear estimate.

Finally, we note that we show a different asymptotic behavior under less restrictive assumptions than those found in [20]. Moreover our results and proofs are different from the above referenced works.

This contribution is organized as follows. Section 2 introduces the notation and defines the estimator. Section 3 gives the asymptotic behavior of the proposed estimator. A brief simulation study on finite sample comparison is presented in Sect. 4. The interested reader is referred to Sect. 6 where we have gathered the technical proofs.

2 Preliminaries

2.1 Notations

Let $(X_1, Y_1), \ldots, (X_n, Y_n)$ be n independent copies of the pair of random variables (X, Y) with values in $\mathbb{R} \times \mathbb{R}$. We suppose that the explanatory variable X has probability density f and model the dependence of the response variable Y to the explanatory variable X through the nonparametric regression model

$$Y = m(X) + \varepsilon. \quad (1)$$

We assume that the regression function $m(\cdot)$ is smooth and that the disturbance ε is a mean zero random variable with finite variance σ^2 that is independent of the covariate X. Consider the linear smoothers for the regression function $m(x)$ which we write as

$$\hat{m}(x) = \sum_{j=1}^{n} \omega_j(x; h) Y_j,$$

where the weight functions $\omega_j(x; h)$ depend on a bandwidth h. If the weight functions are such that $\sum_{j=1}^{n} \omega_j(x; h) = 1$ and $\sum_{j=1}^{n} \omega_j(x; h)^2 = (nh)^{-1}\tau^2$, and if the disturbances satisfy the Lindeberg's condition, then the linear smoother obeys the central limit theorem

$$\sqrt{nh}\left(\hat{m}(x) - \sum_{j=1}^{n} w_j(x; h) m(X_j)\right) \xrightarrow{d} \mathcal{N}(0, \tau^2) \quad \text{as} \quad n \to \infty. \tag{2}$$

We can use (2) to construct asymptotic pointwise confidence intervals for the unknown regression function $m(x)$. But unless the limit of the scaled bias

$$b(x) = \lim_{n \longrightarrow \infty} \sqrt{nh}\left(\sum_{j=1}^{n} w_j(x; h) m(X_j) - m(x)\right),$$

which we call the asymptotic bias, is zero, the confidence interval

$$\left[\hat{m}(x) - Z_{1-\alpha/2}\sqrt{nh}\tau, \hat{m}(x) + Z_{1-\alpha/2}\sqrt{nh}\tau\right]$$

will not cover asymptotically the true regression function $m(x)$ at the nominal $1-\alpha$ level ($Z_{1-\alpha/2}$ stands for the $(1-\alpha/2)$-quantile of the $\mathcal{N}(0, 1)$ distribution). The construction of valid pointwise $1-\alpha$ confidence intervals for regression smoothers is another motivation for developing estimators with zero asymptotic bias.

2.2 *Multiplicative Bias Reduction*

Given a pilot smoother with bandwidth h_0 for the regression function $m(x)$,

$$\tilde{m}_n(x) = \sum_{j=1}^{n} \omega_j(x; h_0) Y_j,$$

consider the ratio $V_j = \frac{Y_j}{\tilde{m}_n(X_j)}$. That ratio is a noisy estimate of the inverse relative estimation error of the smoother $\tilde{m}_n$ at each of the observations, $m(X_j)/\tilde{m}_n(X_j)$. Smoothing V_j using a second linear smoother, say

$$\widehat{\alpha}_n(x) = \sum_{j=1}^{n} \omega_j(x; h_1)V_j,$$

produces an estimate for the inverse of the relative estimation error that can be used as a multiplicative correction of the pilot smoother. This leads to the (nonlinear) smoother

$$\widehat{m}_n(x) = \widehat{\alpha}_n(x)\tilde{m}_n(x). \tag{3}$$

The estimator (3) was first studied for fixed design by [19] and extended to the random design by [20]. In both cases, they assumed that the regression function had four continuous derivatives, and show an improvement in the convergence rate of the bias corrected Nadaraya-Watson kernel smoother. The idea of multiplicative bias reduction can be traced back to [9, 10], who proposed a parametrically guided local linear smoother that extended a parametric pilot regression estimate with a local polynomial smoother. It is showed that the resulting regression estimate improves on the naïve local polynomial estimate as soon as the pilot captures some of the features of the regression function.

3 Theoretical Analysis of Multiplicative Bias Reduction

In this section, we show that the multiplicative smoother has smaller bias with essentially no cost to the variance, assuming only two derivatives of the regression function. While the derivation of our results is for local linear smoothers, the technique used in the proofs can be easily adapted for other linear smoothers, and the conclusions remain essentially unchanged.

3.1 Assumptions

We make the following assumptions:

1. The regression function is bounded and strictly positive, that is, $b \geq m(x) \geq a > 0$ for all x.
2. The regression function is twice continuously differentiable everywhere.

3. The density of the covariate is strictly positive on the interior of its support in the sense that $f(x) \geq b(\mathcal{K}) > 0$ over every compact $\mathcal{K}$ contained in the support of f.
4. ε has finite fourth moments and has a symmetric distribution around zero.
5. Given a bounded symmetric probability density $K(\cdot)$, consider the weights $\omega_j(x; h)$ associated to the local linear smoother. That is, denote by $K_h(\cdot) = K(\cdot/h)/h$ the scaled kernel by the bandwidth h and define for $k = 0, 1, 2, 3$ the sums

$$S_k(x) \equiv S_k(x; h) = \sum_{j=1}^{n} (X_j - x)^k K_h(X_j - x).$$

Then

$$\omega_j(x; h) = \frac{S_2(x; h) - (X_j - x) S_1(x; h)}{S_2(x; h) S_0(x; h) - S_1^2(x; h)} K_h(X_j - x).$$

We set

$$\omega_{0j}(x) = \omega_j(x; h_0) \quad \text{and} \quad \omega_{1j}(x) = \omega_j(x; h_1).$$

6. The bandwidths h_0 and h_1 are such that

$$h_0 \to 0, \quad h_1 \to 0, \quad nh_0 \to \infty, \quad nh_1^3 \to \infty, \quad \frac{h_1}{h_0} \to 0 \quad \text{as} \quad n \to \infty.$$

The positivity assumption (assumption 1) on $m(x)$ is classical when we perform a multiplicative bias correction. It allows to avoid that the terms $Y_j/\tilde{m}_n(X_j)$ blow up. Of course, the regression function might cross the x-axis. For such a situation, [10] proposes to shift all response data Y_i a distance a, so that the new regression function $m(x) + a$ does not any more intersect with the x-axis. Such a method can also be performed here. Assumptions 2–4 are standard to obtain rate of convergence for nonparametric estimators. Assumption 5 means that we conduct the theory for the local linear estimate. The results can be generalized to other linear smoothers. Assumption 6 is not restrictive since it is satisfied for a wide range of values of h_0 and h_1.

3.2 A Technical Aside

The proof of the main results rests on establishing a stochastic approximation of estimator (3) in which each term can be directly analyzed.

Proposition 1 *We have*

$$\widehat{m}_n(x) = \mu_n(x) + \sum_{j=1}^{n} \omega_{1j}(x)A_j(x) + \sum_{j=1}^{n} \omega_{1j}(x)B_j(x) + \sum_{j=1}^{n} \omega_{1j}(x)\xi_j,$$

where $\mu_n(x)$, conditionally on $X_1, \ldots, X_n$ is a deterministic function, A_j, B_j, and ξ_j are random variables. Under condition $nh_0 \to \infty$, the remainder ξ_j converges to 0 in probability and we have

$$\widehat{m}_n(x) = \mu_n(x) + \sum_{j=1}^{n} \omega_{1j}(x)A_j(x) + \sum_{j=1}^{n} \omega_{1j}(x)B_j(x) + O_P\left(\frac{1}{nh_0}\right).$$

Remark 1 A technical difficulty arises because even though ξ_j may be small in probability, its expectation may not be small. We resolve this problem by showing that we only need to modify ξ_j on a set of vanishingly small probability to guarantee that its expectation is also small.

Definition 1 Given a sequence of real numbers a_n, we say that a sequence of random variables $\xi_n = o_p(a_n)$ if for all fixed $t > 0$,

$$\limsup_{n \longrightarrow \infty} \mathbb{P}[|\xi_n| > ta_n] = 0.$$

We will need the following Lemma.

Lemma 1 *If $\xi_n = o_p(a_n)$, then there exists a sequence of random variables $\xi_n^\star$ such that*

$$\limsup_{n \longrightarrow \infty} \mathbb{P}[\xi_n^\star \neq \xi_n] = 0 \quad \textit{and} \quad \mathbb{E}[\xi_n^\star] = o(a_n).$$

We shall use the following notation:

$$\mathbb{E}[\xi_n] = \mathbb{E}[\xi_n^\star].$$

3.3 *Asymptotic Behavior*

We deduce from Proposition 1 and Lemma 1 the following theorem.

Theorem 1 *Under the assumptions (1)-(6), the estimator $\widehat{m}_n$ satisfies:*

$$\mathbb{E}\left(\widehat{m}_n(x)|X_1, \ldots, X_n\right) = \mu_n(x) + O_p\left(\frac{1}{n\sqrt{h_0h_1}}\right) + O_p\left(\frac{1}{nh_0}\right)$$

and

$$\mathbb{V}_\star(\widehat{m}_n(x)|X_1,\ldots,X_n) = \sigma^2 \sum_{j=1}^{n} w_{1j}^2(x) + \mathrm{O}_p\left(\frac{1}{nh_0}\right) + \mathrm{o}_p\left(\frac{1}{nh_1}\right).$$

We deduce from Theorem 1 that if the bandwidth h_0 of the pilot estimator converges to zero much slower than h_1, then $\widehat{m}_n$ has exactly the same asymptotic variance as the local linear smoother of the original data with bandwidth h_1. However, for finite samples, the two step local linear smoother can have a slightly larger variance depending on the choice of h_0. For the bias term, a limited Taylor expansion of $\mu_n(x)$ leads to the following result.

Theorem 2 *Under the assumptions (1)-(6), the estimator $\widehat{m}_n$ satisfies:*

$$\mathbb{E}\left(\widehat{m}_n(x)|X_1,\ldots,X_n\right) = m(x) + \mathrm{o}_p(h_1^2).$$

Remark 2 Note that we only assume that the regression function is twice continuously differentiable. We do not add smoothness assumptions to improve the convergence rate from $\mathrm{O}_p(h_1^2)$ to $\mathrm{o}_p(h_1^2)$. In that manner, our analysis differs from that of [20] who assumed m to be four times continuously differentiable to conclude that the bias corrected smoother converged at the $\mathrm{O}_p(h_1^4)$ rate. For a study of the local linear estimate in the presence of jumps in the derivative, we refer the reader to [5].

Remark 3 Under similar smoothness assumptions, [8, 10, 21] have provided a comprehensive asymptotic behavior for the multiplicative bias corrected estimator with a parametric guide. They obtain the same asymtptotic variance as the local linear estimate and a bias reduction provided the parametric guide captures some of the features of the regression function. We obtain a similar result when the rate of decay of the bandwidth of the pilot estimate is carefully chosen.

Combining Theorems 1 and 2, we conclude that the multiplicative adjustment performs a bias reduction on the pilot estimator without increasing the asymptotic variance. The asymptotic behavior of the bandwidths h_0 and h_1 is constrained by assumption 6. However, it is easily seen that this assumption is satisfied for a large set of values of h_0 and h_1. For example, the choice $h_1 = c_1 n^{-1/5}$ and $h_0 = c_0 n^{-\alpha}$ for $0 < \alpha < 1/5$ leads to

$$\mathbb{E}_\star\left(\widehat{m}_n(x)|X_1,\ldots,X_n\right) - m(x) = \mathrm{o}_p(n^{-2/5})$$

and

$$\mathbb{V}_\star(\widehat{m}_n(x)|X_1,\ldots,X_n) = \mathrm{O}_p\left(n^{-4/5}\right).$$

Remark 4 Estimators with bandwidths of order $O(n^{-\alpha})$ for $0 < \alpha < 1/5$ are oversmoothing the true regression function, and as a result, the magnitude of their biases is of larger than the magnitude of their standard deviations. We conclude that the multiplicative adjustment performs a bias reduction on the pilot estimator.

4 Numerical Examples

Results presented in the previous sections show that our procedure allows to reduce the bias of nonparametric smoothers at no cost for the asymptotic variance. The simulation study in this section shows that the practical benefits of this asymptotic behavior already emerge at modest sample sizes.

4.1 Local Study

To illustrate numerically the reduction in the bias and associate (limited) increase of the variance achieved by the multiplicative bias correction, consider estimating the regression function

$$m(x) = 5 + 3|x|^{5/2} + x^2 + 4\cos(10x)$$

at $x = 0$ (see Fig. 1). The local linear smoother is known to under-estimate the regression function at local maxima and over-estimate local minima, and hence, this example provides a good example to explore bias-reduction variance-increase trade-off. Furthermore, because the second derivative of this regression function is

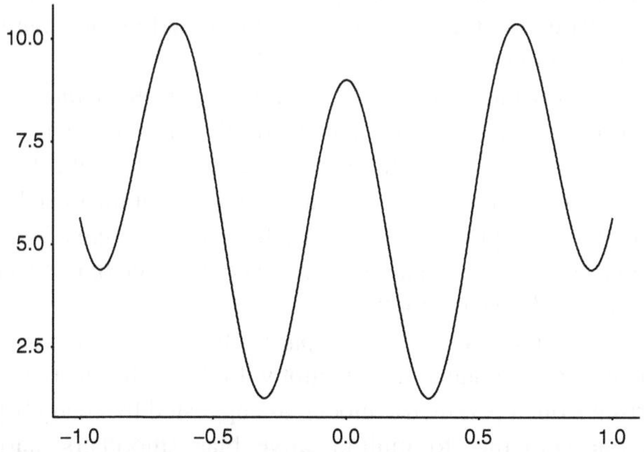

Fig. 1 The regression function to be estimated

continuous but not differentiable at the origin, the results previously obtained by [19] do not apply.

For our Monte-Carlo simulation, the data are generated according to the model

$$Y_i = m(X_i) + \varepsilon_i, \quad i = 1, \ldots, 100,$$

where ε_i are independent $\mathcal{N}(0, 1)$ variables, and the covariates X_i are independent uniform random variables on the interval $[-1, 1]$.

We first consider the local linear estimate and we study its performances over a grid of bandwidths $\mathcal{H} = [0.005, 0.1]$. For the new estimate, the theory recommends to start with an oversmooth pilot estimate. In this regard, we take $h_0 = 0.1$ and study the performance of the multiplicative bias corrected estimate for $h_1 \in \mathcal{H}_1 = [0.005, 0.12]$. To explore the stability of our two-stage estimator with respect to h_0, we also consider the choice $h_0 = 0.02$. For such a choice, the pilot estimate clearly undersmoothes the regression function. For both estimates, we take the Gaussian kernel $K(x) = \exp(-x^2/2)/\sqrt{2\pi}$.

We conduct a Monte Carlo study to estimate bias and variance of each estimate at $x = 0$. To this end, we compute the estimate at $x = 0$ for 1000 samples $(X_i, Y_i), i = 1, \ldots, 100$. The same design $X_i, i = 1, \ldots, 100$ is used for all the sample. The bias at point $x = 0$ is estimated by subtracting $m(0)$ at the mean value of the estimate at $x = 0$ (the mean value is computed over the 1000 replications). Similarly we estimate the variance at $x = 0$ by the variance of the values of the estimate at this point. Figure 2 presents squared bias, variance and mean square error of each estimate for different values of bandwidths h for the local linear smoother and h_1 for our estimate.

Comparing panel (a) and (c) in Fig. 2, we see that if the pilot smoother underestimates the regression function, then the bias is small but the variance is large. For such a pilot smoother, applying a bias correction does not provide any benefit, and the resulting estimator can be worse than a good local linear smoother. Intuitively, the bias of the pilot smoother is already small at the cost of a larger variance, and operating a bias reduction provides little benefit to the bias and can only make the variance worse, leading to a suboptimal smoother.

Comparing panel (a) and (b) in Fig. 2, we note that the squared bias is smaller for the bias corrected smoother over the standard local linear smoother, while the variance of both smoothers is essentially the same. As a result, the mean squared error for the bias corrected smoother is smaller than that of the local linear smoother. This shows that the asymptotic properties outlined in Theorems 1 and 2 emerge for moderate sample sizes. Table 1 quantifies the benefits of the bias corrected smoother over the classical local linear smoother.

We conclude our local study by comparing the multiplicative bias correction smoother starting from a nonparametric pilot with the multiplicative bias correction smoother starting from a parametric model, as suggested by Glad [10]. Specifically, we compare our smoother to multiplicative bias smoothers starting with the following three parametric models:

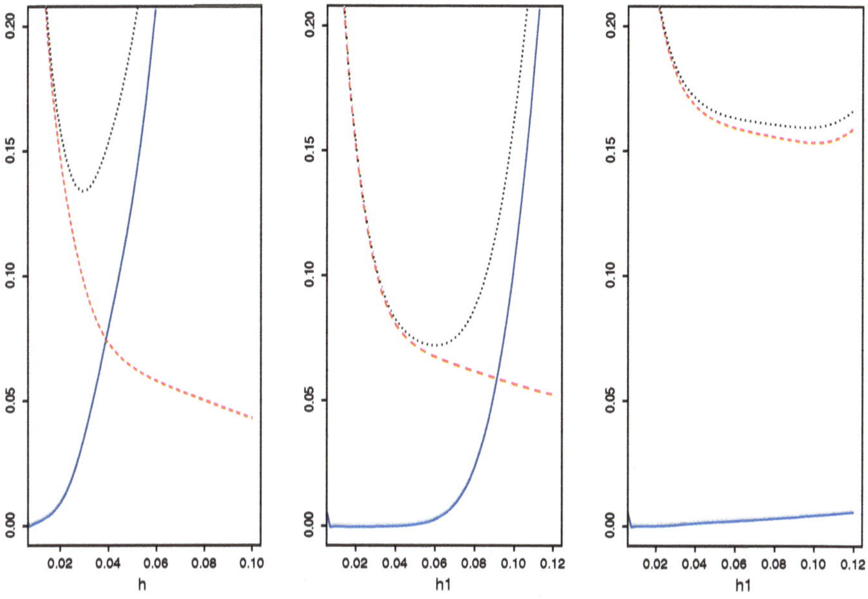

Fig. 2 Mean square error (dotted line), squared bias (solid line), and variance (dashed line) of the local linear estimate (**a**) and multiplicative bias corrected estimate with $h_0 = 0.1$ (**b**) and $h_0 = 0.02$ (**c**) at point $x = 0$

Table 1 Optimal mean square error (MSE) for the local linear estimate (LLE) and the multiplicative bias corrected estimate (MBCE) with $h_0 = 0.1$ at point $x = 0$

	MSE	Bias^2	Variance
LLE	0.130	0.031	0.098
MBCE	0.068	0.003	0.065

- first, the guide is chosen correctly and belong to the true parametric family:

$$\tilde{m}_n^1(x) = \hat{\beta}_0 + \hat{\beta}_1 |x|^{5/2} + \hat{\beta}_2 x^2 + \hat{\beta}_3 \cos(10x);$$

- second, we consider a linear parametric guide (which is obviously wrong):

$$\tilde{m}_n^2(x) = \hat{\beta}_0 + \hat{\beta}_1 x;$$

- finally, we use a more reasonable guide, not correct, but that can reflect some a priori idea on the regression function

$$\tilde{m}_n^3(x) = \hat{\beta}_0 + \hat{\beta}_1 x + \hat{\beta}_2 x^2 + \ldots + \hat{\beta}_8 x^8.$$

All the estimates $\hat{\beta}_j$ stand for the classical least square estimates.

Table 2 Pointwise optimal mean square error at $x = 0$ for the multiplicative bias corrected estimates with parametric starts $\tilde{m}_n^j$, $j = 1, 2, 3$, compared to a multiplicative bias corrected smoother starting with initial bandwidth $h_0 = 0.1$

	MSE	Bias^2	Variance
Start $\tilde{m}_n^1$	0.052	0.000	0.052
Start $\tilde{m}_n^2$	0.129	0.031	0.098
Start $\tilde{m}_n^3$	0.090	0.019	0.071
MBCE	0.068	0.003	0.065

The multiplicative bias correction is performed on these parametric starts using the local linear estimate. The performance of the resulting estimates is measured over a grid of bandwidths $\mathscr{H}_2 = [0.005; 0.4]$. Bias and variance of each estimate are still estimated at $x = 0$. We keep the same setting as above and all the results are averaged over the same 1000 replications. We display in Table 2 the optimal MSE calculated over the grid $\mathscr{H}_2$.

As expected, the performance depends on the choice of the parametric start. Unsurprisingly, the performance of the smoother starting with the parametric guide $\tilde{m}_n^1$ (which belongs to the true model) is best. Table 2 shows that (in terms of MSE) the estimate studied in this work is better than the corrected estimated with parametric start $\tilde{m}_n^2$ and $\tilde{m}_n^3$. This suggests that in practice, when little priori information on the target regression function is available, the method proposed in the present contribution is preferable.

4.2 Global Study

The theory in Sect. 3 does not address the practical issue of bandwidths selection for both the pilot smoother and the multiplicative adjustment. Burr et al. [2] suggests adapting existing automatic bandwidth selection procedures to this problem. There is a large literature on automatic bandwidth selection, including [14, 15]. In this section, we present a numerical investigation of the leave-one-out cross-validation method to select both bandwidths h_0 and h_1 as to minimize the integrated square error of the estimator. The resulting bias smoother is compared with a local polynomial smoother, whose bandwidth is selected in a similar manner.

Our selection of test functions for our investigation relies on the comprehensive numerical study of [18]. We will only compare our multiplicative bias corrected smoother with the classical local linear smoother. In all our examples, we use a Gaussian kernel to construct nonparametric smoothers to estimate the following

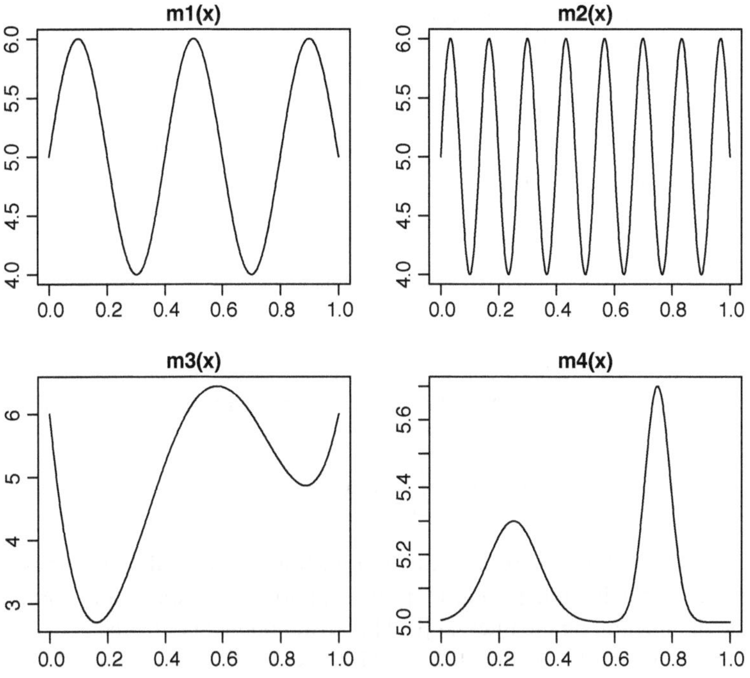

Fig. 3 Regression functions to be estimated

regression functions (see Fig. 3):

(1) $m_1(x) = \sin(5\pi x)$
(2) $m_2(x) = \sin(15\pi x)$
(3) $m_3(x) = 1 - 48x + 218x^2 - 315x^3 + 145x^4$
(4) $m_4(x) = 0.3\exp[-64(x - .25)^2] + 0.7\exp[-256(x - .75)^2]$.

from data $Y_{ji} = m_j(X_i) + \varepsilon_{ji}$, with disturbances $\varepsilon_{j1}, \ldots, \varepsilon_{jn}$ i.i.d. Normal with mean zero and standard deviation $\sigma_j = 0.25\|m_j\|_2$, $j = 1, \ldots, 4$, and $X_1, \ldots, X_n$ i.i.d. Uniform on $[-0.2, 1.2]$.

We use a cross validation device to select both h_0 and h_1 by minimizing simultaneously over a finite grid $\mathscr{H}$ of bandwidths h_0 and h_1 the leave-one-out prediction error. That is, given a grid $\mathscr{H}$, we choose the pair $(\hat{h}_0, \hat{h}_1)$ defined by

$$(\hat{h}_0, \hat{h}_1) = \underset{(h_0,h_1)\in\mathscr{H}\times\mathscr{H}}{\operatorname{argmin}} \frac{1}{n}\sum_{i=1}^{n}(Y_i - \widehat{m}_n^i(X_i))^2.$$

Table 3 Median over 1000 replications of the selected bandwidths and of the integrated square error of the selected estimates

	LLE		MBCE			
	h	ISE (×100)	h_0	h_1	ISE (×100)	R_{ISE}
m_1	0.023	0.957	0.050	0.032	0.735	1.316
m_2	0.011	6.094	0.028	0.012	4.771	1.286
m_3	0.028	2.022	0.071	0.054	1.281	1.591
m_4	0.018	0.087	0.034	0.024	0.074	1.187

LLE and MBCE stands for local linear estimate and multiplicative bias corrected estimate

Here $\widehat{m}_n^i$ stands for the prediction of the bias corrected smoother at X_i, estimated without the observation (X_i, Y_i). We use the Integrated Square Error (ISE)

$$ISE(\widehat{m}) = \int_0^1 (m(x) - \widehat{m}(x))^2 \, dx,$$

to measure the performance of an estimator $\widehat{m}$. Note that even though our estimators are defined on the interval $[-0.2, 1.2]$ (the support of the explanatory variable), we evaluate the integral on the interval $[0, 1]$ to avoid boundary effects.

Table 3 compares the median ISE over 1000 replication, of a standard local linear smoother and our bias corrected smoother from a samples of size $n = 100$. This table further presents the median selected bandwidth, and the ratio of the ISE.

First, in all four cases, the ISE for the MBCE is smaller than that of the LLE. Second, we note that both bandwidths for the multiplicative bias corrected are larger than the optimal bandwidth of the classical local linear smoother. That h_0 is larger is supported by the theory, as the pilot smoother needs to oversmooth. We surmise that larger bandwidth h_1 reflects the fact that the pilot is reasonably close to the true regression function, and hence the multiplicative correction is quite smooth and thus can accommodate a larger bandwidth. Figure 4 displays the boxplots of the integrated square error for each estimate.

Figure 5 presents, for the regression function m_1 with $n = 100$ and 1000 iterations, different estimators on a grid of points. In lines is the true regression function which is unknown. For every point on a fixed grid, we plot, side by side, the mean over 1000 replications of our estimator at that point (left side) and on the right side of that point the mean over 1000 replications of the local polynomial estimator. Leave-one-out cross validation is applied to select the bandwidths h_0 and h_1 for our estimator and the bandwidth h for the local polynomial estimator. We add also the interquartile interval in order to see the fluctuations of the different estimators. In this example, our estimator reduces the bias by increasing the peak and decreasing the valleys. Moreover, the interquartile intervals look similar for both estimator, as predicted by the theory.

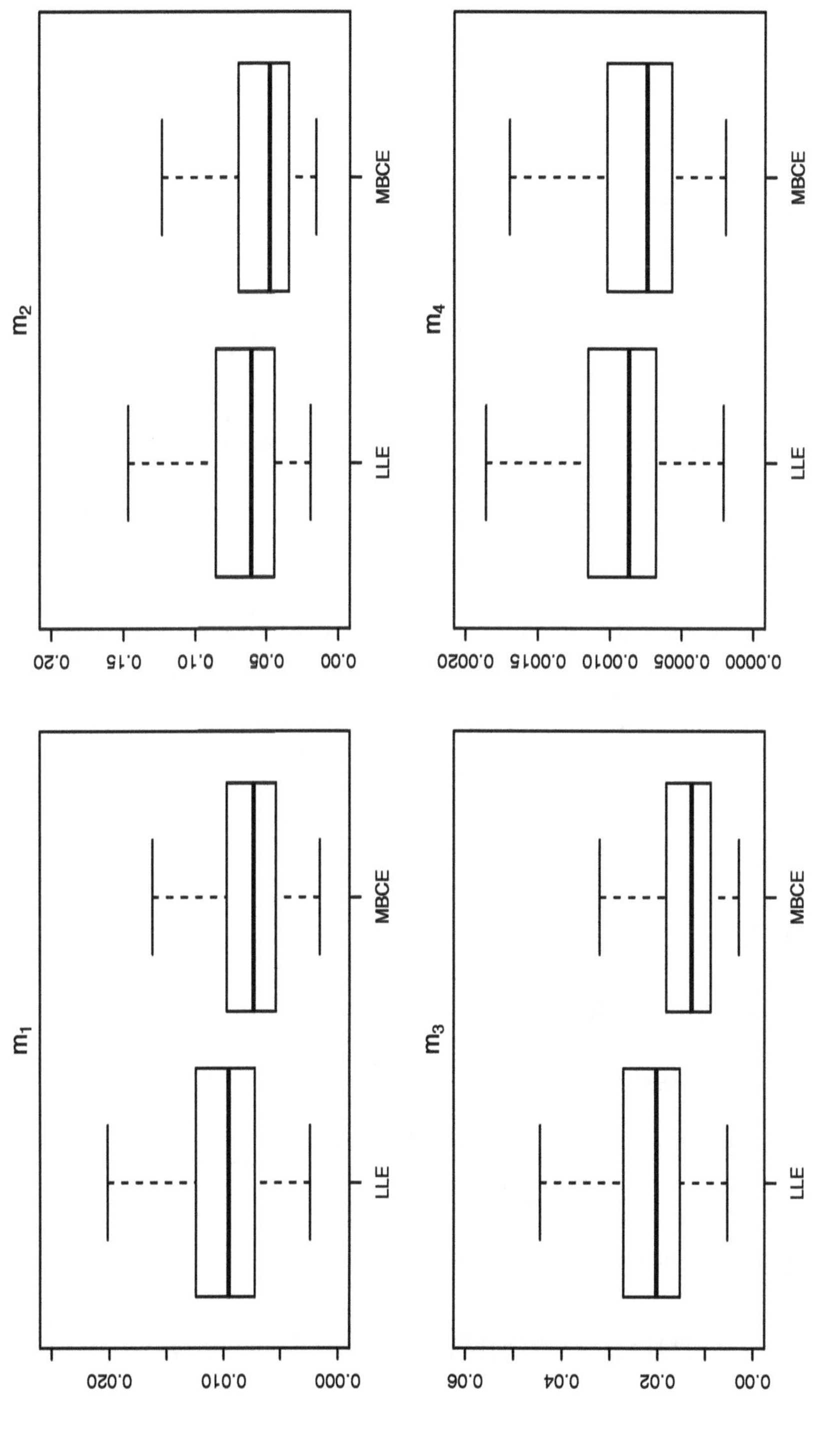

Fig. 4 Boxplot of the integrated square error over the 1000 replications

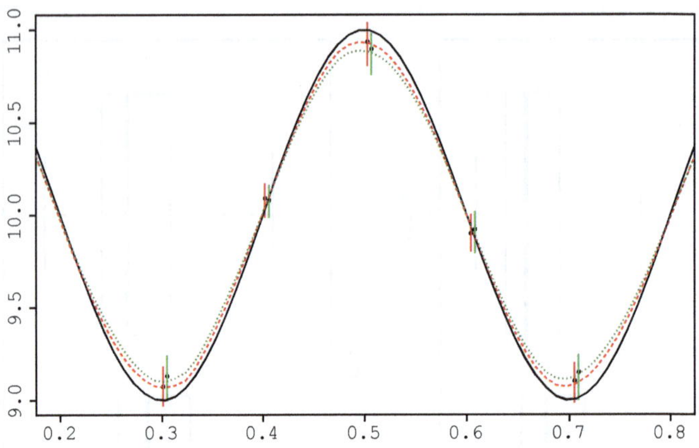

Fig. 5 The solid curve represents the true regression function, our estimator is in dashed line and local linear smoother is dotted

5 Conclusion

This chapter revisits the idea of multiplicative bias reduction under minimal conditions and shows that it is possible to reduce the bias with little effect to the variance. Our theory proves that our proposed estimator has zero asymptotic bias while maintaining the same asymptotic variance than the original smoother. The simulation study in this work shows that this desirable property emerges for even modest sample sizes. The one downside of our estimator is that the computation of data driven "optimal" bandwidths is computationally expensive.

6 Proofs

6.1 Proof of Proposition 1

Write the bias corrected estimator

$$\widehat{m}_n(x) = \sum_{j=1}^{n} \omega_{1j}(x) \frac{\tilde{m}_n(x)}{\tilde{m}_n(X_j)} Y_j = \sum_{j=1}^{n} \omega_{1j}(x) R_j(x) Y_j,$$

and let us approximate the quantity $R_j(x)$. Define

$$\bar{m}_n(x) = \sum_{j=1}^{n} \omega_{0j}(x) m(X_j) = \mathbb{E}\left(\tilde{m}_n(x) | X_1, \ldots, X_n\right),$$

and observe that

$$\begin{aligned}
R_j(x) =& \frac{\tilde{m}_n(x)}{\tilde{m}_n(X_j)} \\
=& \frac{\bar{m}_n(x)}{\bar{m}_n(X_j)} \times \left(1 + \frac{\tilde{m}_n(x) - \bar{m}_n(x)}{\bar{m}_n(x)}\right) \times \left(1 + \frac{\tilde{m}_n(X_j) - \bar{m}_n(X_j)}{\bar{m}_n(X_j)}\right)^{-1} \\
=& \frac{\bar{m}_n(x)}{\bar{m}_n(X_j)} \times [1 + \Delta_n(x)] \times \frac{1}{1 + \Delta_n(X_j)},
\end{aligned}$$

where

$$\Delta_n(x) = \frac{\tilde{m}_n(x) - \bar{m}_n(x)}{\bar{m}_n(x)} = \frac{\sum_{l \le n} \omega_{0l}(x)\varepsilon_l}{\sum_{l \le n} \omega_{0l}(x) m(X_l)}.$$

Write now $R_j(x)$ as

$$R_j(x) = \frac{\bar{m}_n(x)}{\bar{m}_n(X_j)} \left[1 + \Delta_n(x) - \Delta_n(X_j) + r_j(x, X_j)\right]$$

where $r_j(x, X_j)$ is a random variable converging to 0 to be defined later on. Given the last expression and model (1), estimator (3) could be written as

$$\begin{aligned}
\widehat{m}_n(x) =& \sum_{j=1}^{n} \omega_{1j}(x) R_j(x) Y_j \\
=& \sum_{j=1}^{n} \omega_{1j}(x) \frac{\bar{m}_n(x)}{\bar{m}_n(X_j)} m(X_j) + \sum_{j=1}^{n} \omega_{1j}(x) \frac{\bar{m}_n(x)}{\bar{m}_n(X_j)} \left[\varepsilon_j + m(X_j)\left(\Delta_n(x) - \Delta_n(X_j)\right)\right] \\
&+ \sum_{j=1}^{n} \omega_{1j}(x) \frac{\bar{m}_n(x)}{\bar{m}_n(X_j)} \left(\Delta_n(x) - \Delta_n(X_j)\right) \varepsilon_j + \sum_{j=1}^{n} \omega_{1j}(x) \frac{\bar{m}_n(x)}{\bar{m}_n(X_j)} r_j(x, X_j) Y_j \\
=& \mu_n(x) + \sum_{j=1}^{n} \omega_{1j}(x) A_j(x) + \sum_{j=1}^{n} \omega_{1j}(x) B_j(x) + \sum_{j=1}^{n} \omega_{1j}(x) \xi_j.
\end{aligned}$$

which is the first part of the proposition. Under assumption set forth in Sect. 3.1, the pilot smoother $\tilde{m}_n$ converges to the true regression function $m(x)$. Bickel and Rosenblatt [1] shows that this convergence is uniform over compact sets $\mathcal{K}$ contained in the support of the density of the covariate X. As a result, for n large enough $\sup_{x \in \mathcal{K}} |\tilde{m}_n(x) - \bar{m}_n(x)| \le \frac{1}{2}$ with probability 1. So a limited expansion of $(1+u)^{-1}$ yields for $x \in \mathcal{K}$

$$R_j(x) = \frac{\bar{m}_n(x)}{\bar{m}_n(X_j)} \left[1 + \Delta_n(x) - \Delta_n(X_j) + \mathrm{O}_p\left(|\Delta_n(x)\Delta_n(X_j)| + \Delta_n^2(X_j)\right)\right],$$

thus

$$\xi_j = \mathrm{O}_p\left(|\Delta_n(x)\Delta_n(X_j)| + \Delta_n^2(X_j)\right).$$

Under the stated regularity assumptions, we deduce that $\xi_j = \mathrm{O}_p\left(\frac{1}{nh_0}\right)$, leading to the announced result. Proposition 1 is proved.

6.2 *Proof of Lemma 1*

By definition $\limsup_{n\longrightarrow\infty} \mathbb{P}[|\xi_n| > ta_n] = 0$ for all $t > 0$, so that a triangular array argument shows that there exists an increasing sequence $m = m(k)$ such that

$$\mathbb{P}\left[|\xi_n| > \frac{a_n}{k}\right] \leq \frac{1}{k} \qquad \text{for all } n \geq m(k).$$

For $m(k) \leq n \leq m(k+1) - 1$, define

$$\xi_n^\star = \begin{cases} \xi_n & \text{if } |\xi_n| < k^{-1}a_n \\ 0 & \text{otherwise.} \end{cases}$$

It follows from the construction of $\xi_n^\star$ that for $n \in (m(k), m(k+1) - 1)$,

$$\mathbb{P}[\xi_n \neq \xi_n^*] = \mathbb{P}[|\xi_n| > k^{-1}a_n] \leq \frac{1}{k},$$

which converges to zero as n goes to infinity. Finally set $k(n) = \sup\{k : m(k) \leq n\}$, we obtain

$$\mathbb{E}[|\xi_n^\star|] \leq \frac{a_n}{k(n)} = \mathrm{o}(a_n).$$

6.3 *Proof of Theorem 1*

Recall that $\widehat{m}_n(x) = \mu_n(x) + \sum_{j=1}^n \omega_{1j}(x)A_j(x) + \sum_{j=1}^n \omega_{1j}(x)B_j(x) + \mathrm{O}_P\left(\frac{1}{nh_0}\right)$. Focus on the conditional bias, we get

$$\mathbb{E}(\mu_n(x)|X_1, \dots, X_n) = \mu_n(x), \quad \mathbb{E}(A_j(x)|X_1, \dots, X_n) = 0$$

and

$$\mathbb{E}(B_j(x)|X_1,\ldots,X_n) = \frac{\bar{m}_n(x)}{\bar{m}_n(X_j)}\sigma^2\Big(\frac{\omega_{0j}(x)}{\bar{m}_n(x)} - \frac{\omega_{0j}(X_j)}{\bar{m}_n(X_j)}\Big).$$

Since

$$\left|\sum_{j=1}^{n}\omega_{1j}(x)\omega_{0j}(x)\right| \le \sqrt{\sum_{j=1}^{n}\omega_{1j}(x)^2}\sqrt{\sum_{j=1}^{n}\omega_{0j}(x)^2} = \mathrm{O}_p\left(\frac{1}{n\sqrt{h_0h_1}}\right),$$

we deduce that

$$\mathbb{E}\left(\sum_{j=1}^{n}\omega_{1j}(x)B_j(x)\Big|X_1,\ldots,X_n\right) = \mathrm{O}_p\left(\frac{1}{n\sqrt{h_0h_1}}\right).$$

This proves the first part of the theorem. For the conditional variance, we use the following expansion of the two-stage estimator

$$\widehat{m}_n(x) = \sum_{j=1}^{n}\omega_{1j}(x)\frac{\bar{m}_n(x)}{\bar{m}_n(X_j)}Y_j\left(1+\left[\Delta_n(x)-\Delta_n(X_j)\right]\right) + \mathrm{O}_p\left(\frac{1}{nh_0}\right).$$

Using the fact that the residuals have four finite moments and have a symmetric distribution around 0, a moment's thought shows that

$$\mathbb{V}(Y_j\left[\Delta_n(x)-\Delta_n(X_j)\right]|X_1,\ldots,X_n) = \mathrm{O}_p\left(\frac{1}{nh_0}\right)$$

and

$$\mathrm{Cov}(Y_j, Y_j\left[\Delta_n(x)-\Delta_n(X_j)\right]|X_1,\ldots,X_n) = \mathrm{O}_p\left(\frac{1}{nh_0}\right).$$

Hence

$$\mathbb{V}_\star(\widehat{m}_n(x)|X_1,\ldots,X_n) = \mathbb{V}\left(\sum_{j=1}^{n}\omega_{1j}(x)\frac{\bar{m}_n(x)}{\bar{m}_n(X_j)}Y_j\Big|X_1,\ldots,X_n\right) + \mathrm{O}_p\left(\frac{1}{nh_0}\right).$$

Observe that the first term on the right-hand side of this equality can be seen as the variance of the two-stage estimator with a deterministic pilot estimator. It follows from [10] that

$$\mathbb{V}\left(\sum_{j=1}^{n}\omega_{1j}(x)\frac{\bar{m}_n(x)}{\bar{m}_n(X_j)}Y_j\middle|X_1,\ldots,X_n\right)=\sigma^2\sum_{j=1}^{n}\omega_{1j}^2(x)+o_p\left(\frac{1}{nh_1}\right),$$

which proves the theorem.

6.4 Proof of Theorem 2

Recall that

$$\mu_n(x)=\sum_{j\le n}\omega_{1j}(x)\frac{\bar{m}_n(x)}{\bar{m}_n(X_j)}m(X_j).$$

We consider the limited Taylor expansion of the ratio

$$\frac{m(X_j)}{\bar{m}_n(X_j)}=\frac{m(x)}{\bar{m}_n(x)}+(X_j-x)\left(\frac{m(x)}{\bar{m}_n(x)}\right)'+\frac{1}{2}(X_j-x)^2\left(\frac{m(x)}{\bar{m}_n(x)}\right)''(1+o_p(1)),$$

then

$$\mu_n(x)=\bar{m}_n(x)\left\{\frac{m(x)}{\bar{m}_n(x)}\sum_{j=1}^{n}\omega_{1j}(x)+\left(\frac{m(x)}{\bar{m}_n(x)}\right)'\sum_{j=1}^{n}(X_j-x)\omega_{1j}(x)\right.$$
$$\left.+\frac{1}{2}\left(\frac{m(x)}{\bar{m}_n(x)}\right)''\sum_{j=1}^{n}(X_j-x)^2\omega_{1j}(x)(1+o_p(1))\right\}.$$

It is easy to verify that $\sum_{j=1}^{n}\omega_{1j}(x)=1$, $\sum_{j=1}^{n}(X_j-x)\omega_{1j}(x)=0$, and

$$\Sigma_2(x;h_1)=\sum_{j=1}^{n}(X_j-x)^2\omega_{1j}(x)=\frac{S_2^2(x;h_1)-S_3(x;h_1)S_1(x;h_1)}{S_2(x;h_1)S_0(x;h_1)-S_1^2(x;h_1)}.$$

For random designs, we can further approximate (see, e.g., [27])

$$S_k(x,h_1)=\begin{cases}h^k\sigma_K^k f(x)+o_p(h^k) & \text{for } k \text{ even}\\ h^{k+1}\sigma_K^{k+1} f'(x)+o_p(h^{k+1}) & \text{for } k \text{ odd,}\end{cases}$$

where $\sigma_K^k = \int u^k K(u)\,\mathrm{d}u$. Therefore

$$\begin{aligned}\Sigma_2(x;h_1) &= h_1^2 \int u^2 K(u)\,\mathrm{d}u + \mathrm{o}_p(h_1^2)\\ &= \sigma_K^2 h_1^2 + \mathrm{o}_p(h_1^2),\end{aligned}$$

so that we can write $\mu_n(x)$ as

$$\begin{aligned}\mu_n(x) =& \bar{m}_n(x)\left\{\frac{m(x)}{\bar{m}_n(x)} + \frac{\sigma_K^2 h_1^2}{2}\left(\frac{m(x)}{\bar{m}_n(x)}\right)'' + \mathrm{o}_p(h_1^2)\right\}\\ =& m(x) + \frac{\sigma_K^2 h_1^2}{2}\bar{m}_n(x)\left(\frac{m(x)}{\bar{m}_n(x)}\right)'' + \mathrm{o}_p(h_1^2).\end{aligned}$$

Moreover

$$\begin{aligned}\left(\frac{m(x)}{\bar{m}_n(x)}\right)'' =& \frac{\bar{m}_n^2(x)m''(x)}{\bar{m}_n^3(x)} - 2\frac{\bar{m}_n(x)\bar{m}_n'(x)m'(x)}{\bar{m}_n^3(x)}\\ &- \frac{m(x)\bar{m}_n(x)\bar{m}_n''(x)}{\bar{m}_n^3(x)} + 2\frac{m(x)(\bar{m}_n'(x))^2}{\bar{m}_n^3(x)}\end{aligned}$$

and applying the usual approximations, we conclude that

$$\left(\frac{m(x)}{\bar{m}_n(x)}\right)'' = \mathrm{o}_p(1).$$

Putting all pieces together, we obtain

$$\mathbb{E}(\widehat{m}_n(x)|X_1,\ldots,X_n) - m(x) = \mathrm{o}_p(h_1^2) + \mathrm{O}_p\left(\frac{1}{n\sqrt{h_0 h_1}}\right) + \mathrm{O}_p\left(\frac{1}{nh_0}\right).$$

Since $nh_1^3 \longrightarrow \infty$ and $\frac{h_1}{h_0} \longrightarrow 0$, we conclude that the bias is of order $\mathrm{o}_p(h_1^2)$.

References

1. Bickel, P. J., & Rosenblatt, M. (1973). On some global measures of the deviations of density function estimates. *The Annals of Statistics, 1*, 1071–1095.
2. Burr, T., Hengartner, N., Matzner-Løber, E., Myers, S., & Rouvière, L. (2010). Smoothing low resolution gamma spectra. *IEEE Transactions on Nuclear Science, 57*, 2831–2840.
3. Cornillon, P. A., Hengartner, N., Jegou, N., & Matzner-Løber, E. (2013). Iterative bias reduction: a comparative study. *Statistics and Computing, 23*(6), 777–791.
4. Cornillon, P. A., Hengartner, N., & Matzner-Løber, E. (2014). Recursive bias estimation for multivariate regression smoothers. *ESAIM: Probability and Statistics, 18*, 483–502.

5. Desmet, L., & Gijbels, I. (2009). Local linear fitting and improved estimation near peaks. *The Canadian Journal of Statistics, 37*, 473–475.
6. Di Marzio, M., & Taylor, C. (2007). Multistep kernel regression smoothing by boosting. http://www1.maths.leeds.ac.uk/~charles/boostreg.pdf
7. Fan, J., & Gijbels, I. (1996). *Local Polynomial Modeling and its Application, Theory and Methodologies*. New York: Chapman and Hall.
8. Fan, J., Wu, Y., & Feng, Y. (2009). Local quasi-likelihood with a parametric guide. *Annals of Statistics, 37*, 4153–4183.
9. Glad, I. (1998). A note on unconditional properties of parametrically guided Nadaraya-Watson estimator. *Statistics and Probability Letters, 37*, 101–108.
10. Glad, I. (1998). Parametrically guided non-parametric regression. *Scandinavian Journal of Statistics, 25*, 649–668.
11. Gustafsson, J., Hagmann, M., Nielsen, J. P., & Scaillet, O. (2009). Local transformation kernel density estimation of loss distributions. *Journal of Business and Economic Statistics, 27*, 161–175.
12. Hagmann, M., & Scaillet, O. (2007). Local multiplicative bias correction for asymmetric kernel density estimators. *Journal of Econometrics, 141*, 213–249.
13. Hengartner, N., & Matzner-Løber, E. (2009). Asymptotic unbiased density estimators. *ESAIM: Probability and Statistics, 13*, 1–14.
14. Hengartner, N., & Wegkamp, M. (2001). Estimation and selection procedures in regression: an l_1 approach. *Canadian Journal of Statistics, 29*(4), 621–632.
15. Hengartner, N., Wegkamp, M., & Matzner-Løber, E. (2002). Bandwidth selection for local linear regression smoothers. *Journal of the Royal Statistical Society: Series B, 64*, 1–14.
16. Hirukawa, M. (2006). A modified nonparametric prewhitened covariance estimator. *Journal of Time Series Analysis, 27*, 441–476.
17. Hirukawa, M. (2010). Nonparametric multiplicative bias correction for kernel-type density estimation on the unit interval. *Computational Statistics and Data Analysis, 54*, 473–495.
18. Hurvich, C., Simonoff, G., & Tsai, C. L. (1998). Smoothing parameter selection in nonparametric regression using and improved akaike information criterion. *Journal of the Royal Statistical Society: Series B, 60*, 271–294.
19. Jones, M. C., Linton, O., & Nielsen, J. P. (1994). A multiplicative bias reduction method for nonparametric regression. *Statistics and Probability Letters, 19*, 181–187.
20. Jones, M. C., Linton, O., & Nielsen, J. P. (1995). A simple and effective bias reduction method for kernel density estimation. *Biometrika, 82*, 327–338.
21. Martins-Filho, C., Mishra S., & Ullah, A. (2008). A class of improved parametrically guided nonparametric regression estimators. *Econometric Reviews, 27*, 542–573.
22. Mishra, S., Su, L., & Ullah, A. (2010). Semiparametric estimator of time series conditional variance. *Journal of Business and Economic Statistics, 28*, 256–274.
23. Nielsen, J. P. (1998). Multiplicative bias correction in kernel hazard estimation. *Scandinavian Journal of Statistics, 25*, 541–553.
24. Nielsen J. P., & Tanggaard, C. (2001) Boundary and bias correction in kernel hazard estimation. *Scandinavian Journal of Statistics, 28*, 675–698.
25. Scott, D. (1992). *Multivariate Density Estimation: Theory, Practice, and Visualization*. New-York: Wiley.
26. Simonoff, J. S. (1996). *Smoothing methods in statistics*. New York: Springer.
27. Wand, M. P., & Jones, M. C. (1995). *Kernel smoothing*. London: Chapman and Hall.
28. Xiao, Z., & Linton, O. (2002) A nonparametric prewhitened covariance estimator. *Journal of Time Series Analysis, 23*, 215–250.

Efficiency of the V-Fold Model Selection for Localized Bases

F. Navarro and A. Saumard

Abstract Many interesting functional bases, such as piecewise polynomials or wavelets, are examples of localized bases. We investigate the optimality of V-fold cross-validation and a variant called V-fold penalization in the context of the selection of linear models generated by localized bases in a heteroscedastic framework. It appears that while V-fold cross-validation is not asymptotically optimal when V is fixed, the V-fold penalization procedure is optimal. Simulation studies are also presented.

1 Introduction

V-fold cross-validation type procedures are extremely used in statistics and machine learning, with however a rather small set of theoretical results on it [3]. This chapter aims at investigating from the theoretical point of view and on simulations, the efficiency of two V-fold strategies for model selection in a heteroscedastic regression setting, with random design. On the one hand, we investigate the behaviour of the classical V-fold cross-validation to select, among other examples, linear models of wavelets. As pointed out in the case of histogram selection in [2], this procedure is not asymptotically optimal when V is fixed, as it is the case in practice where V is usually taken to be equal to 5 or 10. On the other hand, we study the V-fold penalization proposed by Arlot [2] and show its efficiency in our general context.

More precisely, the present contribution is devoted to an extension of some results obtained in [16] related to efficiency of cross-validation type procedures. Indeed, as remarked in [16] (see Remark 5.1 therein) our results obtained for the selection of linear models endowed with a strongly localized basis (see Definition (**Aslb**), Section 2.1 of [16]) can be extended to more general and more classical

F. Navarro (✉) · A. Saumard
CREST-ENSAI-UBL, Bruz, France
e-mail: fabien.navarro@ensai.fr; adrien.saumard@ensai.fr

P. Bertail et al. (eds.), *Nonparametric Statistics*, Springer Proceedings in Mathematics & Statistics 250, https://doi.org/10.1007/978-3-319-96941-1_4

localized bases, at the price of considering only models with sufficiently small dimensions. Rigorous proofs are given here and further simulation studies are explored.

This chapter is organized as follows. In Sect. 2, we describe our model selection setting. Then V-fold cross-validation is considered in Sect. 3, while the efficiency of V-fold penalization is tackled in Sect. 4. A simulation study is reported in Sect. 5. The proofs are exposed in Sect. 6.

2 Model Selection Setting

Assume that we observe n independent pairs of random variables $\xi_i = (X_i, Y_i) \in \mathcal{X} \times \mathbb{R}$ with common distribution P. For convenience, we also denote by $\xi = (X, Y)$ a random pair, independent of the sample $(\xi_1, \ldots, \xi_n)$, following the same distribution P. The set $\mathcal{X}$ is called the feature space and we assume $\mathcal{X} \subset \mathbb{R}^d$, $d \geq 1$. We denote by P^X the marginal distribution of the design X. We assume that the following regression relation is valid,

$$Y = s_* (X) + \sigma (X) \varepsilon ,$$

with $s_* \in L_2 \left(P^X\right)$ the regression function that we aim at estimating. Conditionally to X, the residual ε is normalized, i.e. it has mean zero and variance one. The function $\sigma : \mathcal{X} \to \mathbb{R}_+$ is a heteroscedastic noise level, assumed to be unknown.

To produce an estimator of s_*, we are given a finite collection of models $\mathcal{M}_n$, with cardinality depending on the amount n of data. Each model $m \in \mathcal{M}_n$ is taken to be a finite-dimensional vector space, of linear dimension D_m. We will further detail in a few lines the analytical structure of the models.

We set $\|s\|_2 = \left(\int_{\mathcal{X}} s^2 dP^X\right)^{1/2}$ the quadratic norm in $L_2 \left(P^X\right)$ and s_m the orthogonal—with respect to the quadratic norm—projection of s_* onto m. For a function $f \in L_1 (P)$, we write $P(f) = Pf = \mathbb{E}[f (\xi)]$. We call the least squares contrast a functional $\gamma : L_2 \left(P^X\right) \to L_1 (P)$, defined by

$$\gamma (s) : (x, y) \mapsto (y - s (x))^2 , \quad s \in L_2 \left(P^X\right) .$$

Using these notations, the regression function s_* is the unique minimizer of the risk,

$$s_* = \arg \min_{s \in L_2(P^X)} P (\gamma (s)) .$$

The projections s_m are also characterized by

$$s_m = \arg\min_{s \in m} P (\gamma (s)) .$$

To each model $m \in \mathcal{M}_n$, we associate a least squares estimator $\hat{s}_m$, defined by

$$\begin{aligned}\hat{s}_m &\in \arg\min_{s \in m} \{P_n (\gamma (s))\} \\ &= \arg\min_{s \in m} \left\{ \frac{1}{n} \sum_{i=1}^{n} (Y_i - s(X_i))^2 \right\} ,\end{aligned}$$

where $P_n = n^{-1} \sum_{i=1}^{n} \delta_{\xi_i}$ is the empirical measure associated to the sample.

The accuracy of estimation is tackled through the excess loss of the estimators,

$$\ell \left(s_*, \hat{s}_m\right) := P \left(\gamma \left(\hat{s}_m\right) - \gamma (s_*)\right) = \left\| \hat{s}_m - s_* \right\|_2^2 .$$

The following "bias-variance" decomposition holds,

$$\ell \left(s_*, \hat{s}_m\right) = \ell (s_*, s_m) + \ell \left(s_m, \hat{s}_m\right) ,$$

where

$$\begin{aligned}\ell (s_*, s_m) &:= P (\gamma (s_m) - \gamma (s_*)) = \| s_m - s_* \|_2^2 \\ \ell \left(s_m, \hat{s}_m\right) &:= P \left(\gamma \left(\hat{s}_m\right) - \gamma (s_m)\right) \geq 0 .\end{aligned}$$

The deterministic quantity $\ell (s_*, s_m)$ is called the bias of the model m, while the random variable $\ell \left(s_m, \hat{s}_m\right)$ is called the excess loss of the least squares estimator $\hat{s}_m$ on the model m. By the Pythagorean Theorem, we have

$$\ell \left(s_m, \hat{s}_m\right) = \left\| \hat{s}_m - s_m \right\|_2^2 .$$

From the collection of models $\mathcal{M}_n$, we aim at proposing an estimator that is as close as possible in terms of excess loss to an oracle model m_*, defined by

$$m_* \in \arg\min_{m \in \mathcal{M}_n} \left\{ \ell \left(s_*, \hat{s}_m\right) \right\} .$$

We choose to select an estimator from the collection $\left\{\hat{s}_m \ ; \ m \in \mathcal{M}_n\right\}$. Hence, the selected model is denoted by $\hat{m}$. The goal is to ensure that the selected estimator achieves an oracle inequality of the form

$$\ell \left(s_*, \hat{s}_{\hat{m}}\right) \leq C \times \inf_{m \in \mathcal{M}_n} \ell \left(s_*, \hat{s}_m\right) ,$$

for a constant $C \geq 1$ as close as possible to one and on an event of probability close to one.

3 V-Fold Cross-Validation

For convenience, let us denote in the following $\widehat{s}_m(P_n)$ the least squares estimator built from the empirical distribution $P_n = 1/n \sum_{i=1}^{n} \delta_{(X_i,Y_i)}$. To perform the V-fold cross-validation (VFCV) procedure, we consider a partition $\left(B_j\right)_{1\leq j\leq V}$ of the index set $\{1,\dots,n\}$ and set

$$P_n^{(j)} = \frac{1}{\operatorname{Card}\left(B_j\right)} \sum_{i\in B_j} \delta_{(X_i,Y_i)} \quad \text{and} \quad P_n^{(-j)} = \frac{1}{n-\operatorname{Card}\left(B_j\right)} \sum_{i\notin B_j} \delta_{(X_i,Y_i)} .$$

We assume that the partition $\left(B_j\right)_{1\leq j\leq V}$ is regular: for all $j \in \{1,\dots,V\}$, $\operatorname{Card}\left(B_j\right) = n/V$. It is worth noting that it is always possible to define our partition such $\sup_j \left|\operatorname{Card}\left(B_j\right) - n/V\right| < 1$ so that the assumption of regular partition is only a slight approximation of the general case. Let us write $\widehat{s}_m^{(-j)} = \widehat{s}_m\left(P_n^{(-j)}\right)$ the estimators built from the data in the block B_j. Now, the selected model $\widehat{m}_{\mathrm{VFCV}}$ is taken equal to any model optimizing the V-fold criterion,

$$\widehat{m}_{\mathrm{VFCV}} \in \arg\min_{n\in\mathcal{M}_n} \left\{\operatorname{crit}_{\mathrm{VFCV}}(m)\right\} , \tag{1}$$

where

$$\operatorname{crit}_{\mathrm{VFCV}}(m) = \frac{1}{V}\sum_{j=1}^{V} P_n^{(j)}\gamma\left(\widehat{s}_m^{(-j)}\right) . \tag{2}$$

Let us now detail the set of assumptions under which we will investigate the accuracy of VFCV.

Set of assumptions: (SA)

(P1) Polynomial complexity of $\mathcal{M}_n$: there exist some constants $c_{\mathcal{M}}, \alpha_{\mathcal{M}} > 0$ such that $\operatorname{Card}(\mathcal{M}_n) \leq c_{\mathcal{M}} n^{\alpha_{\mathcal{M}}}$.

(Alb) There exists a constant $r_{\mathcal{M}}$ such that for each $m \in \mathcal{M}_n$ one can find an orthonormal basis $(\varphi_k)_{k=1}^{D_m}$ satisfying, for all $(\beta_k)_{k=1}^{D_m} \in \mathbb{R}^{D_m}$,

$$\left\|\sum_{k=1}^{D_m} \beta_k\varphi_k\right\|_{\infty} \leq r_{\mathcal{M}}\sqrt{D_m}\,|\beta|_{\infty} , \tag{3}$$

where $|\beta|_{\infty} = \max\{|\beta_k|\,;\, k \in \{1,\dots,D_m\}\}$.

(P2) Upper bound on dimensions of models in $\mathcal{M}_n$: there exists a positive constant $A_{\mathcal{M},+}$ such that for every $m \in \mathcal{M}_n$, $D_m \leq A_{\mathcal{M},+} n^{1/3} (\ln n)^{-2}$.

(Ab) A positive constant A exists that bounds the data and the projections s_m of the target s_* over the models m of the collection $\mathcal{M}_n$: $|Y_i| \leq A < \infty$, $\|s_m\|_\infty \leq A < \infty$ for all $m \in \mathcal{M}_n$.

(An) Uniform lower-bound on the noise level: $\sigma(X_i) \geq \sigma_{\min} > 0$ $a.s.$

(Ap$_u$) The bias decreases as a power of D_m: there exist $\beta_+ > 0$ and $C_+ > 0$ such that

$$\ell(s_*, s_m) \leq C_+ D_m^{-\beta_+} .$$

Assumption (**Alb**) refers to the classical concept of localized basis (Birgé and Massart [6]). It is proved in [5], Section 3.2.1, that linear models of piecewise polynomials with bounded degree on a regular partition of a bounded domain of $\mathbb{R}^d$ are endowed with a localized basis. It is also proved that compactly supported wavelet expansions are also fulfilled with a localized basis on $\mathbb{R}^d$. However, the Fourier basis is not a localized basis. For some sharp concentration results related to the excess loss of least squares estimators built from the Fourier basis, we refer to [19].

The assumption (**Alb**) is more general than the assumption of strongly localized basis used in [16], but the price to pay for such generality is that, according to (**P2**) we can only consider models with dimensions $D_m << n^{1/3}$.

Assumption (**P1**) states that the collection as a polynomial cardinality with respect to the sample size, allowing in particular to consider a collection of models built from a basis expansion.

Then Assumption (**Ab**) is related to boundedness of the data and enables in particular to use Talagrand's type concentration inequalities for the empirical process. Going beyond the bounded setting would in particular bring much more technicalities that might darken our work. For an example of results in an unbounded setting, see, for instance, [4], dealing with optimal selection of regressograms (histograms being a very particular case of our general framework). Assumption (**An**) is essentially a technical assumption that allows to obtain sharp lower bounds for the excess losses of the estimators. Condition (Ap$_u$) is a very classical assumption in the model selection literature, specifying a rate of decay for the biases of the models. This assumption is classically satisfied for piecewise polynomials when the regression function belongs to a Sobolev space and for wavelet models whenever the target belongs to some Besov space (see, for instance, [5] for more details). The specific value of β_+ parameter will only affect the value of the constants in the derived oracle inequalities.

Theorem 1 *Assume that (SA) holds. Let* $r \in (2, +\infty)$ *and* $V \in \{2, \ldots, n-1\}$ *satisfying* $1 < V \leq r$. *Define the V-fold cross-validation procedure as the model selection procedure given by (1). Then, for all* $n \geq n_0((\mathbf{SA}), r)$, *with probability at least* $1 - L_{(\mathbf{SA}),r} n^{-2}$,

$$\ell\left(s_*, s_{\widehat{m}_{VFCV}}\right) \leq \left(1 + \frac{L_{(\mathbf{SA}),r}}{\sqrt{\ln n}}\right) \inf_{m \in \mathcal{M}_n} \left\{\ell\left(s_*, \widehat{s}_m^{(-1)}\right)\right\} + L_{(\mathbf{SA}),r} \frac{(\ln n)^3}{n} .$$

In Theorem 1, we prove an oracle inequality with principal constant tending to one when the sample size goes to infinity. This inequality bounds from above the excess loss of the selected estimator by the excess loss of the oracle learned with a fraction $1 - V^{-1}$ of the original data. Ideally, one would, however, expect from an optimal procedure to recover the oracle built from the entire data. The next section is devoted to this task.

Parameter V (or r) is considered in Theorem 1 as a constant, essentially for ease of presentation. Actually, the value of V may be allowed to depend on n but also on the dimensions D_m, meaning that we may take different values of V according to the different models of the collection. More precisely, it can be seen from the arguments in the proofs (especially from Theorem 8 in [18]) that for each model $m \in \mathcal{M}_n$, it suffices to have $V \leq \max\left\{D_m (\ln n)^{-\tau}; 2\right\}$ where τ is any number in $(1, 3)$ to ensure an oracle inequality with leading constant tending to one when the amount of data tends to infinity. In this case, r cannot be considered as a parameter independent from the sample size anymore, but it can be checked that for the latter constraints on V, the constants $n_0\left(\left(\mathbf{SA}\right), r\right)$ and $L_{(\mathbf{SA}),r}$ do not explode but are still uniformly bounded with respect to n and thus can still be considered as independent from n.

4 V-Fold Penalization

Now we investigate the behaviour of a penalization procedure proposed by Arlot [2] and called V-fold penalization,

$$\widehat{m}_{\mathrm{penVF}} \in \arg\min_{n \in \mathcal{M}_n} \left\{\mathrm{crit}_{\mathrm{penVF}}(m)\right\} ,$$

where

$$\mathrm{crit}_{\mathrm{penVF}}(m) = P_n\left(\gamma\left(\widehat{s}_m\right)\right) + \mathrm{pen}_{\mathrm{VF}}(m) ,$$

with

$$\mathrm{pen}_{\mathrm{VF}}(m) = \frac{V-1}{V} \sum_{j=1}^{V} \left[P_n \gamma\left(\widehat{s}_m^{(-j)}\right) - P_n^{(-j)} \gamma\left(\widehat{s}_m^{(-j)}\right)\right] . \tag{4}$$

The property underlying the V-fold penalization is that the V-fold penalty $\mathrm{pen}_{\mathrm{VF}}$ is an unbiased estimate of the ideal penalty $\mathrm{pen}_{\mathrm{id}}$, the latter allowing to identify the oracle m_*,

$$\begin{aligned} m_* &\in \arg\min_{m \in \mathcal{M}_n} \left\{P\left(\gamma\left(\widehat{s}_m\right)\right)\right\} \\ &= \arg\min_{m \in \mathcal{M}_n} \left\{P_n\left(\gamma\left(\widehat{s}_m\right)\right) + \mathrm{pen}_{\mathrm{id}}(m)\right\} , \end{aligned}$$

where

$$\mathrm{pen}_{\mathrm{id}}(m) = P(\gamma(\widehat{s}_m)) - P_n(\gamma(\widehat{s}_m)) .$$

The following theorem states the asymptotic optimality of the V-fold penalization procedure for a fixed V.

Theorem 2 *Assume that (**SA**) holds. Let $r \in (2, +\infty)$ and $V \in \{2, \ldots, n-1\}$ satisfying $1 < V \leq r$. Define the V-fold cross-validation procedure as the model selection procedure given by*

$$\widehat{m}_{penVF} \in \arg\min_{n \in \mathcal{M}_n} \left\{ P_n(\gamma(\widehat{s}_m)) + \mathrm{pen}_{VF}(m) \right\} .$$

Then, for all $n \geq n_0((\boldsymbol{SA}), r)$, with probability at least $1 - L_{(\boldsymbol{SA}),r} n^{-2}$,

$$\ell\left(s_*, \widehat{s}_{\widehat{m}_{penVF}}\right) \leq \left(1 + \frac{L_{(\boldsymbol{SA}),r}}{\sqrt{\ln n}}\right) \inf_{m \in \mathcal{M}_n} \{\ell(s_*, \widehat{s}_m)\} + L_{(\boldsymbol{SA}),r} \frac{(\ln n)^3}{n} .$$

As for Theorem 1 above, parameter V (or r) is considered in Theorem 2 as a constant but in fact, the value of V may be allowed to depend on n and even on the dimensions D_m, this case corresponding to possibly different choices V according to the models of the collection. As for Theorem 1, it is allowed to have $V \leq \max\left\{D_m (\ln n)^{-\tau}; 2\right\}$ where τ is any number in $(1, 3)$ to ensure an oracle inequality with leading constant tending to one when the amount of data tends to infinity.

5 Simulation Study

In order to assess the numerical performances of the model selection procedures we have discussed, a short simulation study was conducted. Particularly, to illustrate the theory developed above for the selection of linear estimators using the V-fold cross-validation and V-fold penalization, linear wavelet models were considered.

Despite the fact that a linear wavelet estimator is not as flexible, or potentially as powerful, as a nonlinear one, it still preserves the computational efficiency of wavelet methods and can provide comparative results to thresholding estimator, particularly when the unknown function is sufficiently smooth (see [1]).

The simulations were carried out using Matlab and the wavelet toolbox Wavelab850 [10]. The codes used to replicate the numerical results presented here will be available at https://github.com/fabnavarro. For more details on the numerical simulations and comparisons with other model selection procedures, we refer the reader to [19].

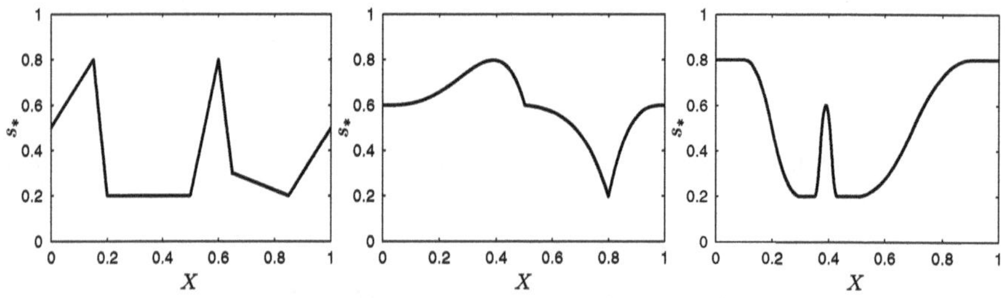

Fig. 1 The three test functions used in the simulation study (from left to right *Angle, Corner* and *Parabolas*)

The simulated data were generated according to $Y_i = s_*(X_i) + \sigma(X_i)\varepsilon_i, \quad i = 1, \ldots, n,$ where $n = 4096$, X_i's are uniformly distributed on [0, 1], ε_i's are independent $\mathcal{N}(0, 1)$ variables and independent of X_i's. The heteroscedastic noise level $\sigma(x) = |\cos(10x)|/10$. Daubechies' compactly supported wavelet with 8 vanishing moments was used. Three standard regression functions with different degrees of smoothness (*Angle, Corner*, and *Parabolas*, see [7, 14]) were considered. They are plotted in Fig. 1 and a visual idea of the noise level is given in Fig. 2b.

The computation of wavelet-based estimators is straightforward and fast in the fixed design case, thanks to Mallat's pyramidal algorithm [13]. In the case of random design, the implementation requires some changes and several strategies have been developed in the literature (see, e.g., [8, 11]). In the regression with uniform design [9] has examined convergence rates when the unknown function is in a Hölder class. They showed that the standard equispaced wavelet method with universal thresholding can be directly applied to the nonequispaced data (without a loss in the rate of convergence). We have followed this approach since it preserves the computational simplicity and efficiency of the equispaced algorithm. In the context of wavelet regression in random design with heteroscedastic dependent errors [12] has also adopted this approach. Thus, the wavelet coefficients of the collection of models is computed by a simple application of Mallat's algorithm using the ordered Y_i's as input variables. The collection is then constructed by successively adding whole resolution levels of wavelet coefficients. Thus, the considered dimensions are $\{D_m, m \in \mathcal{M}_n\} = \{2^j, j = 1, \ldots, J-1\}$, where $J = \log 2(n)$ (the finest resolution level). Finally, the selected model is obtained by minimizing (2) and (4) over the set $m \in \mathcal{M}_n$. Note that these linear models operate in a global fashion since whole levels of coefficients are suppressed as opposed to thresholding methods.

For choosing the threshold parameter in wavelet shrinkage Nason [15] adjusted the usual 2FCV method—which cannot be applied directly to wavelet estimation. In order to implement its strategy in a linear context, we test, for every model of the collection, an interpolated wavelet estimator learned from the (ordered)

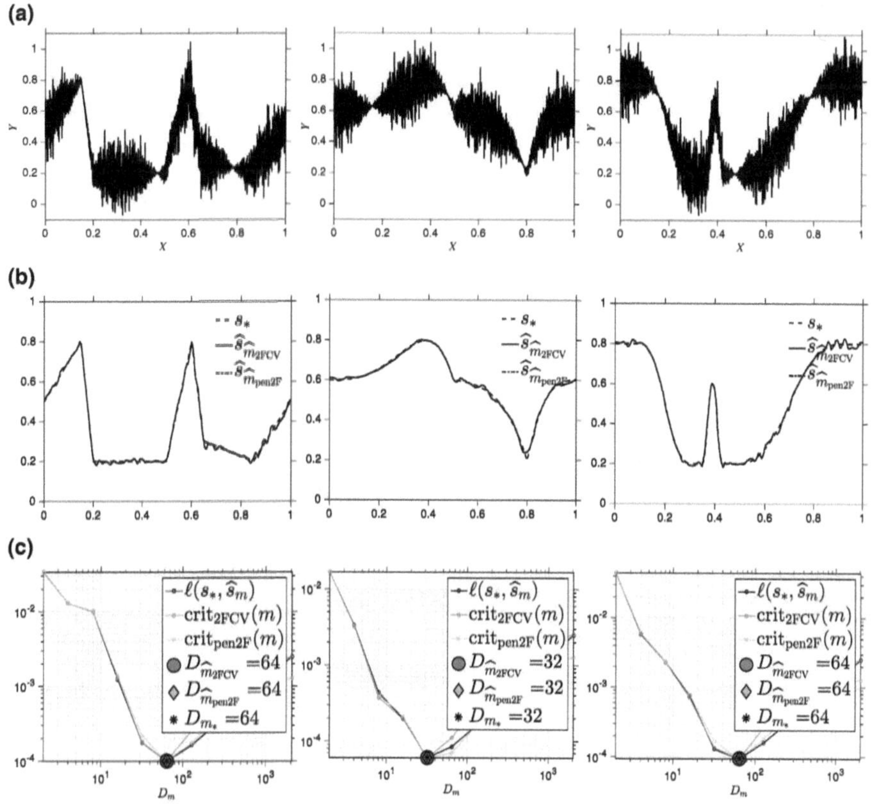

Fig. 2 (**a**) Noisy observations. (**b**) Typical reconstructions from a single simulation with $n = 4096$. Dashed line indicates the true function s_*, solid line corresponds to the estimates $\widehat{s}_{\widehat{m}_{\mathrm{2FCV}}}$ and dashed-dotted line to $\widehat{s}_{\widehat{m}_{\mathrm{pen2F}}}$. (**c**) Graph of the excess risk $\ell(s_*, \widehat{s}_m)$ (black) against the dimension D_m and (rescaled) $\mathrm{crit}_{\mathrm{2FCV}}(m)$ (gray) and $\mathrm{crit}_{\mathrm{pen2F}}(m)$ (light-gray) (in a log-log scale). The gray circle represents the global minimizer $\widehat{m}_{\mathrm{2FCV}}$ of $\mathrm{crit}_{\mathrm{2FCV}}(m)$, the light-gray diamond corresponds to the global minimizer $\widehat{m}_{\mathrm{2FCV}}$ of $\mathrm{crit}_{\mathrm{pen2F}}(m)$ and the black star the oracle model m_*

even-indexed data against the odd-indexed data and vice versa. More precisely, considering the data X_i are ordered, the selected model $\widehat{m}_{\mathrm{2FCV}}$ (resp. $\widehat{m}_{\mathrm{pen2F}}$) is obtained by minimizing (2) (resp. (4)) with $V = 2$, $B_1 = \{2, 4, \dots, n\}$ and $B_2 = \{1, 3, \dots, n-1\}$.

For one Monte Carlo simulation with a sample size $n = 4096$, we display the estimation results in Fig. 2b. Plots of the excess risk $\ell(s_*, \widehat{s}_m)$ against the dimension D_m are plotted in Fig. 2c. The curve $\mathrm{crit}_{\mathrm{2FCV}}(m)$ and $\mathrm{crit}_{\mathrm{pen2F}}(m)$ are also displayed in Fig. 2c. It can be observed that $\mathrm{crit}_{\mathrm{2FCV}}(m)$ and $\mathrm{crit}_{\mathrm{pen2F}}(m)$ give very reliable estimate for the risk $\ell(s_*, \widehat{s}_m)$, and in turn, also a high-quality estimate of the optimal model. Indeed, in this case, both methods consistently select the oracle model m_*.

6 Proofs

As a preliminary result, let us first prove the consistency in sup-norm of our least squares estimators. This is in fact the main change compared to the strongly localized case treated in [16].

Theorem 3 *Let $\alpha > 0$. Assume that m is a linear vector space satisfying Assumption **(Alb)** and use the notations given in the statement of **(Alb)**. Assume also that Assumption **(Ab)** holds. If there exists $A_+ > 0$ such that*

$$D_m \leq A_+ \frac{n^{1/3}}{(\ln n)^2},$$

then there exists a positive constant $L_{A,r_{\mathcal{M}},\alpha}$ such that, for all $n \geq n_0(r_{\mathcal{M}},\alpha)$,

$$\mathbb{P}\left(\left\|\hat{s}_m - s_m\right\|_\infty \geq L_{A,r_{\mathcal{M}},\alpha}\sqrt{\frac{D_m \ln n}{n}}\right) \leq n^{-\alpha}.$$

Proof (Proof of Theorem 3) Let $C > 0$. Set

$$\mathcal{F}_C^\infty := \{s \in m \; ; \; \|s - s_m\|_\infty \leq C\}$$

and

$$\mathcal{F}_{>C}^\infty := \{s \in m \; ; \; \|s - s_m\|_\infty > C\} = m \backslash \mathcal{F}_C^\infty .$$

Take an orthonormal basis $(\varphi_k)_{k=1}^{D_m}$ of $(m, \|\cdot\|_2)$ satisfying (**Alb**). By Lemma 19 of [16], we get that there exists $L_{A,r_m,\alpha}^{(1)} > 0$ such that, by setting

$$\Omega_1 = \left\{ \max_{k\in\{1,\dots,D_m\}} \left|(P_n - P)(\psi_m \cdot \varphi_k)\right| \leq L_{A,r_m,\alpha}^{(1)} \sqrt{\frac{\ln n}{n}} \right\},$$

we have for all $n \geq n_0(A_+)$, $\mathbb{P}(\Omega_1) \geq 1 - n^{-\alpha}$. Moreover, we set

$$\Omega_2 = \left\{ \max_{(k,l)\in\{1,\dots,D_m\}^2} \left|(P_n - P)(\varphi_k \cdot \varphi_l)\right| \leq L_{\alpha,r_m}^{(2)} \min\left\{\|\varphi_k\|_\infty \; ; \; \|\varphi_l\|_\infty\right\} \sqrt{\frac{\ln n}{n}} \right\},$$

where $L_{\alpha,r_m}^{(2)}$ is defined in Lemma 18 of [16]. By Lemma 18 of [16], we have that for all $n \geq n_0(A_+)$, $\mathbb{P}(\Omega_2) \geq 1 - n^{-\alpha}$ and so, for all $n \geq n_0(A_+)$,

$$\mathbb{P}\left(\Omega_1 \bigcap \Omega_2\right) \geq 1 - 2n^{-\alpha}.$$

We thus have for all $n \geq n_0(A_+)$,

$$
\begin{aligned}
&\mathbb{P}\left(\|s_n - s_m\|_\infty > C\right) \\
\leq &\mathbb{P}\left(\inf_{s \in \mathcal{F}^\infty_{>C}} P_n\left(\gamma(s) - \gamma(s_m)\right) \leq \inf_{s \in \mathcal{F}^\infty_{C}} P_n\left(\gamma(s) - \gamma(s_m)\right)\right) \\
= &\mathbb{P}\left(\sup_{s \in \mathcal{F}^\infty_{>C}} P_n\left(\gamma(s_m) - \gamma(s)\right) \geq \sup_{s \in \mathcal{F}^\infty_{C}} P_n\left(\gamma(s_m) - \gamma(s)\right)\right) \\
\leq &\mathbb{P}\left(\left\{\sup_{s \in \mathcal{F}^\infty_{>C}} P_n\left(\gamma(s_m) - \gamma(s)\right) \geq \sup_{s \in \mathcal{F}^\infty_{C/2}} P_n\left(\gamma(s_m) - \gamma(s)\right)\right\} \bigcap \Omega_1 \bigcap \Omega_2\right) + 2n^{-\alpha} .
\end{aligned}
\tag{5}
$$

Now, for any $s \in m$ such that

$$
s - s_m = \sum_{k=1}^{D_m} \beta_k \varphi_k, \ \beta = (\beta_k)_{k=1}^{D_m} \in \mathbb{R}^{D_m},
$$

we have

$$
\begin{aligned}
&P_n\left(\gamma(s_m) - \gamma(s)\right) \\
&= (P_n - P)\left(\psi_m \cdot (s_m - s)\right) - (P_n - P)\left((s - s_m)^2\right) - P\left(\gamma(s) - \gamma(s_m)\right) \\
&= \sum_{k=1}^{D_m} \beta_k (P_n - P)\left(\psi_m \cdot \varphi_k\right) - \sum_{k,l=1}^{D_m} \beta_k \beta_l (P_n - P)\left(\varphi_k \cdot \varphi_l\right) - \sum_{k=1}^{D_m} \beta_k^2 .
\end{aligned}
$$

We set for any $(k, l) \in \{1, \ldots, D_m\}^2$,

$$
R^{(1)}_{n,k} = (P_n - P)\left(\psi_m \cdot \varphi_k\right) \ \text{ and } \ R^{(2)}_{n,k,l} = (P_n - P)\left(\varphi_k \cdot \varphi_l\right) .
$$

Moreover, we set a function h_n, defined as follows:

$$
h_n : \beta = (\beta_k)_{k=1}^{D_m} \longmapsto \sum_{k=1}^{D_m} \beta_k R^{(1)}_{n,k} - \sum_{k,l=1}^{D_m} \beta_k \beta_l R^{(2)}_{n,k,l} - \sum_{k=1}^{D_m} \beta_k^2 .
$$

We thus have for any $s \in m$ such that $s - s_m = \sum_{k=1}^{D_m} \beta_k \varphi_k$, $\beta = (\beta_k)_{k=1}^{D_m} \in \mathbb{R}^{D_m}$,

$$
P_n\left(\gamma(s_m) - \gamma(s)\right) = h_n(\beta) . \tag{6}
$$

In addition we set for any $\beta = (\beta_k)_{k=1}^{D_m} \in \mathbb{R}^{D_m}$,

$$|\beta|_{m,\infty} = r_m \sqrt{D_m}\, |\beta|_\infty \ .$$

It is straightforward to see that $|\cdot|_{m,\infty}$ is a norm on $\mathbb{R}^{D_m}$, proportional to the sup-norm. We also set for a real $D_m \times D_m$ matrix B, its operator norm $\|A\|_m$ associated to the norm $|\cdot|_{m,\infty}$ on the D_m-dimensional vectors. More explicitly, we set for any $B \in \mathbb{R}^{D_m \times D_m}$,

$$\|B\|_m := \sup_{\beta\in\mathbb{R}^{D_m},\ \beta\neq 0} \frac{|B\beta|_{m,\infty}}{|\beta|_{m,\infty}} = \sup_{\beta\in\mathbb{R}^{D_m},\ \beta\neq 0} \frac{|B\beta|_{\infty}}{|\beta|_{\infty}} \ .$$

We have, for any $B = \left(B_{k,l}\right)_{k,l=1,\dots,D_m} \in \mathbb{R}^{D_m \times D_m}$, the following classical formula

$$\|B\|_m = \max_{k\in\{1,\dots,D_m\}} \left\{ \left\{ \sum_{l\in\{1,\dots,D_m\}} \left|B_{k,l}\right| \right\} \right\} .$$

Notice that by inequality (3) of (**Alb**), it holds

$$\mathcal{F}_{>C}^{\infty} \subset \left\{ s \in m \ ; \ s - s_m = \sum_{k=1}^{D_m} \beta_k \varphi_k \ \ \& \ \ |\beta|_{m,\infty} \geq C \right\} \tag{7}$$

and

$$\mathcal{F}_{C/2}^{\infty} \supset \left\{ s \in m \ ; \ s - s_m = \sum_{k=1}^{D_m} \beta_k \varphi_k \ \ \& \ \ |\beta|_{m,\infty} \leq C/2 \right\} . \tag{8}$$

Hence, from (5), (6), (8), and (7) we deduce that if we find on $\Omega_1 \bigcap \Omega_2$ a value of C such that

$$\sup_{\beta\in\mathbb{R}^{D_m},\ |\beta|_{m,\infty}\geq C} h_n(\beta) < \sup_{\beta\in\mathbb{R}^{D_m},\ |\beta|_{m,\infty}\leq C/2} h_n(\beta) \ ,$$

then we will get

$$\mathbb{P}\left(\|\widehat{s}_m - s_m\|_\infty > C\right) \leq 2n^{-\alpha} \ .$$

Taking the partial derivatives of h_n with respect to the coordinates of its arguments, it then holds for any $(k,l) \in \{1,\dots,D_m\}^2$ and $\beta = (\beta_i)_{i=1}^{D_m} \in \mathbb{R}^{D_m}$,

$$\frac{\partial h_n}{\partial \beta_k}(\beta) = R_{n,k}^{(1)} - 2\sum_{i=1}^{D_m} \beta_i R_{n,k,i}^{(2)} - 2\beta_k \tag{9}$$

We look now at the set of solutions β of the following system,

$$\frac{\partial h_n}{\partial \beta_k}(\beta) = 0\,, \forall k \in \{1, \ldots, D_m\}\ . \tag{10}$$

We define the $D_m \times D_m$ matrix $R_n^{(2)}$ to be

$$R_n^{(2)} := \left(R_{n,k,l}^{(2)}\right)_{k,l=1,\ldots,D_m}$$

and by (9), the system given in (10) can be written

$$2\left(I_{D_m} + R_n^{(2)}\right)\beta = R_n^{(1)}\,, \tag{S}$$

where $R_n^{(1)}$ is a D_m-dimensional vector defined by

$$R_n^{(1)} = \left(R_{n,k}^{(1)}\right)_{k=1,\ldots,D_m}\ .$$

Let us give an upper bound of the norm $\left\| R_n^{(2)} \right\|_m$, in order to show that the matrix $I_{D_m} + R_n^{(2)}$ is nonsingular. On Ω_2 we have

$$\begin{aligned}
\left\| R_n^{(2)} \right\|_m &= \max_{k\in\{1,\ldots,D_m\}}\left\{\left\{\sum_{l\in\{1,\ldots,D_m\}} \left|(P_n - P)(\varphi_k\cdot\varphi_l)\right|\right\}\right\} \\
&\leq L_{\alpha,r_m}^{(2)} \max_{k\in\{1,\ldots,D_m\}}\left\{\left\{\sum_{l\in\{1,\ldots,D_m\}} \min\left\{\|\varphi_k\|_\infty\,;\,\|\varphi_l\|_\infty\right\}\sqrt{\frac{\ln n}{n}}\right\}\right\} \\
&\leq r_m L_{\alpha,r_m}^{(2)}\sqrt{\frac{D_m^3 \ln n}{n}}
\end{aligned} \tag{11}$$

Hence, from (11) and the fact that $D_m \leq A_+\frac{n^{1/3}}{(\ln n)^2}$, we get that for all $n \geq n_0(r_m, \alpha)$, it holds on Ω_2,

$$\left\| R_n^{(2)} \right\|_m \leq \frac{1}{2}$$

and the matrix $\left(I_d + R_n^{(2)}\right)$ is nonsingular, of inverse $\left(I_d + R_n^{(2)}\right)^{-1} = \sum_{u=0}^{+\infty}\left(-R_n^{(2)}\right)^u$. Hence, the system (**S**) admits a unique solution $\beta^{(n)}$, given by

$$\beta^{(n)} = \frac{1}{2}\left(I_d + R_n^{(2)}\right)^{-1} R_n^{(1)}\ .$$

Now, on Ω_1 we have

$$\left|R_n^{(1)}\right|_{m,\infty} \leq r_m \sqrt{D_m} \max_{k \in \{1,\ldots,D_m\}} \left|(P_n - P)(\psi_m \cdot \varphi_k)\right| \leq r_m L_{A,r_m,\alpha}^{(1)} \sqrt{\frac{D_m \ln n}{n}}$$

and we deduce that for all $n_0(r_m, \alpha)$, it holds on $\Omega_2 \bigcap \Omega_1$,

$$\left|\beta^{(n)}\right|_{m,\infty} \leq \frac{1}{2} \left\| \left(I_d + R_n^{(2)}\right)^{-1} \right\|_m \left|R_n^{(1)}\right|_{m,\infty} \leq r_m L_{A,r_m,\alpha}^{(1)} \sqrt{\frac{D_m \ln n}{n}} . \tag{12}$$

Moreover, by the formula (6) we have

$$h_n(\beta) = P_n(\gamma(s_m)) - P_n \left(Y - \sum_{k=1}^{D_m} \beta_k \varphi_k\right)^2$$

and we thus see that h_n is concave. Hence, for all $n_0(r_m, \alpha)$, we get that on Ω_2, $\beta^{(n)}$ is the unique maximum of h_n and on $\Omega_2 \bigcap \Omega_1$, by (12), concavity of h_n and uniqueness of $\beta^{(n)}$, we get

$$h_n\left(\beta^{(n)}\right) = \sup_{\beta \in \mathbb{R}^{D_m},\, |\beta|_{m,\infty} \leq C/2} h_n(\beta) > \sup_{\beta \in \mathbb{R}^{D_m},\, |\beta|_{m,\infty} \geq C} h_n(\beta) ,$$

with $C = 2 r_m L_{A,r_m,\alpha}^{(1)} \sqrt{\frac{D_m \ln n}{n}}$, which concludes the proof.

From Theorem 2 of [17] and Theorem 3 above, we deduce the following excess risks bounds.

Theorem 4 *Let $A_+, A_-, \alpha > 0$. Assume that m is a linear vector space of finite dimension D_m satisfying (**Alb(m)**) and use notations of (**Alb(m)**). Assume, moreover, that the following assumption holds:*

(Ab(m)) There exists a constant $A > 0$, such that $\|s_m\|_\infty \leq A$ and $|Y| \leq A$ a.s.

If it holds

$$A_- (\ln n)^2 \leq D_m \leq A_+ \frac{n^{1/3}}{(\ln n)^2} ,$$

then a positive constant A_0 exists, only depending on α, A_- and on the constants A, $\sigma_{\min}$ and r_m such that by setting

$$\varepsilon_n = A_0 \max\left\{ \left(\frac{\ln n}{D_m}\right)^{1/4}, \left(\frac{D_m \ln n}{n}\right)^{1/4} \right\} ,$$

we have for all $n \geq n_0 (A_-, A_+, A, r_m, \sigma_{\min}, \alpha)$,

$$\mathbb{P}\left[(1-\varepsilon_n)\frac{\mathscr{C}_m}{n} \leq \ell(s_m, \widehat{s}_m) \leq (1+\varepsilon_n)\frac{\mathscr{C}_m}{n}\right] \geq 1 - 10n^{-\alpha},$$

$$\mathbb{P}\left[\left(1-\varepsilon_n^2\right)\frac{\mathscr{C}_m}{n} \leq \ell_{\mathrm{emp}}(\widehat{s}_m, s_m) \leq \left(1+\varepsilon_n^2\right)\frac{\mathscr{C}_m}{n}\right] \geq 1 - 5n^{-\alpha},$$

where $\mathscr{C}_m = \sum_{k=1}^{D_m} \operatorname{var}((Y - s_m(X)) \cdot \varphi_k(X))$.

Having at hand Theorem 4, the proofs of Theorems 1 and 4 follow from the exact same lines as the proofs of Theorems 6 and 7 of [16]. To give a more precise view of the ideas involved, let us detail the essential arguments of the proof of Theorem 1.

We set

$$\mathrm{crit}^0_{\mathrm{VFCV}}(m) = \mathrm{crit}_{\mathrm{VFCV}}(m) - \frac{1}{V}\sum_{j=1}^{V} P_n^{(j)}(\gamma(s_*)).$$

The difference between $\mathrm{crit}^0_{\mathrm{VFCV}}(m)$ and $\mathrm{crit}_{\mathrm{VFCV}}(m)$ being a quantity independent of $m \in \mathscr{M}_n$, the procedure defined by $\mathrm{crit}^0_{\mathrm{VFCV}}$ gives the same result as the VFCV procedure defined by $\mathrm{crit}_{\mathrm{VFCV}}$.

We get for all $m \in \mathscr{M}_n$,

$$\begin{aligned}
\mathrm{crit}^0_{\mathrm{VFCV}}(m) &= \frac{1}{V}\sum_{j=1}^{V} P_n^{(j)}\left(\gamma\left(\widehat{s}_m^{(-j)}\right) - \gamma(s_*)\right) \\
&= \frac{1}{V}\sum_{j=1}^{V}\Big[P_n^{(j)}\left(\gamma\left(\widehat{s}_m^{(-j)}\right) - \gamma(s_m)\right) \\
&\quad + \left(P_n^{(j)} - P\right)(\gamma(s_m) - \gamma(s_*)) + P(\gamma(s_m) - \gamma(s_*))\Big] \\
&= \ell\left(s_*, \widehat{s}_m^{(-1)}\right) + \Delta_V(m) + \bar{\delta}(m) \qquad (13)
\end{aligned}$$

where

$$\Delta_V(m) = \frac{1}{V}\sum_{j=1}^{V} P_n^{(j)}\left(\gamma\left(\widehat{s}_m^{(-j)}\right) - \gamma(s_m)\right) - P\left(\gamma\left(\widehat{s}_m^{(-1)}\right) - \gamma(s_m)\right),$$

and

$$\bar{\delta}(m) = \frac{1}{V}\sum_{j=1}^{V}\left(P_n^{(j)} - P\right)(\gamma(s_m) - \gamma(s_*))$$

Now, we have to show that $\Delta_V(m)$ and $\bar{\delta}(m)$ are negligible in front of $\ell\left(s_*, \widehat{s}_m^{(-1)}\right)$. For $\bar{\delta}(m)$, this is done by using Bernstein's concentration inequality (see Lemma 7.5 of [16]). To control $\Delta_V(m)$, we also make use of Bernstein's concentration inequality, but by conditioning successively on the data used to learn the estimators $\widehat{s}_m^{(-j)}$, $j = 1, \ldots, V$ (see Lemma 7.3 and Corollary 7.4 of [16]).

References

1. Antoniadis, A., Gregoire, G., & McKeague, I. (1994). Wavelet methods for curve estimation. *Journal of the American Statistical Association, 89*(428), 1340–1353.
2. Arlot, S. (2008). V-fold cross-validation improved: V-fold penalization. ArXiv:0802.0566v2. http://hal.archives-ouvertes.fr/hal-00239182/en/
3. Arlot, S., & Célisse, A. (2010). A survey of cross-validation procedures for model selection. *Statistical Surveys, 4*, 40–79.
4. Arlot, S., & Massart, P. (2009) Data-driven calibration of penalties for least-squares regression. *Journal of Machine Learning Research, 10*, 245–279 (electronic).
5. Barron, A., Birgé, L., & Massart, P. (1999). Risk bounds for model selection via penalization. *Probability Theory and Related Fields, 113*(3), 301–413.
6. Birgé, L., & Massart, P. (1998). Minimum contrast estimators on sieves: exponential bounds and rates of convergence. *Bernoulli, 4*(3), 329–375.
7. Cai, T. (1999). Adaptive wavelet estimation: a block thresholding and oracle inequality approach. *Annals of Statistics, 27*(3), 898–924.
8. Cai, T., & Brown, L. (1998). Wavelet shrinkage for nonequispaced samples. *Annals of Statistics, 26*, 1783–1799.
9. Cai, T., & Brown, L. (1999). Wavelet estimation for samples with random uniform design. *Statistics and Probability Letters, 42*(3), 313–321.
10. Donoho D., Maleki, A., & Shahram, M. (2006). Wavelab 850. http://statweb.stanford.edu/~wavelab/
11. Hall, P., & Turlach, B. (1997). Interpolation methods for nonlinear wavelet regression with irregularly spaced design. *Annals of Statistics, 25*(5), 1912–1925.
12. Kulik, R., & Raimondo, M. (2009). Wavelet regression in random design with heteroscedastic dependent errors. *Annals of Statistics, 37*(6A), 3396–3430.
13. Mallat, S. (2008). *A wavelet tour of signal processing: The sparse way*. New York: Academic.
14. Marron, J., Adak, S., Johnstone, I., Neumann, M., & Patil, P. (1998). Exact risk analysis of wavelet regression. *Journal of Computational and Graphical Statistics, 7*(3), 278–309.
15. Nason, G. (1996). Wavelet shrinkage using cross-validation. *Journal of the Royal Statistical Society, Series B, 58*, 463–479.
16. Navarro, F., & Saumard, A. (2017). Slope heuristics and V-Fold model selection in heteroscedastic regression using strongly localized bases. *ESAIM: Probability and Statistics, 21*, 412–451.
17. Saumard, A. (2012). Optimal upper and lower bounds for the true and empirical excess risks in heteroscedastic least-squares regression. *Electronic Journal of Statistics, 6*(1–2):579–655.
18. Saumard, A. (2013). Optimal model selection in heteroscedastic regression using piecewise polynomial functions. *Electronic Journal of Statistics, 7*, 1184–1223.
19. Saumard, A. (2017). On optimality of empirical risk minimization in linear aggregation. *Bernoulli* (to appear). arXiv:1605.03433

Non-parametric Lower Bounds and Information Functions

S. Y. Novak

Abstract We argue that common features of non-parametric estimation appear in parametric cases as well if there is a deviation from the classical regularity condition. Namely, in many non-parametric estimation problems (as well as some parametric cases) unbiased finite-variance estimators do not exist; neither estimator converges locally uniformly with the optimal rate; there are no asymptotically unbiased with the optimal rate estimators; etc.

We argue that these features naturally arise in particular parametric subfamilies of non-parametric classes of distributions. We generalize the notion of regularity of a family of distributions and present a general regularity condition, which leads to the notions of the information index and the information function.

We argue that the typical structure of a continuity modulus explains why unbiased finite-variance estimators cannot exist if the information index is larger than two, while in typical non-parametric situations neither estimator converges locally uniformly with the optimal rate. We present a new result on impossibility of locally uniform convergence with the optimal rate.

1 Introduction

It was observed by a number of authors that in many non-parametric estimation problems the accuracy of estimation is worse than in the case of a regular parametric family of distributions, estimators depend on extra tuning "parameters," unbiased estimators are not available, the weak convergence of normalized estimators to the limiting distribution is not uniform at the optimal rate, no estimator is uniformly consistent in the considered class of distributions. These features have been observed, e.g., in the problems of non-parametric density, regression curve, and tail index estimation (cf. [9, ch. 13], and the references therein).

S. Y. Novak (✉)
Middlesex University, London, UK
e-mail: S.Novak@mdx.ac.uk

P. Bertail et al. (eds.), *Nonparametric Statistics*, Springer Proceedings in Mathematics & Statistics 250, https://doi.org/10.1007/978-3-319-96941-1_5

Our aim in this study is to develop a rigorous treatment of these features through a generalization of the notion of regularity of a family of probability distributions. We argue that features mentioned above (which might have been considered accidental drawbacks of particular estimation procedures) in reality are inevitable consequences of the "richness" of the non-parametric class of distributions under consideration.

We argue that the degree of "richness" of the class of distributions determines the accuracy of estimation. The interplay between the degree of "richness" and the accuracy of estimation can be revealed via the non-parametric lower bounds. In some situations the lower bound to the accuracy of estimation is bounded away from zero, meaning consistent estimation is impossible.

2 Regularity Conditions and Lower Bounds

In a typical estimation problem one wants to estimate a quantity of interest a_P from a sample $X_1, \ldots, X_n$ of independent and identically distributed (i.i.d.) observations, where the unknown distribution $P = Ł(X_1)$ belongs to a particular class $\mathscr{P}$.

If there are reasons to assume that the unknown distribution belongs to a *parametric* family $\mathscr{P} = \{P_\theta, \theta \in \Theta\}$, $\Theta \subset \mathscr{X}$, where $\mathscr{X}$ is $\mathbb{R}^m$ or a Hilbert space, then it is natural to choose $a_{P_\theta} = \theta$. Other examples include $a_P = f_P$, the density of P with respect to a given measure μ (assuming every $P \in \mathscr{P}$ has a density with respect to μ), the tail index of a distribution form the class of regularly varying distributions, etc.

Let

$$d_H, \; d_\chi \;\; and \;\; d_{TV}$$

denote Hellinger, χ^2, and the total variation distances, respectively.

In the case of a parametric family of probability distributions a typical regularity condition states/implies that

$$d_H^2(P_\theta; P_{\theta+h}) \sim \|h\|^2 I_\theta/8 \text{ or } d_\chi^2(P_\theta; P_{\theta+h}) \sim \|h\|^2 I_\theta \tag{1}$$

as $h \to 0$, $\theta \in \Theta, \theta+h \in \Theta$, where I_θ is "Fisher's information." If one of regularity conditions (1) holds, estimator $\hat\theta$ is unbiased, and function $\theta \to \mathbb{E}_\theta \|\hat\theta - \theta\|^2$ is continuous, then

$$\mathbb{E}_\theta \|\hat\theta - \theta\|^2 \geq 1/nI_\theta \qquad (\forall \theta \in \Theta). \tag{2}$$

This is the celebrated Fréchet–Rao–Cramér inequality. Thus, if an unbiased estimator with a finite second moment exists, then the optimal unbiased estimator is the one that turns lower bound (2) into equality.

However, the assumption of existence of unbiased estimators may be unrealistic even in parametric estimation problems. For instance, Barankin [1] gives an example of a parametric estimation problem where an unbiased estimator with a finite second moment does not exist.

Below we suggest a generalization of the regularity condition for a family of probability distributions, and introduce the notion of an information index. We then present a non-parametric generalization of the Fréchet–Rao–Cramér inequality. We give reasons why in typical non-parametric estimation problems (as well as in certain parametric ones) unbiased estimators with a finite second moment do not exist.

Notation Below $a_n \sim b_n$ means $a_n = b_n(1+o(1))$ as $n \to \infty$. We write

$$a_n \gtrsim b_n \tag{*}$$

if $a_n \geq b_n(1+o(1))$ as $n \to \infty$.

Recall the definitions of the Hellinger distance d_H and the χ^2-distance d_χ . If the distributions P_1 and P_2 have densities f_1 and f_2 with respect to a measure μ, then

$$d_H^2(P_1;\, P_2) = \frac{1}{2}\int (f_1^{1/2} - f_2^{1/2})^2 d\mu = 1 - \int \sqrt{f_1 f_2}\, d\mu\,,$$

$$d_\chi^2(P_1;\, P_2) = \int (f_2/f_1 - 1)^2 dP_1\,,$$

In the definition of d_χ we presume that $\operatorname{supp} P_1 \supseteq \operatorname{supp} P_2$.

Definition 1 We say the parametric family $\mathscr{P} = \{P_\theta,\ \theta \in \Theta\}$, $\Theta \subset \mathscr{X}$, obeys the regularity condition $(R_{t,H})$ if there exist $\nu > 0$ and $I_{t,H} > 0$ such that

$$d_H^2(P_t;\, P_{t+h}) \sim I_{t,H}\|h\|^\nu \tag{$R_{t,H}$}$$

as $h \to 0$, $t \in \Theta$, $t+h \in \Theta$.

Family $\mathscr{P}$ obeys the regularity condition (R_H) if there exist $\nu > 0$ and function $I_{\cdot,H} > 0$ such that $(R_{t,H})$ holds for every $t \in \Theta$.

Definition 2 We say family $\mathscr{P}$ obeys the regularity condition $(R_{t,\chi})$ if there exist $\nu > 0$ and $I_{t,\chi} > 0$ such that

$$d_\chi^2(P_t;\, P_{t+h}) \sim I_{t,\chi}\|h\|^\nu \tag{$R_{t,\chi}$}$$

as $h \to 0$, $t \in \Theta, t+h \in \Theta$.

Family $\mathscr{P}$ obeys the regularity condition (R_χ) if there exist $\nu > 0$ and function $I_{\cdot,\chi} > 0$ such that $(R_{t,\chi})$ holds for every $t \in \Theta$.

Definitions 1 and 2 extend the notion of regularity of a parametric family of distributions.

A variant of these definitions has $\sim$ replaced with $\leq$.

We are not aware of natural examples where dependence of $d_H^2(P_t; P_{t+h})$ or $d_\chi^2(P_t; P_{t+h})$ on h is more complex. However, if such examples appear, then $(R_{t,H})$ and $(R_{t,\chi})$ can be generalized by replacing $\|h\|^\nu$ in the right-hand sides with $\psi(h)$ for a certain function ψ.

Definition 3 If (R_H) or (R_χ) holds, then we call ν the *"information index"* and $I_{\cdot,H}$ and/or $I_{\cdot,\chi}$ the *"information functions."*

It is known (see, e.g., [12] or [9, ch. 14]) that

$$d_H^2 \leq d_{TV} \leq \sqrt{2} d_H \leq d_\chi . \tag{3}$$

If both (R_H) and (R_χ) are in force, then inequality $2d_H^2 \leq d_\chi^2$ entails

$$2I_{t,H} \leq I_{t,\chi} .$$

In Example 1 below $I_{t,\chi} = 2I_{t,H}$. In the case of a family $\{P_t = \mathcal{N}(t; 1), t \in \mathbb{R}\}$ of normal random variables (r.v.s) one has

$$d_H^2(P_0; P_t) = 1 - e^{-t^2/8} , \; d_\chi^2(P_0; P_t) = e^{t^2} - 1,$$

hence $I_{t,\chi} = 8I_{t,H}$ (cf. [9, ch. 14.4]).

Information index ν indicates how "rich" or "poor" the class $\mathscr{P}$ is. In the case of a regular parametric family of distributions (i.e., a family obeying (1)) one has

$$\nu = 2.$$

"Irregular" parametric families of distributions may obey (R_H) and (R_χ) with $\nu < 2$ (cf. Example 1 and [9, ch. 13]).

Example 1 Let $\mathscr{P} = \{P_t, t > 0\}$, where $P_t = \mathrm{U}[0; t]$ is the uniform distribution on $[0; t]$. Then

$$d_H^2(P_{t+h}; P_t) = 1 - (1+|h|/t)^{-1/2} \sim h/2t \qquad (t \geq h \searrow 0),$$
$$d_\chi^2(P_{t+h}; P_t) = h/t, \;\; d_{TV}(P_{t+h}; P_t) = h/(t+h) \quad (t \geq h > 0).$$

Hence family $\mathscr{P}$ is not regular in the traditional sense (cf. (1)). Yet (R_H) and (R_χ) hold with

$$\nu = 1 , \;\; I_{t,H} = 1/2t , \;\; I_{t,\chi} = 1/t .$$

The optimal estimator $t_n^* = \max\{X_1, \ldots, X_n\}(n+1)/n$ is unbiased, and

$$\mathbb{E}_t(t_n^* - t)^2 = t^2/n(n+2). \qquad \square$$

Parametric subfamilies of non-parametric classes typically obey (R_H) and (R_χ) with $\nu > 2$ (cf. Example 3 and [9, ch. 13]).

We present now lower bounds to the accuracy of estimation when (R_H) or (R_χ) holds. Theorem 1 below indicates that the accuracy of estimation is determined by the information index and the information function.

Definition 4 We say that set Θ obeys property (A_ε) if for every $t \in \Theta$ there exists $t' \in \Theta$ such that $\|t' - t\| = \varepsilon$. Property (A) holds if (A_ε) is in force for all small enough $\varepsilon > 0$.

We say that estimator $\hat{\theta}$ with a finite first moment has "regular" bias if for every $t \in \Theta$ there exists $c_t > 0$ such that

$$\|\mathbb{E}_{t+h}\hat{\theta} - \mathbb{E}_t\hat{\theta}\| \sim c_t\|h\| \qquad (h \to 0). \tag{4}$$

An unbiased estimator obeys (4) with $c_t \equiv 1$. If Θ is an interval, then (A) trivially holds.

Theorem 1 ([9]) *Assume property (A), and suppose that estimator $\hat{t}_n$ obeys (4).*

If (R_χ) holds with $\nu \in (0; 2)$, then, as $n \to \infty$,

$$\sup_{t \in \Theta}(nI_{t,\chi})^{2/\nu}\mathbb{E}_t\|\hat{t}_n - t\|^2/c_t^2 \gtrsim y_\nu^{2/\nu}/(e^{y_\nu} - 1), \tag{5}$$

where y_ν is the positive root of the equation $2(1 - e^{-y}) = \nu y$.

If the function $t \to \mathbb{E}_t\|\hat{t} - t\|^2$ is continuous, then, as $n \to \infty$,

$$(nI_{t,\chi})^{2/\nu}\mathbb{E}_t\|\hat{t}_n - t\|^2/c_t^2 \gtrsim y_\nu^{2/\nu}/(e^{y_\nu} - 1) \qquad (\forall t \in \Theta). \tag{5*}$$

If (R_χ) holds with $\nu > 2$, then $\mathbb{E}_t\|\hat{t}_n\|^2 = \infty$ $(\exists t \in \Theta)$.

The result holds with (R_χ) replaced by (R_H) if $I_{t,\chi}$ is replaced with $I_{t,H}$ and the right-hand side of (5) is replaced with $(\ln 4/3)^{2/\nu}/4$.

According to (5), the rate of the accuracy of estimation for estimators with regular bias cannot be better than $n^{-1/\nu}$. Moreover, (5) establishes that the natural normalizing sequence for $\hat{t}_n - t$ depends in a specific way on n, ν, and the information function.

Theorem 1 supplements the Fréchet–Rao–Cramér inequality that deals with the case $\nu = 2$. Note that (5*) formally extends to the case $\nu = 2$ with $y_2 := 0$ and the right-hand side of (5*) treated as $\lim_{y \to 0} y/(e^y - 1) = 1$.

According to Theorem 1, an estimator $\hat{t}_n$ cannot be unbiased or have a regular bias if (R_χ) or (R_H) holds with $\nu > 2$ and $\mathbb{E}_t\|\hat{t}_n\|^2 < \infty$ for every $t \in \Theta$.

Lower bounds involving continuity moduli are presented in the next section.

3 Lower Bounds Based on Continuity Moduli

We consider now a general situation where one cannot expect regularity conditions to hold (cf. Example 3).

Let $\mathscr{P}$ be an arbitrary class of probability distributions, and let the quantity of interest a_P be an element of a metric space $(\mathscr{X}, d)$. Given $\varepsilon > 0$, we denote by

$$\mathscr{P}_H(P, \varepsilon) = \{Q \in \mathscr{P} : d_H(P; Q) \le \varepsilon\}$$

the neighborhood of distribution $P \in \mathscr{P}$. We call

$$w_H(P, \varepsilon) = \sup_{Q \in \mathscr{P}_H(P,\varepsilon)} d(a_Q; a_P)/2 \quad \text{and} \quad w_H(\varepsilon) = \sup_{P \in \mathscr{P}} w_H(P, \varepsilon)$$

the moduli of continuity.

For instance, if $\mathscr{P} = \{P_t, t \in \Theta\}$, $a_{P_t} = t$ and $d(x; y) = |x - y|$, then

$$2w_H(P_t, \varepsilon) = \sup\{|h| : d_H(P_t; P_{t+h}) \le \varepsilon\}$$

and $w_H(\varepsilon) = \sup_t w_H(P_t, \varepsilon)$.

Similarly we define $\mathscr{P}_\chi(P, \varepsilon)$, $\mathscr{P}_{TV}(P, \varepsilon)$, $w_\chi(\cdot)$ and $w_{TV}(\cdot)$ using the χ^2-distance d_χ and the total variation distance d_{TV}. For instance, if $a_P \in \mathbb{R}$ and $d(x; y) = |x - y|$, then

$$w_{TV}(P, \varepsilon) = \sup_{Q \in \mathscr{P}_{TV}(P,\varepsilon)} |a_Q - a_P|/2.$$

The notion of continuity moduli has been available in the literature on non-parametric estimation for a while (cf. Donoho & Liu [3] and Pfanzagl [10, 11]). It helps to quantify the interplay between the degree of "richness" of class $\mathscr{P}$ and the accuracy of estimation.

Lemma 1 ([9]) *For any estimator* $\hat{a}$ *and every* $P_0 \in \mathscr{P}$,

$$\sup_{P \in \mathscr{P}_H(P_0,\varepsilon)} P(d(\hat{a}_n; a_P) \ge w_H(P_0, \varepsilon)) \ge (1 - \varepsilon^2)^{2n}/4, \tag{6}$$

$$\sup_{P \in \mathscr{P}_\chi(P_0,\varepsilon)} P(d(\hat{a}_n; a_P) \ge w_\chi(P_0, \varepsilon)) \ge [1 + (1 + \varepsilon^2)^{n/2}]^{-2}. \tag{7}$$

Let R be a loss function. Lemma 1 and Chebyshev's inequality yield a lower bound to $\sup_{P \in \mathscr{P}_H(P_0,\varepsilon)} \mathbb{E}_P R(d(\hat{a}_n; a_P))$. For example, (6) with $R(x) = x^2$ yields

$$\sup_{P \in \mathscr{P}_H(P_0,\varepsilon)} \mathbb{E}_P^{1/2} d^2(\hat{a}_n; a_P) \ge w_H(P_0, \varepsilon)(1 - \varepsilon^2)^n/2. \tag{8}$$

A (8)-type result for asymptotically unbiased estimators has been presented by Pfanzagl [11]. Note that Lemma 1 does not impose any extra assumptions.

The best possible rate of estimation can be found by maximizing the right-hand side of (8) in ε. For instance, if

$$w_H(P,\varepsilon) \gtrsim J_{H,P}\,\varepsilon^{2r} \qquad (\varepsilon \to 0) \tag{9}$$

for some $J_{H,P} > 0$, then the rate of the accuracy of estimation cannot be better than n^{-r}.

If (R_H) and/or (R_χ) hold for a parametric subfamily of $\mathscr{P}$, then

$$2w_H(P_t,\varepsilon) \sim (\varepsilon^2/I_{t,H})^{1/\nu} \ \text{ and/or } \ 2w_\chi(P_t,\varepsilon) \sim (\varepsilon^2/I_{t,\chi})^{1/\nu}, \tag{10}$$

yielding (9) with $r = 1/\nu$. Hence the best possible rate of the accuracy of estimation is $n^{-1/\nu}$.

The drawback of this approach is the difficulty of calculating the continuity moduli.

Example 2 Consider the parametric family $\mathscr{P}$ of distributions P_θ with densities

$$f_\theta(x) = \varphi(x-\theta)/2 + \varphi(x+\theta)/2 \qquad (\theta \in \mathbb{R}),$$

where φ is the standard normal density. Set

$$a_{P_\theta} = \theta,\ d(\theta_1;\theta_2) = |\theta_1 - \theta_2|.$$

Then

$$d_H(P_0;\,P_h) \sim h^2/4\,.$$

Thus, $(R_{0,H})$ holds with

$$\nu = 4,\ I_{0,H} = 1/16,$$

$w_H(P_0,\varepsilon) \sim \sqrt{\varepsilon}$ as $\varepsilon \to 0$; there is no asymptotically unbiased with the optimal rate finite-variance estimator; the rate of the accuracy of estimation in a neighborhood of the standard normal distribution P_0 cannot be better than $n^{-1/4}$ (cf. Liu and Brown [5]). An application of (13.8) in [9] yields

$$\sup_{0\le\theta\le\varepsilon} \mathbb{E}_{P_\theta}|\hat\theta_n - \theta|^2 \gtrsim 1/2\sqrt{en} \qquad (n \to \infty) \tag{11}$$

for an arbitrary estimator $\hat\theta_n$ and any $\varepsilon > 0$. □

Put $\varepsilon^2 = c^2/n$ in (8). Then

$$\sup_{P \in \mathscr{P}_H(P_0,\varepsilon)} \mathbb{E}_P^{1/2} d(\hat{a}_n; a_P)^2 \gtrsim e^{-c^2} w_H(P_0, c/\sqrt{n})/2 . \tag{8*}$$

Thus, the rate of the accuracy of estimation of a_P in a neighborhood of P_0 cannot be better than that of

$$w_H(P_0, 1/\sqrt{n})$$

(cf. Donoho & Liu [3]). More specifically, if (9) holds, then

$$\sup_{P \in \mathscr{P}_H(P_0,\varepsilon)} \mathbb{E}_P^{1/2} d^2(\hat{a}_n; a_P) \gtrsim e^{-c^2} J_{H,P_0} c^{2r} n^{-r}/2 . \tag{12}$$

If $J_{H,\cdot}$ is uniformly continuous on $\mathscr{P}$, then (12) with $c^2 = r$ yields the *non-uniform* lower bound

$$\sup_{P \in \mathscr{P}} J_{H,P}^{-1} \mathbb{E}_P^{1/2} d^2(\hat{a}_n; a_P) \gtrsim (r/e)^r n^{-r}/2 . \tag{13}$$

Lower bound (13) is non-uniform because of the presence of the term depending on P in the left-hand side of (13). Note that the traditional approach would be to deal with $\sup_{P \in \mathscr{P}} \mathbb{E}_P\, d^2(\hat{a}_n; a_P)$ (cf. [12]); the latter can in some cases be meaningless while $\sup_{P \in \mathscr{P}} J_{H,P}^{-1} \mathbb{E}_P\, d^2(\hat{a}_n; a_P)$ is finite (cf. (14)).

Example 1 (continued) Let $a_{P_t} = t$, $d(t; s) = |t - s|$. Then

$$w_H(P_t, \varepsilon) = t\varepsilon^2(1 - \varepsilon^2/2)/(1 - \varepsilon^2)^2 \geq t\varepsilon^2 ,$$

and (9) holds with $r = 1$, $J_{H,P_t} = t$. According to (8) with $\varepsilon^2 = 1/n$,

$$\sup_{P_s \in \mathscr{P}_H(P_t,\varepsilon)} \mathbb{E}_s^{1/2} |\hat{t}_n - s|^2 \geq t/2en$$

for any estimator $\hat{t}_n$. Hence $\sup_{t>0} \mathbb{E}_t |\hat{t}_n - t|^2 = \infty$, while the non-uniform bound is

$$\sup_{t>0} \mathbb{E}_t^{1/2} |\hat{t}_n/t - 1|^2 \geq 1/2en(1+2/n). \tag{14}$$

Remark In typical non-parametric situations the rate of the accuracy of estimation is worse than $n^{-1/2}$. However, an interesting fact is that if we choose $a_P = P$ and $d = d_H$, then $w_H(P, \varepsilon) = \varepsilon/2$ for all P, (9) holds with $r = 1/2$, $J_{H,P} = 1/2$, hence

$$\sup_{P \in \mathscr{P}} \mathbb{E}_P\, d_H^2(\hat{a}_n; a_P) \gtrsim 1/32en. \tag{15}$$

4 On Unbiased Estimation

It is not difficult to notice that in most estimation problems concerning non-parametric classes of distributions the available estimators are biased. The topic was studied by a number of authors (see Pfanzagl [11] and the references therein). Examples include non-parametric density, regression curve, hazard function (failure rate), and tail index estimation.

We notice in [6] that the sample autocorrelation is a non-negatively biased estimator of the autocorrelation function (the bias is positive unless the distribution of the sample elements is symmetric).

Theorem 1 suggests a way of showing that there are no unbiased finite-variance estimators for a given class $\mathscr{P}$ of distributions if the class contains a parametric family of distributions obeying the regularity condition (R_H) or (R_χ) with $\nu > 2$.

Example 3 Let $\mathscr{P}_b$, where $b > 0$, be the class of distributions P such that

$$\sup_{0<x\leq 1} \left|x^{-\alpha_P} P(X<x) - 1\right| x^{-b\alpha_P} < \infty \qquad (\exists \alpha_P > 0)$$

(the Hall class). Note that $F(x) \equiv P(X < x) = x^\alpha(1+O(x^{b\alpha}))$ as $x \to 0$ if $P \in \mathscr{P}_b$. We consider the problem of estimating index $\alpha \equiv \alpha_P$ from a sample of independent observations when the unknown distribution belongs to $\mathscr{P}_b$.

Let $P_{\alpha,0}$ and $P_{\alpha,\gamma}$ be the distributions with distribution functions (d.f.s)

$$F_{\alpha,0}(y) = y^\alpha \mathbb{1}\{0 < y \leq 1\},$$
$$F_{\alpha,\gamma}(y) = \delta^{-\gamma} y^{\alpha+\gamma} \mathbb{1}\{0 < y \leq \delta\} + y^\alpha \mathbb{1}\{\delta < y \leq 1\},$$

where $\delta = \gamma^{1/b\alpha}$, $\gamma \in (0; 1)$. One can check that

$$d_H^2(P_{\alpha,0}; P_{\alpha,\gamma}) = \gamma^{1/b}\left[1 - \sqrt{1+\gamma/\alpha}\Big/(1+\gamma/2\alpha)\right] \leq \gamma^{1/r}/8\alpha^2, \qquad (16)$$

$$d_\chi^2(P_{\alpha,0}; P_{\alpha,\gamma}) = \gamma^{1/r}\alpha^{-2}(1+\gamma/2\alpha)^{-1} \leq \gamma^{1/r}/\alpha^2, \qquad (17)$$

where $r = b/(1+2b)$. Thus, $(R_{t,H})$ and $(R_{t,\chi})$ hold with $\nu = 2+1/b$.

According to Theorem 1, there are no unbiased finite-variance estimators of index α.

Note that

$$d_H^2(P_{\alpha,0}; P_{\alpha,h}) \sim h^{2+1/b}/8\alpha^2 \qquad (h \to 0)$$

for the parametric family $\{P_{\alpha,h}, 0 \leq h < 1\} \subset \mathscr{P}_b$, while

$$d_H^2(P_{\alpha,0}; P_{\alpha+h,0}) \sim h^2/8\alpha^2 \qquad (h \to 0)$$

for the parametric family $\{P_{\alpha+h,0}, 0 \leq h < 1\} \subset \mathscr{P}_b$ (cf. [9, p. 293]). □

The next theorem shows that in typical non-parametric situations there are no asymptotically unbiased with the rate estimators.

Let $\{\mathscr{P}_n, n \geq 1\}$ be a non-increasing sequence of neighborhoods of a particular distribution P_0, and let $\{z_n\}$ be a sequence of positive numbers. Pfanzagl [11] calls estimator $\{\hat{a}_n\}$ *asymptotically unbiased* uniformly in $\mathscr{P}_n$ with the rate $\{z_n\}$ if

$$\limsup_{u\to\infty} \limsup_{n\to\infty} \sup_{P\in\mathscr{P}_n} |\mathbb{E}_P K_u((\hat{a}_n - a_P)/z_n)| = 0,$$

where $K_u(x) = x\mathbb{1}\{|x| \leq u\}$.

Denote $\mathscr{P}_{n,\varepsilon} = \mathscr{P}_\chi(P_0, \varepsilon/\sqrt{n})$, where $\varepsilon > 0$.

Theorem 2 ([11]) *Suppose that*

$$\limsup_{\varepsilon\to 0} \varepsilon^{-1} \liminf_{n\to\infty} w_\chi(P_0, \varepsilon/\sqrt{n})/z_n = \infty, \tag{18}$$

$$\lim_{u\to\infty} \liminf_{n\to\infty} P_0(|\hat{a}_n - a_{P_0}|/z_n \leq u) > 0, \tag{19}$$

$$\lim_{u\to\infty} \liminf_{n\to\infty} \mathbb{E}_{P_0} K_u^2((\hat{a}_n - a_{P_0})/z_n) < \infty. \tag{20}$$

Then estimator $\{\hat{a}_n\}$ cannot be asymptotically unbiased with the rate $\{z_n\}$ uniformly in $\mathscr{P}_{n,\varepsilon}$ for some $\varepsilon > 0$.

Pfanzagl [11] showed that in a number of particular non-parametric estimation problems

$$\inf_{\varepsilon>0} \varepsilon^{-c} \liminf_{n\to\infty} w_\chi(P_0, \varepsilon/\sqrt{n})/z_n > 0 \qquad (\exists\, c \in (0; 1)) \tag{21}$$

(cf. (10)). Note that (21) entails (18).

5 On Consistent Estimation

The rate of the accuracy of estimation can be very poor if the class $\mathscr{P}$ of distributions is "rich." In utmost cases the lower bound is bounded away from zero meaning neither estimator is consistent uniformly in $\mathscr{P}$. We present below few such examples.

Example 4 Let $\mathscr{F}$ be a class of distributions with absolutely continuous distribution functions on $\mathbb{R}$ such that $\int |f(x+y) - f(x)|dx \leq |y|$. Ibragimov and Khasminskiy [4] have shown that

$$\sup_{f\in\mathscr{F}} \mathbb{E}_f \int |\hat{f}_n - f| \geq 2^{-9} \qquad (n \geq 1)$$

for any estimator $\hat{f}_n$ of density f (see Devroye [2] for a related result). □

Example 5 Consider the problem of non-parametric regression curve estimation. Given a sample of i.i.d. pairs $(X_1, Y_1), \ldots, (X_n, Y_n)$, one wants to estimate the regression function

$$\psi(x) = \mathbb{E}\{Y|X=x\}.$$

There is no uniformly consistent estimator if the only assumption about $\mathcal{L}(X, Y)$ is that function ψ is continuous.

Let $\mathscr{P}$ be a class of distributions of random pairs (X, Y) taking values in $\mathbb{R}^2$ such that function $\psi(\cdot) = \mathbb{E}\{Y|X = \cdot\}$ is continuous. Set

$$f_0(x, y) = \mathbb{1}\{|x|\varepsilon|y| \le 1/2\}, \; f_1(x, y) = f_0(x, y) + hg(xh^{-c})g(y),$$

where $c>0$, $h\in(0; 1)$ and $g(x) = \sin(2\pi x)\mathbb{1}\{|x|\le 1/2\}$. These are the densities of two distributions of a random pair (X, Y).

Let ψ_k, $k\in\{0; 1\}$, denote the corresponding regression curves. Then

$$\psi_0 \equiv 0, \; \psi_1(x) = 2\pi^{-2}h\sin(2\pi h^{-c}x)\mathbb{1}\{|x|\le h^c/2\}.$$

Hence $\|\psi_0 - \psi_1\| = 2\pi^{-2}h$.

Note that $d_\chi^2(f_0; f_1) \le h^{2+c}/4$. Applying Lemma 13.1 [9], we derive

$$\max_{i\in\{0,1\}} \mathbb{P}_i(\|\hat{\psi}_n - \psi\| \ge h/\pi^2) \ge \left(1+d_\chi^2\right)^{-n}/4 \ge \exp(-nh^{2+c}/4)/4$$

for any regression curve estimator $\hat{\psi}_n$. With $c = n-2$ and $h = n^{-1/n}$, we get

$$\sup_{P\in\mathscr{P}} \mathbb{P}(\|\hat{\psi}_n - \psi\| \ge 1/9) \ge 1/4e^{1/4}. \tag{22}$$

Hence no estimator is consistent uniformly in $\mathscr{P}$. □

Example 6 Consider the problem of non-parametric estimation of the distribution function of the sample maximum. No uniformly consistent estimator exists in a general situation. Indeed, it is shown in [7, 8] that for any estimator $\{\hat{F}_n\}$ of the distribution function of the sample maximum there exist a d.f. F such that

$$\limsup_{n\to\infty} \mathbb{P}_F\left(\|\hat{F}_n - F^n\| \ge 1/9\right) \ge 1/3.$$

Moreover, one can construct d.f.s F_0 and F_1 such that

$$\max_{i\in\{0;1\}} \mathbb{P}_{F_i}(\|\hat{F}_n - F_i^n\| \ge 1/4) \ge 1/4 \qquad (n\ge 1),$$

where F_0 is uniform on [0; 1] and $F_1 \equiv F_{1,n} \to F_0$ everywhere as $n \to \infty$.

An estimator $\tilde{a}_n(\cdot) \equiv \tilde{a}_n(\cdot, X_1, \ldots, X_n)$ is called *shift-invariant* if

$$\tilde{a}_n(x, X_1, \ldots, X_n) = \tilde{a}_n(x+c, X_1+c, \ldots, X_n+c)$$

for every $x \in \mathbb{R}$, $c \in \mathbb{R}$. An estimator $\tilde{a}_n(\cdot)$ is called *scale-invariant* if

$$\tilde{a}_n(x, x_1, \ldots, x_n) = \tilde{a}_n(cx, cx_1, \ldots, cx_n) \qquad (\forall c > 0)$$

for all $x, x_1, \ldots, x_n$.

Examples of shift- and scale-invariant estimators of F^n include F_n^n, where F_n is the empirical distribution function, and the "blocks" estimator

$$\tilde{F}_n = \Big(\sum_{i=1}^{[n/r]} \mathbb{1}\{M_{i,r} < x\}/[n/r] \Big)^n,$$

where $M_{i,r} = \max\{X_{(i-1)r+1}, \ldots, X_{ir}\}$ $(1 \le r \le n)$.

For any shift- or scale-invariant estimator $\{\tilde{F}_n\}$ of the distribution function of the sample maximum there holds

$$\mathbb{P}_{F_0}\Big(\|\tilde{F}_n - F_0^n\| \ge 1/4 \Big) \ge 1/4 \qquad (n \ge 1). \tag{23}$$

Thus, consistent estimation of the distribution function of the sample maximum is only possible under certain assumptions on the class of unknown distributions. □

6 On Uniform Convergence

We saw that the rate of the accuracy of estimation cannot be better than $w_H(P, 1/\sqrt{n})$. According to Donoho and Liu [3], if a_P is linear and class $\mathscr{P}$ of distributions is convex, then there exists an estimator $\hat{a}_n$ attaining this rate.

We show now that in typical non-parametric situations neither estimator converges locally uniformly with the optimal rate.

Definition 5 Let $\mathscr{P}'$ be a subclass of $\mathscr{P}$. We say that estimator $\hat{a}_n$ converges to a_P with the rate z_n *uniformly* in $\mathscr{P}'$ if there exists a non-defective distribution P^* such that

$$\lim_{n\to\infty} \sup_{P \in \mathscr{P}'} \big| P((\hat{a}_n - a_P)/z_n \in A) - P^*(A) \big| = 0 \tag{24}$$

for every measurable set A with $P^*(\partial A) = 0$.

Note that for every $P \in \mathscr{P}'$ (24) yields the weak convergence $(\hat{a}_n - a_P)/v_n \Rightarrow P^*$.

The following result on impossibility of locally uniform convergence with the optimal rate is due to Pfanzagl [10]. It involves a continuity modulus based on the total variation distance.

Let $\mathscr{X} = \mathbb{R}$. Denote $\mathscr{P}_{TV}^{(n)}(P_0, \varepsilon) = \{P \in \mathscr{P} : d_{TV}(P^n; P_0^n) \leq \varepsilon\}$, and recall that

$$w_{TV}^{(n)}(P_0, \varepsilon) = \sup_{P \in \mathscr{P}_{TV}^{(n)}(P_0,\varepsilon)} |a_P - a_{P_0}|/2.$$

Theorem 3 ([10]) *Suppose that*

$$\lim_{\varepsilon \downarrow 0} \varepsilon^{-1} \limsup_{n\to\infty} w_{TV}^{(n)}(P_0, \varepsilon)/z_n = \infty. \tag{25}$$

Then neither estimator can converge to a_P with the rate z_n uniformly in $\mathscr{P}_{TV}^{(n)}(P_0, \varepsilon)$ for some $\varepsilon \in (0; 1)$.

Example 7 Let $\mathscr{P}_b^+$, where $b > 0$, be the non-parametric class of distributions on $(0; 1]$ with densities

$$f(x) = C_{\alpha,b} x^{\alpha-1}(1+r(x)),$$

where $\sup_{0<x\leq 1} |r(x)|x^{-\alpha b} < \infty$. We consider the problem of estimating index α.

Denote $r = b/(1+2b)$. Pfanzagl [10] showed that

$$\varepsilon^{-2r} \liminf_{n\to\infty} n^r w_{TV}^{(n)}(P_0, \varepsilon) > 0 \qquad (\forall \varepsilon \in (0; 1)). \tag{26}$$

Since $r < 1/2$, (25) and (26) entail that neither estimator of index α can converge to α uniformly in $\mathscr{P}_{TV}^{(n)}(P_0, \varepsilon)$ with the rate $z_n = n^{-b/(1+2b)}$. □

The next theorem presents a result on impossibility of locally uniform convergence with the optimal rate involving the modulus of continuity w_H based on the Hellinger distance. The Hellinger distance may be preferable to the total variation distance in identifying the optimal rate of the accuracy of estimation as there are cases where

$$d_{TV}(P_0; P_1) \gg d_H^2(P_0; P_1)$$

for "close" distributions P_0 and P_1. For instance, consider family $\mathscr{P} = \{P_{\alpha,\gamma}\}_{\gamma\geq 0}$, where distributions $\{P_{\alpha,\gamma}\}$ have been defined in Example 3. Then

$$d_{TV}(P_{\alpha,o}; P_{\alpha,\gamma}) \sim \frac{\gamma^{1/r-1}}{\alpha e} \gg d_H^2(P_{\alpha,o}; P_{\alpha,\gamma}) \sim \frac{\gamma^{1/r}}{8\alpha^2} \qquad (\gamma \to 0).$$

Theorem 4 *If (9) holds for a particular $P \in \mathscr{P}$ with $r < 1/2$, then neither estimator converges to a_P with the rate n^{-r} uniformly in $\mathscr{P}_H(P, 1/\sqrt{n})$.*

Theorem 4 generalizes Theorem 13.9 in [9] by relaxing the assumption that there exists a positive continuous derivative of distribution P^* with respect to the Lebesgue measure.

Proof of Theorem 4 Let $\hat{a}_n$ be an arbitrary estimator of a_P. Denote

$$w_{n,\varepsilon} = 2w_H(P, \varepsilon/\sqrt{n}) \qquad (\varepsilon > 0).$$

Let $\varepsilon \in (0; 1]$. For any $c > 0$ one can find $P' \in \mathscr{P}_H(P, \varepsilon/\sqrt{n})$ such that $a_{P'} - a_P \geq w_{n,\varepsilon} - c$. Then for an arbitrary $x \in \mathbb{R}$

$$\begin{aligned} 1 &\leq P(a_{P'} - \hat{a} > -x) + P(\hat{a}_n - a_P \geq x + w_{n,\alpha} - c) \\ &\leq P'(\hat{a} - a_{P'} < x) + P(\hat{a} - a_P \geq x + w_{n,\alpha} - c) + d_{TV}(P'^n; P^n). \end{aligned}$$

According to (3), $d_{TV}(P'^n; P^n) \leq \sqrt{2n}\, d_H(P'; P)$. Hence

$$P'(\hat{a}_n - a_{P'} \geq x) \leq P(\hat{a} - a_P \geq x + w_{n,\varepsilon} - c) + \sqrt{2n}\, d_H(P'; P).$$

Since $d_H(P'; P) \leq \varepsilon/\sqrt{n}$ and $w_{n,\varepsilon} \geq J_{H,P}\kappa_n\varepsilon^{2r}/n^r$ by (9), where $\kappa_n \to 1$ as $n \to \infty$, we can apply the monotone convergence theorem in order to derive that

$$\inf_{P' \in \mathscr{P}_H(P, \varepsilon/\sqrt{n})} P'(\hat{a}_n - a_{P'} \geq x) \leq P(\hat{a}_n - a_P \geq x + J_{H,P}\kappa_n\varepsilon^{2r}/n^r) + \varepsilon\sqrt{2} \qquad (27)$$

for any $\varepsilon \in [0; 1]$.

Suppose that estimator $\hat{a}_n$ converges to a_P uniformly in $\mathscr{P}_H(P, 1/\sqrt{n})$ with the rate $z_n = n^{-r}$. Then there exists a non-defective distribution P^* such that (24) holds with $\mathscr{P}' = \mathscr{P}_H(P, 1/\sqrt{n})$. We will show that this assumption leads to a contradiction.

Let η be an r.v. with the distribution $\mathcal{L}(\eta) = P^*$, and set

$$x = yn^{-r} \qquad (y \in \mathbb{R}).$$

The assumption implies that (24) holds also with $\mathscr{P}' = \mathscr{P}_H(P, \varepsilon/\sqrt{n})$ for every $\varepsilon \in [0; 1]$. Taking into account (24) and (27), we derive

$$\mathbb{P}(\eta \geq y) \leq \mathbb{P}(\eta \geq y + J_{H,P}\kappa_n\varepsilon^{2r}) + \varepsilon\sqrt{2}.$$

Hence $\mathbb{P}(y \leq \eta < y + J_{H,P}\kappa_n\varepsilon^{2r}) \leq \varepsilon\sqrt{2}$. Thus,

$$\mathbb{P}(y \leq \eta \leq y+1) \leq (1 + [1/J_{H,P}\kappa_n\varepsilon^{2r}])\varepsilon\sqrt{2} \leq \varepsilon\sqrt{2} + \varepsilon^{1-2r}\sqrt{2}/J_{H,P}\kappa_n.$$

Since $r < 1/2$, letting $\varepsilon \to 0$, we get

$$\mathbb{P}(y \leq \eta \leq y+1) = 0 \qquad (\forall y \in \mathbb{R}),$$

i.e., P^* is a defective distribution.

The contradiction obtained proves the theorem. □

Acknowledgements The author is grateful to the anonymous reviewer for helpful comments.

References

1. Barankin, E. W. (1949). Locally best unbiased estimates. *Annals of Mathematical Statistics, 20*, 477–501.
2. Devroye, L. (1995). Another proof of a slow convergence result of Birgé. *Statistics & Probability Letters, 23*(1), 63–67.
3. Donoho D. L., & Liu, R. C. (1991). Geometrizing rates of convergence II, III. *Annals of Statistics, 19*(2), 633–667, 668–701.
4. Ibragimov, I. A., & Khasminskii, R. Z. (1980). Estimation of distribution density. *Zap. Nauch. Sem. LOMI, 98*, 61–85.
5. Liu, R. C., & Brown, L. D. (1993). Nonexistence of informative unbiased estimators in singular problems. *Annals of Statistics, 21*(1), 1–13.
6. Novak, S. Y. (2006). A new characterization of the normal law. *Statistics & Probability Letters, 77*(1), 95–98.
7. Novak, S. Y. (2010). Impossibility of consistent estimation of the distribution function of a sample maximum. *Statistics, 44*(1), 25–30.
8. Novak, S. Y. (2010). Lower bounds to the accuracy of sample maximum estimation. *Theory of Stochastic Processes, 15(31)*(2), 156–161.
9. Novak, S. Y. (2011). *Extreme value methods with applications to finance*. London: Taylor & Francis/CRC. ISBN 978-1-43983-574-6.
10. Pfanzagl, J. (2000). On local uniformity for estimators and confidence limits. *Journal of Statistical Planning and Inference, 84*, 27–53.
11. Pfanzagl, J. (2001). A nonparametric asymptotic version of the Cramér-Rao bound. In *State of the art in probability and statistics. Lecture notes in monograph series* (Vol. 36, pp. 499–517). Beachwood, OH: Institute of Statistical Mathematics.
12. Tsybakov, A. B. (2009). *Introduction to nonparametric estimation*. New York: Springer.

Modification of Moment-Based Tail Index Estimator: Sums Versus Maxima

N. Markovich and M. Vaičiulis

Abstract In this contribution, we continue the investigation of the SRCEN estimator of the extreme-value index γ (or the tail index $\alpha = 1/\gamma$) proposed in McElroy and Politis (J Statist Plan Infer 137:1389–1406, 2007) for $\gamma > 1/2$. We propose a new estimator based on the local maximum. This, in fact, is a modification of the SRCEN estimator to the case $\gamma > 0$. We establish the consistency and asymptotic normality of the newly proposed estimator for i.i.d. data. Additionally, a short discussion on the comparison of the estimators is included.

1 Introduction and Main Results

Let $X_k,\ k \geq 1$ be non-negative independent, identically distributed (i.i.d.) random variables (r.v.s) with the distribution function (d.f.) F. Suppose that F belongs to the domain of attraction of the Fréchet distribution

$$\Phi_\gamma(x) = \begin{cases} 0, & x \leq 0, \\ \exp\{-x^{-1/\gamma}\}, & x > 0, \end{cases} \qquad \Phi := \Phi_1,$$

which means that there exist normalizing constants $a_m > 0$ such that

$$\lim_{m\to\infty} \mathrm{P}\left(\frac{L_m}{a_m} \leq x\right) = \lim_{m\to\infty} F^m(a_m x) = \Phi_\gamma(x), \tag{1}$$

for all $x > 0$, where $L_{u,v} = \max\{X_u, \ldots, X_v\}$ for $1 \leq u \leq v$ and $L_v = L_{1,v}$. The parameter $\gamma > 0$ is referred to as positive extreme-value index in the statistical literature.

N. Markovich
V.A. Trapeznikov Institute of Control Sciences of Russian Academy of Sciences, Moscow, Russia

M. Vaičiulis (✉)
Institute of Mathematics and Informatics, Vilnius University, Vilnius, Lithuania

P. Bertail et al. (eds.), *Nonparametric Statistics*, Springer Proceedings in Mathematics & Statistics 250, https://doi.org/10.1007/978-3-319-96941-1_6

Meerschaert and Scheffler [13] introduced the estimator for $\gamma \geq 1/2$, which is based on the growth rate of the logged sample variance of N observations $X_1, \ldots, X_N$:

$$\hat{\gamma}_N = \frac{1}{2\ln(N)} \ln_+ \left(N s_N^2\right),$$

where $s_N^2 = N^{-1} \sum_{i=1}^N \left(X_i - \bar{X}_N\right)^2$, $\bar{X}_N = (X_1 + \cdots + X_N)/N$ and $\ln_+(x) = 0 \vee \ln x$.

McElroy and Politis [12] divided the observations $X_1, \ldots, X_N$ into non-intersecting blocks $\{X_{(k-1)m^2+1}, \ldots, X_{km^2}\}$, $1 \leq k \leq [N/m^2]$ of the width m^2, while each such block was divided into non-intersecting sub-blocks of the width m. To estimate $\gamma > 1/2$ the so-called SRCEN estimator was proposed as the sample mean over all blocks:

$$\hat{\gamma}_N^{(1)}(m) = \frac{1}{[N/m^2]} \sum_{i=1}^{[N/m^2]} \xi_i(m),$$

where

$$\xi_i(m) = \frac{\ln\left(\sum_{j=(i-1)m^2+1}^{im^2} X_j^2\right)}{2\ln(m)} - \frac{1}{m}\sum_{k=1}^{m} \frac{\ln\left(\sum_{j=(k-1)m^2+(k-1)m+1}^{(k-1)m^2+km} X_j^2\right)}{2\ln(m)}, \tag{2}$$

and $[\cdot]$ denotes the integer part. In applications a simple heuristic rule for the choice of sub-block width $m = [N^{1/3}]$, provided in [12], works quite well, see the Monte-Carlo simulation studies in [12, 17] and [18].

Using the inequality of arithmetic and geometric means we obtain that for sample $X_1, \ldots, X_N$, $\hat{\gamma}_N^{(1)}(m) \geq 1/2$ holds with equality if and only if $X^2_{(i-1)m^2+1} = \cdots = X^2_{im^2}$, $1 \leq i \leq [N/m^2]$.

In this chapter we provide an estimator similar to the SRCEN estimator but one that can be used for $\gamma > 0$, not only for $\gamma > 1/2$. Namely, we replace the sums in (2) by corresponding maxima and introduce the new estimator

$$\hat{\gamma}_N^{(2)}(m) = \frac{1}{[N/m^2]} \sum_{i=1}^{[N/m^2]} \widetilde{\xi}_i(m)$$

where

$$\widetilde{\xi}_i(m) = \frac{\ln\left(L_{(i-1)m^2+1, im^2}\right)}{\ln(m)} - \frac{1}{m}\sum_{j=1}^{m} \frac{\ln\left(L_{(i-1)m^2+(j-1)m+1, (i-1)m^2+jm}\right)}{\ln(m)}.$$

In fact, the estimator $\hat{\gamma}_N^{(2)}(m)$ is based on the convergence $\mathbb{E}\ln(L_m)/\ln(m) \to \gamma$ as $m \to \infty$, which implies

$$2\mathbb{E}\left(\frac{\ln(L_{m^2})}{\ln(m^2)}\right) - \mathbb{E}\left(\frac{\ln(L_m)}{\ln(m)}\right) \to \gamma, \quad m \to \infty. \tag{3}$$

Thus, the estimator $\hat{\gamma}_N^{(2)}(m)$ is nothing else, but a moment-type estimator for the left-hand side in (3).

Note that $\hat{\gamma}_N^{(2)}(m)$ and $\hat{\gamma}_N^{(1)}(m)$ are scale-free, i.e., they do not change when X_j is replaced by cX_j with $c > 0$.

Typically, the estimators, whose constructions are based on the grouping of the observations into the blocks, are well suited for recursive on-line calculations. In particular, if $\hat{\gamma}_N^{(1)}(m) = \hat{\gamma}_N^{(1)}(m; X_1, \ldots, X_N)$ denotes the estimate of γ obtained from observations $X_1, \ldots, X_N$ and we get the next group of updates $X_{N+1}, \ldots, X_{N+m^2}$, then we obtain

$$\hat{\gamma}_N^{(1)}(m; X_1, \ldots, X_{N+m^2}) = \frac{1}{\tilde{N}+1}\sum_{i=1}^{\tilde{N}+1} \xi_i(m) = \frac{1}{\tilde{N}+1}\left(\tilde{N}\hat{\gamma}_N^{(1)}(m) + \xi_{\tilde{N}+1}(m)\right),$$

denoting $\tilde{N} = [N/m^2]$. After getting L additional groups $\{X_{N+(k-1)m^2+1}, \ldots, X_{N+km^2}\}$, $k = 1, \ldots, L$, we have

$$\hat{\gamma}_N^{(1)}(m; X_1, \ldots, X_{N+Lm^2}) = \frac{1}{\tilde{N}+L}\sum_{i=1}^{\tilde{N}+L} \xi_i(m)$$

$$= \frac{1}{\tilde{N}+L}\left(\tilde{N}\hat{\gamma}_N^{(1)}(m) + \xi_{\tilde{N}+1}(m) + \ldots + \xi_{\tilde{N}+L}(m)\right).$$

It is important that $\hat{\gamma}_N^{(1)}(m; X_1, \ldots, X_{N+Lm^2})$ is obtained using $\hat{\gamma}_N^{(1)}(m)$ after $O(1)$ calculations. The same is valid for $\widehat{\gamma}_N^{(2)}(m)$ substituting $\xi_i(m)$ by $\widetilde{\xi}_i(m)$. The discussion on on-line estimation of the parameter $\gamma > 0$ can be found in Section 1.2.3 of [11].

There are situations when data can be divided naturally into blocks but only the largest observations within blocks (the block-maxima) are available. Several such examples are mentioned in [15], see also [1], where battle deaths in major power wars between 1495 and 1975 were analyzed. Then the estimator $\hat{\gamma}_N^{(2)}(m)$ can be applied while the estimators $\hat{\gamma}_N$ and $\hat{\gamma}_N^{(1)}(m)$ are not applicable.

We will formulate our assumptions in terms of a so-called quantile function V of the d.f. F, which is defined as the left continuous generalized inverse:

$$V(t) := \inf\left\{x \geq 0 : -\frac{1}{\ln F(x)} \geq t\right\}.$$

The domain of attraction condition (1) can be stated in the following way in terms of V: regarding the d.f. F, (1) holds if and only if for all $x > 0$,

$$\lim_{t\to\infty} \frac{V(tx)}{V(t)} = x^{\gamma}, \tag{4}$$

i.e., the function V varies regularly at infinity with the index $\gamma > 0$ (written $V \in RV_{\gamma}$), see, e.g., [3, p.34].

First our result states that $\hat{\gamma}_N^{(2)}(m)$ is a weakly consistent estimator for $\gamma > 0$. For the sake of completeness we include a corresponding result (as a direct consequence of Prop. 1 in [12]) for the SRCEN estimator $\hat{\gamma}_N^{(1)}(m)$.

Theorem 1 *Let observations $X_1, \ldots, X_N$ be i.i.d. r.v.s with d.f. F.*

(i) Suppose F satisfies the first-order condition (4) with $\gamma > 1/2$. Suppose, in addition, that the probability density function $p(x)$ of F exists and is bounded, and also that $p(x)/x$ is bounded in a neighborhood of zero. Then for the sequence $m = m(N)$ satisfying

$$m(N) \to \infty, \quad \frac{N \ln^2 m}{m^2} \to \infty, \quad N \to \infty, \tag{5}$$

it holds

$$\hat{\gamma}_N^{(1)}(m) \xrightarrow{\text{P}} \gamma, \tag{6}$$

where $\xrightarrow{\text{P}}$ denotes convergence in probability.

(ii) Suppose F satisfies (4) with $\gamma > 0$. Suppose, in addition,

$$F(\delta) = 0 \tag{7}$$

for some $\delta > 0$. Then for the sequence $m = m(N)$ satisfying (5) it holds

$$\hat{\gamma}_N^{(2)}(m) \xrightarrow{\text{P}} \gamma. \tag{8}$$

As usual, in order to get asymptotic normality for estimators the so-called second-order regular variation condition in some form is assumed. We recall that the function V is said to satisfy the second-order condition if for some measurable function $A(t)$ with the constant sign near infinity, which is not identically zero, and $A(t) \to 0$ as $t \to \infty$,

$$\lim_{t\to\infty} \frac{\frac{V(tx)}{V(t)} - x^{\gamma}}{A(t)} = x^{\gamma} \frac{x^{\rho} - 1}{\rho} \tag{9}$$

holds for all $x > 0$ with $\rho < 0$, which is a second-order parameter. The function $A(t)$ measures the rate of convergence of $V(tx)/V(t)$ towards x^γ in (4), and $|A(t)| \in RV_\rho$, see [8].

In this work, we assume a second-order condition stronger than (9). Namely, we assume that we are in Hall's class of models (see [9]), where

$$V(t) = Ct^\gamma \left(1 + \rho^{-1} A(t)\,(1 + o(1))\right), \quad t \to \infty \tag{10}$$

with $A(t) = \gamma\beta t^\rho$, where $C > 0$, $\beta \in \mathbb{R} \setminus \{0\}$ and $\rho < 0$. The relation (10) is equivalent to

$$F(x) = \exp\left\{-\left(\frac{x}{C}\right)^{-1/\gamma}\left(1 + \frac{\beta}{\rho}\left(\frac{x}{C}\right)^{\rho/\gamma} + o\left(x^{\rho/\gamma}\right)\right)\right\}, \quad x \to \infty. \tag{11}$$

Theorem 2 *Let the observations $X_1, \ldots, X_N$ be i.i.d. r.v.s with d.f. F.*

(i) Suppose F satisfies the second-order condition (11) with $\gamma > 1/2$ and, in addition, that the probability density function $p(x)$ of F exists and it is bounded, and also that $p(x)/x$ is bounded in a neighborhood of zero. Then for the sequence $m = m(N)$ satisfying $m \to \infty$ and

$$N^{1/2} m^{-2\vee(-1+\rho)\vee(-2\gamma)} \ln(m) \to 0, \quad \text{if } -1 \vee \rho \neq 1 - 2\gamma,$$
$$N^{1/2} m^{-2\gamma} \ln^2(m) \to 0, \quad \text{if } -1 \vee \rho = 1 - 2\gamma,$$

$$\frac{N^{1/2}\ln(m)}{m}\left(\hat{\gamma}_N^{(1)}(m) - \gamma\right) \xrightarrow{\mathrm{d}} \mathcal{N}\left(0, \frac{\left(\gamma^2 - (1/4)\right)\pi^2}{6}\right), \quad N \to \infty, \tag{12}$$

holds, where $\xrightarrow{\mathrm{d}}$ stands for the convergence in distribution.

(ii) Suppose F satisfies (7) and (11) with $\gamma > 0$. Then, for the sequence $m = m(N)$ satisfying (5) and

$$\frac{N^{1/2}}{m} A(m) \to \nu \in (-\infty, +\infty), \tag{13}$$

it follows

$$\frac{N^{1/2}\ln(m)}{m}\left(\hat{\gamma}_N^{(2)}(m) - \gamma\right) \xrightarrow{\mathrm{d}} \mathcal{N}\left(-\frac{\nu\Gamma(1-\rho)}{\rho}, \frac{\gamma^2\pi^2}{6}\right), \quad N \to \infty. \tag{14}$$

The remainder of this chapter is organized as follows. In the next section we investigate the asymptotic mean squared error (AMSE) of the introduced estimator, and compare this estimator with several classical estimators, using the same methodology as in [4]. The last section contains the proofs of the results.

2 Comparison

The AMSE of the estimator $\hat{\gamma}_N^{(2)}(m)$ is given by

$$\text{AMSE}\left(\hat{\gamma}_N^{(2)}(m)\right) := \frac{1}{\ln^2(m)}\left\{\frac{\Gamma^2(1-\rho)A^2(m)}{\rho^2} + \frac{\gamma^2\pi^2 m^2}{6N}\right\}. \tag{15}$$

Regular variation theory, provided in [5] (see also [4]), allows us to perform the minimization of the sum in the curly brackets of (15). Namely, under the choice

$$\bar{m}(N) = \left(\frac{6\Gamma^2(1-\rho)\beta^2}{-\rho\pi^2}\right)^{1/(2(1-\rho))} N^{1/(2(1-\rho))}\left(1+o(1)\right), \quad N \to \infty,$$

we have

$$\text{AMSE}\left(\hat{\gamma}_N^{(2)}(\bar{m})\right) \sim \Gamma^2(-\rho)\beta^2\left(\frac{6\beta^2\Gamma^2(1-\rho)}{\pi^2(-\rho)}\right)^{1/(1-\rho)} \frac{N^{\rho/(1-\rho)}}{\ln^2(N)}, \quad N \to \infty.$$

Probably, the Hill's estimator

$$\gamma_N^{(H)}(k) = \frac{1}{k}\sum_{j=0}^{k-1} \ln\left(\frac{X_{N-j,N}}{X_{N-k,N}}\right),$$

is the most popular, [10]. Here, $1 \le k \le N$ is a tail sample fraction, while $X_{1,N} \le X_{2,N} \le \cdots \le X_{N,N}$ are order statistics from a sample $X_1, \ldots, X_N$. Let us denote $r = -1 \vee \rho$ and

$$\upsilon = \begin{cases} \beta, & -1 < \rho < 0, \\ \beta + (1/2), & \rho = -1, \\ 1/2, & \rho < -1. \end{cases}$$

From [4] it follows that the minimal AMSE of the Hill's estimator under assumption (11) satisfies the relation

$$\text{AMSE}\left(\gamma_N^{(H)}\left(\bar{k}\right)\right) \sim \frac{1-2r}{-2r}\left(\frac{-2r\upsilon^2\gamma^{2-4r}}{(1-r)^2}\right)^{1/(1-2r)} N^{2r/(1-2r)}, \quad N \to \infty,$$

where

$$\bar{k}(N) = \left(\frac{(1-r)^2}{-2r\upsilon^2}\right)^{1/(1-2r)} N^{-2r/(1-2r)}\left(1+o(1)\right), \quad N \to \infty.$$

Now we can compare the estimators $\hat{\gamma}_N^{(2)}(\bar{m})$ and $\gamma_N^{(H)}\left(\bar{k}\right)$. Denote the relative minimal AMSE in the same way as in [4]:

$$\mathrm{RMAMSE}(\gamma, \beta, \rho) = \lim_{N\to\infty} \frac{\mathrm{AMSE}\left(\gamma_N^{(H)}\left(\bar{k}\right)\right)}{\mathrm{AMSE}\left(\hat{\gamma}_N^{(2)}(\bar{m})\right)}.$$

Following [4] we may conclude that $\gamma_N^{(H)}\left(\bar{k}\right)$ dominates $\hat{\gamma}_N^{(2)}(\bar{m})$ at the point (γ, β, ρ) if $\mathrm{RMAMSE}(\gamma, \beta, \rho) < 1$ holds. Note that $\mathrm{RMAMSE}(\gamma, \beta, \rho) = 0$ holds for $-2 < \rho < 0$, i.e. $\gamma_N^{(H)}\left(\bar{k}\right)$ dominates $\hat{\gamma}_N^{(2)}(\bar{m})$, while for $\rho \le -2$ we have $\mathrm{RMAMSE}(\gamma, \beta, \rho) = \infty$ and thus, $\hat{\gamma}_N^{(2)}(\bar{m})$ outperforms $\gamma_N^{(H)}\left(\bar{k}\right)$ in this region of the parameter ρ. It is worth to note that the same conclusion holds if we replace Hill's estimator by another estimator investigated in [4].

Unfortunately, it is impossible to compare the performance of $\hat{\gamma}_N^{(1)}(m)$ and other estimators taking the AMSE as a measure. By taking $\nu = 0$ in (14) one can compare the estimators $\hat{\gamma}_N^{(1)}(m)$ and $\hat{\gamma}_N^{(2)}(m)$ under the same block width m^2. By comparing variances in the limit laws (12) and (14) we conclude that $\hat{\gamma}_N^{(1)}(m)$ outperforms $\hat{\gamma}_N^{(2)}(m)$ for $\gamma > 1/2$.

3 Proofs

Let us firstly provide preliminary results that are useful in our proofs.

Lemma 1 *Let $X_1, \ldots, X_N$ be i.i.d. r.v.s with d.f. F. Suppose F satisfies (4) with $\gamma > 0$ and (7). Then*

$$\lim_{m\to\infty} \mathbb{E} \ln\left(\frac{L_m}{V(m)}\right) = \chi\gamma, \tag{16}$$

$$\lim_{m\to\infty} \mathbb{E} \ln^2\left(\frac{L_m}{V(m)}\right) = \gamma^2\left(\chi^2 + \frac{\pi^2}{6}\right), \tag{17}$$

$$\lim_{m\to\infty} \mathbb{E} \ln^4\left(\frac{L_m}{V(m)}\right) = \gamma^4\left(\chi^4 + \chi^2\pi^2 + \frac{3\pi^4}{20} + 8\chi\zeta(3)\right), \tag{18}$$

$$\lim_{m\to\infty} \mathbb{E}\left(\ln\left(\frac{L_{m^2}}{V(m^2)}\right)\ln\left(\frac{L_m}{V(m)}\right)\right) = \chi^2\gamma^2, \tag{19}$$

holds, where $\chi \approx 0.5772$ is the Euler–Mascheroni constant defined by $\chi = -\int_0^\infty \ln(t)\exp\{-t\}t$, while $\zeta(t)$ denotes the Riemann zeta function, $\zeta(3) \approx 1.202$.

Proof of Lemma 1 We shall prove (16). Let Y be an r.v. with d.f. Φ. It is easy to check that it holds

$$\ln\left(\frac{L_m}{V(m)}\right) \stackrel{\mathrm{d}}{=} \ln\left(\frac{V(mY)}{V(m)}\right).$$

By Theorem B.1.9 in [3], the assumption $V \in RV_\gamma$, $\gamma > 0$ implies that for arbitrary $\epsilon_1 > 0$, $\epsilon_2 > 0$ there exists $m_0 = m_0(\epsilon_1, \epsilon_2)$ such that for $m \geq m_0$, $my \geq m_0$,

$$(1-\epsilon_1)y^\gamma \min\left\{y^{\epsilon_2}, y^{-\epsilon_2}\right\} < \frac{V(my)}{V(m)} < (1+\epsilon_1)y^\gamma \max\left\{y^{\epsilon_2}, y^{-\epsilon_2}\right\}$$

holds. Whence we get that under restriction $0 < \epsilon_1 < 1$ it follows

$$\ln(1-\epsilon_1) + (\gamma - u(y))\ln(y) < \ln\left(\frac{V(my)}{V(m)}\right) < \ln(1+\epsilon_1) + (\gamma + u(y))\ln(y), \tag{20}$$

where $u(y) = -\epsilon_2 I\{y < 1\} + \epsilon_2 I\{y \geq 1\}$ and $I\{\cdot\}$ denotes the indicator function.

We write for $m > m_0$,

$$\mathbb{E}\left(\ln\left(\frac{V(mY)}{V(m)}\right)\right) = J_{1,m} + J_{2,m},$$

where

$$J_{1,m} = \int_0^{m_0/m} \ln\left(\frac{V(my)}{V(m)}\right) \mathrm{d}\Phi(y), \quad J_{2,m} = \int_{m_0/m}^{\infty} \ln\left(\frac{V(my)}{V(m)}\right) \mathrm{d}\Phi(y).$$

The statement (16) follows from

$$\lim_{m\to\infty} J_{1,m} = 0, \tag{21}$$

$$\lim_{m\to\infty} J_{2,m} = \chi\gamma. \tag{22}$$

Substituting $my = t$ we get

$$\begin{aligned} \left|J_{1,m}\right| &\leq \int_0^{m_0} \left|\ln\left(\frac{V(t)}{V(m)}\right)\right| \mathrm{d}\Phi(t/m) \\ &= \int_0^{m_0} \left|\ln V(t)\right| \mathrm{d}\Phi(t/m) + \Phi(m_0/m)\left|\ln V(m)\right|. \end{aligned}$$

By using $\mathrm{d}\Phi(t/m) = m\Phi\,(t/(m-1))\,\mathrm{d}\Phi(t)$ we obtain

$$\left|J_{1,m}\right| \le m\Phi\,(m_0/(m-1)) \int_0^{m_0} |\ln V(t)|\,\mathrm{d}\Phi(t) + \Phi(m_0/m)\,|\ln V(m)|\,.$$

Assumption (7) ensures $V(0) \ge \delta$, which implies $\int_0^{m_0} |\ln V(t)|\,\mathrm{d}\Phi(t) < \infty$. Since the sequence $V(n)$ is of a polynomial growth and $\Phi(m_0/m) = \exp\{-m/m_0\}$ tends to zero exponentially fast, then relation (21) follows.

To prove (22) we use inequality (20). Then we obtain

$$\left|J_{2,m} - \chi\gamma\right| \le \max\{-\ln(1-\epsilon_1), \ln(1+\epsilon_1)\} + \epsilon_2 \mathbb{E}\,|\ln(Y)| + \gamma \int_0^{m_0/m} |\ln(y)|\,\mathrm{d}\Phi(y).$$

One can check that $\mathbb{E}\,|\ln(Y)| = \chi - 2\mathrm{Ei}(-1)$, where $\mathrm{Ei}(x)$, $x \in \mathbb{R}\setminus\{0\}$ denotes the exponential integral function, $\mathrm{Ei}(-1) \approx -0.219384$.

Since $\epsilon_1 > 0$ and $\epsilon_2 > 0$ may be taken arbitrary small, the proof of relation (22) will be finished if we show that $\int_0^{m_0/m} |\ln(y)|\,\mathrm{d}\Phi(y) \to 0$, $m \to \infty$. Substituting $t = my$ we get

$$\begin{aligned}
\int_0^{m_0/m} |\ln(y)|\,\mathrm{d}\Phi(y) &= \int_0^{m_0} |\ln(t/m)|\,\mathrm{d}\Phi(t/m) \\
&= m \int_0^{m_0} |\ln(t/m)|\,\Phi(t/(m-1))\mathrm{d}\Phi(t) \\
&\le m\Phi(m_0/(m-1))\,(\ln(m) + \mathbb{E}\,|\ln(Y)|) \to 0,
\end{aligned}$$

as $m \to \infty$. This completes the proof of (22), and also of relation (16).

Proofs of relations (17) and (18) are similar and thus are skipped. It remains to prove (19). We note that L_m and L_{m+1,m^2} are independent r.v.s and $L_{m^2} = L_m \vee L_{m+1,m^2}$. Let Y_1 and Y_2 are independent r.v.s with d.f. Φ. Then it holds

$$\ln\left(\frac{L_{m^2}}{V(m^2)}\right)\ln\left(\frac{L_m}{V(m)}\right) \stackrel{\mathrm{d}}{=} \ln\left(\frac{V(mY_1) \vee V(m(m-1)Y_2)}{V(m^2)}\right)\ln\left(\frac{V(mY_1)}{V(m)}\right),$$

and consequently,

$$\mathbb{E}\left(\ln\left(\frac{L_{m^2}}{V(m^2)}\right)\ln\left(\frac{L_m}{V(m)}\right)\right) = \mathbb{E}\left(\ln\left(\frac{V(mY_1) \vee V(m(m-1)Y_2)}{V(m^2)}\right)\ln\left(\frac{V(mY_1)}{V(m)}\right)\right).$$

Let us recall that $V(t)$, $t \ge 0$ is a non-decreasing function, see, e.g., Prop. 2.3 in [6]. By using this property we obtain

$$\mathbb{E}\left(\ln\left(\frac{V(mY_1) \vee V(m(m-1)Y_2)}{V(m^2)}\right)\ln\left(\frac{V(mY_1)}{V(m)}\right)\right) = J_{3,m} + J_{4,m} + J_{5,m},$$

where

$$J_{3,m} = \mathbb{E}\left(\ln\left(\frac{V(mY_1)}{V(m^2)}\right)\ln\left(\frac{V(mY_1)}{V(m)}\right)I\{Y_1 > (m-1)Y_2\}\right),$$

$$J_{4,m} = \mathbb{E}\left(\ln\left(\frac{V(m(m-1)Y_2)}{V(m^2)}\right)\right)\mathbb{E}\left(\ln\left(\frac{V(mY_1)}{V(m)}\right)\right),$$

$$J_{5,m} = \mathbb{E}\left(\ln\left(\frac{V(m(m-1)Y_2)}{V(m^2)}\right)\ln\left(\frac{V(mY_1)}{V(m)}\right)I\{Y_1 > (m-1)Y_2\}\right).$$

Let us rewrite quantity $J_{4,m}$ as follows:

$$J_{4,m} = \left\{\ln\left(\frac{V(m(m-1))}{V(m^2)}\right) + \mathbb{E}\ln\left(\frac{L_{m(m-1)}}{V(m(m-1))}\right)\right\}\mathbb{E}\ln\left(\frac{L_m}{V(m)}\right).$$

For any $\epsilon > 0$ there exists natural $\tilde{m}_0$ such that $1/m < \epsilon$ for $m \geq m_0$. Then $V(m^2(1-\epsilon))/V(m^2) \leq V\left(m^2(1-1/m)\right)/V(m^2) \leq 1$. By (4) we get $V(m^2(1-\epsilon))/V(m^2) \to (1-\epsilon)^\gamma$, $m \to \infty$. Since $\epsilon > 0$ can be taken arbitrary small, the relation $V(m(m-1))/V(m^2) \to 1$, $m \to \infty$ holds. By using the last relation and (16) we deduce that $J_{4,m} \to \chi^2\gamma^2$ holds as $m \to \infty$.

Next, we have

$$J_{3,m} = \mathbb{E}\left(\ln^2\left(\frac{V(mY_1)}{V(m)}\right)I\{Y_1 > (m-1)Y_2\}\right)$$
$$+\ln\left(\frac{V(m)}{V(m^2)}\right)\mathbb{E}\left(\ln\left(\frac{V(mY_1)}{V(m)}\right)I\{Y_1 > (m-1)Y_2\}\right).$$

We apply the Hölder's inequality to get

$$\left|J_{3,m}\right| \leq \left\{\mathbb{E}\ln^4\left(\frac{L_m}{V(m)}\right)\right\}^{1/2}\{\mathrm{P}(Y_1 > (m-1)Y_2)\}^{1/2}$$
$$+\left|\ln\left(\frac{V(m)}{V(m^2)}\right)\right|\left\{\mathbb{E}\ln^2\left(\frac{L_m}{V(m)}\right)\right\}^{1/2}\{\mathrm{P}(Y_1 > (m-1)Y_2)\}^{1/2}.$$

We find that $\mathrm{P}(Y_1 > (m-1)Y_2) = 1/m$ holds. Let us recall the well-known property of regularly varying functions: if $V \in RV_\gamma$, then

$$\lim_{m\to\infty}\frac{\ln V(m)}{\ln(m)} = \gamma, \tag{23}$$

see, e.g., Prop. B.1.9 in [3]. By using (23) we obtain $\ln\left(V(m^2)/V(m)\right) \sim \gamma\ln(m)$, $m \to \infty$. Thus, keeping in mind (17) and (18) we obtain $\left|J_{3,m}\right| = O\left(m^{-1/2}\ln(m)\right)$,

$m \to \infty$. By a similar argument we obtain $\left|J_{5,m}\right| = O\left(m^{-1/2}\right)$, $m \to \infty$. This finishes the proof of (19) and Lemma 1.

Proof of Theorem 1 First we prove (8). Let us rewrite

$$\hat{\gamma}_N^{(2)}(m) = \gamma + \left\{\mathbb{E}\hat{\gamma}_N^{(2)}(m) - \gamma\right\} + S_N(m), \tag{24}$$

where

$$\begin{aligned}\mathbb{E}\hat{\gamma}_N^{(2)}(m) - \gamma = &\left\{\frac{\ln V(m^2) - \ln V(m)}{\ln(m)} - \gamma\right\} \\ &+\frac{1}{\ln(m)}\left(\mathbb{E}\ln\left(\frac{L_{m^2}}{V(m^2)}\right) - \mathbb{E}\ln\left(\frac{L_m}{V(m)}\right)\right)\end{aligned} \tag{25}$$

and

$$\begin{aligned}S_N(m) = \frac{1}{[N/m^2]\ln(m)} \sum_{i=1}^{[N/m^2]} &\left\{\left\{\ln\left(\frac{L_{(i-1)m^2+1,im^2}}{V(m^2)}\right) - \mathbb{E}\ln\left(\frac{L_{m^2}}{V(m^2)}\right)\right\}\right. \\ &\left.-\frac{1}{m}\sum_{j=1}^{m}\left\{\ln\left(\frac{L_{(i-1)m^2+(j-1)m+1,(i-1)m^2+jm}}{V(m)}\right) - \mathbb{E}\ln\left(\frac{L_m}{V(m)}\right)\right\}\right\}.\end{aligned}$$

By combining (16) and (23) we deduce that $\mathbb{E}\hat{\gamma}_N^{(2)}(m) - \gamma \to 0$, $m \to \infty$. Thus, it is enough to prove that $S_N(m) \xrightarrow{\mathrm{P}} 0$ as $N \to \infty$. By Chebyshev's inequality, for any $\epsilon > 0$ it holds $\mathrm{P}\left(|S_N(m)| > \epsilon\right) \le \epsilon^{-2}\mathbb{E}\left(S_N(m)\right)^2$. We have

$$\begin{aligned}\mathbb{E}\left(S_N(m)\right)^2 = &\frac{1}{[N/m^2]\ln^2(m)}\left\{\mathrm{Var}\left(\ln\left(\frac{L_{m^2}}{V(m^2)}\right)\right)\right. \\ &\left.-2\mathrm{Cov}\left(\ln\left(\frac{L_{m^2}}{V(m^2)}\right), \ln\left(\frac{L_m}{V(m)}\right)\right) + \frac{1}{m}\mathrm{Var}\left(\ln\left(\frac{L_m}{V(m)}\right)\right)\right\}.\end{aligned} \tag{26}$$

Use (16)–(17) and (19) to deduce that the sum in the curly brackets has a finite limit as $m \to \infty$. Thus, assumption (5) ensures $\mathbb{E}\left(S_N(m)\right)^2 \to 0$, $m \to \infty$. This finishes the proof of (8).

Consider now (6), where the restriction $\gamma > 1/2$ holds. Assumption (4) is equivalent to $1 - F \in RV_{-1/\gamma}$. By the Representation Theorem (see Thm. B.1.6. in [3]), there exists a function $\ell \in RV_0$, such that

$$1 - F(x^{1/2}) = x^{-1/(2\gamma)}\ell\left(x^{1/2}\right), \quad x \to \infty. \tag{27}$$

Following the Mijnheer Theorem (see Thm. 1.8.1 in [16]), we determine the norming function $a(m) \in RV_{2\gamma}$ from

$$\lim_{m\to\infty} \frac{m\ell\left(a^{1/2}(m)\right)}{(a(m))^{1/(2\gamma)}} = d(\gamma), \quad d(\gamma) = \Gamma(1-1/(2\gamma))\cos(\pi/(4\gamma)). \tag{28}$$

Put $Q(m) = (X_1^2 + \cdots + X_m^2)/a_m$. Then $Q(m) \stackrel{d}{\to} Z$, as $m \to \infty$, where Z is totally skewed to the right $1/(2\gamma)$-stable r.v. with characteristic function

$$\mathbb{E}\exp\{i\theta Z\} = \exp\left\{-|\theta|^{1/(2\gamma)}\left(1 - \mathrm{i}\,\mathrm{sgn}(\theta)\tan\left(\frac{\pi}{4\gamma}\right)\right)\right\}. \tag{29}$$

Similarly to (24) we use the decomposition

$$\hat{\gamma}_N^{(1)}(m) = \gamma + \left\{\mathbb{E}\hat{\gamma}_N^{(1)}(m) - \gamma\right\} + \tilde{S}_N(m),$$

where

$$\tilde{S}_N(m) = \frac{1}{2[N/m^2]\ln(m)} \sum_{i=1}^{[N/m^2]} \left\{\ln\left(\sum_{j=(i-1)m^2+1}^{im^2} \frac{X_j^2}{a(m^2)}\right) - \mathbb{E}\ln Q(m^2)\right\}$$
$$-\frac{1}{m}\sum_{j=1}^{m}\left\{\ln\left(\sum_{j=(i-1)m^2+(i-1)m+1}^{(i-1)m^2+im} \frac{X_j^2}{a(m)}\right) - \mathbb{E}\ln Q(m)\right\}.$$

The bias of the estimator $\hat{\gamma}_N^{(1)}(m)$ is given by $\mathbb{E}\hat{\gamma}_N^{(1)}(m) - \gamma = \Delta(m^2) - (1/2)\Delta(m)$, where

$$\Delta(m) = \frac{\ln a(m)}{\ln m} - 2\gamma + \frac{1}{\ln m}\{\mathbb{E}\ln Q(m) - \mathbb{E}\ln Z\}.$$

In Prop. 1–2 of [12] it is proved

$$\mathbb{E}\ln Q(m) \to \mathbb{E}\ln Z, \quad \mathbb{E}\ln^2 Q(m) \to \mathbb{E}\ln^2 Z, \tag{30}$$
$$\mathrm{Cov}\left(\ln Q(m^2), \ln Q(m)\right) \to 0, \quad m \to \infty. \tag{31}$$

It is worth to note that the moments $\mathbb{E}\ln Z$ and $\mathbb{E}\ln^2 Z$ can be found explicitly. Indeed, there is a direct connection between moments of order $r < 1/(2\gamma)$ and log-moments of order $k \in \mathbb{N}$:

$$\mathbb{E}\ln^k Z = \frac{\mathrm{d}^k}{\mathrm{d}r^k}\mathbb{E}Z^r\bigg|_{r=0}, \tag{32}$$

see [19]. Regarding the moments $\mathbb{E}Z^r$, the following relation is proved in Section 8.3 of [14]:

$$\mathbb{E}Z^r = \frac{\Gamma(1-2\gamma r)}{\Gamma(1-r)}\left(1+\tan^2\left(\frac{\pi}{4\gamma}\right)\right)^{\gamma r}, \quad -1 < r < 1/(2\gamma). \tag{33}$$

By using (32) and (33) we obtain

$$\mathbb{E}\ln Z = -\chi + 2\chi\gamma + \gamma\ln\left(\tan^2\left(\frac{\pi}{4\gamma}\right)+1\right), \tag{34}$$

$$\mathbb{E}\ln^2 Z = \chi^2 - \frac{\pi^2}{6} + 4\chi^2\gamma^2 - 4\chi^2\gamma + \frac{2\pi^2\gamma^2}{3} + \gamma^2\log^2\left(\tan^2\left(\frac{\pi}{4\gamma}\right)+1\right)$$
$$+4\chi\gamma^2\log\left(\tan^2\left(\frac{\pi}{4\gamma}\right)+1\right) - 2\chi\gamma\log\left(\tan^2\left(\frac{\pi}{4\gamma}\right)+1\right). \tag{35}$$

We combine (23) and the first relation in (30) to deduce that $\Delta(m) \to 0$, $m \to \infty$, which implies $\mathbb{E}\hat{\gamma}_N^{(1)}(m) - \gamma \to 0$, $m \to \infty$. Thus, relation (6) will be proved if we show that under assumptions (5), $\mathbb{E}\left(\tilde{S}_N(m)\right)^2 \to 0$. The last relation can be verified by using (30) and (31), and

$$\mathbb{E}\left(\tilde{S}_N(m)\right)^2 = \frac{\mathrm{Var}\left(\ln Q(m^2)\right) - 2\mathrm{Cov}\left\{\ln Q(m^2), \ln Q(m)\right\} + m^{-1}\mathrm{Var}\left(\ln Q(m)\right)}{4[N/m^2]\ln^2(m)}. \tag{36}$$

This completes the proof of Theorem 1.

Proof of Theorem 2 In view of decomposition (24), the assertion (14) follows from

$$\mathbb{E}\left(S_N(m)\right)^2 \sim \frac{\pi^2\gamma^2 m^2}{6N\ln^2(m)}, \tag{37}$$

$$\left\{\mathbb{E}\left(S_N(m)\right)^2\right\}^{-1/2} S_N(m) \xrightarrow{\mathrm{d}} \mathcal{N}(0,1), \tag{38}$$

$$\frac{N^{1/2}\ln(m)}{m}\left(\mathbb{E}\hat{\gamma}_N^{(1)}(m) - \gamma\right) \to -\frac{\nu\Gamma(1-\rho)}{\rho}, \quad N \to \infty, \tag{39}$$

where ν is the same as in (13).

Relation (37) follows from (26) by applying (16)–(17) and (19). To prove (38), by using (16)–(19) we check the 4-th order Lyapunov condition for i.i.d. random variables forming a triangular array. We skip standard details.

By using (10) we obtain

$$\frac{\ln V(m)}{\ln(m)} - \gamma = \frac{1}{\ln(m)}\left\{\ln(C) + \frac{A(m)}{\rho}(1+o(1))\right\}, \quad m \to \infty.$$

Following the proof of Lemma 2 in [18] one can obtain

$$\mathbb{E}\ln\left(\frac{L_m}{V(m)}\right) - \chi\gamma = \frac{\Gamma(1-\rho)-1}{\rho}A(m)\,(1+o(1))\,, \quad m\to\infty.$$

We combine the last two relations, assumption (13) and decomposition (25) to verify (39).

Let us discuss the proof of (12) now. Relations (30), (31), (34)–(36) imply $\mathbb{E}\left(\tilde{S}_N(m)\right)^2 \sim m^2 N^{-1}\ln^{-2}(m)\left(\gamma^2-(1/4)\right)\pi^2/6$, $N\to\infty$. In view of the last relation it is enough to prove that

$$\left\{\mathrm{Var}\left(\tilde{S}_N(m)\right)\right\}^{-1/2}\tilde{S}_N(m) \overset{\mathrm{d}}{\to} \mathcal{N}(0,1)\,, \tag{40}$$

$$\mathbb{E}\hat{\gamma}_N^{(1)}(m)-\gamma = \begin{cases} O\left(m^{-1\vee\rho\vee(1-2\gamma)}\right), & -1\vee\rho\neq 1-2\gamma, \\ O\left(m^{1-2\gamma}\ln(m)\right), & -1\vee\rho = 1-2\gamma. \end{cases} \tag{41}$$

We skip a standard proof of (40) and focus on the investigation of the bias $\mathbb{E}\hat{\gamma}_N^{(1)}(m)-\gamma$. Firstly, we prove that

$$\frac{\ln a(m^2)-\ln a(m)}{2\ln(m)} - \gamma = O\left(\frac{m^{-1\vee\rho}}{\ln(m)}\right), \quad m\to\infty. \tag{42}$$

The relation (11) can be written in the form $1-F(x)=x^{-1/\gamma}\ell(x)$, $x\to\infty$, where function $\ell\in RV_0$ has the form

$$\ell(x) = C^{1/\gamma}\left(1+\tilde{C}(\beta,\rho)\,(x/C)^{(-1\vee\rho)/\gamma}+o\left(x^{(-1\vee\rho)/\gamma}\right)\right), \quad x\to\infty, \tag{43}$$

where

$$\tilde{C}(\beta,\rho) = \begin{cases} \beta/\rho, & -1<\rho<0, \\ -(2\beta-1)/\rho, & \rho=-1, \beta\neq 1/2, \\ -1/2, & \rho<-1. \end{cases}$$

Now, by using (28), one can find that under assumption (11) the norming function satisfies the asymptotic relation

$$a(m) = \left(C^{1/\gamma}/d(\gamma)\right)^{2\gamma} m^{2\gamma}\left(1+2\gamma\tilde{C}(\beta,\rho)d^{-(-1\vee\rho)}(\gamma)m^{-1\vee\rho}+o\left(m^{-1\vee\rho}\right)\right)$$

as $m\to\infty$, while the last relation implies (42).

We claim that

$$\frac{\mathbb{E}\ln Q(m) - \mathbb{E}\ln Z}{\ln m} = \begin{cases} O\left(m^{-1\vee\rho\vee(1-2\gamma)}\right), & -1\vee\rho \neq 1-2\gamma, \\ O\left(m^{1-2\gamma}\ln(m)\right), & -1\vee\rho = 1-2\gamma \end{cases} \tag{44}$$

as $m \to \infty$.

Then terms $\ln^{-1}(m^2)\left\{\mathbb{E}\ln Q(m^2) - \mathbb{E}\ln Z\right\}$ and $(2\ln(m))^{-1}\left\{\ln a(m^2) - \ln a(m)\right\} - \gamma$ are negligible with respect to $\ln^{-1}(m)\left\{\mathbb{E}\ln Q(m) - \mathbb{E}\ln Z\right\}$ and thus, the relation (41) follows.

To verify (44) we use the similar decomposition $\mathbb{E}\ln Q(m) - \mathbb{E}\ln Z = R_{1,m} - R_{2,m} - R_{3,m}$ as in the proof of Prop. 3 in [12], where

$$R_{1,m} = \int_0^{\infty} \{\mathrm{P}(\ln Q(m) > x) - \mathrm{P}(\ln Z > x)\}\,\mathrm{d}x,$$

$$R_{2,m} = \int_{-\ln m}^{0} \{\mathrm{P}(\ln Q(m) < x) - \mathrm{P}(\ln Z < x)\}\,\mathrm{d}x,$$

$$R_{3,m} = \int_{-\infty}^{-\ln m} \{\mathrm{P}(\ln Q(m) < x) - \mathrm{P}(\ln Z < x)\}\,\mathrm{d}x.$$

By using substitution $t = \exp\{x\}$ we obtain

$$R_{1,m} = \int_1^{\infty} t^{-1}\{\mathrm{P}(Q(m) > t) - \mathrm{P}(Z > t)\}\,\mathrm{d}t.$$

Similarly we get $R_{2,m} = \int_{1/m}^{1} t^{-1}\{\mathrm{P}(Q(m) < t) - \mathrm{P}(Z < t)\}\,\mathrm{d}t$. From Corollary 2 in [2] it follows

$$\sup_{t\geq 0} f_\gamma(t)\,|\mathrm{P}(Q(m) > t) - \mathrm{P}(Z > t)| = O\left(\lambda\left(m^{2\gamma}\right) + m^{-2\gamma}\right), \quad m \to \infty,$$

where $f_\gamma(t) = 1 + t^{2\gamma}\ln^{-2}(\mathrm{e} + t)$ and $\lambda(R) = \lambda_1(R) + R^{-1+1/(2\gamma)}\lambda_2(R)$, $R > 0$, where

$$\lambda_1(R) = \sup_{u\geq R} u^{1/(2\gamma)}\left|\mathrm{P}(X_1^2 > u) - \mathrm{P}(Z > u)\right|,$$

$$\lambda_2(R) = \int_0^R \left|\mathrm{P}(X_1^2 > u) - \mathrm{P}(Z > u)\right|\mathrm{d}u.$$

It is well-known that $P(Z > x) = C_1 x^{-1/(2\gamma)}\left(1 + C_2 x^{-1/(2\gamma)} + o\left(x^{-1/(2\gamma)}\right)\right)$, $x \to \infty$ holds, where $C_k = C_k(\gamma)$ are some constants. The asymptotic of $\mathrm{P}(X_1^2 > u)$ is given in (27), where a function ℓ slowly varying at infinity is given in (43). Recall that $\hat{\gamma}_N^{(1)}(m)$ is a scale-free estimator. Thus, without loss of generality, we may assume that the scale parameter C in (43) satisfies $C^{1/\gamma} = C_1$. Then we have

$$\mathrm{P}(X_1^2 > x) - \mathrm{P}(Z > x) = Dx^{(-2\vee(\rho-1))/(2\gamma)} + o\left(x^{(-2\vee(\rho-1))/(2\gamma)}\right), \quad x \to \infty, \tag{45}$$

where $D \neq 0$ is some constant. By applying (45) we obtain immediately $\lambda_1\left(m^{2\gamma}\right) = O\left(m^{-1\vee\rho}\right)$, $m \to \infty$. If $-2 \vee (\rho - 1) > -2\gamma$, by ex. 1.2 in [7], a relation $f(x) \sim x^r$, $x \to \infty$ implies

$$\int_0^x f(t)\mathrm{d}t \sim \begin{cases} x^{r+1}/(r+1), & r > -1, \\ \ln(x), & r = -1, \end{cases} \quad x \to \infty \tag{46}$$

and thus we obtain $m^{1-2\gamma}\lambda_2(m^{2\gamma}) = O\left(m^{-1\vee\rho}\right)$, $m \to \infty$. In the case $-2 \vee (\rho - 1) = -2\gamma$, by applying (46) one more time we get $m^{1-2\gamma}\lambda_2(m^{2\gamma}) = O\left(m^{1-2\gamma}\ln(m)\right)$, $m \to \infty$. As for the case $-2 \vee (\rho - 1) < -2\gamma$, we have $m^{1-2\gamma}\lambda_2(m^{2\gamma}) = O\left(m^{1-2\gamma}\right)$, $m \to \infty$. By putting the obtained results together we get

$$\sup_{t\geq 0} f_\gamma(t)\left|\mathrm{P}\left(Q(m) > t\right) - \mathrm{P}\left(Z > t\right)\right| = \begin{cases} O\left(m^{-1\vee\rho\vee(1-2\gamma)}\right), & -1 \vee \rho \neq 1 - 2\gamma, \\ O\left(m^{1-2\gamma}\ln(m)\right), & -1 \vee \rho = 1 - 2\gamma \end{cases}$$

as $m \to \infty$.

Applying the last asymptotic relation we obtain immediately

$$\left|R_{2,m}\right| = \begin{cases} O\left(m^{-1\vee\rho\vee(1-2\gamma)}\ln(m)\right), & -1 \vee \rho \neq 1 - 2\gamma, \\ O\left(m^{1-2\gamma}\ln^2(m)\right), & -1 \vee \rho = 1 - 2\gamma \end{cases}$$

and $\left|R_{1,m}\right| = o\left(\left|R_{2,m}\right|\right)$ as $m \to \infty$. Since the relation $\left|R_{3,m}\right| = O\left(m^{-1}\right) = o\left(\left|R_{2,m}\right|\right)$, $m \to \infty$ holds (see proof of Prop. 3 in [12]), the statement of Theorem 2 follows.

References

1. Cederman, L.-E., Warren , T. C., & Sornette, D. (2011). Testing Clausewitz: Nationalism, mass mobilization, and the severity of war. *International Organization, 65*, 605–638.
2. Daugavet, A. I. (1987). Estimate of the rate of convergence of number characterises in limit theorems with a stable limit law. *Teoriya Veroyatnostei i ee Primeneniya, 32*, 585–589.
3. De Haan, L., & Ferreira, A. (2006). *Extreme value theory: An introduction*. New York: Springer.
4. De Haan, L., & Peng, L. (1998). Comparison of tail index estimators. *Statist. Nederlandica, 52*, 60–70.
5. Dekkers, A. L. M., & de Haan, L. (1993). Optimal choice of sample fraction in extreme-value estimation. *Journal of Multivariate Analysis, 47*, 173–195.

6. Embrechts, P., & Hofert, P. (2013). A note on generalized inverses. *Mathematical Methods of Operations Research, 77*, 423–432.
7. Fedoruk, M. V. (1987). *Asymptotics: Integrals and series*. Moscow: Nauka.
8. Geluk, J., & de Haan, L. (1987). *Regular variation, extensions and Tauberian theorems. CWI Tract* (Vol. 40). Amsterdam: Center for Mathematics and Computer Sciences.
9. Hall, P., & Welsh, A. H. (1985). Adaptive estimates of parameters of regular variation. *Annals of Statistics, 13*, 331–341.
10. Hill, B. M. (1975). A simple general approach to inference about the tail of a distribution. *Annals of Statistics, 3*, 1163–1174.
11. Markovich, N. M. (2007). *Nonparametric analysis of univariate heavy-tailed data*. Chichester: Wiley.
12. McElroy, T., & Politis, D. N. (2007). Moment-based tail index estimation. *Journal of Statistical Planning and Inference, 137*, 1389–1406.
13. Meerschaert, M., & Scheffler, H. (1998). A simple robust estimator for the thickness of heavy tails. *Journal of Statistical Planning and Inference, 71*, 19–34.
14. Paolella, M. S. (2007). *Intermediate probability: A computational approach*. Chichester: Wiley.
15. Qi, Y. (2010). On the tail index of a heavy tailed distribution. *Annals of the Institute of Statistical Mathematics, 62*, 277–298.
16. Samorodnitsky, G., & Taqqu, M. S. (1994). Stable non-Gaussian random processes. New York: Chapman and Hall.
17. Vaičiulis, M. (2009). An estimator of the tail index based on increment ratio statistics. *Lithuanian Mathematical Journal, 49*, 222–233.
18. Vaičiulis, M. (2014). Local maximum based tail index estimator. *Lithuanian Mathematical Journal, 54*, 503–526.
19. Zolotarev, V. M. (1986). *One-dimensional stable distributions. Translations of Mathematical Monographs* (Vol. 65). Providence: American Mathematical Society.

Constructing Confidence Sets for the Matrix Completion Problem

A. Carpentier, O. Klopp, and M. Löffler

Abstract In the present contribution we consider the problem of constructing honest and adaptive confidence sets for the matrix completion problem. For the Bernoulli model with known variance of the noise we provide a method with polynomial time complexity for constructing confidence sets that adapt to the unknown rank of the true matrix.

1 Introduction

In recent years, there has been a considerable interest in statistical inference for high-dimensional matrices. One particular problem is matrix completion where one observes only a small number $n \ll m_1 m_2$ of the entries of a high-dimensional $m_1 \times m_2$ matrix M_0 of unknown rank r; it aims at inferring the missing entries. The problem of matrix completion comes up in many areas including collaborative filtering, multi-class learning in data analysis, system identification in control, global positioning from partial distance information and computer vision, to mention some of them. For instance, in computer vision, this problem arises as many pixels may be missing in digital images. In collaborative filtering, one wants to make automatic predictions about the preferences of a user by collecting information from many users. So, we have a data matrix where rows are users and columns are items. For each user, we have a partial list of his preferences. We would like to predict the missing ones in order to be able to recommend items that he may be interested in.

A. Carpentier
IMST, Otto von Guericke University Magdeburg, Magdeburg, Germany
e-mail: alexandra.carpentier@ovgu.de

O. Klopp (✉)
IDS, ESSEC Business School, Cergy, France
e-mail: kloppolga@math.cnrs.fr

M. Löffler
StatsLab, University of Cambridge, Cambridge, UK
e-mail: m.loffler@statslab.cam.ac.uk

P. Bertail et al. (eds.), *Nonparametric Statistics*, Springer Proceedings in Mathematics & Statistics 250, https://doi.org/10.1007/978-3-319-96941-1_7

In general, recovery of a matrix from a small number of observed entries is impossible, but, if the unknown matrix has low rank, then accurate and even exact recovery is possible. In the noiseless setting, [3, 5, 6] established the following remarkable result: assuming that it satisfies a low coherence condition, M_0 can be recovered exactly by constrained nuclear norm minimization with high probability from only $n \gtrsim \text{rank}(M_0)(m_1 \vee m_2)\log^2(m_1 \vee m_2)$ entries observed uniformly at random.

What makes low-rank matrices special is that they depend on a number of free parameters that is much smaller than the total number of entries. Taking the singular value decomposition of a matrix $A \in \mathbb{R}^{m_1 \times m_2}$ of rank r, it is easy to see that A depends upon $(m_1+m_2)r - r^2$ free parameters. This number of free parameters gives us a lower bound for the number of observations needed to complete the matrix.

A situation, common in applications, corresponds to the noisy setting in which the few available entries are corrupted by noise. Noisy matrix completion has been extensively studied recently (e.g., [2, 8, 13, 16]). Here we observe a relatively small number of entries of a data matrix

$$Y = M_0 + E$$

where $M_0 = ((M_0)_{ij}) \in \mathbb{R}^{m_1 \times m_2}$ is the unknown matrix of interest and $E = (\varepsilon_{ij}) \in \mathbb{R}^{m_1 \times m_2}$ is a matrix of random errors. It is an important issue in applications to be able to say from the observations how well the recovery procedure has worked or, in the sequential sampling setting, to be able to give data-driven stopping rules that guarantee the recovery of the matrix M_0 at a given precision. This fundamental statistical question was recently studied in [7] where two statistical models for matrix completion are considered: the *trace regression model* and the *Bernoulli model* (for details see Sect. 2). In particular, in [7], the authors show that in the case of unknown noise variance, the information-theoretic structure of these two models is fundamentally different. In the trace regression model, even if only an upper bound for the variance of the noise is known, a honest and rank adaptive Frobenius-confidence set whose diameter scales with the minimax optimal estimation rate exists. In the Bernoulli model however, such sets do not exist.

Another major difference is that, in the case of known variance of the noise, [7] provides a realizable method for constructing confidence sets for the trace regression model whereas for the Bernoulli model only the existence of adaptive and honest confidence sets is demonstrated. The proof uses the duality between the problem of testing the rank of a matrix and the existence of honest and adaptive confidence sets. In particular, the construction in [7] is based on an infimum test statistic which cannot be computed in polynomial time. This is not feasible in practice. Thus, in the present note we develop an alternative method of constructing a confidence set in the Bernoulli model which is computable with a polynomial time algorithm.

2 Notation, Assumptions, and Some Basic Results

Let A, B be matrices in $\mathbb{R}^{m_1 \times m_2}$. We define the *matrix scalar product* as $\langle A, B\rangle = \mathrm{tr}(A^T B)$. The trace norm of a matrix $A = (a_{ij})$ is defined as $\|A\|_* := \sum \sigma_j(A)$, the operator norm as $\|A\| := \sigma_1(A)$ and the Frobenius norm as $\|A\|_2^2 := \sum_i \sigma_i^2 = \sum_{i,j} a_{ij}^2$ where $(\sigma_j(A))_j$ are the singular values of A ordered decreasingly. $\|A\|_\infty = \max\limits_{i,j}|a_{ij}|$ denotes the largest absolute value of any entry of A. In what follows, we use symbols C, c for a generic positive constant, which is independent of n, m_1, m_2, rank and may take different values at different places. We denote by $a \vee b = \max(a, b)$. We let $\mathbb{P}_M$ denote the distribution of the data when the parameter is M. For a set of matrices $\mathscr{C}$ we denote by $\|\mathscr{C}\|_2$ its diameter measured in Frobenius norm.

We assume that each entry of Y is observed independently of the other entries with probability $p = n/(m_1 m_2)$. More precisely, if $n \le m_1 m_2$ is given and B_{ij} are i.i.d. Bernoulli random variables of parameter p independent of the ε_{ij}'s, we observe

$$Y_{ij} = B_{ij}\left((M_0)_{ij} + \varepsilon_{ij}\right), \quad 1 \le i \le m_1, 1 \le j \le m_2. \tag{1}$$

This model for the matrix completion problem is usually called the *Bernoulli model*. Another model often considered in the matrix completion literature is the trace regression model (e.g., [2, 10, 13, 16]). Let $k_0 = \mathrm{rank}(M_0) \vee 1$.

In many of the most cited applications of the matrix completion problem, such as recommendation systems or the problem of global positioning from the local distances, the noise is bounded but not necessarily identically distributed. This is the assumption which we adopt in the present chapter. More precisely, we assume that the noise variables are independent, homoscedastic, bounded, and centered:

Assumption 1 *For any* $(ij) \in [m_1]\times[m_2]$ *we assume that* $\mathbb{E}(\varepsilon_{ij}) = 0$, $\mathbb{E}(\varepsilon_{ij}^2) = \sigma^2$ *and that there exists a positive constant* $u > 0$ *such that*

$$\max_{ij}\left|\varepsilon_{ij}\right| \le u.$$

Let $m = \min(m_1, m_2)$, $d = m_1 + m_2$. For any $l \in \mathbb{N}$ we set $[l] = \{1, \dots, l\}$. For any integer $0 \le k \le m$ and any $\mathbf{a} > 0$, we define the parameter space of rank k matrices with entries bounded by $\mathbf{a}$ in absolute value as

$$\mathscr{A}(k, \mathbf{a}) = \left\{M \in \mathbb{R}^{m_1 \times m_2} : \mathrm{rank}(M) \le k,\ \|M\|_\infty \le \mathbf{a}\right\}. \tag{2}$$

For constants $\beta \in (0, 1)$ and $c = c(\sigma, \mathbf{a}) > 0$ we have that

$$\inf_{\widehat{M}} \sup_{M_0 \in \mathscr{A}(k,\mathbf{a})} \mathbb{P}_{M_0,\sigma}\left(\frac{\|\widehat{M} - M_0\|_2^2}{m_1 m_2} > c\frac{kd}{n}\right) \ge \beta$$

where $\simeq$ denotes two-sided inequality up to a universal constant and $\widehat{M}$ is an estimator of M_0 (see, e.g., [11]). This bound is valid uniformly over $M_0 \in \mathscr{A}(k, \mathbf{a})$. It has been also shown in [11] that an iterative soft thresholding estimator $\widehat{M}$ satisfies with $\mathbb{P}_{M_0}$-probability at least $1 - 8/d$

$$\frac{\|\widehat{M} - M_0\|_F^2}{m_1 m_2} \leq C\frac{(\mathbf{a} + \sigma)^2 k_0 d}{n} \quad \text{and} \quad \|M_0 - \widehat{M}\|_\infty \leq 2\mathbf{a} \tag{3}$$

for a constant $C > 0$. These lower and upper bounds imply that for the Frobenius loss the minimax risk for recovering a matrix $M_0 \in \mathscr{A}(k_0, \mathbf{a})$ is of order $\sqrt{\frac{(\sigma + \mathbf{a})^2 k_0 d m_1 m_2}{n}}$.

For $k \in [m]$ we set

$$r_k = C\frac{(\sigma + \mathbf{a})^2 dk}{n},$$

where C is the numerical constant in (3). We use the following definition of honest and adaptive confidence sets:

Definition 1 Let $\alpha, \alpha' > 0$ be given. A set $\mathscr{C}_n = \mathscr{C}_n((Y_{ij}, B_{ij}), \alpha) \subset \mathscr{A}(m, \mathbf{a})$ is a honest confidence set at level α for the model $\mathscr{A}(m, \mathbf{a})$ if

$$\liminf_n \inf_{M \in \mathscr{A}(m,\mathbf{a})} \mathbb{P}_M(M \in \mathscr{C}_n) \geq 1 - \alpha.$$

Furthermore, we say that $\mathscr{C}_n$ is adaptive for the sub-model $\mathscr{A}(k, \mathbf{a})$ at level α' if there exists a constant $C = C(\alpha, \alpha') > 0$ such that

$$\sup_{M \in \mathscr{A}(k,\mathbf{a})} \mathbb{P}_M \left(\|\mathscr{C}_n\|_2 > C r_k\right) \leq \alpha'$$

while still retaining

$$\sup_{M \in \mathscr{A}(m,\mathbf{a})} \mathbb{P}_M \left(\|\mathscr{C}_n\|_2 > C r_m\right) \leq \alpha'.$$

The lim inf in this definition is to be understood in a high-dimensional sense, i.e. m_1, m_2 both depend on n and grow to ∞ as $n \to \infty$ such that $n \leq m_1(n) m_2(n)$.

3 A Non-asymptotic Confidence Set for the Matrix Completion Problem

Let $\widehat{M}$ be an estimator of M_0 based on the observations (Y_{ij}, B_{ij}) from the Bernoulli model (1) such that $\|\widehat{M}\|_\infty \leq \mathbf{a}$. Assume that for some $\beta > 0$, $\widehat{M}$ satisfies the following bound:

$$\sup_{M_0 \in \mathscr{A}(k_0, \mathbf{a})} \mathbb{P}\left(\frac{\|\widehat{M} - M_0\|_2^2}{m_1 m_2} \leq C\frac{(\sigma + \mathbf{a})^2 k_0 d}{n}\right) \geq 1 - \beta. \tag{4}$$

We can take, for example, the thresholding, estimator considered in [11] which attains (4) with $\beta = 8/d$. Our construction is based on Lepski's method [15]. We denote by $\widehat{M}_k$ the projection of $\widehat{M}$ on the set $\mathscr{A}(k, \mathbf{a})$ of matrices of rank k with sup-norm bounded by $\mathbf{a}$:

$$\widehat{M}_k \in \operatorname*{argmin}_{A \in \mathscr{A}(k,\mathbf{a})} \|\widehat{M} - A\|_2.$$

We set

$$S = \{k : \|\widehat{M} - \widehat{M}_k\|_2^2 \leq r_k\} \qquad \text{and} \qquad \hat{k} = \min\{k \in S\}$$

and we will use $\widetilde{M} = \widehat{M}_{\hat{k}}$ to center the confidence set with diameter controlled by the residual sum of squares statistic $\hat{r}_n$:

$$\hat{r}_n = \frac{1}{n} \sum_{ij} (Y_{ij} - B_{ij}\widetilde{M}_{ij})^2 - \sigma^2. \tag{5}$$

Given $\alpha > 0$, we denote

$$\bar{z} = \frac{p}{256}\|M - \widetilde{M}\|_2^2 + z(uc^*)^2 d\hat{k} \quad \text{and} \quad \xi_{\alpha,u} = 2u^2\sqrt{\log(\alpha^{-1})} + \frac{4u^2\log(\alpha^{-1})}{3\sqrt{n}}.$$

Here z is a sufficiently large numerical constant to be chosen later on and $c^* \geq 2$ is a universal constant in Corollary 3.12 [1]. We define the confidence set as follows:

$$C_n = \left\{M \in \mathbb{R}^{m_1 \times m_2} : \frac{\|M - \widetilde{M}\|_2^2}{m_1 m_2} \leq 128\left(\hat{r}_n + \frac{\mathbf{a}^2 z d\hat{k} + \bar{z}}{n} + \frac{\xi_{\alpha,u}}{\sqrt{n}}\right)\right\}. \tag{6}$$

Theorem 1 *Let* $\alpha > 0$, $d > 16$ *and suppose that* $\widehat{M}$ *attains the bound* (4) *with probability at least* $1 - \beta$. *Let* C_n *be given by (6). Assume that* $\|M_0\|_\infty \leq \mathbf{a}$ *and that*

Assumption 1 is satisfied. Then, for every $n \geq m\log(d)$, *we have*

$$\mathbb{P}_{M_0}(M_0 \in C_n) \geq 1 - \alpha - \exp(-cd). \tag{7}$$

Moreover, with probability at least $1 - \beta - \exp(-cd)$

$$\frac{\|C_n\|_2^2}{m_1 m_2} \leq C\frac{(\sigma + \mathbf{a})^2 d k_0}{n}. \tag{8}$$

Theorem 1 implies that C_n is a honest and adaptive confidence set:

Corollary 1 *Let* $\alpha > 0$, $d > 16$ *and suppose that* $\widehat{M}$ *attains the risk bound* (4) *with probability at least* $1 - \beta$. *Let* C_n *be given by (6). Assume that Assumption 1 is satisfied. Then, for* $n \geq m\log(d)$, C_n *is a* $\alpha + \exp(-cd)$ *honest confidence set for the model* $\mathscr{A}(m, \mathbf{a})$ *and adapts to every sub-model* $\mathscr{A}(k, a)$, $1 \leq k \leq m$, *at level* $\beta + \exp(-cd)$.

Remark The procedure for building a confidence set consists of

1. Building an adaptive estimator satisfying Eq. (4), that is computable in polynomial time, e.g. the thresholding estimator from [11] or the matrix lasso [13]
2. Projecting the estimator on the smallest possible model which is coherent with the estimator and amounts to computing the SVD of $\widehat{M}$ once. This has complexity of smaller order than the complexity of the matrix lasso.
3. Computing $\hat{r}_n$ for which on average n terms have to be considered

Summarizing, the computational cost of computing the adaptive confidence set is of smaller order than the computational complexity of constructing an adaptive estimator.

Proof (Proof of Theorem 1) For $1 \leq k \leq m_1 \wedge m_2$ we consider the following sets

$$\mathscr{C}(k, \mathbf{a}) = \left\{ A \in \mathbb{R}^{m_1 \times m_2} \, : \, \|A\|_\infty \leq \mathbf{a}, \, \|M_0 - A\|_2^2 \geq \frac{256(\mathbf{a} \vee u)^2 zd}{p} \text{ and } \operatorname{rank}(A) \leq k \right\}$$

and write

$$\mathscr{C} = \cup_{k=1}^{m} \mathscr{C}(k, \mathbf{a}). \tag{9}$$

When $\left\| M_0 - \widetilde{M} \right\|_2^2 \leq \dfrac{256(\mathbf{a} \vee u)^2 zd}{p}$ we have that $M_0 \in C_n$. So, we only need to consider the case $\left\| M_0 - \widetilde{M} \right\|_2^2 \geq \dfrac{256(\mathbf{a} \vee u)^2 zd}{p}$. In this case we have that $\widetilde{M} \in \mathscr{C}$. We introduce the observation operator $\mathscr{X}$ defined as follows:

$$\mathscr{X} : \mathbb{R}^{m_1 \times m_2} \to \mathbb{R}^{m_1 \times m_2} \qquad \text{with} \qquad \mathscr{X}(A) = (B_{ij} a_{ij})_{ij}.$$

and set $\|A\|_{L_2(\Pi)}^2 = \mathbb{E}\|\mathscr{X}(A)\|_2^2 = p\|A\|_2^2$. We can decompose

$$\hat{r}_n = n^{-1}\|\mathscr{X}(\widetilde{M} - M_0)\|_2^2 + 2n^{-1}\langle \mathscr{X}(E), M_0 - \widetilde{M}\rangle + n^{-1}\|\mathscr{X}(E)\|_2^2 - \sigma^2.$$

Then we can bound the probability $\mathbb{P}_{M_0}(M_0 \notin C_n)$ by the sum of the following probabilities:

$$I := \mathbb{P}_{M_0}\left(\frac{\|\widetilde{M} - M_0\|_{L_2(\Pi)}^2}{128} > \|\mathscr{X}(\widetilde{M} - M_0)\|_2^2 + z\mathbf{a}^2 d\hat{k}\right),$$

$$II := \mathbb{P}_{M_0}\left(-2\langle \mathscr{X}(E), M_0 - \widetilde{M}\rangle > \bar{z}\right),$$

$$III := \mathbb{P}_{M_0}\left(-\|\mathscr{X}(E)\|_2^2 + n\sigma^2 > \sqrt{n}\xi_{\alpha,u}\right).$$

By Lemma 1, the first probability is bounded by $8\exp(-4d)$ for $z \geq (27c^*)^2$. For the second term we use Lemma 7 which implies that $II \leq \exp(-cd)$ for $z \geq 6240$. Finally, for the third term, Bernstein's inequality implies

$$\mathbb{P}\left\{-\|\mathscr{X}(E)\|_2^2 + n\sigma^2 > t\right\} \leq \exp\left(-\frac{t^2}{2\sigma^2 nu^2 + \frac{2}{3}u^2 t}\right).$$

Taking $t = 2u^2\sqrt{n\log(\alpha^{-1})} + \frac{4}{3}u^2\log(\alpha^{-1})$ we obtain that $III \leq \alpha$ by definition of $\xi_{\alpha,u}$. This completes the proof of (7).

To prove (8), using Lemmas 1 and 7, we can bound the square Frobenius norm diameter of our confidence set C_n defined in (6) as follows:

$$\frac{\|C_n\|_2^2}{m_1 m_2} \lesssim \frac{\|\widetilde{M} - M_0\|_2^2}{m_1 m_2} + \left(r_{\hat{k}} + \frac{\xi_{\alpha,u}}{\sqrt{n}}\right).$$

This bound holds on an event of probability at least $1 - \exp(-cd)$. Now we restrict to the event where $\widehat{M}$ attains the risk bound in (4) which happens with probability at least $1 - \beta$. On this event, $M_0 \in \mathscr{A}(k_0, \mathbf{a})$ implies $\|\widehat{M} - \widehat{M}_{k_0}\|_2^2 \leq r_{k_0}$. So, $k_0 \in S$ and $\hat{k} \leq k_0$. Now, the triangle inequality and $r_{\hat{k}} \leq r_{k_0}$ imply that on the intersection of those two events we have that

$$\|\widetilde{M} - M_0\|_2^2 \lesssim m_1 m_2\left(r_{k_0} + r_{\hat{k}}\right) \lesssim m_1 m_2 r_{k_0}.$$

This, together with the definition of $\xi_{\alpha,u}$ and the condition $n \leq m_1 m_2$, completes the proof of (8).

4 Technical Lemmas

Lemma 1 *With probability larger than* $1 - 8\exp(-4d)$ *we have that*

$$\sup_{A\in\mathscr{C}} \frac{\left|\|\mathscr{X}(M_0 - A)\|_2 - \|M_0 - A\|_{L_2(\Pi)}\right| - \frac{7}{8}\|M_0 - A\|_{L_2(\Pi)}}{\mathbf{a}\sqrt{(\mathrm{rank}(A) \vee 1)d}} \leq 27\,c^*$$

where c^* *is a universal numerical constant and* $\mathscr{C}$ *is defined in* (9).

Proof We have that

$$\begin{aligned}
&\mathbb{P}\left(\sup_{A\in\mathscr{C}} \frac{\left|\|\mathscr{X}(M_0 - A)\|_2 - \|M_0 - A\|_{L_2(\Pi)}\right| - \frac{7}{8}\|M_0 - A\|_{L_2(\Pi)}}{\mathbf{a}\sqrt{(\mathrm{rank}(A) \vee 1)d}} \geq 27\,c^*\right)\\
&\leq \sum_{k=1}^{k_0} \underbrace{\mathbb{P}\left(\sup_{A\in\mathscr{C}(k,\mathbf{a})} \left|\|\mathscr{X}(M_0 - A)\|_2 - \|M_0 - A\|_{L_2(\Pi)}\right| - \frac{7}{8}\|M_0 - A\|_{L_2(\Pi)} \geq 27\,c^*\mathbf{a}\sqrt{kd}\right)}_{\mathrm{I}}.
\end{aligned} \tag{10}$$

In order to upper bound I, we use a peeling argument. We set $\alpha = 7/6$ and $\nu^2 = \dfrac{188\mathbf{a}^2 zd}{p}$. Moreover, for $l \in \mathbb{N}$ set

$$S_l = \left\{A \in \mathscr{C}(k,\mathbf{a}) \;:\; \alpha^l \nu \leq \|A - M_0\|_2 \leq \alpha^{l+1}\nu\right\}.$$

Then

$$\begin{aligned}
\mathrm{I} &\leq \sum_{l=1}^{\infty} \mathbb{P}\left(\sup_{A\in S_l} \left|\|\mathscr{X}(M_0 - A)\|_2 - \|M_0 - A\|_{L_2(\Pi)}\right| \geq 27\,c^*\mathbf{a}\sqrt{kd} + \frac{7}{8}\alpha^l\mathbf{a}\sqrt{188zd}\right)\\
&\leq \sum_{l=1}^{\infty} \underbrace{\mathbb{P}\left(\sup_{A\in\mathscr{C}(k,\mathbf{a},\alpha^{l+1}\nu)} \left|\|\mathscr{X}(M_0 - A)\|_2 - \|M_0 - A\|_{L_2(\Pi)}\right| \geq 27\,c^*\mathbf{a}\sqrt{kd} + \frac{7}{8}\alpha^l\mathbf{a}\sqrt{188zd}\right)}_{\mathrm{II}}
\end{aligned}$$

where $\mathscr{C}(k, \mathbf{a}, T) = \{A \in \mathscr{C}(k, \mathbf{a}) \;:\; \|M_0 - A\|_2 \leq T\}$. The following lemma gives an upper bound on II:

Lemma 2 *Consider the following set of matrices:*

$$\mathscr{C}(k, \mathbf{a}, T) = \{A \in \mathscr{C}(k, \mathbf{a}) \;:\; \|M_0 - A\|_2 \leq T\}$$

and set

$$Z_T = \sup_{A\in\mathscr{C}(k,\mathbf{a},T)} \left|\|\mathscr{X}(M_0 - A)\|_2 - \|M_0 - A\|_{L_2(\Pi)}\right|.$$

Then, we have that

$$\mathbb{P}\left(Z_T \geq \frac{3}{4}\sqrt{p}T + 27\,c^*\mathbf{a}\sqrt{kd}\right) \leq 4e^{-c_1\,p\,T^2/\mathbf{a}^2}$$

with $c_1 \geq (512)^{-1}$.

Lemma 2 implies that $\mathrm{II} \leq 4\exp(-c_1\,p\,\alpha^{2l}\nu^2/\mathbf{a}^2)$ and we obtain

$$\mathrm{I} \leq 4\sum_{l=1}^{\infty}\exp(-c_1\,p\,\alpha^{2l+2}\nu^2/\mathbf{a}^2) \leq 4\sum_{l=1}^{\infty}\exp\left(-2c_1\,p\,\nu^2\,\log(\alpha)\,l/\mathbf{a}^2\right)$$

where we used $e^x \geq x$. We finally compute for $\nu^2 = 188\mathbf{a}^2 zdp^{-1}$

$$\mathrm{I} \leq \frac{4\exp\left(-2c_1\,p\,\nu^2\,\log(\alpha)/\mathbf{a}^2\right)}{1-4\exp\left(-2c_1\,p\,\nu^2\,\log(\alpha)/\mathbf{a}^2\right)} \leq 8\exp\left(-376\,c_1 zd\log(7/6)\right) \leq \exp(-5d)$$

where we take $z \geq (27c^*)^2$. Using (10) and $d \geq \log(m)$ we obtain the statement of Lemma 1.

Proof (Proof of Lemma 2) This proof is close to the proof of Theorem 1 in [16]. We start by applying a discretization argument. Let $\{G_\delta^1, \dots G_\delta^{N(\delta)}\}$ be a δ−covering of $\mathscr{C}(k, \mathbf{a}, T)$ given by Lemma 3. Then, for any $A \in \mathscr{C}(k, \mathbf{a}, T)$ there exist an index $i \in \{1, \dots, N(\delta)\}$ and a matrix Δ with $\|\Delta\|_2 \leq \delta$ such that $A = G_\delta^i + \Delta$. Using the triangle inequality we thus obtain that

$$\begin{aligned}\big|\|M_0 - A\|_{L_2(\Pi)} - \|\mathscr{X}(M_0 - A)\|_2\big| \leq &\left|\left\|\mathscr{X}(M_0 - G_\delta^i)\right\|_2 - \|M_0 - G_\delta^i\|_{L_2(\Pi)}\right| \\ &+ \|\mathscr{X}\Delta\|_2 + \sqrt{p}\delta.\end{aligned}$$

Lemma 3 implies that $\Delta \in \mathscr{D}_\delta(2k, 2\mathbf{a}, 2T)$ where

$$\mathscr{D}_\delta(k, \mathbf{a}, T) = \{A \in \mathbb{R}^{m_1\times m_2} \;:\; \|A\|_\infty \leq \mathbf{a},\ \|A\|_2 \leq \delta \quad \text{and} \quad \|A\|_* \leq \sqrt{k}T\}.$$

Then,

$$Z_T \leq \max_{i=1,\dots,N(\delta)}\left|\left\|\mathscr{X}(M_0 - G_\delta^i)\right\|_2 - \|M_0 - G_\delta^i\|_{L_2(\Pi)}\right| + \sup_{\Delta\in\mathscr{D}_\delta(2k,2\mathbf{a},2T)}\|\mathscr{X}\Delta\|_2 + \sqrt{p}\delta.$$

Now we take $\delta = T/8$ and use Lemmas 5 and 6 to obtain that

$$Z_T \leq \sqrt{p}\delta + 8\,c^*\mathbf{a}\sqrt{kd} + 19\mathbf{a}\,c^*\sqrt{2kd} + \sqrt{p}T/2 + \sqrt{p}\delta \leq 27\,c^*\mathbf{a}\sqrt{kd} + 6\sqrt{p}T/8.$$

with probability at least $1 - 8\exp\left(-\frac{pT^2}{512\mathbf{a}^2}\right)$.

Lemma 3 *Let $\delta = T/8$. There exists a set of matrices $\{G_\delta^1, \dots G_\delta^{N(\delta)}\}$ with $N_\delta \le \left(\frac{18T}{\delta}\right)^{2(d+1)k}$ and such that*

(i) For any $A \in \mathscr{C}(k, \mathbf{a}, T)$ there exists a $G_\delta^A \in \{G_\delta^1, \dots G_\delta^{N(\delta)}\}$ satisfying

$$\|A - G_\delta^A\|_2 \le \delta \quad \text{and} \quad (A - G_\delta^A) \in \mathscr{D}_\delta(2k, 2\mathbf{a}, 2T).$$

(ii) Moreover, $\|G_\delta^j - M_0\|_\infty \le 2\mathbf{a}$ and $\|G_\delta^j - M_0\|_2 \le 2T$ for any $j = 1, \dots, N_\delta$.

Proof We use the following result (see Lemma 3.1 in [4] and Lemma A.2 in [18]):

Lemma 4 *Let $S(k, T) = \{A \in \mathbb{R}^{m_1 \times m_2} : \mathrm{rank}(A) \le k \text{ and } \|A\|_2 \le T\}$. Then, there exists an ϵ-net $\bar{S}(k, T)$ for the Frobenius norm obeying*

$$\left|\bar{S}(k, T)\right| \le (9T/\epsilon)^{(m_1+m_2+1)k}.$$

Let $S_{M_0}(k, T) = \{A \in \mathbb{R}^{m_1 \times m_2} : \mathrm{rank}(A) \le k \text{ and } \|A - M_0\|_2 \le T\}$ and take a $X_0 \in \mathscr{C}(k, \mathbf{a}, T)$. We have that $S_{M_0}(k, T) - X_0 \subset S(2k, 2T)$. Let $\bar{S}(2k, 2T)$ be an δ-net given by Lemma 4. Then, for any $A \in S_{M_0}(k, T)$ there exists a $\bar{G}_\delta^A \in \bar{S}(2k, 2T)$ such that $\|A - X_0 - \bar{G}_\delta^A\|_2 \le \delta$. Let $G_\delta^j = \Pi(\bar{G}_\delta^j) + X_0$ for $j = 1, \dots, \left|\bar{S}(2k, 2T)\right|$ where Π is the projection operator under Frobenius norm into the set $\mathscr{D}(2k, 2\mathbf{a}, 2T) = \{A \in \mathbb{R}^{m_1 \times m_2} : \|A\|_\infty \le 2\mathbf{a}, \text{ and } \|A\|_* \le 2\sqrt{2k}T\}$. Note that as $\mathscr{D}(2k, 2\mathbf{a}, 2T)$ is convex and closed, Π is non-expansive in Frobenius norm. For any $A \in \mathscr{C}(k, \mathbf{a}, T) \subset S_{M_0}(k, T)$, we have that $A - X_0 \in \mathscr{D}(2k, 2\mathbf{a}, 2T)$ which implies

$$\|A - X_0 - \Pi(\bar{G}_\delta^A)\|_2 = \|\Pi(A - X_0 - \bar{G}_\delta^A)\|_2 \le \|A - \bar{G}_\delta^A - X_0)\|_2 \le \delta$$

and we have that $(A - \Pi(\bar{G}_\delta^A) - X_0) \in \mathscr{D}_\delta(2k, 2\mathbf{a}, 2T)$ which completes the proof of (i) of Lemma 3. To prove (ii), note that by the definition of Π we have that $\|G_\delta^j - M_0\|_\infty = \|\Pi(\bar{G}_\delta^j) + X_0 - M_0\|_\infty = \|\Pi(\bar{G}_\delta^j + X_0 - M_0)\|_\infty \le 2\mathbf{a}$ and $\|G_\delta^j - M_0\|_2 \le 2T$.

Lemma 5 *Let $\delta = T/8$ and assume that $n \ge m\log(m)$. We have that with probability at least $1 - 4\exp\left(-\frac{pT^2}{512\mathbf{a}^2}\right)$*

$$\sup_{\Delta \in \mathscr{D}_\delta(2k, 2a, 2T)} \|\mathscr{X}\Delta\|_2 \le 19\mathbf{a}\, c^*\sqrt{2kd} + \sqrt{p}T/2.$$

Proof Let $X_T = \sup_{\Delta \in \mathscr{D}_\delta(2k, 2\mathbf{a}, 2T)} \|\mathscr{X}\Delta\|_2$. We use the following concentration inequality by Talagrand:

Theorem 1 *Suppose that* $f : [-1,1]^N \to \mathbb{R}$ *is a convex Lipschitz function with Lipschitz constant L. Let* $\Xi_1, \dots \Xi_N$ *be independent random variables taking value in* $[-1,1]$. *Let* $Z := f(\Xi_1, \dots, \Xi_n)$. *Then for any* $t \geq 0$,

$$\mathbb{P}\left(|Z - \mathbb{E}(Z)| \geq 16L + t\right) \leq 4e^{-t^2/2L^2}.$$

For a proof see [17, Theorem 6.6.] and [8, Theorem 3.3]. Let $f(x_{11}, \dots, x_{m_1m_2}) := \sup\limits_{\Delta \in \mathscr{D}_\delta(2k, 2\mathbf{a}, 2T)} \sqrt{\sum_{(i,j)} x_{ij}^2 \Delta_{ij}^2}$. It is easy to see that $f(x_{11}, \dots, x_{m_1m_2})$ is a Lipschitz function with Lipschitz constant $L = 2\mathbf{a}$. Indeed,

$$\begin{aligned}
&\left|f(x_{11}, \dots, x_{m_1m_2}) - f(z_{11}, \dots, z_{m_1m_2})\right| \\
&\quad = \left| \sup_{\Delta \in \mathscr{D}_\delta(2k, 2\mathbf{a}, 2T)} \sqrt{\sum_{(i,j)} x_{ij}^2 \Delta_{ij}^2} - \sup_{\Delta \in \mathscr{D}_\delta(2k, 2\mathbf{a}, 2T)} \sqrt{\sum_{(i,j)} z_{ij}^2 \Delta_{ij}^2} \right| \\
&\quad \leq \sup_{\Delta \in \mathscr{D}_\delta(2k, 2\mathbf{a}, 2T)} \sqrt{\sum_{(i,j)} (x_{ij} - z_{ij})^2 \Delta_{ij}^2} \leq 2\mathbf{a}\|x - z\|_2
\end{aligned}$$

where $x = (x_{11}, \dots, x_{m_1m_2})$ and $z = (z_{11}, \dots, z_{m_1m_2})$. Now, Theorem 1 implies

$$\mathbb{P}\left(X_T \geq \mathbb{E}(X_T) + 32\mathbf{a} + t\right) \leq 4\exp\left(-\frac{t^2}{8\mathbf{a}^2}\right). \tag{11}$$

Next, we bound the expectation $\mathbb{E}(X_T)$. Applying Jensen's inequality, a symmetrization argument and the Ledoux-Talagrand contraction inequality (e.g., [12, Theorem 2.2]) we obtain

$$\begin{aligned}
(\mathbb{E}(X_T))^2 &\leq \mathbb{E}\left(\sup_{\Delta \in \mathscr{D}_\delta(2k, 2\mathbf{a}, 2T)} \sum_{(i,j)} B_{ij}\Delta_{ij}^2 \right) \\
&\leq \mathbb{E}\left(\sup_{\Delta \in \mathscr{D}_\delta(2k, 2\mathbf{a}, 2T)} \sum_{(i,j)} B_{ij}\Delta_{ij}^2 - \mathbb{E}\left(B_{ij}\Delta_{ij}^2\right) \right) + p\delta^2 \\
&\leq 2\,\mathbb{E}\left(\sup_{\Delta \in \mathscr{D}_\delta(2k, 2\mathbf{a}, 2T)} \left| \sum_{(i,j)} \eta_{ij} B_{ij}\Delta_{ij}^2 \right| \right) + p\delta^2 \\
&\leq 8\mathbf{a}\,\mathbb{E}\left(\sup_{\Delta \in \mathscr{D}_\delta(2k, 2\mathbf{a}, 2T)} \left| \sum_{(i,j)} \eta_{ij} B_{ij}\Delta_{ij} \right| \right) + p\delta^2 \\
&= 8\mathbf{a}\,\mathbb{E}\left(\sup_{\Delta \in \mathscr{D}_\delta(2k, 2\mathbf{a}, 2T)} |\langle \Sigma_R, \Delta\rangle| \right) + p\delta^2 \leq 16\,\mathbf{a}\sqrt{2k}\,T\,\mathbb{E}\left(\|\Sigma_R\|\right) + p\delta^2
\end{aligned}$$

where $\{\eta_{ij}\}$ is an i.i.d. Rademacher sequence, $\Sigma_R = \sum_{(i,j)} B_{ij}\eta_{ij}X_{ij}$ with $X_{ij} = e_i(m_1)e_j^T(m_2)$ and $e_k(l)$ are the canonical basis vectors in $\mathbb{R}^l$. Lemma 4 in [11] and $n \geq m\log(m)$ imply that

$$\mathbb{E}\,\|\Sigma_R\| \leq c^*\sqrt{pd} \tag{12}$$

where $c^* \geq 2$ is a universal numerical constant. Using (12), $\sqrt{x+y} \leq \sqrt{x} + \sqrt{y}$, $2xy \leq x^2 + y^2$ and $\delta = T/8$ we compute

$$\mathbb{E}(X_T) \leq 4\left(\mathbf{a}\,c^*\sqrt{2kpd}\,T\right)^{1/2} + \sqrt{p}\delta \leq 16\mathbf{a}\,c^*\sqrt{2kd} + 3\sqrt{p}T/8.$$

Taking in (11) $t = \sqrt{p}T/8$ we obtain the statement of Lemma 5.

Lemma 6 *Let $\delta = T/8$, $d > 16$ and $(G_\delta^1, \ldots, G_\delta^{N(\delta)})$ be the collection of matrices given by Lemma 3. We have that*

$$\max_{k=1,\ldots,N(\delta)} \left| \left\| \mathscr{X}(M_0 - G_\delta^k) \right\|_2 - \|M_0 - G_\delta^k\|_{L_2(\Pi)} \right| \leq \sqrt{p}\delta + 8\,c^*\mathbf{a}\sqrt{kd}$$

with probability at least $1 - 4\exp\left(-\frac{p\delta^2}{8\mathbf{a}^2}\right)$.

Proof For any fixed $A \in \mathbb{R}^{m_1\times m_2}$ satisfying $\|A\|_\infty \leq 2a$ we have that

$$\|\mathscr{X} A\|_2 = \sqrt{\sum_{ij} B_{ij}A_{ij}^2} = \sup_{\|u\|_2=1} \sum_{ij} \left(u_{ij}B_{ij}A_{ij}\right).$$

Then we can apply Theorem 1 with $f(x_{11}, \ldots, x_{m_1m_2}) := \sup\limits_{\|u\|_2=1} \sum_{ij}\left(u_{ij}x_{ij}\right)$ to obtain

$$\mathbb{P}\left(|\|\mathscr{X} A\|_2 - \mathbb{E}\|\mathscr{X} A\|_2| > t + 32\mathbf{a}\right) \leq 4\exp\left\{-\frac{t^2}{8\mathbf{a}^2}\right\}. \tag{13}$$

On the other hand, let $Z = \sup\limits_{\|u\|_2=1} \sum_{ij}\left(u_{ij}B_{ij}A_{ij}\right)$. Applying Corollary 4.8 from [14] we obtain that $\mathrm{Var}(Z) = \|A\|_{L_2(\Pi)}^2 - (\mathbb{E}\|\mathscr{X} A\|_2)^2 \leq 16^2\mathbf{a}^2$ which together with (13) implies

$$\mathbb{P}\left(\left|\|\mathscr{X} A\|_2 - \|A\|_{L_2(\Pi)}\right| > t + 48\mathbf{a}\right) \leq 4\exp\left\{-\frac{t^2}{8\mathbf{a}^2}\right\}. \tag{14}$$

Now Lemma 6 follows from Lemma 3, (14) with $t = \sqrt{p}\delta + 5c^*\mathbf{a}\sqrt{kd}$ and the union bound.

Lemma 7 *We have that*

$$\sup_{A\in\mathscr{C}} \frac{|\langle \mathscr{X}(E), A - M_0\rangle| - \frac{1}{256}\|M_0 - A\|^2_{L_2(\Pi)}}{d\,\mathrm{rank}(A)} \leq 6240(uc^*)^2.$$

with probability larger than $1 - \exp(-cd)$ *with* $c \geq 0.0003$.

Proof Following the lines of the proof of Lemma 1 with $\alpha = \sqrt{65/64}$ and $\nu^2 = \dfrac{252(\mathbf{a} \vee u)^2 zd}{p}$ we obtain

$$\mathbb{P}\left(\sup_{A\in\mathscr{C}} \frac{|\langle \mathscr{X}(E), A - M_0\rangle| - \frac{1}{256}\|M_0 - A\|^2_{L_2(\Pi)}}{d\,\mathrm{rank}(A)} \geq 6240(uc^*)^2\right)$$

$$\leq \sum_{k=1}^{k_0}\sum_{l=1}^{\infty} \mathbb{P}\left(\sup_{A\in\mathscr{C}(k,\mathbf{a},\alpha^{l+1}\nu)} |\langle \mathscr{X}(E), A - M_0\rangle| \geq 6240(uc^*)^2 dk + \frac{p\alpha^{2l}\nu^2}{256}\right)$$

$$\leq 4\sum_{k=1}^{k_0}\sum_{l=1}^{\infty} \exp\left(-\frac{p\alpha^{2l+2}\nu^2}{c_2(\mathbf{a}\vee u)^2}\right) \leq \exp\left(-cd\right)$$

where we use the following lemma:

Lemma 8 *Consider the following set of matrices*

$$\mathscr{C}(k, \boldsymbol{a}, T) = \{A \in \mathscr{C}(k, \boldsymbol{a}) \,:\, \|M_0 - A\|_2 \leq T\}$$

and set

$$\widetilde{Z}_T = \sup_{A\in\mathscr{C}(k,a,T)} |\langle \mathscr{X}(E), A - M_0\rangle| \,.$$

We have that

$$\mathbb{P}\left(\widetilde{Z}_T \geq 6240(c^*u)^2 dk + pT^2/260\right) \leq 4\exp\left(-pT^2/c_2(\mathbf{a}\vee u)^2\right)$$

with $c_2 \leq 12(1560)^2$.

Proof (Proof of Lemma 8) Fix an $X_0 \in \mathscr{C}(k, \mathbf{a}, T)$. For any $A \in \mathscr{C}(k, \mathbf{a}, T)$, we set $\Delta = A - X_0$ and we have that $\mathrm{rank}(\Delta) \leq 2k$ and $\|\Delta\|_2 \leq 2T$. Then using $|\langle \mathscr{X}(E), A - M_0\rangle| \leq |\langle \mathscr{X}(E), X_0 - M_0\rangle| + |\langle \mathscr{X}(E), \Delta\rangle|$ we obtain that

$$\widetilde{Z}_T \leq |\langle \mathscr{X}(E), X_0 - M_0\rangle| + \sup_{\Delta\in\mathscr{T}(2k,2\mathbf{a},2T)} |\langle \mathscr{X}(E), \Delta\rangle|$$

where

$$\mathscr{T}(2k, 2\mathbf{a}, 2T) = \{A \in \mathbb{R}^{m_1 \times m_2} : \|A\|_\infty \leq 2\mathbf{a},\ \|A\|_2 \leq 2T \quad \text{and} \quad \operatorname{rank}(A) \leq 2k\}.$$

Bernstein's inequality and $\|X_0 - M_0\|_2 \leq T$ imply that

$$\mathbb{P}\{|\langle \mathscr{X}(E), X_0 - M_0\rangle| > t\} \leq 2\exp\left(-\frac{t^2}{2\sigma^2 pT^2 + \frac{4}{3}u\mathbf{a}t}\right).$$

Taking $t = pT^2/520$ we obtain

$$\mathbb{P}\left\{|\langle \mathscr{X}(E), X_0 - M_0\rangle| > pT^2/520\right\} \leq 2\exp\left(-\frac{pT^2}{c_2(a \vee u)^2}\right). \tag{15}$$

On the other hand, Lemma 9 implies that with probability at least $1 - 2\exp\left(-\frac{pT^2}{c_3(\mathbf{a} \vee u)^2}\right)$

$$\sup_{\Delta \in \mathscr{T}(2k, 2\mathbf{a}, 2T)} |\langle \mathscr{X}(E), \Delta\rangle| \leq 6240(c^* u)^2 kd + pT^2/520$$

which together with (15) implies the statement of Lemma 8.

Lemma 9 *Assume that* $n \geq m\log(m)$. *We have that with probability at least* $1 - 2\exp\left(-\frac{pT^2}{c_3(a \vee u)^2}\right)$

$$\sup_{\Delta \in \mathscr{T}(2k, 2a, 2T)} |\langle \mathscr{X}(E), \Delta\rangle| \leq 6240(c^* u)^2 kd + pT^2/520$$

where $c_3 \leq 12(1560)^2$ *is a numerical constant.*

Proof Let $\widetilde{X}_T = \sup_{\Delta \in \mathscr{T}(2k, 2\mathbf{a}, 2T)} |\langle \mathscr{X}(E), \Delta\rangle| = \sup_{\Delta \in \mathscr{T}(2k, 2\mathbf{a}, 2T)} \langle \mathscr{X}(E), \Delta\rangle$. First we bound the expectation $\mathbb{E}(\widetilde{X}_T)$:

$$\mathbb{E}(\widetilde{X}_T) \leq \mathbb{E}\left(\sup_{\Delta \in \mathscr{T}(2k, 2\mathbf{a}, 2T)} \left|\sum_{(i,j)} \varepsilon_{ij} B_{ij} \Delta_{ij}\right|\right) = \mathbb{E}\left(\sup_{\Delta \in \mathscr{T}(2k, 2\mathbf{a}, 2T)} |\langle \Sigma, \Delta\rangle|\right)$$
$$\leq 2\sqrt{2k}\, T\, \mathbb{E}(\|\Sigma\|)$$

where $\Sigma = \sum_{(i,j)} B_{ij}\varepsilon_{ij} X_{ij}$ with $X_{ij} = e_i(m_1)e_j^T(m_2)$ and $e_k(l)$ are the canonical basis vectors in $\mathbb{R}^l$. Using $n \geq m\log(m)$ Lemma 4 in [11] and Corollary 3.3 in [1] implies that

$$\mathbb{E}\,\|\boldsymbol{\Sigma}\| \leq c^{*}u\sqrt{pd}. \tag{16}$$

where $c^{*} \geq 2$ is a universal numerical constant. Using (16) we obtain

$$\mathbb{E}(\widetilde{X}_T) \leq 2c^{*}u\sqrt{2kpd}\,T \leq 3120(c^{*}u)^2kd + pT^2/1560. \tag{17}$$

Now we use Theorem 3.3.16 in [9] (see also Theorem 8.1 in [7]) to obtain

$$\begin{aligned}\mathbb{P}\left(\widetilde{X}_T \geq \mathbb{E}(\widetilde{X}_T) + t\right) &\leq \exp\left(-\frac{t^2}{4u\mathbf{a}\mathbb{E}(\widetilde{X}_T) + 4\sigma^2 pT^2 + 9u\mathbf{a}t}\right) \\ &\leq \exp\left(-\frac{t^2}{8\mathbf{a}u^2c^{*}\sqrt{2kpd}T + 4\sigma^2 pT^2 + 9u\mathbf{a}t}\right) \end{aligned} \tag{18}$$

Taking in (18) $t = pT^2/1560 + 2uc^{*}\sqrt{2kpd}T$, together with (17) we obtain the statement of Lemma 9.

Acknowledgements The work of A. Carpentier is supported by the DFG's Emmy Noether grant MuSyAD (CA 1488/1-1) and by the DFG CRC 1294 on Data Assimilation and by the DFG GRK 2297 MathCoRe. The work of O. Klopp was conducted as part of the project Labex MME-DII (ANR11-LBX-0023-01). The work of M. Löffler was supported by the European Research Council (ERC) grant No. 647812.

References

1. Bandeira, A. S., & van Handel, R. (2016). Sharp nonasymptotic bounds on the norm of random matrices with independent entries. *The Annals of Probability, 44*(4), 2479–2506.
2. Cai, T., & Zhou, W. X. (2016). Matrix completion via max-norm constrained optimization. *Electronic Journal of Statistics, 10*(1), 1493–1525.
3. Candès, E. J., & Plan, Y. (2009). Matrix completion with noise. *Proceedings of IEEE, 98*(6), 925–936.
4. Candès, E. J., & Plan, Y. (2010). Tight oracle bounds for low-rank matrix recovery from a minimal number of random measurements. *CoRR*. abs/1001.0339.
5. Candès, E. J., & Recht, B. (2009). Exact matrix completion via convex optimization. *Fondations of Computational Mathematics, 9*(6), 717–772.
6. Candès, E. J., & Tao, T. (2010). The power of convex relaxation: Near-optimal matrix completion. *IEEE Transactions on Information Theory, 56*(5), 2053–2080.
7. Carpentier, A., Klopp, O., Löffler, M., & Nickl, R. (2018). Adaptive confidence sets for matrix completion. *Bernoulli, 24*, 2429–2460.
8. Chatterjee, S. (2015). Matrix estimation by universal singular value thresholding. *Annals of Statistics, 43*(1), 177–214.
9. Giné E., & Nickl, R. (2015). *Mathematical foundations of infinite-dimensional statistical models. Cambridge series in statistical and probabilistic mathematics*. Cambridge: Cambridge University Press.

10. Klopp, O. (2014). Noisy low-rank matrix completion with general sampling distribution. *Bernoulli, 20*(1), 282–303.
11. Klopp, O. (2015). Matrix completion by singular value thresholding: Sharp bounds. *Electronic Journal of Statistics, 9*(2), 2348–2369.
12. Koltchinskii, V. (2011). *Oracle inequalities in empirical risk minimization and sparse recovery problems. Lecture notes in mathematics* (Vol. 2033). Heidelberg: Springer. Lectures from the 38th Probability Summer School held in Saint-Flour, 2008, École d'Été de Probabilités de Saint-Flour.
13. Koltchinskii, V., Lounici, K., & Tsybakov, A. B. (2011). Nuclear-norm penalization and optimal rates for noisy low-rank matrix completion. *Annals of Statistics, 39*(5), 2302–2329.
14. Ledoux, M. (2001). *The concentration of measure phenomenon. Mathematical surveys and monographs* (Vol. 89). Providence: American Mathematical Society.
15. Lepskii, O. V. (1992). On problems of adaptive estimation in white Gaussian noise. *Advances in Soviet Mathematics, 12*, 87–106.
16. Negahban, S., & Wainwright, M. J. (2012). Restricted strong convexity and weighted matrix completion: Optimal bounds with noise. *Journal of Machine Learning Research, 13*, 1665–1697.
17. Talagrand, M. (1996). A new look at independence. *The Annals of Probability, 24*(1), 1–34.
18. Wang, Y. X., & Xu, H. (2012). Stability of matrix factorization for collaborative filtering. In *Proceedings of the 29th International Conference on Machine Learning, ICML 2012*, Edinburgh, Scotland, UK, June 26–July 1, 2012.

A Nonparametric Classification Algorithm Based on Optimized Templates

J. Kalina

Abstract This contribution is devoted to a classification problem into two groups. A novel algorithm is proposed, which is based on a distance of each observation from the centroid (prototype, template) of one of the groups. The general procedure is described on the particular task of mouth localization in facial images, where the centroid has the form of a mouth template. While templates are most commonly constructed as simple averages of positive examples, the novel optimization criterion allows to improve the separation between observations of one group (images of mouths) and observations of the other group (images of non-mouths). The separation is measured by means of the weighted Pearson product-moment correlation coefficient. On the whole, the new classification method can be described as conceptually simple and at the same time powerful.

1 Introduction

In this chapter, a novel nonparametric classification method to two groups is proposed, which is based on a centroid (prototype, template) of one of the groups. The method does not consider any distributional assumptions, allows a clear interpretation, and optimizes the centroid without any parametric model, as it is common in the nonparametric regression context.

The method is explained and illustrated on a particular classification task in the context of 2D images, namely a mouth detection in images of faces. Nevertheless, the method does not use any specific properties of mouths and not even of images and thus can be described as a general classification method suitable for high-dimensional data. Thus, the optimization algorithm may bring improvement in a wide variety of applications in different fields, e.g. in medicine or forensic science.

J. Kalina (✉)
Institute of Computer Science of the Czech Academy of Sciences, Prague, Czech Republic

National Institute of Mental Health, Klecany, Czech Republic
e-mail: kalina@cs.cas.cz

P. Bertail et al. (eds.), *Nonparametric Statistics*, Springer Proceedings in Mathematics & Statistics 250, https://doi.org/10.1007/978-3-319-96941-1_8

The concept of centroids in classification tasks has been recently investigated within various classification [9, 26] or unsupervised [6] learning tasks with recommendations to replace means by other (shrunken, regularized) centroids to improve the classification performance. Searching for a suitable centroid in various applications (e.g., image analysis or gene expression data) may be complicated i.e. by strongly correlated variables or by a high dimensionality of the problem, especially if the number of variables is large compared to the number of training data [14].

Centroid-based classification is popular in image analysis for being simple, powerful, and comprehensible. In the context of images, a centroid is called a template and is interpreted as a typical form, a virtual object with ideal appearance or shape, or image model [26], while the most common classification method based on templates is known as template matching. It is commonly performed by measuring similarity (most often by means of Pearson product-moment correlation coefficient) between the template and every rectangular part of the image, which has the same size as the template. Such area of the image with the largest similarity with a given mouth template is classified as the part of the image corresponding to the template.

Template matching has established applications, e.g. in person recognition, computer vision, forensic science, or archeology [2, 5, 21, 25] even recently. Unflagging attention is paid to templates as tools within more complicated approaches (e.g., sets of landmarks are used as templates in geometric morphometrics [27]), but also to self-standing local or global 2D or 3D templates, geometrical descriptors in the form of rigid templates [4], allowing to model the covariance structure of the data and to distinguish the intrapersonal variation from noise. All such approaches, just like any likelihood-based approaches to detection of eyes [24], faces [19] or humans [20], can be however described as parametric with all possible disadvantages following from violations of the probabilistic assumptions and models.

To the best of our knowledge, there have been no recommendations on a sophisticated construction of templates, which is surprising with regard to their simplicity and applicability. Commonly, a template is constructed as the average of several different positive examples. This ignores the requirement that templates should be very different from all possible negative examples, which are defined as parts of the image not corresponding to any mouth. If a mouth detection task is considered as an example, positive examples are mouths and negative will be called non-mouths. At the same time, we are not aware of any procedures for optimizing the discrimination between two images.

Our previous work in the task of face detection attempted to improve the correlation coefficient [11] exploiting its robust versions [13, 15], while only the current contribution is focused on improving and optimizing templates. Section 2 of this chapter describes a nonparametric method for optimizing templates together with an approximate algorithm. Results of computations over a particular data set of images of faces are presented in Sect. 3. Section 4 discusses the results as well as advantages of the proposed approach. Finally, Sect. 5 concludes the work.

2 Methods

2.1 Optimization Criterion

We propose a minimax optimization procedure for a nonparametric construction of an optimal centroid (template). The approach, although rather general, will be explained on the example of locating the mouth using a single template. Let $r_w(\mathbf{r}, \mathbf{s})$ denote the weighted (Pearson, i.e. product-moment) correlation coefficient between two data vectors $\mathbf{r}$ and $\mathbf{s}$ with given weights. Without any prior dimensionality reduction, r_w will be used as a measure of similarity between the template and the image throughout this chapter. Our aim is to optimize the classification rule over grey values of the template for the weighted correlation coefficient. The template is improved over the training data set of images, starting with an initial template. Fixed weights are used throughout the whole procedure, which express the importance of particular pixels for the classification task.

Let us consider a given template $\mathbf{t}$ and let us use the notation, e.g. $\mathbf{t} = \mathsf{vec}(\mathbf{t}) = (t_1, \ldots, t_n)^T$ so that the matrix is converted to a vector. Denoting n the number of its pixels, we consider a particular mouth $\mathbf{x} = (x_1, \ldots, x_n)^T$ and a particular non-mouth $\mathbf{z} = (z_1, \ldots, z_n)^T$ as vectors of the same size as the template. We also consider given non-negative weights $\mathbf{w} = (w_1, \ldots, w_n)^T$, which fulfil

$$\sum_{i=1}^{n} w_i = 1. \tag{1}$$

The function

$$f(\mathbf{x}, \mathbf{z}, \mathbf{t}, \mathbf{w}) = \frac{r_w^F(\mathbf{x}, \mathbf{t})}{r_w^F(\mathbf{z}, \mathbf{t})} \tag{2}$$

will be considered, where the monotone Fisher transformation

$$r_w^F(\mathbf{x}, \mathbf{y}) = \frac{1}{2} \log \frac{1 + r_w(\mathbf{x}, \mathbf{y})}{1 - r_w(\mathbf{x}, \mathbf{y})} \tag{3}$$

is applied on r_w extending its values to the whole real line $(-\infty, \infty)$ and improving the separation between mouths and non-mouths by putting emphasis on the worst non-mouths with the largest weighted correlation coefficient with the template. Besides, (3) can be characterized as a variance-stabilizing transformation. The value of (2) exceeds 1 if and only if the particular mouth $\mathbf{x}$ is classified correctly, i.e. it is well discriminated from the non-mouth. The larger (2), the better separation between $\mathbf{x}$ and $\mathbf{z}$.

The template optimization proposed below maximizes the worst separation, i.e. the separation for the worst case, which is defined as the mouth and non-mouth in a particular image with the worst separation over the whole training data set. For a particular image i in the data set of images $\mathscr{I}$, let us consider the set I of all rectangular areas in image i, which have the same size as the template. If there is exactly one mouth in every image, then I consists of the mouth and of the set I_z corresponding to the remaining areas (non-mouths). Retaining given weights $\mathbf{w}$, the optimal template is found as

$$\operatorname*{argmax}_{t_1,\ldots,t_n \in \mathbb{R}} \min_{i \in \mathscr{I}} \min_{\mathbf{z} \in I_z} f(\mathbf{x}, \mathbf{z}, \mathbf{t}, \mathbf{w}), \tag{4}$$

considering all non-mouths $\mathbf{z}$ in every particular image $i \in \mathscr{I}$ of the training data set.

Let us now consider more mouths in a given image. Then, we consider I to contain also the set I_x with several (rather than one) areas corresponding to the mouth. Then, we consider the worst of the mouths and thus instead of (4),

$$\operatorname*{argmax}_{t_1,\ldots,t_n \in \mathbb{R}} \min_{i \in \mathscr{I}} \min_{\mathbf{z} \in I_z} \min_{\mathbf{x} \in I_x} f(\mathbf{x}, \mathbf{z}, \mathbf{t}, \mathbf{w}) \tag{5}$$

will be searched for.

2.2 *Optimizing Templates*

A linear approximation to the solution of the nonlinear problem (5) will be now used to simplify the optimization, while results will be presented later in Sect. 3. Let us use the notation (2) for the separation between a particular mouth $\mathbf{x}$ and a non-mouth $\mathbf{z}$ in the worst case over all images, given the template $\mathbf{t}$ and weights $\mathbf{w}$. The task of minimizing (5) will be replaced by a linear approximation obtained from the Taylor series as

$$f(\mathbf{x}, \mathbf{z}, \mathbf{t} + \boldsymbol{\delta}, \mathbf{w}) \approx f(\mathbf{x}, \mathbf{z}, \mathbf{t}, \mathbf{w}) + \sum_{i=1}^{n} \delta_i \frac{\partial f(\mathbf{x}, \mathbf{z}, \mathbf{t}, \mathbf{w})}{\partial t_i}. \tag{6}$$

Small constants $\delta_1, \ldots, \delta_n$, where n is the size of the template, will be added to the grey intensities of the initial template with the aim to increase the worst separation. Formally, the linear problem

$$\max_{\delta_1,\ldots,\delta_n \in \mathbb{R}} \sum_{i=1}^{n} \delta_i \frac{\partial f(\mathbf{x}, \mathbf{z}, \mathbf{t}, \mathbf{w})}{\partial t_i} \tag{7}$$

will be solved under constraints

- $\sum_{i=1}^{n} \delta_i = 0$,
- $0 \leq t_i + \delta_i \leq C$ with a given $C > 0$, $i = 1, \ldots, n$,
- $-\varepsilon \leq \delta_i \leq \varepsilon$ with a given $\varepsilon > 0$, $i = 1, \ldots, n$.

The linear problem for the separation in the worst case should be formulated and solved repeatedly within an iterative algorithm. The simplex algorithm will be used to solve the linear problem. In some applications, e.g. mouth localization, it is reasonable to accompany the set of constraints by a requirement on the optimal template to be symmetric along its vertical axis, which at the same time reduces the dimensionality of the problem.

As the worst separation (5) increases, sooner or later it reaches the level of the second worst case. Therefore, we introduce additional constraints to improve the separation for several cases simultaneously. Let us now consider all non-mouths together with the mouths from the same images, which have the separation larger than the very worst case by less than (say) $p = 0.01$. For each one of these worst cases with the mouth denoted by $\mathbf{x}^*$ and the non-mouth $\mathbf{z}^*$, let us require

$$\sum_{i=1}^{n} \delta_i \frac{\partial f(\mathbf{x}, \mathbf{z}, \mathbf{t}, \mathbf{w})}{\partial t_i} = \sum_{i=1}^{n} \delta_i \frac{\partial f(\mathbf{x}^*, \mathbf{z}^*, \mathbf{t}, \mathbf{w})}{\partial t_i} \tag{8}$$

as additional constraints for (7). The classification is based on mouths and non-mouths near the boundary between these two classes and the whole iterative computation is described in Algorithm 1. The value of (5) as a function of a given template and weights will be denoted as $M(\mathbf{t}, \mathbf{w})$.

Remark 1 A repeated evaluation of (5) requires to find the worst case over all images repeatedly, as it may be different from the worst case in the previous iteration.

Algorithm 1 Linear approximation for optimizing a given (initial) template

Require: Symmetric initial template $\mathbf{t}_0$, symmetric weights $\mathbf{w}_0$, $p = 0.01$
Ensure: Optimal template $\mathbf{t}^*$

```
k := 0
repeat
    Consider the linear problem (7) for t = t_k.
    Find such mouths and non-mouths, which have the separation using t_k smaller than
    M(t_k, w_0) + p.
    Formulate constraints (8) for all such mouths and non-mouths for the linear problem.
    Solve the constrained linear problem by linear programming to obtain δ_1, ..., δ_n.
    k := k + 1
    t_k := (t_{k-1,1} + δ_1, ..., t_{k-1,n} + δ_n)^T
until M(t_k, w_0) ≥ M(t_{k-1}, w_0)
t* := t_{k-1}
```

3 Results

3.1 Description of the Data

The novel classification method, i.e. template matching with the template obtained with the approach of Sect. 2.2, will be now illustrated on the task of mouth localization in a data set of 212 raw grey-scale images of faces of size 192 × 256 pixels. The data set, which was created within projects BO 1955/2-1 and WU 314/2-1 of the German Research Council (DFG), was acquired at the Institute of Human Genetics, University of Duisburg-Essen [1, 11]. A grey intensity in the interval [0, 1] corresponds to each pixel. Each image contains a face of one individual sitting straight in front of the camera under standardized conditions, i.e. looking straight in the camera with open eyes without glasses, without hair covering the face or other nuisance effects. The size or rotation of the head differs only slightly.

First, the data set is divided to a training data set of 124 images and an independent validation data set of the remaining 88 images. The reason for this division was pragmatic as the 88 images were acquired later but still under the same standardized conditions fulfilling assumptions of independent validation.

In order to compare results of various methods, we manually localized the position of the midpoint of the mouth in every image of both data sets. The localization is considered successful if the midpoint of the detected mouth has the distance from the manually localized midpoint less or equal to three pixels (cf. Sect. 2.1). In this way, every mouth contains the middle parts of the lips, but reaches neither the nostrils nor the chin. All further computations are performed on raw images without a prior reduction of dimensionality or feature extraction. We used C++ for programming the entire code and R software for visualization of the results.

3.2 Locating the Mouth: Initial Results

For the given database of images of faces, the average mouth computed from the whole training data set contains a clear mouth without any inkling of moustache, but has rather a weak performance if used as a template. It localizes the mouth correctly in 85% of the images in the training data set. Instead, we attempted to construct various mouth templates of the same fixed size 26 × 56 pixels as averages of groups (clusters) of mouths mutually resembling each other. The size was selected to cover each mouths together with a small neighborhood. We consider only this fixed template size, while data driven approaches to selecting a suitable template size are much more complicated [17]. Several templates constructed in such a way yield a better performance than the overall average. The template with the best mouth localization performance among 10 such templates is the bearded template of Fig. 1, constructed as the average of 7 mouths of bearded men.

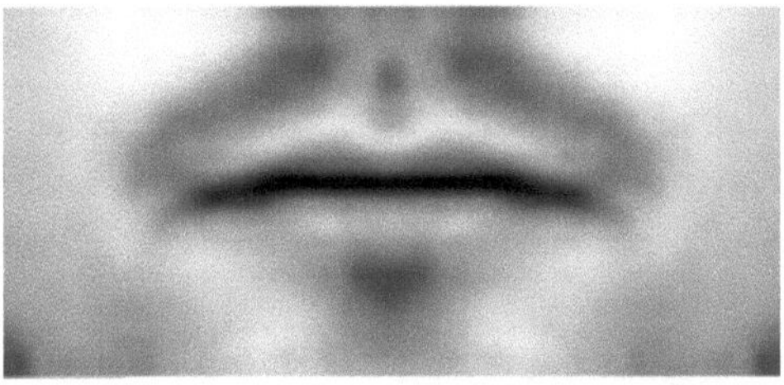

Fig. 1 Bearded template of size 26 × 56 pixels

The classification performance of the templates turns out to be improved if radial weights (Fig. 2, second row left) for the template are used. They are inversely proportional to the distance of each pixel from the template midpoint, stressing the central area of the template rather than its boundary parts. If a pixel with coordinates $[i, j]$ is considered, radial weights in this pixel are inversely proportional to its distance from the midpoint $[i_0, j_0]$, formally for even n_1 and n_2 defined as

$$w_{ij}^R = \frac{1}{\sqrt{(i - i_0)^2 + (j - j_0)^2}} \tag{9}$$

and a standardization (1) is used.

Table 1 presents values of the worst separation (5) obtained with various templates on the training and validation data sets. The optimal template was always constructed over the training data set. Its classification performance was subsequently evaluated over the training as well as the validation data sets.

Let us also compare the results obtained with template matching with those obtained with standard algorithms. The approaches of [22] and [28] were applied on raw images over the considered data set. To compare the results also with several standard classification methods, we performed the following manual pre-processing. Out of the training data set, a set of 124 manually selected mouth images of size 26 × 56 pixels and 124 non-mouths of the same size, where each non-mouth comes from one image, was created. Here, such non-mouth was selected from each image which has the largest correlation coefficient with the bearded template. The results are shown in Table 2 for various general classification methods as well as specific procedures for object detection in 2D images. However, the separation measure (2) cannot be applied to any of these standard methods.

3.3 *Optimal Mouth Template*

The template optimization of Sect. 2.2 will be now used to further improve the results of template matching in terms of separation between mouths and non-mouths. The optimization starts always with the bearded initial template but considers various choices of the weights. Symmetry of the optimal mouth template

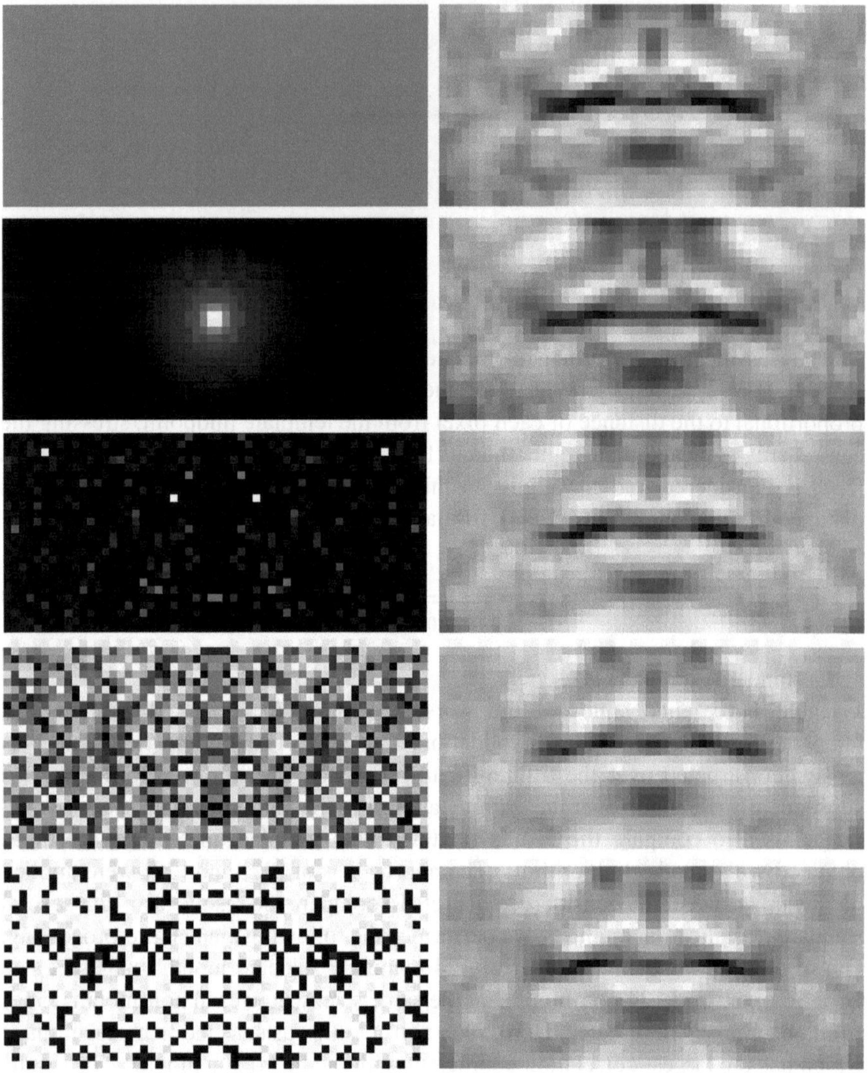

Fig. 2 Results of template optimization. Left: choice of the weights. Middle: optimal template obtained for these weights and for the bearded initial template. Right: optimal weights for this optimal template, obtained by the two-stage search with the upper bounds $c_1 = c_2 = 0.005$

is required to reduce the computational complexity. In the computations, the choice $C = 0.005$ was made to prevent overfitting. This is rather a precaution because templates resulting from the optimization commonly reach C only in a small percentage of pixels.

The computation based on the linear approximation requires several hundreds of iterations in small steps. Concerning the value of ε, we started with $\varepsilon = 0.000020$

Table 1 Performance of mouth localization evaluated by means of (5) in the training and validation data sets for equal or radial weights

	Equal weights		Radial weights	
Template	Training set	Validation set	Training set	Validation set
Average	0.66	0.69	0.85	0.82
Bearded	0.78	0.55	1.13	0.94
Optimal ($C = 0.005$)	2.12	1.81	1.79	1.52

Different templates include the average across the training data set, a bearded template (Fig. 1), and the solution of the linear approximation of Sect. 2.2, starting with the bearded initial template

Table 2 Percentages of correctly classified images using different standard methods for classification or object detection

	Results over the		
Method	Training set	Validation set	Software
Viola-Jones [22]	1.00	1.00	MATLAB, package vision
Zhu [28]	1.00	1.00	Online supplement of [28]
Support vector machines	0.90	0.85	R, package e1071
Classification tree	0.97	0.90	R, package tree
Multilayer perceptron	1.00	1.00	R, package neural
SCRDA	0.98	0.92	R, package rda

The classification rule was learned over the training data set and its performance was subsequently evaluated over both data sets

Table 3 Performance of mouth localization evaluated by means of (5)

Weights	Initial template		Optimal template	
(row of Fig. 2)	Training set	Validation set	Training set	Validation set
1	0.78	0.75	2.12	1.53
2	1.13	1.03	1.79	1.43
3	0.77	0.72	2.05	1.36
4	0.80	0.81	2.06	1.48
5	0.82	0.75	2.05	1.41

Different weights are used from different rows of Fig. 2 (left). For the initial bearded template, the performance is evaluated for the training (T) and validation (V) data sets. The optimal template was constructed over the training data set and its performance was subsequently evaluated over both data sets

but later iterations required a decrease to $\varepsilon = 0.000001$ in order to continue improving the worst separation. Also, there happen to be as many as 10 worst cases from different images during the last iterations. To speed the computation, we have a good experience with violating Remark 1 and finding the worst case over the whole data set only in each fifth iteration.

The resulting optimal templates are shown in Fig. 2 for different choices of fixed weights. The performance of the optimal templates for these various choices of weights is presented in Table 3. As we can see, the optimal templates in all cases contain clear lips but no beard any more (Fig. 2).

4 Discussion

In this chapter, a novel classification method is proposed, which can be characterized as a nonparametric classifier to two groups based on optimizing the centroid (template) of one of the groups. It is illustrated on a particular task of optimal construction of a mouth template for the automatic mouth localization in 2D grey-scale images of faces. We use a weighted Pearson product-moment correlation coefficient as the measure of similarity between the image and the template. The procedure may find applications in a broad scope of classification problems not limited to template matching, which itself is acknowledged as one of standard, powerful, comprehensible, and simple methods useful for object detection in single images. Still, the topic of optimal construction of templates has not attracted sufficient attention.

If the average mouth is used as a single template, the performance of template matching is rather weak over the given dataset. A simple bearded template yields the best performance with the mouth localization results to be correct in 100% of images if radial weights are used. While some of the standard algorithms of machine learning as well as specific approaches of image analysis are able to reach the 100% performance as well, the advantages of the new approach include the possibility to measure directly the separation (2) in the worst case. Such measure is tailor-made for template matching and allows to search for an optimal template, while it cannot be even evaluated for any other classification procedure.

The optimization criterion is formulated to separate the mouths and non-mouths in the worst case across the whole data set. The optimization task was solved exploiting a linear approximation to the high-dimensional optimization task and depends on a small number (not more than ten) of non-mouths with the largest resemblance to the mouths. The optimization is able to remarkably improve the initial classification performance and the improvement is retained if verified on an independent validation data set. This contradicts the popular belief that the average of mouths of different people as a very suitable template. We have a good experience with the bearded initial template, although the beard seems to disappear from the optimized templates. Additional computations also reveal the resulting template not to be very sensitive to the choice of the initial template.

The procedure can also be described as a nonparametric search for a shrinkage version of the centroid of one group (cf. [23]). Numerous classification procedures for high-dimensional data are based on shrinkage estimators of the population mean, which reduce the mean square error compared to the classical mean for multivariate data [10, 16]. In the classification task, the mean of each group is commonly shrunken towards the overall mean [18] or towards zero [9]. However, all such approaches require to consider the prototype (e.g., regularized mean) also of the non-mouths, while the population of non-mouths is substantially more heterogeneous (diverse) than that of mouths.

The novel method works reliably on the considered data without any initial reduction of dimensionality, which allows a clear interpretation and represents also the difference from numerous habitual approaches to image analysis which require a prior feature extraction. It may be, however, used after a prior dimensionality reduction (feature extraction) as well. The classifier does not seem to have a tendency to overfitting in spite of the high dimensionality of the task, when the number of pixels $n = 1456$ largely exceeds the number of images.

5 Conclusion

In this chapter, a novel general nonparametric approach to classification to two groups is proposed and implemented. It is based on measuring the weighted correlation coefficient between an observation and a centroid of one of the two groups, for which an optimization criterion tailor-made for the classification task is formulated. The novel method does not require any distributional assumptions and does not evaluate any form of a likelihood. It may be applied to classification tasks for high-dimensional multivariate data in various fields, while it is common to use the arithmetic mean to play the role of a centroid (prototype, estimator of the population mean) of the groups in classification tasks, e.g. in the framework of linear discriminant analysis [10, 12].

Principles of the new nonparametric classifier, although rather general, are explained and illustrated on a particular classification task in images. In such context, the centroid can be denoted as a template and we may speak of a nonparametric construction of optimal templates exploiting all benefits of a nonparametric approach.

The optimization criterion of the new method is based on improving the separation only for a small set of the worst cases. These are in our case mouths with the worst separation from non-mouths, i.e. non-mouths with the largest resemblance to a mouth. This is a common feature of various nonparametric optimization approaches, e.g. kernel-based methods [7, 10] or support vector machines, where the latter are based only on selected observations (support vectors) near the boundary between the classes. In our examples, the novel template optimization brings improvements in the separation between positive and negative examples, as verified on an independent validation data set.

The resulting classification procedure can be perceived as a comprehensible method allowing to interpret which variables contribute the most to the similarity between the template and the corresponding part of the image. This is an advantage over competing image analysis procedures, which commonly contain a large number of parameters with a great impact on the result but with a too difficult interpretability (e.g., [22]).

Limitations of the new method include its computational intensity due to its nonparametric character. Still, the method may be suitable, e.g. for applications in medicine or forensic science, which do not require a fast computation. The demanding computation is performed however only in the optimization (i.e., learning) of the template, while assigning a new observation to one of the groups (i.e., the template matching itself) can be computed quickly. Other disadvantages include the suitability of template matching only for standardized images, although the method itself is general and uses specific properties of neither faces nor images. Further, we did not invent a special solution for images with the mouth located at the boundary of the image; it may be worthwhile to replace r_w by a robust correlation coefficient [15]. While template matching is common as an elementary tool within more complicated computational pipelines, it must be admitted that templates themselves cannot compete with image analysis approaches invariant to illumination changes [3]. The rigid character of the template represents another restriction, while deformable 2D templates represent a more flexible alternative, allowing to model the deformation of the object from the ideal template.

A future research is intended to be devoted to the following tasks, which are arranged from the simplest applications to more complicated extensions.

- Localizing other objects (e.g., eyes) in facial images by optimal templates in 2D or 3D images.
- Optimizing also the weights for the weighted correlation coefficient.
- Applying the new method to nonparametric classification of data which are not images, especially data which are high-dimensional and not normally distributed (e.g., molecular genetic measurements). While this chapter considers mouths and non-mouths in blocks (i.e., within images), gene expression data have an analogous structure in pairs, while a sample of a patient and a sample of a control individual both are measured within a pair within a microarray.
- Optimizing deformable templates, which are obtained by a shape alteration (distortion, warping) of rigid templates [8].

Acknowledgements This work was financially supported by the Czech Health Research Council project NV15-29835A and the Neuron Fund for Support of Science. The author is thankful to Prof. Dr. Laurie Davies and Dr. Ctirad Matonoha for valuable suggestions.

Reference

1. Böhringer, S., Vollmar, T., Tasse, C., Würtz, R. P., Gillessen-Kaesbach, G., Horsthemke, B., et al. (2006). Syndrome identification based on 2D analysis software. *European Journal of Human Genetics, 14*, 1082–1089.
2. Chen, J. H., Chen, C. S., & Chen, Y. S. (2003). Fast algorithm for robust template matching with M-estimators. *IEEE Transactions on Signal Processing, 51*, 230–243.
3. Chong, H. Y., Gortler, S. J., & Zickler, T. (2008). A perception-based color space for illumination-invariant image processing. *ACM Transactions on Graphics, 27*, Article 61.
4. Dalal, N., & Triggs, B. (2005). Histograms of oriented gradients for human detection. In *IEEE Computer Society Conference on Computer Vision and Pattern Recognition CVPR 2005* (pp. 886–893).

5. Damas, S., Cordón, O., Ibáñez, O., Santamaria, J., Alemán, I., Botella, M., & Navarro, F. (2011). Forensic identification by computer-aided craniofacial superimposition: A survey. *ACM Computing Survey, 43*, Article 27.
6. Gao, J., & Hitchcock, D. B. (2010). James-Stein shrinkage to improve k-means cluster analysis. *Computational Statistics & Data Analysis, 54*, 2113–2127.
7. Girosi, F., Jones, M., & Poggio, T. (1995). Regularization theory and neural networks architectures. *Neural Computation 7*, 219–269.
8. Grenander, U. (1993). *General pattern theory. A mathematical study of regular structures*. Oxford: Oxford University Press.
9. Guo, Y., Hastie, T., & Tibshirani, R. (2007). Regularized discriminant analysis and its application in microarrays. *Biostatistics, 8*, 86–100.
10. Hastie, T., Tibshirani, R., & Friedman, J. (2009). *The elements of statistical learning*. (2nd ed.) New York: Springer.
11. Kalina, J. (2012). Facial symmetry in robust anthropometrics. *Journal of Forensic Sciences, 57*(3), 691–698.
12. Kalina, J. (2012). Highly robust statistical methods in medical image analysis. *Biocybernetics and Biomedical Engineering, 32*(2), 3–16.
13. Kalina, J. (2015). Three contributions to robust regression diagnostics. *Journal of Applied Mathematics, Statistics and Informatics, 11*(2), 69–78.
14. Kalina, J., & Schlenker, A. (2015). A robust supervised variable selection for noisy high-dimensional data. *BioMed Research International, 2015*, Article 320385, 1–10.
15. Shevlyakov, G. L., & Oja, H. (2016). *Robust correlation: Theory and applications*. New York: Wiley.
16. Stein, C. (1956). Inadmissibility of the usual estimator for the mean of a multivariate normal distribution. In *Proceedings of the Third Berkeley Symposium on Mathematical Statistics and Probability* (Vol. 1, pp. 197–206). Berkeley: University of California Press.
17. Tang, F., & Tao, H. (2007). Fast multi-scale template matching using binary features. In *IEEE Workshop on Applications of Computer Vision WACV'07*, 36.
18. Tibshirani, R., Hastie, T., & Narasimhan, B. (2003). Class prediction by nearest shrunken centroids, with applications to DNA microarrays. *Statistical Science, 18*, 104–117.
19. Torralba, A., Murphy, K. P., & Freeman, W. T. (2007). Sharing visual features for multiclass and multiview object detection. *IEEE Transactions on Pattern Analysis and Machine Intelligence, 5*, 854–869.
20. Tuzel, O., Porikli, F., & Meer, P. (2007). Human detection via classification on Riemannian manifolds. In *IEEE Computer Society Conference on Computer Vision and Pattern Recognition CVPR 2007* (pp. 1–8).
21. Vanderbei, R. J. (2009). *Linear programming: Foundations and extensions* (3rd ed.). New York: Springer.
22. Viola, P., & Jones, M. J. (2004). Robust real-time face detection. *International Journal of Computer Vision, 57*, 137–154.
23. Wang, C., Tong, T., Cao, L., & Miao, B. (2014). Non-parametric shrinkage mean estimation for quadratic loss functions with unknown covariance matrices. *Journal of Multivariate Analysis, 125*, 222–232.
24. Wang, X., & Tang, X. (2005). Subspace analysis using random mixture models. In *IEEE Computer Society Conference on Computer Vision and Pattern Recognition CVPR 2005* (pp. 574–580).
25. Wei, L., Yu, W., & Li, M. (2011). Skull assembly and completion using template-based surface matching. In *Proceedings International Conference on 3D Imaging, Modeling, Processing, Visualization and Transmission 3DIMPVT 2011* (pp. 413–420).
26. Yang, J., Han, F., Irizarry, R. A., & Liu, H. (2014). Context aware group nearest shrunken centroids in large-scale genomic studies. *Journal of Machine Learning Research, 33*, 1051–1059.

27. Zelditch, M., Swiderski, D., Sheets, D. H., & Fink, W. (2012). *Geometric morphometrics for biologists* (2nd ed.). London: Elsevier.
28. Zhu, X., & Ramanan, D. (2012). Face detection, pose estimation, and landmark localization in the wild. In *IEEE Conference on Computer Vision and Pattern Recognition 2012* (pp. 2879–2886). New York: IEEE.

PAC-Bayesian Aggregation of Affine Estimators

L. Montuelle and E. Le Pennec

Abstract Aggregating estimators using exponential weights depending on their risk appears optimal in expectation but not in probability. We use here a slight overpenalization to obtain oracle inequality in probability for such an explicit aggregation procedure. We focus on the fixed design regression framework and the aggregation of linear estimators and obtain results for a large family of linear estimators under a non-necessarily independent sub-Gaussian noise assumptions.

1 Introduction

We consider here a classical fixed design regression model

$$\forall i \in \{1, \ldots, n\},\ Y_i = f_0(x_i) + W_i$$

with f_0 an unknown function, x_i the fixed design points, and $W = (W_i)_{i \leq n}$ a centered sub-Gaussian noise. We assume that we have at hand a family of linear estimate $\{\hat{f}_t(Y) = A_t Y | A_t \in \mathscr{S}_n^+(\mathbb{R}), b_t \in \mathbb{R}^n, t \in \mathscr{T}\}$, for instance a family of projection estimator, of linear ordered smoother in a basis or in a family of basis. The most classical way to use such a family is to select one of the estimates according to the observations, for instance using a penalized empirical risk principle. A better way is to combine linearly those estimates with weights depending on the observation. A simple strategy is the Exponential Weighting Average in which all those estimates are averaged with a weight proportional to $\exp\left(-\frac{\widetilde{r}_t}{\beta}\right) \pi(t)$ where $\widetilde{r}_t$ is a (penalized) estimate of the risk of $\hat{f}_t$. This strategy is not new nor optimal

L. Montuelle
RTE, La Défense, France
e-mail: lucie.montuelle@rte-france.com

E. Le Pennec (✉)
CMAP/XPOP, École Polytechnique, Palaiseau, France
e-mail: erwan.le-pennec@polytechnique.edu

P. Bertail et al. (eds.), *Nonparametric Statistics*, Springer Proceedings in Mathematics & Statistics 250, https://doi.org/10.1007/978-3-319-96941-1_9

as explained below but is widely used in practice. In this chapter, we analyze the performance of this simple EWA estimator by providing oracle inequalities in probability under mild sub-Gaussian assumption on the noise.

Our aim is to obtain the best possible estimate of the function f_0 at the grid points. This setting is probably one of the most common in statistics and many regression estimators are available in the literature. For non-parametric estimation, Nadaraya-Watson estimator [39, 52] and its fixed design counterpart [26] are widely used, just like projection estimators using trigonometric, wavelet [24] or spline [51] basis, for example. In the parametric framework, least squares or maximum likelihood estimators are commonly employed, sometimes with minimization constraints, leading to LASSO [47], ridge [34], elastic net [60], AIC [1], or BIC [45] estimates.

Facing this variety, the statistician may wonder which procedure provides the best estimation. Unfortunately, the answer depends on the data. For instance, a rectangular function is well approximated by wavelets but not by trigonometric functions. Since the best estimator is not known in advance, our aim is to mimic its performances in terms of risk. This is theoretically guaranteed by an oracle inequality:

$$R(f_0, \tilde{f}) \leq C_n \inf_{t \in \mathcal{T}} R(f_0, \hat{f}_t) + \epsilon_n$$

comparing the risk of the constructed estimator $\tilde{f}$ to the risk of the best available procedure in the collection $\{\hat{f}_t, t \in \mathcal{T}\}$. Our strategy is based on convex combination of these preliminary estimators and relies on PAC-Bayesian aggregation to obtain a single adaptive estimator. We focus on a wide family, commonly used in practice : affine estimators $\{\hat{f}_t(Y) = A_t(Y - b) + b + b_t | A_t \in \mathcal{S}_n^+(\mathbb{R}), b_t \in \mathbb{R}^n, t \in \mathcal{T}\}$ with $b \in \mathbb{R}^n$ a common recentering.

Aggregation procedures have been introduced by Vovk [50], Littlestone and Warmuth [37], Cesa-Bianchi et al. [13], Cesa-Bianchi and Lugosi [14]. They are a central ingredient of bagging [9], boosting [25, 44], or random forest ([3] or [10]; or more recently [6–8, 27]).

The general aggregation framework is detailed in [40] and studied in [11, 12] through a PAC-Bayesian framework as well as in [53–59]. See, for instance, [49] for a survey. Optimal rates of aggregation in regression and density estimation are studied by Tsybakov [48], Lounici [38], Rigollet and Tsybakov [42], Rigollet [41] and Lecué [35].

A way to translate the confidence of each preliminary estimate is to aggregate according to a measure exponentially decreasing when the estimate's risk rises. This widely used strategy is called exponentially weighted aggregation. More precisely, as explained before, the weight of each element $\hat{f}_t$ in the collection is proportional to $\exp\left(-\frac{\widetilde{r}_t}{\beta}\right)\pi(t)$ where $\widetilde{r}_t$ is a (penalized) estimate of the risk of $\hat{f}_t$, β is a positive parameter, called the temperature, that has to be calibrated and π is a prior measure over $\mathcal{T}$. The key property of exponential weights is that they explicitly minimize the aggregated risk penalized by the Kullback-Leibler divergence to the prior measure π [12]. Our aim is to give sufficient conditions on the risk estimate $\widetilde{r}_t$ and the temperature β to obtain an oracle inequality for the risk of the aggregate. Note that when the family $\mathscr{T}$ is countable, the exponentially weighted aggregate is a weighted sum of the preliminary estimates.

This procedure has shown its efficiency, offering lower risk than model selection because we bet on several estimators. Aggregation of projections has already been addressed by Leung and Barron [36]. They have proved, by the mean of an oracle inequality, that the aggregate performs almost as well, in expectation, as the best projection in the collection. Those results have been extended to several settings and noise conditions [5, 18, 19, 21–23, 29, 30, 43, 46] under a *frozen* estimator assumption: they should not depend on the observed sample. This restriction, not present in the work by Leung and Barron [36], has been removed by Dalalyan and Salmon [20] within the context of affine estimator and exponentially weighted aggregation. Nevertheless, they make additional assumptions on the matrices A_t and the Gaussian noise to obtain an optimal oracle inequality in expectation for affine estimates. Very sharp results have been obtained in [15, 31] and [32]. Those papers, except the last one, study a risk in expectation.

Indeed, the Exponential Weighting Aggregation is not optimal anymore in probability. Dai et al. [17] have indeed proved the sub-optimality in deviation of exponential weighting, not allowing to obtain a sharp oracle inequality in probability. Under strong assumptions and independent noise, [4] provides a sharp oracle inequality with optimal rate for another aggregation procedure called Q-aggregation. It is similar to exponential weights but the criterion to minimize is modified and the weights no longer are explicit. Results for the original EWA scheme exist nevertheless but with a constant strictly larger than 1 in the oracle inequality. Dai [16] obtain, for instance, a result under a Gaussian white noise assumption by penalizing the risk in the weights and taking a temperature at least 20 times greater than the noise variance. Golubev and Ostobski [32] does not use an overpenalization but assumes some ordered structure on the estimate to obtain a result valid even for low temperature. An unpublished work, by Gerchinovitz [28], provides also weak oracle inequality with high probability for projection estimates on non-linear models. Alquier and Lounici [2] consider *frozen* and bounded preliminary estimators and obtain a sharp oracle inequality in deviation for the excess risk under a sparsity assumption, if the regression function is bounded, with again a modified version of exponential weights.

In this work, we will play on both the temperature and the penalization. We will be able to obtain oracle inequalities for the Exponential Weighting Aggregation under a general sub-Gaussian noise assumption that does not require a coordinate independent setting. We conduct an analysis of the relationship between the choice of the penalty and the minimal temperature. In particular, we show that there is a continuum between the usual noise based penalty and a sup norm type one allowing a *sharp* oracle inequality.

2 Framework and Estimate

Recall that we observe

$$\forall i \in \{1, \dots, n\},\ Y_i = f_0(x_i) + W_i$$

with f_0 an unknown function and x_i the fixed grid points. Our only assumption will be on the noise. We do not assume any independence between the coordinates W_i but only that $W = (W_i)_{i \le n} \in \mathbb{R}^n$ is a centered sub-Gaussian variable. More precisely, we assume that $\mathbb{E}(W) = 0$ and there exists $\sigma^2 \in \mathbb{R}^+$ such that

$$\forall \alpha \in \mathbb{R}^n,\ \mathbb{E}\left[\exp\left(\alpha^\top W\right)\right] \le \exp\left(\frac{\sigma^2}{2}\|\alpha\|_2^2\right),$$

where $\|.\|_2$ is the usual euclidean norm in $\mathbb{R}^n$. If W is a centered Gaussian vector with covariance matrix Σ, then σ^2 is nothing but the largest eigenvalue of Σ.

The quality of our estimate will be measured through its error at the design points. More precisely, we will consider the classical euclidean loss, related to the squared norm

$$\|g\|_2^2 = \sum_{i=1}^n g(x_i)^2.$$

Thus, our unknown is the vector $(f_0(x_i))_{i=1}^n$ rather than the function f_0.

As announced, we will consider affine estimators $\hat{f}_t(Y) = A_t(Y - b) + b + b_t$ corresponding to affine smoothed projection.

We will assume that

$$\hat{f}_t(Y) = A_t(Y - b) + b + b_t = \sum_{i=1}^n \rho_{t,i}\langle Y - b, g_{t,i}\rangle g_{t,i} + b + b_t$$

where $(g_{t,i})_{i=1}^n$ is an orthonormal basis, $(\rho_{t,i})_{i=1}^n$ a sequence of non-negative real numbers, and $b_t \in \mathbb{R}^n$. By construction, A_t is thus a symmetric positive semi-definite real matrix. We assume furthermore that the matrix collection $\{A_t\}_{t \in \mathcal{T}}$ is such that $\sup_{t \in \mathcal{T}} \|A_t\|_2 \le 1$. For the sake of simplicity, we only use the notation $\hat{f}_t(Y) = A_t(Y - b) + b + b_t$ in the following.

To define our estimate from the collection $\{\hat{f}_t(Y) = A_t Y + b_t | A_t \in \mathcal{S}_n^+(\mathbb{R}), b_t \in \mathbb{R}^n, t \in \mathcal{T}\}$, we specify the estimate $\widetilde{r_t}$ of the (penalized) risk of the estimator $\hat{f}_t(Y)$, choose a prior probability measure π over $\mathcal{T}$ and a temperature $\beta > 0$. We define the exponentially weighted measure ρ_{EWA}, a probability measure over $\mathcal{T}$, by

$$d\rho_{EWA}(t) = \frac{\exp\left(-\frac{1}{\beta}\widetilde{r_t}\right)}{\int \exp\left(-\frac{1}{\beta}\widetilde{r_{t'}}\right) d\pi(t')} d\pi(t)$$

and the exponentially weighted aggregate f_{EWA} by $f_{EWA} = \int \hat{f}_t \, d\rho_{EWA}(t)$. If $\mathcal{T}$ is countable, then

$$f_{EWA} = \sum_{t \in \mathcal{T}} \frac{e^{-\widetilde{r_t}/\beta} \pi_t}{\sum_{t' \in \mathcal{T}} e^{-\widetilde{r_{t'}}/\beta} \pi_{t'}} \hat{f}_t .$$

This construction naturally favors low risk estimates. When the temperature goes to zero, this estimator becomes very similar to the one minimizing the risk estimate while it becomes an indiscriminate average when β grows to infinity. The choice of the temperature appears thus to be crucial and a low temperature seems to be desirable.

Our choice for the risk estimate $\widetilde{r_t}$ is to use the classical Stein unbiased estimate, which is sufficient to obtain optimal oracle inequalities in expectation,

$$r_t = \|Y - \hat{f}_t(Y)\|_2^2 + 2\sigma^2 \mathrm{Tr}(A_t) - n\sigma^2$$

and add a penalty $\mathrm{pen}(t)$. We will consider simultaneously the case of a penalty independent of f_0 and the one where the penalty may depend on an upper bound of (kind of) sup norm.

More precisely, we allow the use, at least in the analysis, of an upper bound $\|\widetilde{f_0 - b}\|_\infty$ which can be thought as the supremum of the sup norm of the coefficients of f_0 in any basis appearing in $\mathcal{T}$. Indeed, we define $\|\widetilde{f_0 - b}\|_\infty$ as the smallest non-negative real number C such that for any $t \in \mathcal{T}$,

$$\|A_t(f_0 - b)\|_2^2 \leq C^2 \mathrm{Tr}(A_t^2).$$

By construction, $\|\widetilde{f_0 - b}\|_\infty$ is smaller than the sup norm of any coefficients of $f_0 - b$ in any basis appearing in the collection of estimators. Note that $\|\widetilde{f_0 - b}\|_\infty$ can also be upper bounded by $\|f_0 - b\|_1$, $\|f_0 - b\|_2$ or $\sqrt{n}\|f_0 - b\|_\infty$ where the ℓ_1 and sup norm can be taken in any basis.

Our aim is to obtain sufficient conditions on the penalty $\mathrm{pen}(t)$ and the temperature β so that an oracle inequality of type

$$\begin{aligned} \|f_0 - f_{EWA}\|_2^2 \leq \inf_{\mu \in \mathcal{M}_+^1(\mathcal{T})} & (1+\epsilon) \int \|f_0 - \hat{f}_t\|_2^2 d\mu(t) \\ & + (1+\epsilon') \left(\int \mathrm{price}(t) d\mu(t) + 2\beta \mathrm{KL}(\mu, \pi) + \beta \ln \frac{1}{\eta} \right) \end{aligned}$$

holds either in probability or in expectation. Here, ϵ and ϵ' are some small non-negative numbers possibly equal to 0 and $\mathrm{price}(t)$ a loss depending on the choice of $\mathrm{pen}(t)$ and β. When $\mathcal{T}$ is countable, such an oracle proves that the risk of our aggregate estimate is of the same order as the one of the best estimate in the collection as it implies

$$\|f_0 - f_{EWA}\|_2^2 \leq \inf_{t\in\mathcal{T}} \left\{(1+\epsilon)\|f_0 - \hat{f}_t\|_2^2 + (1+\epsilon')\left(\text{price}(t) + \beta \ln \frac{1}{\pi(t)^2\eta}\right)\right\}.$$

Before stating our more general result, which is in Sect. 4, we provide a comparison with some similar results in the literature on the countable $\mathcal{T}$ setting.

3 Penalization Strategies and Preliminary Results

The most similar result in the literature is the one from [16] which holds under a Gaussian white noise assumption and uses a penalty proportional to the known variance σ^2:

Proposition 3.1 ([16]) *If* $\text{pen}(t) = 2\sigma^2\text{Tr}(A_t)$, *and* $\beta \geq 4\sigma^2 16$, *then for all* $\eta > 0$, *with probability at least* $1-\eta$,

$$\|f_0 - f_{EWA}\|^2 \leq \min_t \left\{\left(1 + \frac{128\sigma^2}{3\beta}\right)\|f_0 - \hat{f}_t\|^2 + 8\sigma^2\text{Tr}(A_t) + 3\beta \ln \frac{1}{\pi_t} + 3\beta \ln \frac{1}{\eta}\right\}.$$

Our result generalizes this result to the non-necessarily independent sub-Gaussian noise. We obtain

Proposition 3.2 *If* $\beta \geq 20\sigma^2$, *there exists* $\gamma \in [0, 1/2)$, *such that if* $\text{pen}(t) \geq \frac{4\sigma^2}{\beta-4\sigma^2}\text{Tr}(A_t^2)\sigma^2$, *for any* $\eta > 0$, *with probability at least* $1-\eta$,

$$\|f_0 - f_{EWA}\|^2 \leq \inf_t \left\{\left(1 + \frac{4\gamma}{1-2\gamma}\right)\|f_0 - \hat{f}_t\|^2 + \left(1 + \frac{2\gamma}{1-2\gamma}\right)\left(\text{pen}(t) + 2\sigma^2\text{Tr}(A_t) + 2\beta \ln \frac{1}{\pi_t} + \beta \ln \frac{1}{\eta}\right)\right\}.$$

The parameter γ is explicit and satisfies $\epsilon = O(\frac{\sigma^2}{\beta})$. We recover thus a similar weak oracle inequality under a weaker assumption on the noise. It should be noted that [4] obtains a sharp oracle inequality for a slightly different aggregation procedure but only under the very strong assumption that $\text{Tr}(A_t) \leq \ln \frac{1}{\pi(t)}$.

Following [33], a lower bound on the penalty that involves the sup norm of f_0, can be given. In that case, the oracle inequality is sharp as $\epsilon = \epsilon' = 0$. Furthermore, the parameter γ is not necessary and the minimum temperature is lower.

Proposition 3.3 *If* $\beta > 4\sigma^2$, *and*

$$\mathrm{pen}(t) \geq \frac{4\sigma^2}{\beta - 4\sigma^2}\left(\sigma^2 \mathrm{Tr}(A_t^2) + 2\left[\widetilde{\|f_0 - b\|_\infty^2} \mathrm{Tr}(A_t^2) + \|b_t\|_2^2\right]\right),$$

then for any $\eta > 0$, *with probability at least* $1 - \eta$,

$$\|f_0 - f_{EWA}\|^2 \leq \inf_t \Bigg\{ \|f_0 - \hat{f}_t\|^2 + 2\sigma^2 \mathrm{Tr}(A_t) + \frac{8\sigma^2}{\beta - 4\sigma^2}\left[\widetilde{\|f_0 - b\|_\infty^2} \mathrm{Tr}(A_t^2) + \|b_t\|_2^2\right] + \mathrm{pen}(t) + 2\beta \ln \frac{1}{\pi_t} + \beta \ln \frac{1}{\eta} \Bigg\}.$$

We are now ready to state the central result of this contribution, which gives an explicit expression for γ and introduce an optimization parameter $\nu > 0$, from which this theorem can be deduced.

4 A General Oracle Inequality

We consider now the general case for which $\mathscr{T}$ is not necessarily countable. Recall that we have defined the exponentially weighted measure ρ_{EWA}, a probability measure over $\mathscr{T}$, by

$$d\rho_{EWA}(t) = \frac{\exp\left(-\frac{1}{\beta}\widetilde{r_t}\right)}{\int \exp\left(-\frac{1}{\beta}\widetilde{r_{t'}}\right) d\pi(t')} d\pi(t)$$

and the exponentially weighted aggregate f_{EWA} by $f_{EWA} = \int \hat{f}_t \, d\rho_{EWA}(t)$. Propositions 3.2 and 3.3 will be obtained as straightforward corollaries.

Our main contribution is the following two similar theorems:

Theorem 4.1 *For any* $\beta \geq 20\sigma^2$, *let*

$$\gamma = \frac{\beta - 12\sigma^2 - \sqrt{\beta - 4\sigma^2}\sqrt{\beta - 20\sigma^2}}{16\sigma^2}.$$

If for any $t \in \mathscr{T}$,

$$\mathrm{pen}(t) \geq \frac{4\sigma^2}{\beta - 4\sigma^2}\sigma^2 \mathrm{Tr}(A_t^2),$$

then

- *for any* $\eta \in (0, 1]$, *with probability at least* $1 - \eta$,

$$\|f_0 - f_{EWA}\|_2^2 \le \inf_{\mu \in \mathcal{M}_+^1(\mathcal{T})} \left(1 + \frac{4\gamma}{1-2\gamma}\right) \int \|f_0 - \hat{f}_t\|_2^2 d\mu(t)$$
$$+ \left(1 + \frac{2\gamma}{1-2\gamma}\right) \int \text{pen}(t) + 2\sigma^2 \text{Tr}(A_t) d\mu(t)$$
$$+ \beta \left(1 + \frac{2\gamma}{1-2\gamma}\right) \left(2\text{KL}(\mu, \pi) + \ln \frac{1}{\eta}\right).$$

- *Furthermore*

$$\mathbb{E}\|f_0 - f_{EWA}\|_2^2 \le \inf_{\mu \in \mathcal{M}_+^1(\mathcal{T})} \left(1 + \frac{4\gamma}{1-2\gamma}\right) \int \mathbb{E}\|f_0 - \hat{f}_t\|_2^2 d\mu(t)$$
$$+ \left(1 + \frac{2\gamma}{1-2\gamma}\right) \int \text{pen}(t) + 2\sigma^2 \text{Tr}(A_t) d\mu(t) + 2\beta \left(1 + \frac{2\gamma}{1-2\gamma}\right) \text{KL}(\mu, \pi).$$

and

Theorem 4.2 *For any* $\delta \in [0, 1]$, *if* $\beta > 4\sigma^2$, *If for any* $t \in \mathcal{T}$,

$$\text{pen}(t) \ge \frac{4\sigma^2}{\beta - 4\sigma^2} \left(\sigma^2 \text{Tr}(A_t^2) + 2\left[\widetilde{\|f_0 - b\|_\infty^2} \text{Tr}(A_t^2) + \|b_t\|_2^2\right]\right),$$

then

- *for any* $\eta \in (0, 1]$, *with probability at least* $1 - \eta$,

$$\|f_0 - f_{EWA}\|_2^2 \le \inf_{\mu \in \mathcal{M}_+^1(\mathcal{T})} \int \|f_0 - \hat{f}_t\|_2^2 d\mu(t)$$
$$+ \int \text{pen}(t) + 2\sigma^2 \text{Tr}(A_t) + \frac{8\sigma^2}{\beta - 4\sigma^2} \left[\widetilde{\|f_0 - b\|_\infty^2} \text{Tr}(A_t^2) + \|b_t\|_2^2\right] d\mu(t)$$
$$+ \beta \left(2\text{KL}(\mu, \pi) + \ln \frac{1}{\eta}\right).$$

- *Furthermore*

$$\mathbb{E}\|f_0 - f_{EWA}\|_2^2 \le \inf_{\mu \in \mathcal{M}_+^1(\mathcal{T})} \left(1 + \frac{4\gamma}{1-2\gamma}\right) \int \mathbb{E}\|f_0 - \hat{f}_t\|_2^2 d\mu(t) + \int \text{pen}(t)$$
$$+ 2\sigma^2 \text{Tr}(A_t) + \frac{8\sigma^2}{\beta - 4\sigma^2} \left[\widetilde{\|f_0 - b\|_\infty^2} \text{Tr}(A_t^2) + \|b_t\|_2^2\right] d\mu(t) + 2\beta \text{KL}(\mu, \pi).$$

When $\mathscr{T}$ is discrete, one can replace the minimization over all the probability measures in $\mathscr{M}_+^1(\mathscr{T})$ by the minimization overall all Dirac measures δ_{f_t} with $t \in \mathscr{T}$. Propositions 3.2 and 3.3 are then straightforward corollaries. Note that the result in expectation is obtained with the same penalty, which is known not to be necessary, at least in the Gaussian case, as shown by Dalalyan and Salmon [20].

If we assume the penalty is given

$$\mathrm{pen}(t) = \kappa \mathrm{Tr}(A_t^2)\sigma^2,$$

one can rewrite the assumption in terms of κ. The weak oracle inequality holds for any temperature greater than $20\sigma^2$ as soon as $\kappa \geq \frac{4\sigma^2}{\beta - 4\sigma^2}$. While an exact oracle inequality holds for any vector f_0 and any temperature β greater than $4\sigma^2$ as soon as

$$\frac{\beta - 4\sigma^2}{4\sigma^2}\kappa - 1 \geq \frac{\widetilde{\|f_0 - b\|_\infty^2} + \|b_t\|^2/\mathrm{Tr}(A_t^2)}{\sigma^2}.$$

For fixed κ and β, this corresponds to a low peak signal to noise ratio $\frac{\widetilde{\|f_0 - b\|_\infty^2}}{\sigma^2}$ up to the $\|b_t\|^2$ term which vanishes when $b_t = 0$. Note that similar results hold for a penalization scheme but with much larger constants and some logarithmic factor in n.

Finally, the minimal temperature of $20\sigma^2$ can be replaced by some smaller value if one further restricts the smoothed projections used. As it appears in the proof, the temperature can be replaced by $8\sigma^2$ or even $6\sigma^2$ when the smoothed projections are, respectively, classical projections and projections on the same basis. The question of the minimality of such temperature is still open. Note that in this proof, there is no loss due to the sub-Gaussianity assumption, since the same upper bound on the exponential moment of the deviation as in the Gaussian case is found, providing the same penalty and bound on temperature.

The two results can be combined in a single one producing weak oracle inequalities for a wider range of temperatures than Theorem 4.1. Our proof is available in an extended version of this contribution in which, we prove that a continuum between those two cases exists: a weak oracle inequality, with smaller leading constant than the one of Theorem 4.1, holds as soon as there exists $\delta \in [0, 1)$ such that $\beta \geq 4\sigma^2(1 + 4\delta)$ and

$$\frac{\beta - 4\sigma^2}{4\sigma^2}\kappa - 1 \geq (1 - \delta)(1 + 2\gamma)^2 \frac{\widetilde{\|f_0 - b\|_\infty^2} + \|b_t\|^2/\mathrm{Tr}(A_t^2)}{\sigma^2},$$

where the signal to noise ratio guides the transition. The temperature required remains nevertheless always above $4\sigma^2$. The convex combination parameter δ measures the account for signal to noise ratio in the penalty.

Note that in practice, the temperature can often be chosen smaller. It is an open question whether the $4\sigma^2$ limit is an artifact of the proof or a real lower bound. In the Gaussian case, [32] have been able to show that this is mainly technical. Extending this result to our setting is still an open challenge.

References

1. Akaike, H. (1973). Information theory and an extension of the maximum likelihood principle. In *Second International Symposium on Information Theory (Tsahkadsor, 1971)* (pp. 267–281). Budapest: Akadémiai Kiadó.
2. Alquier, P., & Lounici, K. (2011). PAC-Bayesian bounds for sparse regression estimation with exponential weights. *Electronic Journal of Statistics, 5*, 127–145.
3. Amit, Y., & Geman, D. (1997). Shape quantization and recognition with randomized trees. *Neural Computation, 9*(7), 1545–1588.
4. Bellec, P. C. (2018). Optimal bounds for aggregation of affine estimators. *The Annals of Statistics, 46*(1), 30–59.
5. Belloni, A., Chernozhukov, V., & Wang, L. (2011). Square-root lasso: pivotal recovery of sparse signals via conic programming. *Biometrika, 98*(4), 791–806.
6. Biau, G. (2012). Analysis of a random forests model. *Journal of Machine Learning Research, 13*, 1063–1095.
7. Biau, G., & Devroye, L. (2010). On the layered nearest neighbour estimate, the bagged nearest neighbour estimate and the random forest method in regression and classification. *Journal of Multivariate Analysis, 101*(10), 2499–2518.
8. Biau, G., Devroye, L., & Lugosi, G. (2008). Consistency of random forests and other averaging classifiers. *Journal of Machine Learning Research, 9*, 2015–2033.
9. Breiman, L. (1996). Bagging predictors. *Machine Learning, 24*(2), 123–140.
10. Breiman, L. (2001). Random forests. *Machine Learning, 45*(1), 5–32.
11. Catoni, O. (2004). *Statistical learning theory and stochastic optimization: Vol. 1851. Lecture notes in mathematics*. Berlin: Springer. Lecture notes from the 31st Summer School on Probability Theory held in Saint-Flour, 8–25 July 2001.
12. Catoni, O. (2007). *Pac-Bayesian supervised classification: The thermodynamics of statistical learning: Vol. 56. Institute of Mathematical Statistics Lecture Notes—Monograph Series*. Beachwood, OH: Institute of Mathematical Statistics.
13. Cesa-Bianchi, N., Freund, Y., Haussler, D., Helmbold, D. P., Schapire, R. E., & Warmuth M. K. (1997). How to use expert advice. *Journal of the ACM, 44*(3), 427–485.
14. Cesa-Bianchi, N., & Lugosi, G. (1999). On prediction of individual sequences. *The Annals of Statistics, 27*(6), 1865–1895.
15. Chernousova, E., Golubev, Y., & Krymova, E. (2013). Ordered smoothers with exponential weighting. *Electronic Journal of Statistics, 7*, 2395–2419.
16. Dai, D., Rigollet, P., Xia, L., & Zhang, T. (2014). Aggregation of affine estimators. *Electronic Journal of Statistics, 8*, 302–327.
17. Dai, D., Rigollet, P., & Zhang, T. (2012). Deviation optimal learning using greedy Q-aggregation. *The Annals of Statistics, 40*(3), 1878–1905.
18. Dalalyan, A. S. (2012). SOCP based variance free Dantzig selector with application to robust estimation. *Comptes Rendus Mathematique Academie des Sciences, Paris, 350*(15–16), 785–788.
19. Dalalyan, A. S., Hebiri,M., Meziani, K., & Salmon, J. (2013). Learning heteroscedastic models by convex programming under group sparsity. *Proceedings of the 30th International Conference on Machine Learning, Proceedings of Machine Learning Research, 28*(3), 379–387.

20. Dalalyan, A. S., & Salmon, J. (2012). Sharp oracle inequalities for aggregation of affine estimators. *The Annals of Statistics, 40*(4), 2327–2355.
21. Dalalyan, A. S., & Tsybakov, A. B. (2007). Aggregation by exponential weighting and sharp oracle inequalities. In N. H. Bshouty & C. Gentile (Eds.), *Learning theory: Vol. 4539. Lecture notes in computer science* (pp. 97–111). Berlin: Springer.
22. Dalalyan, A. S., & Tsybakov, A. B. (2008). Aggregation by exponential weighting, sharp pac-bayesian bounds and sparsity. *Machine Learning, 72*(1–2), 39–61.
23. Dalalyan, A. S., & Tsybakov, A. B. (2012). Sparse regression learning by aggregation and Langevin Monte-Carlo. *Journal of Computer and System Sciences, 78*(5), 1423–1443.
24. Donoho, D. L., Johnstone, I. M., Kerkyacharian, G., & Picard D. (1995). Wavelet shrinkage: Asymptopia? *Journal of the Royal Statistical Society Series B, 57*(2), 301–369.
25. Freund, Y. (1995). Boosting a weak learning algorithm by majority. *Information and Computation, 121*(2), 256–285.
26. Gasser, T., & Müller, H. G. (1984). Estimating regression functions and their derivatives by the kernel method. *Scandinavian Journal of Statistics, 11*(3), 171–185.
27. Genuer, R. (2011). *Forêts aléatoires : aspects théoriques, sélection de variables et applications*. PhD thesis, Université Paris-Sud.
28. Gerchinovitz, S. (2011). *Prediction of Individual Sequences and Prediction in the Statistical Framework : Some Links Around Sparse Regression and Aggregation Techniques*. Thesis, Université Paris Sud.
29. Giraud, C. (2008). Mixing least-squares estimators when the variance is unknown. *Bernoulli, 14*(4), 1089–1107.
30. Giraud, C., Huet, S., & Verzelen, N. (2012). High-dimensional regression with unknown variance. *Statistical Science, 27*(4), 500–518.
31. Golubev, Y. (2012). Exponential weighting and oracle inequalities for projection estimates. *Problems of Information Transmission, 48*, 269–280.
32. Golubev, Y., & Ostobski, D. (2014). Concentration inequalities for the exponential weighting method. *Mathematical Methods of Statistics, 23*(1), 20–37.
33. Guedj, B., & Alquier, P. (2013). PAC-Bayesian estimation and prediction in sparse additive models. *Electronic Journal of Statistics, 7*, 264–291.
34. Hastie, T., Tibshirani, R., & Friedman, J. (2009). *The elements of statistical learning: Data mining, inference, and prediction. Springer series in statistics* (2nd ed.). New York: Springer.
35. Lecué, G. (2007). Optimal rates of aggregation in classification under low noise assumption. *Bernoulli, 13*(4), 1000–1022.
36. Leung, G., & Barron, A. R. (2006). Information theory and mixing least-squares regressions. *IEEE Transactions on Information Theory, 52*(8), 3396–3410.
37. Littlestone, N., & Warmuth, M. K. (1994). The weighted majority algorithm. *Information and Computation, 108*(2), 212–261.
38. Lounici, K. (2007). Generalized mirror averaging and D-convex aggregation. *Mathematical Methods of Statistics, 16*(3), 246–259.
39. Nadaraya, É. (1965). On non-parametric estimates of density functions and regression curves. *Theory of Probability & Its Applications, 10*(1), 186–190.
40. Nemirovski, A. (2000). Topics in non-parametric statistics. In *Lectures on probability theory and statistics (Saint-Flour, 1998): Vol. 1738. Lecture notes in mathematics* (pp. 85–277). Berlin: Springer.
41. Rigollet, P. (2006). *Inégalités d'oracle, agrégration et adaptation*. PhD thesis, Université Pierre et Marie Curie- Paris VI.
42. Rigollet, P., & Tsybakov, A. B. (2007). Linear and convex aggregation of density estimators. *Mathematical Methods of Statistics, 16*(3), 260–280.
43. Rigollet, P., & Tsybakov, A. B. (2012). Sparse estimation by exponential weighting. *Statistical Science, 27*(4), 558–575.
44. Schapire, R. E. (1990). The strength of weak learnability. *Machine Learning, 5*(2), 197–227.
45. Schwarz, G. (1978). Estimating the dimension of a model. *The Annals of Statistics, 6*(2), 461–464.

46. Sun, T., & Zhang, C.-H. (2012). Scaled sparse linear regression. *Biometrika, 99*(4), 879–898.
47. Tibshirani, R. (1994). Regression shrinkage and selection via the lasso. *Journal of the Royal Statistical Society, Series B, 58*, 267–288.
48. Tsybakov, A. B. (2003). Optimal rates of aggregation. In B. Schölkopf & M. K. Warmuth (Eds.), *Learning theory and kernel machines: Vol. 2777. Lecture notes in computer science* (pp. 303–313). Berlin/Heidelberg: Springer.
49. Tsybakov, A. B. (2008). Agrégation d'estimateurs et optimisation stochastique. *Journal de la Société Franēaise de Statistique & Review of Statistics and Its Application, 149*(1), 3–26.
50. Vovk, V. G. (1990). Aggregating strategies. In *Proceedings of the Third Annual Workshop on Computational Learning Theory, COLT '90* (pp. 371–386). San Francisco, CA: Morgan Kaufmann Publishers Inc.
51. Wahba, G. (1990). *Spline models for observational data: Vol. 59. CBMS-NSF regional conference series in applied mathematics*. Philadelphia, PA: Society for Industrial and Applied Mathematics (SIAM).
52. Watson, G. S. (1964). Smooth regression analysis. *Sankhya Series A, 26*, 359–372.
53. Yang, Y. (2000). Adaptive estimation in pattern recognition by combining different procedures. *Statistica Sinica, 10*(4), 1069–1089.
54. Yang, Y. (2000). Combining different procedures for adaptive regression. *Journal of Multivariate Analysis, 74*(1), 135–161.
55. Yang, Y. (2000). Mixing strategies for density estimation. *The Annals of Statistics, 28*(1), 75–87.
56. Yang, Y. (2001). Adaptive regression by mixing. *Journal of the American Statistical Association, 96*(454), 574–588.
57. Yang, Y. (2003). Regression with multiple candidate models: Selecting or mixing? *Statistica Sinica, 13*(3), 783–809.
58. Yang, Y. (2004). Aggregating regression procedures to improve performance. *Bernoulli, 10*(1), 25–47.
59. Yang, Y. (2004). Combining forecasting procedures: some theoretical results. *Econometric Theory, 20*(1), 176–222.
60. Zou, H., & Hastie T. (2005). Regularization and variable selection via the elastic net. *Journal of the Royal Statistical Society, Series B: Statistical Methodology, 67*(2), 301–320.

Light- and Heavy-Tailed Density Estimation by Gamma-Weibull Kernel

L. Markovich

Abstract In our previous papers we focus on the gamma kernel estimators of density and its derivatives on positive semi-axis by dependent data by univariate and multivariate samples. We introduce the gamma product kernel estimators for the multivariate joint probability density function (pdf) with the nonnegative support and its partial derivatives by the multivariate dependent data with a strong mixing. The asymptotic behavior of the estimates and the optimal bandwidths in the sense of minimal mean integrated squared error (MISE) are obtained. However, it is impossible to fit accurately the tail of the heavy-tailed density by pure gamma kernel. Therefore, we construct the new kernel estimator as a combination of the asymmetric gamma and Weibull kernels, i.e. Gamma-Weibull kernel. The gamma kernel is nonnegative and it changes the shape depending on the position on the semi-axis and possesses good boundary properties for a wide class of densities. Thus, we use it to estimate the pdf near the zero boundary. The Weibull kernel is based on the Weibull distribution which can be heavy-tailed and hence, we use it to estimate the tail of the unknown pdf. The theoretical asymptotic properties of the proposed density estimator like the bias and the variance are derived. We obtain the optimal bandwidth selection for the estimate as a minimum of the MISE. The optimal rate of convergence of the MISE for the density is found.

L. Markovich (✉)
Moscow Institute of Physics and Technology , Dolgoprudny, Moscow Region, Russia

Institute for Information Transmission Problems, Moscow, Russia

V. A. Trapeznikov Institute of Control Sciences, Moscow, Russia
e-mail: kimo1@mail.ru

P. Bertail et al. (eds.), *Nonparametric Statistics*, Springer Proceedings
in Mathematics & Statistics 250, https://doi.org/10.1007/978-3-319-96941-1_10

1 Introduction

It is well known that in modeling of a wide range of applications in engineering, signal processing, medical research, quality control, actuarial science, and climatology among others the nonnegatively supported pdfs are widely used. For example, the distributions from the gamma family play a key role in actuarial science. Most total insurance claim distributions are shaped like gamma pdfs [11]: nonnegatively supported, skewed to the right and unimodal. The gamma distributions are also used to model rainfalls [1]. Erlang and χ^2 pdfs are used in modeling insurance portfolios [13]. The pdfs from the exponential class play a prominent role in the optimal filtering in the signal processing and control of nonlinear processes [7]. On the basis of the high popularity of the nonnegatively supported distributions, it is fairly natural to study the estimation methods of such pdfs by finite data samples. One of the most common nonparametric pdf estimation methods are kernel estimators. However most of the known asymmetric kernel estimators are oriented on the univariate nonnegative independent identically distributed (iid) data and the light-tailed distributions. For example, for the iid random variables (r.v.s), the estimators with gamma kernels were proposed in [6]. The gamma kernel estimator was developed for univariate dependent data in [4]. In [3] the gamma kernel estimator of the multivariate pdf for the nonnegative iid r.v.s was introduced. Other asymmetrical kernel estimators for the case of the iid data like inverse Gaussian and reciprocal inverse Gaussian estimators were studied in [20]. The comparison of these asymmetric kernels with the gamma kernel is given in [5]. However, for a real life modeling the multivariate dependent probability models are used. For example, to attempt modeling portfolios of insurance losses the dependent multivariate probability models with gamma distributed univariate margins were used in [11]. Moreover, in the risk theory the pdfs can be heavy-tailed [9]. Modeling the heavy-tailed densities is important to compute and forecast the portfolio value-at-risk when the underlying risk factors have a heavy-tailed distribution [12, 19]. Hence, the need of the multivariate pdf estimation for the nonnegative dependent r.v.s and heavy-tailed pdfs arises. In [16] we introduce the gamma product kernel estimators for the multivariate joint pdf with the nonnegative support and its partial derivatives by the multivariate dependent data. The author develops both the asymptotic behavior of the estimates and the optimal bandwidths in the sense of the minimal mean integrated squared error (MISE). Note that the derivative estimation requires a specific bandwidth different from that for the pdf estimation. The mathematical technic applied for the derivative estimation is similar to the one applied for the pdf. However all formulas became much more complicated particulary because of the special Digamma functions arisen. Thus, one has to find the order by a bandwidth from complicated expressions containing logarithms and the special function. Other asymmetrical kernel estimators like inverse Gaussian and reciprocal inverse Gaussian estimators were studied in [20]. The comparison of these asymmetric kernels with the gamma kernel is given in [5]. The gamma kernel is nonnegative and flexible regarding the shape. This allows to provide a satisfactory fitting of the multi-modal pdfs and their derivatives. Gamma kernel estimators have no boundary bias if $f''(0) = 0$ holds, i.e. when the underlying pdf $f(x)$ has a shoulder at $x = 0$ [22]. This shoulder property is fulfilled, for example, for a wide exponential class of pdfs. Other bias correction methods can be found in [14] for the univariate iid data and in [10] for the multivariate iid data. However, less attention is dedicated to the tail fitting. The main focus of this chapter is on the nonparametric estimation of the heavy-tailed pdfs which are defined on the positive part of the real axis. It is obvious that the known classical estimators cannot be directly applied to the heavy-tailed pdfs. These are characterized by slower decay to zero of heavy tails than that of an exponential rate, the lack of some or all moments of the distribution, and sparse observations at the tail domain of the distribution [9]. The known approaches of the heavy-tailed density estimation are the kernel estimators with the variable bandwidth, the estimators based on the preliminary transform of the initial r.v. to a new one and "piecing-together approach" which provides a certain parametric model for the tail of the pdf and a nonparametric model to approximate the "body" of the pdf [18]. In this contribution, we introduce a new kernel constructed from the gamma and the Weibull kernels. The new Gamma-Weibull kernel has two smoothing parameters (bandwidths) and the third parameter that is the width of the boundary domain of the gamma part of the kernel. A stapling between the gamma and the Weibull parts is provided. The asymptotic behavior of the estimates and the optimal bandwidths in the sense of the minimal MISE are obtained. Normally, the Pareto distribution tail is accepted as a tail model for regularly varying heavy-tailed distributions. We selected the Weibull distribution tail since it does not belong to the latter class of the distributions and can be either heavy-tailed or light-tailed, depending on the shape parameter. This chapter is organized as follows. In Sect. 1.1 we provide a brief overview of the results known for the gamma kernel density and its derivative estimators. In Sect. 2 we introduce the Gamma-Weibull kernel estimator and in Sect. 3 its convergence rate and the optimal bandwidth parameters that minimize its MISE are derived.

1.1 Gamma Kernel

In this section we briefly recall the theory known for gamma kernel estimators. Let $\{X_i; i = 1, 2, \ldots\}$ be a strongly stationary sequence with an unknown probability density function $f(x)$, which is defined on $x \in [0, \infty)$. To estimate $f(x)$ by a known sequence of observations $\{X_i\}$ the non-symmetric gamma kernel estimator was defined in [6] by the formula

$$\widehat{f}_n(x) = \frac{1}{n} \sum_{i=1}^{n} K_{\rho_b(x), b}(X_i). \tag{1}$$

Here

$$K_{\rho_b(x),b}(t) = \frac{t^{\rho_b(x)-1}\exp(-t/b)}{b^{\rho_b(x)}\Gamma(\rho_b(x))} \quad (2)$$

is the kernel function, b is a smoothing parameter (bandwidth) such that $b \to 0$ as $n \to \infty$, $\Gamma(\cdot)$ is a standard gamma function and

$$\rho_b(x) = \begin{cases} x/b, & \text{if} \quad x \geq 2b, \\ (x/(2b))^2 + 1, & \text{if} \quad x \in [0, 2b). \end{cases} \quad (3)$$

The use of gamma kernels is due to the fact that they are nonnegative, change their shape depending on the position on the semi-axis, and possess better boundary bias than symmetrical kernels. The boundary bias becomes larger for multivariate densities. Hence, to overcome this problem the gamma kernels were applied in [3]. Earlier the gamma kernels were only used for the density estimation of identically distributed sequences in [3, 6] and for stationary sequences in [4]. Along with the pdf estimation it is often necessary to estimate the derivative of the pdf. The estimation of the univariate pdf derivative by the gamma kernel estimator was proposed in [17] for iid data and in [15] for a strong mixing dependent data. Our procedure achieves the optimal MISE of order $n^{-4/7}$ when the optimal bandwidth is of order $n^{-2/7}$. In [21] an optimal MISE of the kernel estimate of the first derivative of order $n^{-4/7}$ corresponding to the optimal bandwidth of order $n^{-1/7}$ for symmetrical kernels was indicated. The unknown smoothing parameter b was obtained as the minimum of the mean integrated squared error ($MISE$) which, as known, is equal to

$$MISE(\hat{f}_n(x)) = \mathsf{E}\int_0^{\infty}(f(x) - \hat{f}_n(x))^2 dx.$$

Remark 1 The latter integral can be split into two integrals $\int_0^{2b}$ and $\int_{2b}^{\infty}$. In the case when $x \geq 2b$ the integral $\int_0^{2b}$ tends to zero when $b \to 0$. Hence, we omit the consideration of this integral in contrast to [22]. The first integral has the same order by b as the second one, thus it cannot affect the selection of the optimal bandwidth.

In [21, p. 49], it was indicated an optimal $MISE$ of the first derivative kernel estimate $n^{-\frac{4}{7}}$ with the bandwidth of order $n^{-\frac{1}{7}}$ for symmetrical kernels. Nevertheless, our procedure achieves the same order $n^{-\frac{4}{7}}$ with a bandwidth of order $n^{-\frac{2}{7}}$. Moreover, our advantage concerns the reduction of the bias of the density derivative at the zero boundary by means of asymmetric kernels. Gamma kernels allow us to avoid boundary transformations which is especially important for multivariate cases.

2 Gamma-Weibull Kernel

The term heavy-tailed is used to the class of pdf whose tails are not exponentially bounded, i.e. their tails are heavier than the exponential pdf tail [2, 9]. The examples of such pdfs are Lognormal, Pareto, Burr, Cauchy, Weibull with shape parameter less than 1 among others. Let $\{X_i; i = 1, 2, \ldots\}$ be a strongly stationary sequence with an unknown pdf $f(x)$ which is defined on the nonnegative semi axes $x \in [0, \infty)$. Our objective is to estimate the unknown heavy-tailed pdf by a known sequence of observations $\{X_i\}$. Since the pdf is assumed to be asymmetric and heavy-tailed we cannot use the standard symmetrical kernels. Let us construct a special kernel function which would be both flexible on the domain near the zero boundary and it could estimate the heavy tail of the distribution. For the domain $x \in [0, a]$, $a > 0$ we use the gamma kernel estimator that was defined in [6] by the formula

$$\widehat{f_{G_n}}(x) = \frac{1}{n}\sum_{i=1}^{n} K_{\rho(x,h),\theta}(X_i) = \frac{1}{n}\sum_{i=1}^{n} \frac{X_i^{\rho(x,h)-1} e^{-X_i/\theta}}{\theta^{\rho(x,h)}\Gamma(\rho(x,h))}, \qquad \rho, \theta > 0.$$

Here, $\Gamma(\rho)$ is the gamma function evaluated at ρ and h is the bandwidth of the kernel. The shape parameters ρ, θ will be selected further. For the domain $x > a$ the Weibull kernel estimator is constructed by

$$\widehat{f_{W_n}}(x) = \frac{1}{n}\sum_{i=1}^{n} K_{k(x,b)}(X_i) = \frac{1}{n}\sum_{i=1}^{n} \frac{k(x,b)}{\lambda}\left(\frac{X_i}{\lambda}\right)^{k(x,b)-1} \exp\left(-\left(\frac{X_i}{\lambda}\right)^{k(x,b)}\right),$$

where the shape parameters are $\lambda > 0$, $0 < k < 1$ and b is the bandwidth of the kernel. Hence, the pdf estimator is the following

$$\widehat{f_{GW_n}}(x) = \begin{cases} \widehat{f_{G_n}}(x) & \text{if} \quad x \in [0, a], \\ \widehat{f_{W_n}}(x) & \text{if} \quad x > a. \end{cases} \tag{4}$$

The latter kernel estimator has two bandwidth parameters h and b and one special parameter a. The parameters $\rho(x, h)$, $k(x, b)$, λ and θ can be found from the matching conditions

$$f_G(X, \rho(x, h), \theta)\Big|_{x=a} - f_W(a, k(x, b), \lambda)\Big|_{x=a} = 0, \tag{5}$$

$$f_G'(X, \rho(x, h), \theta)\Big|_{x=a} - f_W'(a, k(x, b), \lambda)\Big|_{x=a} = 0. \tag{6}$$

From the condition (5) we can deduce that the shape parameters of the kernels are

$$\rho(a, h) = k(a, b), \quad \theta = \lambda.$$

From the condition (6) we can deduce that

$$\frac{tial\rho(x,h)}{tialx}\Big|_{x=a} = \frac{tialk(x,b)}{tialx}\Big|_{x=a}. \tag{7}$$

Hence, we can select any variety of $\rho(x,h)$ and $k(x,b)$ that satisfy the latter conditions to get some kernel estimators. Let us select, for example, the following parameters

$$\rho(x,h) = \frac{x + c_1 h}{a}, \quad k(x,b) = \frac{x + c_2 b}{a}. \tag{8}$$

Hence, the bandwidth parameters satisfy the condition $h = bc_2/c_1$. Since $k(x,b) < 1$ for the heavy-tailed Weibull pdf the parameters c_1, c_2 are some negative constants that we will select further. As the measure of error of the proposed estimator (4) we consider the MISE and the unknown smoothing parameters h and b are obtained as the minima of (4).

3 Convergence Rate of the Density Estimator

In this section we obtain the asymptotic properties of the estimator (4). To this end we derive the bias and the variance of the estimates in the following lemmas.

Lemma 1 *If $b \to 0$ as $n \to \infty$, then the bias of the pdf estimate* (4) *is equal to*

$$Bias(\widehat{f}(x)) = \begin{cases} C_1(x,a) + hC_2(x,a,c_1) + o(h) \text{ if } & x \in [0,a], \\ B_1(x,a) + bB_2(x,a,c_2) + o(b) \text{ if } & x > a, \end{cases}, \tag{9}$$

where we denote

$$C_1(x,a) = \frac{xa}{2} f''(x), \quad C_2(x,a,c_1) \equiv c_1 c_2(x,a) = c_1\left(f'(x) + f''(x)\frac{a}{2} + f'''(x)\frac{xa}{2}\right), \tag{10}$$

$$B_1(x,a) = f\left(a\Gamma(t)\right) - f(x) + f''\left(a\Gamma(t)\right)\frac{a^2}{2}\left(\Gamma(t) - \Gamma(r)\right)^2, \tag{11}$$

$$\begin{aligned} B_2(x,a,c_2) \equiv c_2 b_2(x,a) = \frac{a^2 c_2}{x^2}\Big(& - f'\left(a\Gamma(t)\right)\Gamma(t)\Psi(t) \\ & + f''\left(a\Gamma(t)\right)a\Big(\left(\Gamma(t) - \Gamma(r)\right)\left(\Gamma(t)\Psi(t) - 2\Gamma(r)\Psi(r)\right)\Big) \\ & - f'''\left(a\Gamma(t)\right)\frac{a^2}{2}\left(\Gamma(t) - \Gamma(r)\right)^2\Gamma(t)\Psi(t)\Big), \end{aligned} \tag{12}$$

and

$$t = 1 + \frac{a}{x}, \quad r = 1 + \frac{2a}{x}. \tag{13}$$

Lemma 2 *If $b \to 0$ as $n \to \infty$, then the variance of the pdf estimate* (4) *is equal to*

$$Var(\widehat{f}(x)) = \frac{1}{n}\Big(A_1(x,a) - (C_1(x,a) + f(x))^2 + h(A_2(x,a,c_2) \tag{14}$$

$$- 2C_2(x,a,c_1)(C_1(x,a) + f(x)))\Big) + o(h) \quad \textit{if} \quad x \in [0,a],$$

$$Var(\widehat{f}(x)) = \frac{1}{n}\Big(D_1(x,a) - (B_1(x,a) - f(x))^2 + b(D_2(x,a,c_2)$$

$$- 2B_2(x,a,c_2)(B_1(x,a) - f(x)))\Big) + o(b) \quad \textit{if} \quad x > a,$$

where we denote

$$A_1(x,a) = -f\left(x - \frac{a}{2}\right)\frac{\sqrt{x}}{\sqrt{a}(a-2x)}, \tag{15}$$

$$A_2(x,a,c_1) \equiv c_1 a_2(x,a) = -c_1\Bigg(f\left(x - \frac{a}{2}\right)\frac{a+2x}{2\sqrt{ax}(a-2x)^2}$$

$$+ \frac{\sqrt{x}}{\sqrt{a}(a-2x)}\left(f'\left(x - \frac{a}{2}\right) + \frac{a}{4}\left(x - \frac{a}{2}\right)f''\left(x - \frac{a}{2}\right)\right)\Bigg)$$

and

$$D_1(x,a) = \frac{x2^{\frac{3x}{a}-1}}{a^2}\left(f(2a)\left(\frac{x(x-3a)}{2a^2} + 2\right) + f'(2a)(x-a) + f''(2a)2a^2\right), \tag{16}$$

$$D_2(x,a,c_2) \equiv d_{21}(x,a) + c_2 d_{22}(x,a) = \frac{x2^{\frac{3x}{a}-1}}{a^2}\Bigg(f(2a)\Bigg(\frac{(x-2a)(dx^2 + a(c_2 - dx))}{2a^3}$$

$$+ \frac{dx^2 + a(c_2 - dx)}{a} - \frac{x(x-a)(x-2a)(6\gamma - 10 - \ln(4))}{2a^3}\Bigg)$$

$$+ f'(2a)\left(\frac{dx^2 + a(c_2 - dx)}{a} - 2xd(x - 2a + 1) - \frac{x(x-a)(6\gamma - 10 - \ln(4))}{a}\right)$$

$$- f''(2a)\left(2x(d(x-a) + 2a^2 + a(6\gamma - 10 - \ln(4)))\right) + \frac{c_2 2^{\frac{3x}{a}-1}}{a^3}(-x\ln(a) + a + 2x\ln(2))$$

$$\times \left(f(2a)\left(\frac{x(x-3a)}{2a^2} + 2\right) + f'(2a)(x-a) + f''(2a)2a^2\right)\Bigg).$$

The proofs of the latter lemmas are given in Appendix.

3.1 The Optimal Bandwidth Parameters for the Density Estimator

To find the mean integrated squared error (MISE) we use the results of two last sections. Hence, for the domain $x \in (0, a]$ the MSE is equal to

$$\begin{aligned} MSE(\hat{f}(x))_G &= C_1^2(x,a) + h^2 C_2^2(x,a,c_1) + 2hC_1(x,a)C_2(x,a,c_1) \\ &+ \frac{1}{n}\Big(A_1(x,a) - (C_1(x,a) + f(x))^2 + h(A_2(x,a,c_1) \\ &- 2C_2(x,a,c_1)(C_1(x,a) + f(x)))\Big) + o(h). \end{aligned} \tag{17}$$

Thus, from the minima of the latter equation we can obtain the optimal bandwidth parameter for the domain $x \in (0, a]$

$$h_{opt}(x,a,n) = -\frac{C_1(x,a)}{C_2(x,a,c_1)} - \frac{1}{C_2(x,a,c_1)n}\left(\frac{A_2(x,a,c_1)}{2C_2(x,a,c_1)} - C_1(x,a) - f(x)\right). \tag{18}$$

Substituting the latter bandwidth into (17) we get

$$\begin{aligned} MSE(\hat{f}(x))_{Gopt} &= -\frac{1}{n^2}\left(\frac{A_2(x,a,c_1)}{2C_2(x,a,c_1)} - (C_1(x,a) + f(x))\right)^2 \\ &+ \frac{1}{n}\left(A_1(x,a) - \frac{A_2(x,a,c_1)C_1(x,a)}{C_2(x,a,c_1)} + (C_1^2(x,a) - f^2(x))\right). \end{aligned}$$

For the domain $x > a$ the MSE is the following:

$$\begin{aligned} MSE(\hat{f}(x))_W &= B_1^2(x,a) + b^2 B_2^2(x,a,c_2) + 2bB_1(x,a)B_2(x,a,c_2) \\ &+ \frac{1}{n}\Big(D_1(x,a) - (B_1(x,a) - f(x))^2 + b(D_2(x,a,c_2) \\ &- 2B_2(x,a,c_2)(B_1(x,a) - f(x)))\Big) + o(b) \end{aligned} \tag{19}$$

and the optimal bandwidth is

$$b_{opt}(x,a,n) = \frac{-B_1(x,a)}{B_2(x,a,c_2)} - \frac{1}{B_2(x,a,c_2)n}\left(\frac{D_2(x,a,c_2)}{2B_2(x,a,c_2)} - B_1(x,a) + f(x)\right). \tag{20}$$

Substituting the latter bandwidth into (19) we get the following rate:

$$MSE(\hat{f}(x))_{Wopt} = -\frac{1}{n^2}\left(\frac{D_2(x,a,c_2)}{2B_2(x,a,c_2)} - (B_1(x,a)+f(x))\right)^2$$
$$+\frac{1}{n}\left(D_1(x,a) - \frac{D_2(x,a,c_2)B_1(x,a)}{B_2(x,a,c_1)} + (B_1^2(x,a) - f^2(x))\right).$$

To satisfy the condition $h_{opt}(a,a,n) = b_{opt}(a,a,n)c_2/c_1$, we have to find the parameters a, c_1, c_2. Let us select the bandwidth $b_{opt}(a,a,n)$ which is optimal for the tail part of the estimate. Hence, the second bandwidth is $h_{b_{opt}}(a,a,n) = b_{opt}(a,a,n)c_2/c_1$. We can find such constants a, c_1, c_2 that provide

$$\min_{a,c_1,c_2}\{h_{opt}(a,a,n) - h_{b_{opt}}(a,a,n)\}.$$

Hence, substituting the values of the bandwidths (18) and (20) we get the following condition:

$$c_2 = \frac{1}{d_{21}(a,a)}\left(\frac{B_1(a,a)b_2(a,a)}{C_1(a,a)}\left(\frac{a_2(a,a)}{c_2(a,a)} - 2f(a)\right) - 2f(a)b_2(a,a) - d_{22}(a,a)\right)$$

where

$$d_{21}(a,a) = 2^{\frac{3a}{2}}\left(\frac{f'(2a)}{\ln(10)}(a-1)(\gamma\ln(10) - \ln(5))\right.$$
$$\left.+ f''(2a)\left(\gamma - 1 + 2a^2\left(\frac{\ln(4)}{\ln(10)} - 6\gamma + 10\right) + \frac{\ln(2)}{\ln(10)}\right)\right),$$

$$d_{22}(a,a) = 2^{\frac{3a}{2}-1}\left(\frac{f(2a)}{a^2}\left(\frac{2a-1}{2} + 3\left(1 - \frac{1}{\ln(10)(\ln(a) - 2\ln(2))}\right)\right)\right.$$
$$\left.+ \frac{f'(2a)}{a} + f''(2a)\left(1 - \frac{1}{\ln(10)}(\ln(a) - 2\ln(2))\right)\right)$$

and $a_2(a,a)$, $B_1(a,a)$, $b_2(a,a)$, $C_1(a,a)$, and $c_2(a,a)$ are defined in Lemmas 1 and 2. Note that we can select any negative c_1, e.g. $c_1 = -1$. Hence, the selection of a provides the choice of c_2 which gives us the optimal bandwidths (18) and (20) for both domains $x \in (0,a]$ and $x > a$, respectively. In practice, the calculation of c_2 requires a pilot estimation of the pdf $f(x)$. One can use the rule of thumb method with the gamma reference function (see [8]).

4 Conclusion

The new kernel estimator of the heavy-tailed probability density function on the positive semi-axis by iid data is proposed. Our estimator is based on two kernels: the gamma kernel for the boundary domain and the Weibull kernel for the tail domain. Since the Weibull density can be either heavy-tailed or light-tailed, depending on the shape parameter, the proposed kernel estimator can be applied to both types of densities. The Gamma-Weibull kernel is smooth due to the cross-linking condition on the boundary and the introduced cross-linking parameter. We provide the asymptotic properties of the estimator by optimal rates of convergence of its mean integrated squared error. We develop explicit formulas for the optimal smoothing parameters (bandwidths). Further development may concern the investigation of alternative bandwidth selection methods. The results can also be extended to multivariate samples with mixing conditions.

Acknowledgements The study was supported by a Foundation for Basic Research, grant 16-08-01285A.

Appendix

Proof of Lemma 1

To find the bias of the estimate $\widehat{f}(x)$ let us write the expectation of the kernel estimator (4)

$$E(\widehat{f}(x)) = \begin{cases} E_G(\widehat{f}(x)) = \int_0^\infty K_{\rho(x,h),\theta}(y)f(y)dy = E(f(\xi_x)), \text{ if } & x \in [0,a), \\ E_W(\widehat{f}(x)) = \int_0^\infty K_{k(x,b),\lambda}(y)f(y)dy = E(f(\eta_x)), \text{ if } & x \geq a. \end{cases} \tag{21}$$

where ξ_x is the gamma distributed $(\rho(x,h),\theta)$ r.v. with the expectation $\mu_x = \rho(x,h)\theta$ and the variance $Var(\xi_x) = \rho(x,h)\theta^2$ and η_x is the weibull distributed $(k(x,b),\lambda)$ r.v. with the expectation $\widetilde{\mu}_x = \lambda\Gamma(1+\frac{1}{k(x,b)})$ and the variance $\widetilde{Var}(\eta_x) = \lambda^2\left(\Gamma(1+\frac{2}{k(x,b)}) - \Gamma(1+\frac{1}{k(x,b)})\right)^2$. Let us use the parameters (8) and $\theta = \lambda = a$. Hence, using the Taylor series in the point μ_x the expectation for the domain $x \in [0,a]$ can be written as

$$\begin{aligned} E(f(\xi_x)) &= f(\mu_x) + \frac{1}{2}f''(\mu_x)Var(\xi_x) + o(h) \\ &= f(x+c_1h) + \frac{a(x+c_1h)}{2}f''(x+c_1h) + o(h) \\ &= f(x) + f'(x)c_1h + \frac{a(x+c_1h)}{2}\left(f''(x) + f'''(x)c_1h\right) + o(h). \end{aligned} \tag{22}$$

Thus, it is straightforward to verify that the bias of the estimate in the domain $x \in [0, a]$ is

$$Bias_G(\widehat{f}(x)) = C_1(x, a) + C_2(x, a)h + o(h),$$

where we used the notations (10). To find the bias for the domain $x > a$ we need to Taylor expand $E(f(\eta_x))$ in the point $\widetilde{\mu}_x$. However $\widetilde{\mu}_x$ and $\widetilde{Var}(\eta_x)$ contain the gamma function depending on the bandwidth parameter. To find their order on b we need to expand them knowing that $b \to 0$ and $nb \to \infty$ as the $n \to \infty$. Hence, we can write

$$\widetilde{\mu}_x = a\Gamma(t) - b\frac{a^2 c_2}{x^2}\Gamma(t)\Psi(t) + o(b),$$

$$\widetilde{Var}(\eta_x) = a^2(\Gamma(t) - \Gamma(r))^2 + b\frac{2a^3 c_2}{x^2}(\Gamma(t) - \Gamma(r))(\Gamma(t)\Psi(t) - 2\Gamma(r)\Psi(r)) + o(b),$$

where we used the notation (13) and $\Psi(\cdot)$ is a digamma function. Thus, the expectation (21) can be written as

$$\begin{aligned} E(f(\eta_x)) &= f(\widetilde{\mu}_x) + \frac{1}{2}f''(\widetilde{\mu}_x)\widetilde{Var}(\eta_x) + o(h) \qquad (23) \\ &= f(a\Gamma(t)) - f'(a\Gamma(t))\frac{a^2 c_2 \Gamma(t)\Psi(t)}{x^2}b \\ &+ \frac{1}{2}\left(a^2(\Gamma(t) - \Gamma(r))^2 + \frac{2a^3 c_2 b}{x^2}(\Gamma(t) - \Gamma(r))(\Gamma(t)\Psi(t) - 2\Gamma(r)\Psi(r))\right) \\ &\times \left(f''(a\Gamma(t)) - f'''(a\Gamma(t))\frac{a^2 c_2 b}{x^2}\Gamma(t)\Psi(t)\right) + o(b). \end{aligned}$$

Therefore, we can write that the bias of the pdf estimate in the domain $x > a$ is

$$Bias_W(\widehat{f}(x)) = B_1(x, a) + bB_2(x, a) + o(b),$$

where we used the notations (11) and (12).

Proof of Lemma 2

By definition the variance is

$$Var(\hat{f}(x)) = \frac{1}{n}Var(K(x)) = \frac{1}{n}\left(E(K^2(x)) - E^2(K(x))\right). \qquad (24)$$

The second term of the right-hand side of (24) is the square of (22) and (23) for the domains $x \in [0, a]$ and $x > a$, respectively. The first term of the right-hand side of (24) for the domain $x \in [0, a]$ can be represented by

$$E(K_G^2(x)) = \int_0^\infty K_G^2(y)f(y)dy = \int_0^\infty \frac{y^{2\left(\frac{x+c_1h}{a}-1\right)}e^{-2y/a}}{a^{2\frac{x+c_1h}{a}}\Gamma^2\left(\frac{x+c_1h}{a}\right)}f(y)dy = B(x,h,a)E(f(\zeta_x)), \tag{25}$$

where ζ_x is the gamma distributed r.v. (with parameters $\left(\frac{2(x+c_1h)}{a}-1, \frac{a}{2}\right)$). The expectation is $\mu_\zeta = x + c_1h - \frac{a}{2}$ and the variance is $Var(\zeta_x) = (x+c_1h)\frac{a}{2} - \frac{a^2}{4}$. Here we used the following notation:

$$B(x,h,a) = \frac{\Gamma\left(\frac{2(x+c_1h)}{a}-1\right)}{a\Gamma^2\left(\frac{x+c_1h}{a}\right)2^{\frac{2(x+c_1h)}{a}-1}}. \tag{26}$$

Using the Stirling's formula for the gamma function, $x \in (0, a]$ and $h \to 0$ as $n \to \infty$ we can expand (26) as

$$B(x,h,a) = -\frac{\sqrt{x}}{\sqrt{a}(a-2x)} - hc_1\frac{a+2x}{2\sqrt{ax}(a-2x)^2} + o(h).$$

The expectation in (25) can be Taylor expanded similarly to (23) as

$$\begin{aligned}
E(f(\zeta_x)) &= f\left(x+c_1h-\frac{a}{2}\right) + \left((x+c_1h)\frac{a}{4}-\frac{a^2}{8}\right)f''\left(x+c_1h-\frac{a}{2}\right) + o(h) \\
&= f\left(x-\frac{a}{2}\right) + c_1h\left(f'\left(x-\frac{a}{2}\right) + \frac{a}{4}\left(x-\frac{a}{2}\right)f''\left(x-\frac{a}{2}\right)\right) + o(h).
\end{aligned}$$

Hence, the expectation (25) is

$$E(K^2(x)) = A_1(x,a) + hA_2(x,a) + o(h),$$

where we used the notations (15). The variance (24) for the domain $x \in (0, a]$ is

$$Var_G(\hat{f}(x)) = \frac{1}{n}(A_1(x,a) - C_1(x,a) + hc_1(A_2(x,a) - 2C_1(x,a)C_2(x,a))) + o(h).$$

Now we turn our attention to the domain $x > a$. Similarly to the previous part of the proof it can be written that

$$\begin{aligned}
E(K_W^2(x)) &= \int_0^\infty K_W^2(y)f(y)dy = \int_0^\infty \frac{k(x,b)^2}{a^2}\left(\frac{y}{a}\right)^{2(k(x,b)-1)}\exp\left(-2\left(\frac{y}{a}\right)^{k(x,b)}\right)f(y)dy \\
&= \frac{4^{k(x,b)}k(x,b)}{a^{k(x,b)}}E(f(\varsigma_x)\varsigma_x^{k(x,b)-1}),
\end{aligned} \tag{27}$$

where ς_x is the Weibull distributed r.v. with the parameters $\left(k(x,b), 2^{k(x,b)a}\right)$ and the expectation is

$$m_x = 2(a - bxd) + o(b^2), \quad d = \gamma - 1 + \ln(2)$$

and the variance is

$$Var_{\varsigma_x} = 4a^2 - 4bax(6\gamma - 10 + \ln(4)) + o(b^2),$$

where γ is the Euler-Mascherson constant. Hence, the expectation (27) can be written as

$$E(f(\varsigma_x)\varsigma_x^{k(x,b)-1}) = f(m_x)m_x^{k(x,b)-1} + \frac{Var_{\varsigma_x}}{2}\Big(f''(m_x)m_x^{k(x,b)-1}$$

$$+ 2(k(x,b)-1)f'(m_x)m_x^{k(x,b)-2} + (k(x,b)-1)(k(x,b)-2)f(m_x)m_x^{k(x,b)-3}\Big) + o(b)$$

$$= m_x^{k(x,b)-1}\Big(f(m_x) + \frac{Var_{\varsigma_x}}{2}\Big(f''(m_x) + (k(x,b)-1)m_x^{-1}$$

$$\cdot \Big(f'(m_x) + (k(x,b)-2)f(m_x)m_x^{-1}\Big)\Big)\Big).$$

Using the Taylor series, we can write that

$$m_x^{k(x,b)-1} = (2a)^{\frac{x}{a}-1}\Big(1 + \frac{b}{a}\Big(c_2\ln(2a) + xd(1-x)\Big)\Big) + o(b),$$

$$(k(x,b)-1)m_x^{-1} = \frac{x-a}{2a^2} + b\frac{a(c_2 - dx) + dx^2}{2a^3} + o(b)$$

$$(k(x,b)-2)m_x^{-1} = \frac{x-2a}{2a^2} + b\frac{a(c_2 - 2dx) + dx^2}{2a^3} + o(b)$$

$$\frac{4^{k(x,b)}k(x,b)}{a^{k(x,b)}} = x4^{\frac{x}{a}}a^{-\frac{x}{a}-1} + c_2 b 4^{\frac{x}{a}}a^{-\frac{x}{a}-2}(-x\ln(a) + a + x\ln(4)) + o(b).$$

Finally, the variance can be written as follows:

$$Var_W(\hat{f}(x)) = \frac{1}{n}\Big(D_1(x,a) + bD_2(x,a,c_2) - (B_1(x,a) + bB_2(x,a) + f(x))^2\Big),$$

where we used the notations (16).

References

1. Aksoy, H. (2000). Use of gamma distribution in hydrological analysis. *Turkish Journal of Engineering and Environmental Sciences, 24*, 419–428.
2. Asmussen, S. R. (2003). Steady-state properties of GI/G/1. *Applied Probability and Queues. Stochastic Modelling and Applied Probability, 51*, 266–301.
3. Bouezmarnia, T., & Rombouts, J. V. K. (2007). Nonparametric density estimation for multivariate bounded data. *Journal of Statistical Planning and Inference, 140*(1), 139–152.
4. Bouezmarnia, T., & Rombouts, J. V. K. (2010). Nonparametric density estimation for positive times series. *Computational Statistics and Data Analysis, 54*(2), 245–261.
5. Bouezmarnia, T., & Scaillet, O. (2003). Consistency of asymmetric kernel density estimators and smoothed histograms with application to income data. *Econometric Theory, 21*, 390–412.
6. Chen, S. X. (2000). Probability density function estimation using gamma kernels. *Annals of the Institute of Statistical Mathematics, 54*, 471–480.
7. Dobrovidov, A. V., Koshkin, G. M., & Vasiliev, V. A. (2012). *Nonparametric state space models*. Heber, UT: Kendrick Press.
8. Dobrovidov, A. V., Markovich, L.A. (2013). Data-driven bandwidth choice for gamma kernel estimates of density derivatives on the positive semi-axis. In *Proceedings of IFAC International Workshop on Adaptation and Learning in Control and Signal Processing, July 3–5, 2013, Caen, France* (pp. 500–505). https://doi.org/10.3182/20130703-3-FR-4038.00086, arXiv:1401.6801.
9. Embrechts, P., Klüppelberg, C., & Mikosch, T. (1997). *Modelling extremal events for insurance and finance*. Berlin: Springer.
10. Funke, B., & Kawka, R. (2015). Nonparametric density estimation for multivariate bounded data using two non-negative multiplicative bias correction methods. *Computational Statistics and Data Analysis, 92*,148–162.
11. Furman, E. (2008). On a multivariate gamma distribution. *Statistics & Probability Letters, 78*, 2353–2360.
12. Glasserman, P., Heidelberger, P., & Shahabuddin P. (2002). Portfolio value-at-risk with heavy-tailed risk factors. *Mathematical Finance, 12*, 239–269.
13. Hürlimann, W. (2001). Analytical evaluation of economic risk capital for portfolios of gamma risks. *ASTIN Bulletin, 31*, 107–122.
14. Igarashi, G., & Kakizawa, Y. (2015). Bias corrections for some asymmetric kernel estimators. *Journal of Statistical Planning and Inference, 159*, 37–63.
15. Markovich, L. A. (2016). Nonparametric gamma kernel estimators of density derivatives on positive semi-axis by dependent data. *RevStat Statistical Journal, 14*(3), 327–348.
16. Markovich, L. A. (2018). Nonparametric estimation of multivariate density and its derivative by dependent data using gamma kernels. arXiv:1410.2507v3.
17. Markovich, L. A., & Dobrovidov, A. V. (2013). Nonparametric gamma kernel estimators of density derivatives on positive semi-axis. In *Proceedings of IFAC MIM 2013, Petersburg, Russia* (pp. 944–949).
18. Markovich, N. M. (2007). *Nonparametric analysis of univariate heavy-tailed data*. Hoboken: Wiley.
19. Rozga, A., Arneric, J. (2009). Dependence between volatility persistence, kurtosis and degrees of freedom. *Revista Investigacion Operacional, 30*(1), 32–39.
20. Scaillet, O. (2004). Density estimation using inverse and reciprocal inverse gaussian kernels. *Journal of Nonparametric Statistics, 16*, 217–226.
21. Wand, M. P., & Jones, M. C. (1995). *Kernel smoothing*. London: Chapman and Hall.
22. Zhang, S. (2010). A note on the performance of the gamma kernel estimators at the boundary. *Statistics & Probability Letters, 80*, 548–557.

Adaptive Estimation of Heavy Tail Distributions with Application to Hall Model

D. N. Politis, V. A. Vasiliev, and S. E. Vorobeychikov

Abstract The problem of tail index estimation of Hall distribution is considered. We propose the estimators of tail index using the truncated estimation method developed for ratio type functionals. It is shown that the truncated estimator constructed on the sample of fixed size has a guaranteed accuracy in the sense of the L_{2m}-norm, $m \geq 1$. The asymptotic properties of estimators are although investigated. These properties make it possible to find the rates of decreasing of the χ^2 divergence in the almost surely sense between distribution and its adaptive estimator. Simulations confirm theoretical results.

1 Introduction

The models with heavy tail distributions are of interest in many applications connected with financial mathematics, insurance theory [1, 4, 15], telecommunication [16], and physics [2]. Usually it is assumed that the distribution function contains as an unknown multiplier a slowly varying function. The problem of tail index estimation was studied by Hill [9] who proposed the estimators based on the order statistics. The estimator is optimal in mean square sense on the class of distribution functions with heavy tails in presence of unknown slowly varying function. It should be noted that Hill's estimators are unstable and can diverge essentially from the estimated parameter for large sample sizes [4, 17].

Later other approaches to estimation problem were proposed (see, e.g., [6, 10] and the references therein). In [18] a least squares estimator for tail index was proposed which is based on the estimation of parameters in linear regression. The geometric-type estimators of the tail index are proposed and investigated in [2].

D. N. Politis
Department of Mathematics, University of California, San Diego, La Jolla, CA, USA
e-mail: dpolitis@ucsd.edu

V. A. Vasiliev (✉) · S. E. Vorobeychikov
Department of Applied Mathematics and Cybernetics, Tomsk State University, Tomsk, Russia
e-mail: vas@mail.tsu.ru; sev@mail.tsu.ru

P. Bertail et al. (eds.), *Nonparametric Statistics*, Springer Proceedings
in Mathematics & Statistics 250, https://doi.org/10.1007/978-3-319-96941-1_11

Some estimators have the form of ratio statistics, see, e.g., Embrechts et al. [4]. For example, formula (1.7) of Markovich [10] describes a well-known class of ratio estimators which are generalization of Hill's estimator in the sense that an arbitrary threshold level instead of an order statistic is used—see, e.g., Novak [11–14], Resnick and Stărică [17], or Goldie and Smith [5].

In this work, the truncated estimation method of ratio type functionals, proposed by Vasiliev [19], is used to obtain estimators with guaranteed accuracy in the sense of the L_{2m}-norm, $m \geq 1$. The estimators are constructed on the basis of empirical functionals without usage of non-parametric approach in an effort to obtain (or get close to) the parametric optimal rate of convergence. These estimators can be used to construct the adaptive estimators of distribution functions. It allows one to find the rates of decreasing $\varphi_\varepsilon^{-1}(n),\ \varepsilon > 0$ of the χ^2 divergence in the almost surely sense between distributions and their adaptive estimators.

As an example we have found the rate of decreasing for Hall distribution [7, 8] with unknown tail index. Similar results for convergence of the χ^2 divergence in probability are presented, e.g., in [6].

2 Adaptive Distribution Estimation

Let $\mathscr{F} = \{F_\Delta(x),\ x \in G \subseteq \mathscr{R}^1,\ \Delta \in \mathscr{D} \subseteq \mathscr{R}^q\}$ be the parametric family of heavy tail distributions. Here $\mathscr{D}$ is an admissible set of the unknown parameter Δ. Denote Δ_n an estimator of Δ.

Suppose that for every $\Delta \in \mathscr{D}$ the density $f_\Delta(x) = dF_\Delta(x)/dx$ exists. It is easy to verify that the χ^2 divergence between F_Δ and F_{Δ_n} has the form

$$\chi^2(F_\Delta, F_{\Delta_n}) = \int\limits_G \frac{dF_{\Delta_n}(x)}{dF_\Delta(x)} dF_{\Delta_n}(x) - 1 = \int\limits_G \left(\frac{f_{\Delta_n}(x)}{f_\Delta(x)} - 1\right)^2 f_\Delta(x)dx.$$

The problem is to construct estimators F_{Δ_n} of concrete well-known distributions F_Δ on the basis of a special type parameter estimators Δ_n with known rates of decreasing $\varphi_\varepsilon^{-1}(n),\ \varepsilon > 0$ of the χ^2 divergence in the following sense

$$\lim_{n\to\infty} \varphi_\varepsilon(n)\chi^2(F_\Delta, F_{\Delta_n}) = 0 \text{ a.s.} \tag{1}$$

Suppose the following

Assumption (A). Assume there exists the number $\delta_0 > 0$ such that for true value Δ the set $\mathscr{D}_0 = \{\delta : \Delta + \delta \subseteq \mathscr{D},\ ||\delta|| \leq \delta_0\}$ is not empty and

$$\sup_{\delta\in\mathscr{D}_0} \int\limits_G ||\nabla_\Delta f_{\Delta+\delta}(x)||^2 f_\Delta^{-1}(x)dx < \infty.$$

Then, using the Taylor expansion for the function $f_{\Delta_n}(x)$ on the set $\Omega_n = \{\omega : ||\Delta_n - \Delta|| \leq \delta_0\}$ we have

$$\chi^2(F_\Delta, F_{\Delta_n}) = \int\limits_G \left(f_{\Delta_n}(x) - f_\Delta(x)\right)^2 f_\Delta^{-1}(x)dx$$

$$= \int\limits_G \left[\sum_{i=1}^q \frac{\partial f_{\Delta+\alpha(\Delta_n-\Delta)}(x)}{\partial \Delta_i}(\Delta_n - \Delta)_i\right]^2 \cdot f_\Delta^{-1}(x)dx$$

$$\leq ||\Delta_n - \Delta||^2 \cdot \int\limits_G ||\nabla_\Delta f_{\Delta+\alpha(\Delta_n-\Delta)}(x)||^2 f_\Delta^{-1}(x)dx \leq C_0||\Delta_n - \Delta||^2,$$

where $\alpha \in (0, 1)$.

Thus to prove (1) it is enough to find the functions $\varphi_\varepsilon(n)$ and estimators Δ_n such that

$$\lim_{n\to\infty} \varphi_\varepsilon(n)||\Delta_n - \Delta||^2 = 0 \text{ a.s.} \tag{2}$$

The general truncated estimation method presented in [19] makes possible to obtain estimators of tail indexes of various type distributions with the properties

$$E||\Delta_n - \Delta||^{2p} \leq r^{-1}(n, p), \quad n \geq 1, \tag{3}$$

which are fulfilled for every $p \geq 1$ and some functions $r(n, p) \to \infty$ as $n \to \infty$ and/or $p \to \infty$.

Define $\overline{\Omega}_n$ a complement of the set Ω_n. Suppose that there exists a number p_0 such that the series

$$\sum_{n\geq 1} r^{-1}(n, p_0) < \infty.$$

Then using inequalities

$$P(\chi^2(F_\Delta, F_{\Delta_n}) > C_0||\Delta_n - \Delta||^2) \leq P(\overline{\Omega}_n) = P(||\Delta_n - \Delta|| > \delta_0)$$

$$\leq \delta_0^{-2p_0} E||\Delta_n - \Delta||^{2p_0} \leq \delta_0^{-2p_0} r^{-1}(n, p_0),$$

and the Borel-Cantelli lemma we have

$$||\Delta_n - \Delta||^{-2}\chi^2(F_\Delta, F_{\Delta_n}) \to 0 \quad \text{a.s.} \tag{4}$$

Define $\varphi(n, p) = (n^{-2}r(n, p))^{1/p}$. By making use of the Borel-Cantelli lemma for every $p \geq 1$ in particular it follows

$$\lim_{n\to\infty} \varphi(n, p)||\Delta_n - \Delta||^2 = 0 \text{ a.s.} \tag{5}$$

From (4) and (5) we get

$$\varphi(n, p)\chi^2(F_\Delta, F_{\Delta_n}) \to 0 \quad \text{a.s.}$$

and the function $\varphi_\varepsilon(n)$ can be defined as $\varphi_\varepsilon(n) = \varphi(n, p_\varepsilon)$ with an appropriate chosen $p_\varepsilon > 0$.

We will apply this approach to the adaptive estimation problem of the Hall model for distribution function $F_\Delta(x) = 1 - C_1x^{-\gamma} - C_2x^{-1/\alpha}$, $\gamma^{-1} = \alpha + \theta$.

In the next section the estimator $\Delta_n = \hat{\gamma}_n$ of γ with needed properties will be constructed and investigated.

Define $\rho = \theta/\alpha - 1$.

The following theorem presents the main result of this contribution.

Theorem 2.1 *For every $\varepsilon > 0$ there exist numbers p_ε such that the property (1) for the Hall model is fulfilled with*

$$\varphi_\varepsilon(n) = n^{\frac{\rho+1}{\rho+2}-\varepsilon}.$$

3 Estimation of Heavy Tail Index of the Hall Model

The problem is to estimate by i.i.d. observations $X_1, \ldots, X_n$ the parameter $\gamma = 1/\beta$ of the Hall distribution function [7]

$$F_\Delta(x) = 1 - C_1x^{-1/\beta} - C_2x^{-1/\alpha}, \quad x \geq c,$$

where $\beta > 0$, $\alpha > 0$; $\beta = \alpha + \theta \geq \beta_0 > 0$.

Then the tail distribution function

$$P(x) = C_1x^{-\gamma}(1 + C_3x^{-\gamma(\rho+1)}),$$

where $C_3 = C_2/C_1$, $\gamma = 1/\beta$, $\rho = \theta/\alpha - 1$.

The density function has the form

$$f(x) = C_1\gamma x^{-(\gamma+1)} + (C_2/\alpha)x^{-(1/\alpha+1)}$$

and Assumption (A) is fulfilled for $\mathscr{D} = \{\Delta = \gamma,\ \gamma > 0\}$, $\mathscr{D}_0 = \{\delta : |\delta| \leq \gamma/2\}$ and $\delta_0 = \gamma/2$.

To construct the estimator for γ we find its appropriate representation. For some $X = (x_1, x_2)$, $x_1 > x_2 > c$ by the definition of $P(x)$ we have

$$\log P(x_1) = \log C_1 - \gamma \log x_1 + \log(1 + C_3 x_1^{-\gamma(\rho+1)}),$$

$$\log P(x_2) = \log C_1 - \gamma \log x_2 + \log(1 + C_3 x_2^{-\gamma(\rho+1)}).$$

Thus we can find γ as a solution of this system

$$\gamma = \frac{\log(P(x_2)/P(x_1))}{\log(x_1/x_2)} - \log\left[1 + \frac{C_3(x_2^{-\gamma(\rho+1)} - x_1^{-\gamma(\rho+1)})}{1 + C_3 x_1^{-\gamma(\rho+1)}}\right]$$

and it is natural to define the estimators γ_n of γ as follows:

$$\gamma_n(X) = \frac{\log(P_n(x_2)/P_n(x_1))}{\log(x_1/x_2)} \cdot \chi(P_n(x_1) \geq \log^{-1} n), \quad n > 1.$$

Here $P_n(x)$ is the empirical tail distribution function

$$P_n(x) = \frac{1}{n} \sum_{i=1}^{n} \chi(X_i \geq x).$$

To get the estimator γ_n with the optimal rate of convergence (in the sense of L_2-norm see [3, 6, 8]), we put for $p \geq 1$ the sequence $X(n) = (x_1(n), x_2(n))$, where

$$x_1(n) = \mathrm{e} \cdot x_2(n), \qquad x_2(n) = n^{\frac{p}{\gamma[2p(\rho+2)-1]}} \tag{6}$$

The deviation of this estimator has the form

$$\gamma_n(X) - \gamma = \Bigg\{ \log(P_n(x_2(n))/P(x_2(n))) - \log(P_n(x_1(n))/P(x_1(n)))$$

$$- \log\left[1 + \frac{C_3(1 - \mathrm{e}^{-1})}{1 + C_3 x_1^{-\gamma(\rho+1)}(n)} x_2^{-\gamma(\rho+1)}(n)\right]\Bigg\} \cdot \chi(P_n(x_1(n)) \geq \log^{-1} n)$$

$$- \gamma \cdot \chi(P_n(x_1(n)) < \log^{-1} n). \tag{7}$$

For any $m \geq 1$ and $x \geq c$ it follows

$$E(P_n(x) - P(x))^{2m} \leq \frac{2B_m P(x)}{n^m}, \quad n \geq 1, \tag{8}$$

where B_m is a constant from the Burkholder inequality.

We will use the following inequality

$$|\log(P_n(x)/P(x))| = |\log\left(1+\frac{P_n(x)-P(x)}{P(x)}\right)\cdot\chi(P_n(x)-P(x)>0)$$

$$+\log\left(1+\frac{|P_n(x)-P(x)|}{P_n(x)}\right)\cdot\chi(P_n(x)-P(x)\le 0)|$$

$$\le |P_n(x)-P(x)|\cdot\Big[\frac{1}{P(x)}+\frac{1}{P_n(x)}\Big] = |P_n(x)-P(x)|\cdot\Big[\frac{2}{P(x)}+\left(\frac{1}{P_n(x)}-\frac{1}{P(x)}\right)\Big]$$

$$\le \frac{2|P_n(x)-P(x)|}{P(x)}+\frac{(P_n(x)-P(x))^2}{P(x)P_n(x)}.$$

Then using the c_r-inequality and (8) for $i=1,2$ we estimate

$$E\log^{2p}(P_n(x_i)/P(x_i))\cdot\chi\left(P_n(x_1)\ge\log^{-1}n\right)$$

$$\le \frac{C}{n^p P^{2p-1}(x_1)}+\frac{C\log^{2p}n}{n^{2p}P^{2(p-1)}(x_1)}. \tag{9}$$

In what follows, C will denote a generic non-negative constant whose value is not critical (and not always the same).

Further, by the Chebyshev inequality and (8) we have

$$P(P_n(x)<\log^{-1}n) = P(P(x)-P_n(x)>P(x)-\log^{-1}n)$$

$$\le \frac{E[P_n(x)-P(x)]^{4p}}{[P(x)-\log^{-1}n]^{4p}} \le \frac{C}{n^{2p}P^{4p-1}(x)} \le C\frac{x^{(4p-1)\gamma}}{n^{2p}}. \tag{10}$$

From (7), (9), and (10) it follows

$$E(\gamma_n(X(n))-\gamma)^{2p} \le Cr^{-1}(n,p), \tag{11}$$

$$r(n,p) = n^{\frac{2p(\rho+1)}{2p(\rho+1)+2p-1}p}$$

and we can put according to the definition of $\varphi(n,p)=(n^{-2}r(n,p))^{1/p}$ with the $r(n,p)$ defined in (11)

$$p_\varepsilon \ge 2\varepsilon^{-1}, \qquad \varphi_\varepsilon(n) = n^{\frac{\rho+1}{\rho+2}-\varepsilon}.$$

Note that proposed parameter estimation procedure gives estimator γ_n with convergence rate, with optimal (for $p=1$) convergence rate, see[6]. At the same time the sequences (6) in the definition of γ_n depend on the unknown model parameters. Then the adaptive estimation procedure should be constructed, e.g., on the presented scheme using some estimators of γ and ρ. The main aim is to get adaptive estimators with the optimal convergence rate.

Consider, for instance, the case of known ρ. Define the known deterministic sequence $(m_n)_{n\geq 1}$, $m_n = n^{\kappa}$, $\kappa \in (0, 1)$ and pilot estimator $\tilde{\gamma}_n = \tilde{\gamma}_n(\tilde{X}(m_n))$ of γ as follows

$$\tilde{\gamma}_n(\tilde{X}(m_n))$$

$$= \min\left\{\frac{\log(P_{m_n}(\tilde{x}_2(m_n))/P_{m_n}(\tilde{x}_1(m_n)))}{\log(\tilde{x}_1(m_n)/\tilde{x}_2(m_n))} \cdot \chi(P_{m_n}(\tilde{x}_1(m_n)) \geq \log^{-1} m_n), \gamma_0\right\}, \tag{12}$$

where $\tilde{X}(n) = (\tilde{x}_1(n), \tilde{x}_2(n))$,

$$\tilde{x}_1(n) = \mathrm{e} \cdot \tilde{x}_2(n), \qquad \tilde{x}_2(n) = n^{\frac{p}{\gamma_0[2p(\rho+2)-1]}}, \quad \gamma_0 = \beta_0^{-1}. \tag{13}$$

This estimator has the property

$$E(\tilde{\gamma}_n - \gamma)^{2p} \leq C \cdot r_0^{-1}(n, p), \quad p \geq 1,$$

$$r_0(n, p) = n^{\frac{2p\gamma\kappa(\rho+1)}{\gamma_0[2p(\rho+1)+2p-1]}p},$$

which can be proved similar to (11) and is strongly consistent according to the Borel-Cantelli lemma with the following rate

$$n^{\nu}(\tilde{\gamma}_n - \gamma) \to 0 \quad \text{a.s.} \tag{14}$$

for every

$$0 < \nu < \frac{\gamma\kappa(\rho+1)}{\gamma_0(2\rho+3)}.$$

Indeed, for every $a > 0$, ν defined above and p large enough

$$\sum_{n\geq 1} P(n^{\nu}(\tilde{\gamma}_n - \gamma) > a) \leq a^{-2p} \sum_{n\geq 1} n^{2\nu p} E(\tilde{\gamma}_n - \gamma)^{2p} \leq C \sum_{n\geq 1} \frac{n^{2\nu p}}{r_0(n, p)} < \infty.$$

Define the adaptive estimator of γ as follows

$$\hat{\gamma}_n = \frac{\log(\tilde{P}_n(\hat{x}_2(n))/\tilde{P}_n(\hat{x}_1(n)))}{\log(\hat{x}_1(n)/\hat{x}_2(n))}, \tag{15}$$

where $\tilde{P}_n$ is the empirical tail distribution function

$$\tilde{P}_n(x) = \frac{1}{n - m_n} \sum_{k=m_n+1}^{n} \chi(X_k \geq x)$$

and $\hat{X}(n) = (\hat{x}_1(n), \hat{x}_2(n))$,

$$\hat{x}_1(n) = \mathrm{e} \cdot \hat{x}_2(n), \qquad \hat{x}_2(n) = n^{\frac{p}{\tilde{\gamma}_n[2p(\rho+2)-1]}}. \tag{16}$$

The estimator $\hat{\gamma}_n$ has the property

$$E[(\hat{\gamma}_n - \gamma)^{2p} | \mathscr{F}_{m_n}] \leq C \cdot \tilde{r}^{-1}(n, p), \quad p \geq 1,$$

where σ-algebra $\mathscr{F}_{m_n} = \sigma\{X_1, \ldots, X_{m_n}\}$ and

$$\tilde{r}(n, p) = n^{\frac{2p\gamma(\rho+1)}{\tilde{\gamma}_n[2p(\rho+1)+2p-1]}p}.$$

Thus, using the Borel-Cantelli lemma and strong consistency of the pilot estimator $\tilde{\gamma}_n$ it is easy to prove the last property of Theorem 2.1 for the adaptive estimator $P_{\hat{\gamma}}$ in the Hall model.

To prove the strong consistency of $\varphi(n, p)(\hat{\gamma}_n - \gamma)^2$ and, as follows, Theorem 2.1, we establish first the convergence to zero of $\tilde{\varphi}(n, p)(\hat{\gamma}_n - \gamma)^2$, where $\tilde{\varphi}(n, p) = (n^{-2}\tilde{r}(n, p))^{1/p}$:

$$\sum_{n \geq 1} P(\tilde{\varphi}(n, p)(\hat{\gamma}_n - \gamma)^2 > a) \leq a^{-2p} \sum_{n \geq 1} E\tilde{\varphi}^p(n, p) E[(\hat{\gamma}_n - \gamma)^{2p} | \mathscr{F}_{m_n}]$$

$$= a^{-2p} \sum_{n \geq 1} n^2 E\tilde{r}(n, p) E[(\hat{\gamma}_n - \gamma)^{2p} | \mathscr{F}_{m_n}] \leq C \sum_{n \geq 1} \frac{1}{n^2} < \infty.$$

Then as $n \to \infty$

$$\tilde{\varphi}(n, p)(\hat{\gamma}_n - \gamma)^2 \to 0 \quad \text{a.s.}$$

Using the property (14), we have

$$\log \tilde{r}(n, p) r^{-1}(n, p) \sim (\hat{\gamma}_n - \gamma) \log n \to 0 \quad \text{a.s.}$$

and, as follows, for the function $\varphi(n, p)$ defined after formula (11), as $n \to \infty$

$$\varphi(n, p)(\hat{\gamma}_n - \gamma)^2 \to 0 \quad \text{a.s.}$$

Thus Theorem 2.1 is proven with

$$p_\varepsilon > (\rho+1)\max\left[\frac{\gamma_0}{\kappa\gamma(\rho+1)}, \frac{2}{\rho+1-\varepsilon(\rho+2)}\right] \tag{17}$$

if in the Hall distribution estimator we use, according to the notation in Sect. 2 the adaptive parameter estimator $\Delta_n = \hat{\gamma}_n$ of $\Delta = \gamma$, defined in (15).

4 Simulation Results

To establish the convergence of χ^2 divergence (1) one needs to check the condition (3), which is the key point to investigate the properties of estimators. In this section, to present some numerical results we define the quantity Λ_n as L_{2p}-norm of normalized deviation of estimator γ_n from the parameter γ

$$\Lambda_n = [r(n,p)E(\gamma_n-\gamma)^{2p}]^{\frac{1}{2p}} \tag{18}$$

in Hall's model [6]

$$F(x) = 1 - 2x^{-1} + x^{-2.5}, \quad x \geq 1, \ \ \rho = 0.5,$$

as a function of n. The values of Λ_n are given in Figs. 1 and 2 for different sample sizes n. Each coordinate is computed as an empirical average over 1000 Monte Carlo simulations of the experiment (for each value of n).

First the simulation was performed for the case when one can choose the sequences $x_1(n),\ \ x_2(n)$ according (6) to get the estimators γ_n with the rate of convergence close to the optimal one, see [6]. The value of $p = p_\varepsilon$ was chosen as $p_\varepsilon = [2/\varepsilon] + 1$. The results are presented in Fig. 1 for $\varepsilon = 0.1$ and $\varepsilon = 0.05$. One can see that Λ_n remains bounded from above as n increases and therefore the condition (3) is fulfilled. Similar results were obtained for $p > p_\varepsilon$.

The results for adaptive estimator $\hat{\gamma}_n$ (15) with the sequences $\hat{x}_1(n),\ \hat{x}_2(n)$ defined by (16) are given in Fig. 2 for $\varepsilon = 0.1$ and $\varepsilon = 0.05$. The value of β_0 was equal 0.5, the sequences $\tilde{x}_1(n),\ \tilde{x}_2(n)$ were defined by (13) with $m_n = n^\kappa,\ \kappa = 0.8$. The pilot estimator $\tilde{\gamma}_n$ was determined by (12), the power p was defined as the right-hand side of inequality (17). The quantity Λ_n remains bounded from above as n increases as well.

Our numerical simulations in all cases give practical confirmation of the theoretical properties of the proposed estimators.

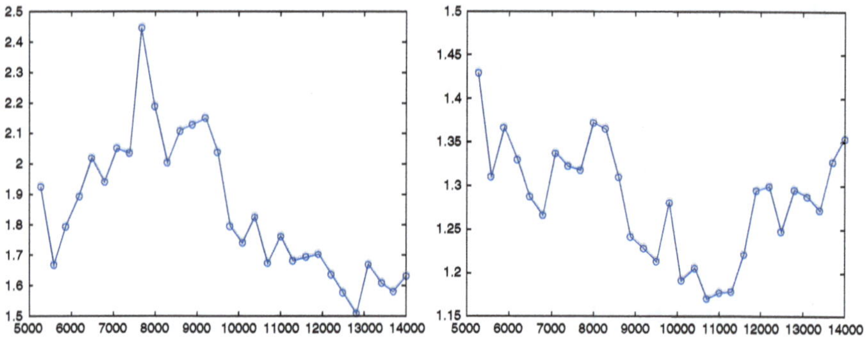

Fig. 1 Λ_n as a function of n in Hall's distribution with $\gamma = 1.0$. Left panel: $\varepsilon = 0.1$, Right panel: $\varepsilon = 0.05$

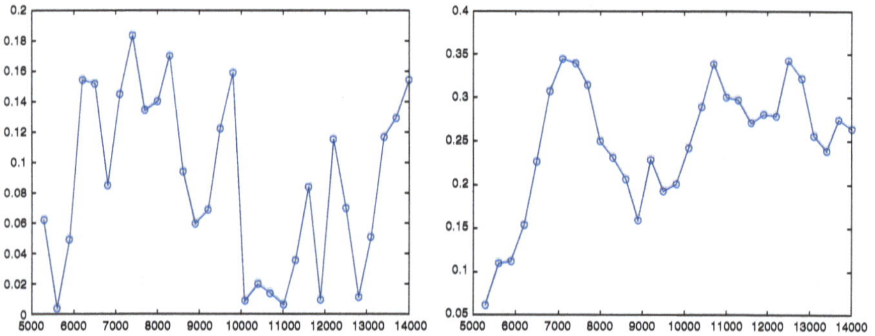

Fig. 2 Λ_n as a function of n in Hall's distribution with $\gamma = 1.0$. Left panel: $\varepsilon = 0.1$, Right panel: $\varepsilon = 0.05$

Acknowledgements This study was supported by the NSF grant DMS 16-13026 and RNF Project No. 17-11-01049.

References

1. Beirlant, J., Teugels, J. L., & Vynckier, P. (1996). *Practical analysis of extreme values*. Leuven: Leuven University Press.
2. Brito, M., Cavalcante, L., & Freitas, A. C. M. (2016). Bias-corrected geometric-type estimators of the tail index. *Journal of Physics A: Mathematical and Theoretical, 49*, 1–38.
3. Drees, H. (1998). Optimal rates of convergence for estimates of the extreme value index. *The Annals of Statistics, 26*, 434–448.
4. Embrechts, P., Kluppelberg, C., & Mikosch, T. (1997). *Modelling extremal events for insurance and finance*. Berlin: Springer.
5. Goldie, C. M., & Smith, R. L. (1987). Slow variation with remainder: Theory and applications. *Quarterly Journal of Mathematics, 38*, 45–71.

6. Grama, I., & Spokoiny, V. (2008). Statistics of extremes by oracle estimation. *The Annals of Statistics, 36*, 1619–1648.
7. Hall, P. (1982). On some simple estimates of an exponent of regular variation. *Journal of the Royal Statistical Society, Series B, 44*, 37–42.
8. Hall, P., & Welsh, A. H. (1984). Best attainable rates of convergence for estimates of parameters of regular variation. *The Annals of Statistics, 12*, 1079–1084.
9. Hill, B. M. (1975). A simple general approach to inference about the tail of a distribution. *The Annals of Statistics, 3*, 1163–1174.
10. Markovich, N. (2007). *Nonparametric analysis of univariate heavy-tailed data research and practice.* Hoboken, NJ: John Wiley and Sons.
11. Novak, S. Y. (1996). On the distribution of the ratio of sums of random variables. *Theory of Probability and Its Applications, 41*, 479–503
12. Novak, S. Y. (2000). On self-normalised sums. *Mathematical Methods of Statistics, 9*, 415–436.
13. Novak, S. Y. (2002). Inference of heavy tails from dependent data. *Siberian Advances in Mathematics, 12*, 73–96.
14. Novak, S. Y. (2011). *Extreme value methods with applications to finance.* London: Chapman and Hall/CRC Press
15. Ramlau-Hansen, H. (1988). A solvency study in non-life insurance. *Scandinavian Actuarial Journal, 80*, 3–34.
16. Resnick, S. (1997). Heavy Tail. Modeling and teletraffic data: Special invited paper. *The Annals of Statistics, 25*(5), 1805–1869.
17. Resnick, S. I., & Stărică, C. (1999). Smoothing the moment estimate of the extreme value parameter. *Extremes, 1*, 263–294.
18. Schultze, J., & Steinebach, J. (1996). On least squares estimates of an exponential tail coefficient. *Statistics & Decisions, 14*, 353–372.
19. Vasiliev, V. (2014). A truncated estimation method with guaranteed accuracy. *Annals of the Institute of Statistical Mathematics, 66*, 141–163.

Extremal Index for a Class of Heavy-Tailed Stochastic Processes in Risk Theory

C. Tillier

Abstract Extreme values for dependent data corresponding to high threshold exceedances may occur in clusters, i.e., in groups of observations of different sizes. In the context of stationary sequences, the so-called extremal index measures the strength of the dependence and may be useful to estimate the average length of such clusters. This is of particular interest in risk theory where public institutions would like to predict the replications of rare events, in other words, to understand the dependence structure of extreme values. In this contribution, we characterize the extremal index for a class of stochastic processes that naturally appear in risk theory under the assumption of heavy-tailed jumps. We focus on Shot Noise type-processes and we weaken the usual assumptions required on the Shot functions. Precisely, they may be possibly random with not necessarily compact support and we do not make any assumption regarding the monotonicity. We bring to the fore the applicability of the result on a Kinetic Dietary Exposure Model used in modeling pharmacokinetics of contaminants.

1 Motivations and Framework

The assessment of major risks in our technological society has become vital because of the economic, environmental, and human impacts of recent industrial disasters. Hence, risk analysis has received an increasing attention over the past years in the scientific literature in various areas, e.g., in dietary risk, hydrology, finance and insurance; see [1, 7, 12], for instance. By nature, risk theory concerns the probability of occurrence of rare events which are functions—sums or products—of heavy-tailed random variables. Hence, stochastic processes provide an appropriate framework for modeling such phenomena through time. For instance, non-life insurance mathematics deal with particular types of *Shot Noise Processes* (SNP)

C. Tillier (✉)
University of Hamburg, SPST, Hamburg, Germany

P. Bertail et al. (eds.), *Nonparametric Statistics*, Springer Proceedings in Mathematics & Statistics 250, https://doi.org/10.1007/978-3-319-96941-1_12

defined as

$$S_1(t) = \sum_{i=0}^{N(t)} W_i h(t - T_i), \qquad t \geq 0, \tag{1}$$

where usually $(W_i)_{i\geq 0}$ are independent and identically distributed (i.i.d.) random variables (r.v.'s), h is a nonincreasing measurable function, and N is a homogeneous Poisson process. In this insurance context, S_1 may be used to represent the amount of aggregate claims that an insurer has to cope with; see [20] for a complete review of non-life insurance mathematics. More generally, this kind of jump processes are useful in many applications to model time series for which sudden jumps occur such as in dietary risk assessment, finance, hydrology or as reference models for intermittent fluctuation in physical systems; see [8] and [24], for instance. The study of the extremal behavior of these stochastic processes leads to risk indicators such as the expected time over a threshold or the expected shortfall, which supply information about the exceedances that give rise to hazardous situations; see [23], for instance.

Besides, since extremal events may occur in clusters, the study of the dependence structure of rare events is a major issue, for example to predict potential replications of earthquakes in environmental sciences. This dependence structure may be captured by the extremal index defined in the seminal contribution [16]. Recall that a stationary sequence $(Z_i)_{i\in\mathbb{Z}}$ has an extremal index $\theta \in [0, 1]$ if for all $\tau > 0$ and all sequence $u_n(\tau)$ such that $\lim_{n\to\infty} n\mathbb{P}(Z_1 \geq u_n(\tau)) = \tau$, it holds that

$$\lim_{n\to\infty} \mathbb{P}\left(\max_{i=1,\ldots,n} Z_i \leq u_n(\tau)\right) = e^{-\theta\tau}. \tag{2}$$

Less formally, the extremal index indicates somehow, how many times in average an extremal event will reproduce. The case $\theta = 1$ (respectively $\theta = 0$) corresponds to independent data, i.e., to extreme values occurring in an isolated fashion (respectively to potentially infinite size clusters).

Authors in [9, 14] and [18] characterize the extremal index in several particular configurations of (1) and study the extremal properties of the process; see also [10, 11] and [13]. More recently, [19] compute the extremal index when the jumps $(W_i)_{i\geq 0}$ form a chain-dependent sequence (the cumulative distribution function (c.d.f.) is linked to a secondary Markov chain) and they assume that h is a bounded positive strictly decreasing function supported on a finite interval.

In this chapter, we continue the investigation of the extremal index for such stochastic processes relaxing the conditions required on h. We focus on an extension of such SNP on the form

$$S(t) = \sum_{i=0}^{\infty} W_i h_i(t - T_i), \qquad t \geq 0, \tag{3}$$

where $t \to h_i(t)$ is a random function. Throughout this chapter, we work under the following three conditions (C1)–(C3).

(C1) *Jumps* $(W_i)_{i\in\mathbb{N}}$ are nonnegative r.v.'s with c.d.f. H such that $\overline{H} = 1 - H$ is regularly varying at infinity with index $-\alpha$, $\alpha > 0$, that is

$$\frac{\overline{H}(wx)}{\overline{H}(x)} \underset{x\to\infty}{\longrightarrow} w^{-\alpha}, \qquad \forall\, w > 0.$$

(C2) *Jumps instants* $(T_i)_{i\in\mathbb{N}}$ are defined for $i \geq 1$ by $T_i = \sum_{k=1}^{i} \Delta T_k$ and $T_0 = 0$ while the *inter-jumps* $(\Delta T_i)_{i\in\mathbb{N}^*}$ are i.i.d. positive r.v.'s with finite expectation.

(C3) For all $i \geq 0$, the random functions h_i are positive, stationary, and independent of T_i.

The condition (C1) is the heavy-tailed distribution assumption on the jumps. We refer to [22] for an exhaustive review of the univariate regular variation theory. The condition (C2) means that $(T_0, T_1, \ldots)$ forms a renewal sequence so that one may define the associated renewal process $\{N(t)\}_{t\geq 0}$ by $N(t) := \#\{i \geq 0 : T_i \leq t\}$ for $t \geq 0$. The remaining part of the manuscript is organized as follows. In Sect. 2, we present the main result regarding the extremal index of the process (3) while an illustrative application is given in Sect. 3. In the Appendix, we recall the main notions involved in the proof of the theorem.

2 The Extremal Index

The extremal index defined in Eq. (2) holds for discrete-time series. The purpose of this work is to investigate the dependence structure of the extreme values of the continuous-time stochastic processes S defined in Eq. (3). Depending on the context, it means that we are interested in the dependence structure either of its maxima or of its minima. In dietary risk assessment, S aims at representing the evolution of a contaminant in the human body through time; see Sect. 3 for more details. Toxicologists determine thresholds from which the exceedance may have some adverse effect for the health of an individual and we are therefore interested in the maxima of S. Similarly, in hydrology, (3) may be used to describe the flow of a river and a hazardous situation—seen as a rare event—arises when the flow exceeds a critical threshold; see [17]. On the other the hand, in most of the applications, the random functions h_i are in essence monotonic for each $i \geq 0$ when t grows. To be convinced, let us go back to the dietary risk assessment. In this context, h_i models the elimination of the contamination and is thus a decreasing function for each $i \geq 0$. For instance, $h_i(t) \equiv e^{-t}, t \geq 0$ has been proposed in [6] and is also used in non-life insurance mathematics; see [20].

Assuming that the random function h_i are monotonic for each $i \geq 0$, it is straightforward to see that the extreme values—the maxima or the minima—occur on the embedded chain, i.e., the process $\{S(t)\}_{t\geq 0}$ sampled at the jump arrivals $T_1, T_2 \ldots$. As a consequence, the dependence structure of the continuous-time stochastic process S may be deduced from the analysis of the dependence structure of the underlying sequence $(S(T_1), S(T_2), \ldots)$.

This is the purpose of this section: to compute the extremal index of the embedded chain of the jump process (3). This means that we focus on the following discrete-time risk process

$$S(T_k) = \sum_{i=1}^{k} W_i h_i(T_k - T_i), \qquad k > 0. \tag{4}$$

Hereinafter, for $i \geq 1$, define $-T_{-i}$ (respectively W_{-i} and h_{-i}) as an independent copy of T_i (respectively of W_i and h_i) so that under (C1) and (C3), $(W_i)_{i=-\infty}^{\infty}$ and $(h_i)_{i=-\infty}^{\infty}$ form, respectively, an i.i.d. and a stationary sequence of positive r.v.'s. Facing with the issue of non-stationarity of the embedded chain (4)—required for the computation of the extremal index—we study a stationary version/modification denoted $(S_k)_{k\in\mathbb{Z}}$ and defined by

$$S_k = \sum_{i=-\infty}^{k} W_i h_i(T_k - T_i), \qquad k > 0. \tag{5}$$

We now introduce the condition (D1), under which the stationary sequence (5) is well defined.

(D1) The random function h_i satisfy

- $\sum_{i=0}^{\infty} \mathbb{E}[h_1(T_i)] < \infty, \quad \alpha < 1.$
- There exists $\epsilon > 0$ such that $\sum_{i=0}^{\infty} \mathbb{E}[h_1^{\alpha-\epsilon}(T_i)] < \infty, \quad \alpha \leq 2.$
- $\sum_{i=0}^{\infty} \mathbb{E}[h_1^2(T_i)] < \infty, \quad \alpha > 2.$

Theorem 1 *Assume Model* (5) *holds. Under Conditions (C1)–(C3) and (D1), the extremal index θ is given by*

$$\theta = \frac{\mathbb{E}\left[\max_{i\geq 0} h_i^{\alpha}(T_i)\right]}{\sum_{i=0}^{\infty} \mathbb{E}[h_i^{\alpha}(T_i)]}. \tag{6}$$

We do not raise the question of the estimation of the extremal index in this work. In many cases, the jump process (3) is a PDMP (Piecewise-Deterministic Markov Process) and [4] propose a robust estimator for the extremal index; see also [5] and the application in Sect. 3 for an illustrative example.

2.1 Proof of Theorem 1

For reader's convenience, the definitions of the main notions, namely the *tail index*, the *anti-clustering*, and the *strong mixing* conditions are postponed in the Appendix. Let us first define the intermediate stationary sequences $\{S_k^{(m)}, k \geq 0\}_{m\geq 0}$ such that

$$S_k^{(m)} = \sum_{i=k-m}^{k} W_i h_i(T_k - T_i). \tag{7}$$

The extremal index θ of the stationary sequence $\{S_k\}_{k\geq 0}$ will be deduced from the extremal index θ_m of $\{S_k^{(m)}, k \geq 0\}_{m\geq 0}$ in the following way. In [3, Theorem 4.5], the authors show that if a jointly regularly varying (see the definition of "jointly regularly varying" in Definition 1 in the Appendix) stationary sequence $(Z_i)_{i\in\mathbb{Z}}$ is strongly mixing and satisfies the anti-clustering condition, then $(Z_i)_{i\in\mathbb{Z}}$ admits an extremal index $\tilde{\theta}$ given by

$$\tilde{\theta} = \mathbb{P}\left(\max_{k\geq 1} Y_k \leq 1\right), \tag{8}$$

where $(Y_i)_{i\in\mathbb{N}}$ is the tail process of $(Z_i)_{i\in\mathbb{Z}}$. Using this result, we obtain the extremal index θ_m for each $m \geq 1$. Next, we show that the assumptions of Proposition 1.4 in [9] hold to conclude that $\lim_{m\to\infty} \theta_m = \theta$. We start by characterizing the tail process of sequence $\{S_k^{(m)}\}_{k\in\mathbb{N}}$ in the following lemma.

Lemma 1 *Assume that Conditions (C1)–(C3) and (D1) hold. For each $m \geq 1$, the tail process of $\{S_k^{(m)}\}_{k\in\mathbb{N}}$ denoted by $\{Y_k^{(m)}\}_{k\in\mathbb{N}}$ is defined by*

$$Y_n^{(m)} = \begin{cases} \frac{h_{N_m}(T_{n+N_m})}{h_{N_m}(T_{N_m})} Y_0^{(m)}, & 0 \leq n \leq m, \\ 0 & \text{for } n > m, \end{cases}$$

with $P\left(Y_0^{(m)} > y\right) = y^{-\alpha}$ and N_m is an integer-valued random variable such that

$$\mathbb{P}(N_m = n) = \frac{\mathbb{E}[h_n^\alpha(T_n)]}{\sum_{i=1}^m \mathbb{E}[h_i^\alpha(T_i)]}, \quad 0 \leq n \leq m.$$

Besides, for any random variable U measurable with respect to $(h_j, T_j)_{j\in\mathbb{Z}}$, we have

$$\mathbb{E}[U \mid N_m = i] = \frac{\mathbb{E}[h_i^\alpha(T_i)U]}{\mathbb{E}[h_i^\alpha(T_i)]}, \quad 0 \leq i \leq m.$$

Proof (Proof of Lemma 1) For clarity of notation, we omit the superscript (m) and we assume that $h_k \equiv 0$ if $k > m$ and we denote $X_1 \overset{d}{=} X_2$ when two random variables X_1, X_2 share the same distribution. Then, for a fixed n, we have

$$\mathbb{P}\left(\max_{i=0,\dots,n} S_i/y_i \le x \mid S_0 > x\right) = 1 - \mathbb{P}\left(\max_{i=0,\dots,n} S_i/y_i > x \mid S_0 > x\right)$$
$$= 1 - \frac{\mathbb{P}\left(\left(\max_{i=0,\dots,n} S_i/y_i\right) \wedge S_0 > x\right)}{\mathbb{P}(S_0 > x)}.$$

Under the two conditions (C1) and (C2), $(W_k)_{k\in\mathbb{Z}}$ is an i.i.d. sequence and $T_i - T_k \overset{d}{=} T_{i-k}$ for $i \ge 0$ and $k \le 0$. It follows that

$$\lim_{x\to\infty} \mathbb{P}\left(\max_{i=0,\dots,n} S_i/y_i > x, S_0 > x\right) \underset{x\to\infty}{\sim} \sum_{k=0}^{\infty} \mathbb{P}\left(W_k\left(\max_{i=0,\dots,n} h_k(T_i - T_k)/y_i\right) \wedge h_k(-T_k) > x\right)$$
$$\underset{x\to\infty}{\sim} \sum_{k=0}^{\infty} \mathbb{E}\left[\bigvee_{i=0}^{n} \frac{h_k^\alpha(T_{k+i})}{y_i} \wedge h_k^\alpha(T_k)\right] \mathbb{P}(W_1 > x),$$

and

$$S_0 = \sum_{i=-\infty}^{0} W_i h_i(T_0 - T_i) \overset{d}{=} \sum_{i=0}^{\infty} W_i h_i(T_i),$$

under (C3). Moreover, since the two sequences $(h_i)_{i\in\mathbb{Z}}$ and $(T_i)_{i\in\mathbb{Z}}$ are mutually independent, the results in Section 3 of [15] imply that the series S_0 is almost surely convergent under Condition (D1) and we have

$$\lim_{x\to\infty} \frac{\mathbb{P}(S_0 > x)}{\overline{H}(x)} = \sum_{i=0}^{\infty} \mathbb{E}\left[h_1^\alpha(T_i)\right] < \infty. \tag{9}$$

From Eq. (9), we have proved that S_0 is regularly varying at infinity with the same index α than the jumps $(W_i)_{i\ge 0}$. It also follows that

$$\lim_{x\to\infty} \mathbb{P}\left(\max_{i=0,\dots,n} S_i/y_i \le x \mid S_0 > x\right) = 1 - \frac{\mathbb{E}\left[\bigvee_{i=0}^{n} \frac{h_k^\alpha(T_{k+i})}{y_i} \wedge h_k^\alpha(T_k)\right]}{\sum_{k=0}^{\infty} \mathbb{E}[h_k^\alpha(T_k)]}.$$

Setting

$$p_k = \frac{\mathbb{E}[h_k^\alpha(T_k)]}{\sum_{j=0}^{\infty} \mathbb{E}[h_j^\alpha(T_j)]},$$

we obtain

$$\lim_{x\to\infty} \mathbb{P}\left(\max_{i=0,\dots,n} S_i/y_i \le x \mid S_0 > x\right) = 1 - \sum_{k=0}^{\infty} p_k \frac{\mathbb{E}\left[\bigvee_{i=0}^{n} h_k^\alpha(T_{k+i})/y_i \wedge h_k^\alpha(T_k)\right]}{\mathbb{E}[h_k^\alpha(T_k)]}$$

$$= 1 - \sum_{k=0}^{\infty} p_k \mathbb{E}\left[\bigvee_{i=0}^{n} \frac{h_N^\alpha(T_{N+i})}{y_i h_N^\alpha(T_N)} \wedge 1 \mid N = k\right]$$

$$= 1 - \mathbb{E}\left[\bigvee_{i=0}^{n} \frac{h_N^\alpha(T_{N+i})}{y_i h_N^\alpha(T_N)} \wedge 1\right]$$

$$= 1 - \mathbb{P}\left(Y_0 \bigvee_{i=0}^{n} \frac{h_N^\alpha(T_{N+i})}{y_i h_N^\alpha(T_N)} > 1\right),$$

where Y_0 is a Pareto random variable independent of $\{h_i, T_i\}_{i\in\mathbb{Z}}$. This proves our claim.

For each $m \ge 1$, the strong mixing condition holds for each sequence $\{S_k^{(m)}, k \ge 0\}$ since it is m-dependent. Indeed, by independence $\alpha_h = 0$ for all $h \ge m + 1$. Likewise, since $\{S_k^{(m)}, k \ge 0\}$ is m-dependent, the anti-clustering condition is satisfied with $r_n \mathbb{P}(Z_1 > a_n) = o(1)$; see Section 4.1 in [2]. As a first consequence, we obtain in the following lemma the expression of the extremal index θ_m of the intermediate sequence $\{S_k^{(m)}\}_{k\ge 0}$.

Lemma 2 *Assume that Conditions (C1)–(C3) and (D1) hold. For each $m \ge 1$, the extremal index θ_m of the intermediate sequence $\{S_k^{(m)}, k \ge 0\}$ defined in (7) is given by*

$$\theta_m = \frac{\mathbb{E}\left[\bigvee_{j=0}^{m} h_j^\alpha(T_j)\right]}{\sum_{i=0}^{m} \mathbb{E}[h_1^\alpha(T_i)]}. \tag{10}$$

Proof (Proof of Lemma 2) Fix $m \ge 1$ throughout the proof. By Eq. (8), since $\mathbb{P}(Y_0 > x) = x^{-\alpha}$, we have

$$\theta_m = \mathbb{P}\left(\max_{k\ge 1} Y_k^{(m)} \le 1\right)$$

$$= \mathbb{P}\left(\max_{k\ge 1} Y_0 \Theta_k^{(m)} \le 1\right)$$

$$= 1 - \mathbb{P}\left(\max_{k\ge 1} Y_0 \Theta_k^{(m)} \ge 1\right)$$

$$= 1 - \mathbb{E}\left[\max_{k\ge 1} \left(\Theta_k^{(m)}\right)^\alpha \wedge 1\right],$$

where $\left(\Theta_k^{(m)}\right)_{k\geq 0}$ refers to the spectral tail process of the intermediate sequence $\{S_k^{(m)}, k \geq 0\}$ defined in [3]. Applying Lemma 1, we obtain

$$\begin{aligned}
\theta_m &= \mathbb{P}\left(Y_0 \max_{1\leq k\leq m} \frac{h_{k+N}(T_{k+N})}{h_N(T_N)} \leq 1\right) \\
&= 1 - \mathbb{E}\left[\max_{1\leq k\leq m} \frac{h^\alpha_{k+N}(T_{k+N})}{h^\alpha_N(T_N)} \bigwedge 1\right] \\
&= 1 - \sum_{n=0}^{m} \mathbb{E}\left[\max_{1\leq k\leq m} \frac{h^\alpha_{k+N}(T_{k+N})}{h^\alpha_N(T_N)} \bigwedge 1 \mid N=n\right] \mathbb{P}(N=n) \\
&= 1 - \frac{\sum_{n=0}^m \mathbb{E}\left[(\max_{1\leq k\leq m} h^\alpha_{n+k}(T_{n+k})) \wedge h^\alpha_n(T_n)\right]}{\sum_{j=0}^m \mathbb{E}[h^\alpha_j(T_j)]} \\
&= \frac{\sum_{n=0}^m \mathbb{E}\left[h^\alpha_n(T_n) - (\max_{1\leq k\leq m} h^\alpha_{n+k}(T_{n+k})) \wedge h^\alpha_n(T_n)\right]}{\sum_{j=0}^m \mathbb{E}[h^\alpha_j(T_j)]}. \qquad (11)
\end{aligned}$$

Observe that using the identity $\max(a, b) = a + b - \min(a, b)$ for any a, b in $\mathbb{R}^+$, one can show that for any sequence $(a_n)_{n\in\mathbb{N}}$ of nonnegative real numbers such that $\sum_{n=0}^\infty a_n < \infty$, we have

$$\max_{n\in\mathbb{N}} a_n = \sum_{n\in\mathbb{N}} \left(a_n - \max_{k\geq 1} a_{n+k} \wedge a_n\right). \qquad (12)$$

Under (D1), from Eq. (9), $\sum_{n=0}^\infty \mathbb{E}[h^\alpha_n(T_n)] < \infty$ and we can apply the relation (12) to the last equality (11). This proves Lemma 2.

Up to now, we have characterized the extremal index θ_m from the tail process of $\{S_k^{(m)}\}_{k\in\mathbb{N}}$. To conclude, it remains to prove that the extremal index of $(S_k)_{k\in\mathbb{Z}}$ is given by $\lim_{m\to\infty} \theta_m = \theta$. For this purpose, we apply Proposition 1.4 of [9]. We must check the following two conditions: for all sequence u_n such that

$$n\mathbb{P}(S_0 > u_n) \to \beta \in (0, \infty),$$

we have

$$\lim_{\epsilon\to 0} \limsup_{n\to\infty} n\mathbb{P}((1-\epsilon)u_n < S_0 \leq (1+\epsilon)u_n) = 0 \qquad (13)$$

and

$$\lim_{m\to\infty} \limsup_{n\to\infty} n\mathbb{P}(|S_0 - S_0^{(m)}| > \epsilon u_n) = 0. \qquad (14)$$

Since we have already proved in Eq. (9) that S_0 is regularly varying with index $-\alpha$, $\alpha > 0$, we have

$$\limsup_{n\to\infty} n\mathbb{P}((1-\epsilon)u_n < S_0 \leq (1+\epsilon)u_n) = \beta((1-\epsilon)^{-\alpha} - (1+\epsilon)^{-\alpha}).$$

Letting $\epsilon \to 0$ proves Eq. (13). Moreover, by the same arguments which lead to the expression for the tail behavior of S_0, we have

$$\lim_{n\to\infty} n\mathbb{P}(|S_0 - S_0^{(m)}| > u_n\epsilon) = \epsilon^{-\alpha}\beta \frac{\sum_{n=m+1}^{\infty} \mathbb{E}[h_1^{\alpha}(T_n)]}{\sum_{n=0}^{\infty} \mathbb{E}[h_1^{\alpha}(T_n)]}.$$

Letting $m \to \infty$ proves Eq. (14). We finally have

$$\theta = \lim_{m\to\infty} \theta_m = \frac{\mathbb{E}\left[\bigvee_{i=0}^{\infty} h_i^{\alpha}(T_i)\right]}{\sum_{i=0}^{\infty} \mathbb{E}[h_1^{\alpha}(T_i)]},$$

which concludes the proof of Theorem 1.

3 Application

For the sake of application of Theorem 1, we consider a specific dietary risk assessment model studied in [6] called KDEM for Kinetic Dietary Exposure Model; see also [7] for the statistical analysis of the model and for more details on dynamic dietary risk processes. For each $i \geq 0$, we assume that the intakes $(W_i)_{i\geq 0}$ are pure Pareto distributed with tail index $\alpha > 0$ and we set $h_i(t) = e^{-\omega_i t}\mathbb{I}_{[0,\infty[}(t)$, where $\mathbb{I}_{[\cdot]}(\cdot)$ is the indicator function. Besides, we consider that $(\omega_i)_{i\in\mathbb{N}}$ is an i.i.d. sequence of nonnegative r.v.'s with finite expectation. In this context, for each $i \geq 0$, h_i is a nonincreasing random elimination function that governs the elimination process of the i-th intake W_i ingested at time T_i up to time t. Then $(\omega_i)_{i\in\mathbb{N}}$ is a random elimination parameter, which permits to take into account fluctuations in the assimilation process. The model may be written as

$$S(t) = \sum_{i=1}^{N(t)} W_i e^{-\omega_i(t-T_i)}, \qquad t > 0, \tag{15}$$

where $t \to N(t) := \#\{i \geq 0 : T_i \leq t\}$ is a renewal process that counts the numbers of intakes that occurred until time $t > 0$. Figure 1 shows how the process (15) evolves through time.

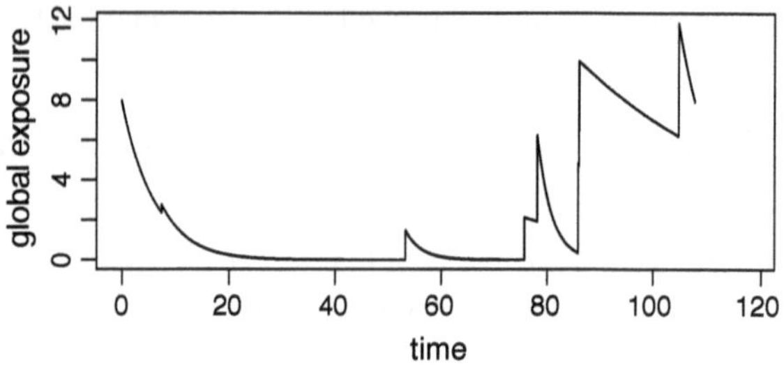

Fig. 1 The elimination driven by the random variable $(\omega_i)_i$ may vary between two intakes. Observe the PDMP-type trajectory. Due to the heavy-tailed distribution of the intakes $(W_i)_i$, some jumps are rather large

To get an explicit result, we consider that the intakes arise regarding a homogeneous Poisson process meaning that the duration between intakes is independent and exponentially distributed. Applying Theorem 1, we get the following explicit formulae of the extremal index.

Proposition 1 *Assume that Model* (15) *holds with positive i.i.d.* $(\omega_i)_{i\in\mathbb{N}}$ *satisfying* $\mathbb{E}[\omega_1] < \infty$. *Assume moreover that* $\overline{H}(x) = x^{-\alpha}, \alpha > 0$ *for all* $x > 0$ *and* N *is a Poisson process with intensity* $\lambda > 0$. *Then we have*

$$\theta = \frac{\alpha}{\alpha + \lambda\mathbb{E}[\omega^{-1}]}. \tag{16}$$

Proof Note first that in this setup, this is straightforward that for all $\alpha > 0$, Assumptions (C1)–(C3) as well as (D1) hold. Indeed, the Pareto distribution is a particular case of such regularly varying random variables so that (C1) is satisfied. (C2) holds as a sum of i.i.d. r.v.'s whose distribution is exponential with mean $1/\lambda$, $\lambda > 0$. Finally, (C3) is satisfied since the random variables $(\omega_i)_{i\in\mathbb{N}}$ are i.i.d., then the random functions $(h_i)_{i\in\mathbb{N}}$ are positive i.i.d. r.v.'s with $0 < \mathbb{E}[\omega_1] < \infty$ implying (D1). Now, observe first that for the numerator, we have

$$\mathbb{E}\left[\max_{i\geq 0}\{h_i^\alpha(T_i)\}\right] = \mathbb{E}\left[\max_{i\geq 0}\{e^{-\alpha\omega_i T_i}\mathbb{I}_{[0,\infty[}(T_i)\}\right] \leq 1.$$

Besides, since $T_0 = 0$ under (C2), we have

$$\mathbb{E}\left[h_0^\alpha(T_0)\right] = \mathbb{E}\left[e^{-\alpha\omega 0}\right] = 1,$$

leading to $\mathbb{E}\left[\max_{i\geq 0}\{h_i^\alpha(T_i)\}\right] = 1$. It also follows that the denominator may be written as

$$\sum_{i=0}^{\infty}\mathbb{E}\left[h_i^\alpha(T_i)\right] = 1 + \sum_{i=1}^{\infty}\mathbb{E}\left[h_i^\alpha(T_i)\right]$$

with

$$\sum_{i=1}^{\infty} \mathbb{E}\left[h_i^{\alpha}(T_i)\right] = \sum_{i=1}^{\infty} \int_0^{\infty} \left(\int_0^{\infty} e^{-\alpha\omega t} dF_{\Delta T}^{*i}(t) \right) dF_W(\omega)$$

$$= \int_0^{\infty} \left(\sum_{i=1}^{\infty} \mathbb{E}\left[e^{-\alpha\omega\Delta T}\right]^i \right) dF_W(\omega)$$

$$= \int_0^{\infty} \frac{\mathbb{E}\left[e^{-\alpha\omega\Delta T}\right]}{1 - \mathbb{E}\left[e^{-\alpha\omega\Delta T}\right]} dF_W(\omega)$$

$$= \int_0^{\infty} \frac{\lambda}{\alpha\omega} dF_W(\omega)$$

$$= \frac{\lambda}{\alpha} \mathbb{E}[\omega^{-1}],$$

where "$*$" refers to the convolution operator. This concludes the proof.

To conclude this part, we briefly discuss the veracity of Proposition 1. In this regard, assume Model (15) holds with the assumptions of Proposition 1. Assume moreover that the elimination parameter $\omega > 0$ is constant. Observe now that its embedded chain, namely

$$S(T_k) = \sum_{i=1}^{k} W_i e^{-\omega(T_k - T_i)}, \qquad k > 0$$

may be expressed as

$$S(T_k) = e^{-\omega\Delta T_k} S(T_{k-1}) + W_k, \qquad k > 0. \tag{17}$$

The latest equation is nothing else than a particular case of the so-called SRE for Stochastic Recurrence Equation. It has been studied for a while. In particular, [21] showed that its extremal index is given by $\theta = 1 - \mathbb{E}[e^{-\alpha\omega\Delta T_1}]$. In the specific setup of Proposition 1 where the $(\Delta T_i)_{i\in\mathbb{N}^*}$ are exponentially distributed, the Laplace transform $\mathbb{E}[e^{-\Delta T_1}]$ is explicit. It follows that the extremal index is given by

$$\theta = \frac{\alpha}{\alpha + \lambda\omega^{-1}}.$$

We retrieve the result of Proposition 1 with constant $\omega > 0$.

Acknowledgements Charles Tillier would like to thank Patrice Bertail and Philippe Soulier, both Professors at Paris Nanterre University for insightful comments and discussions which led to an improvement of this work. Financial supports by the ANR network AMERISKA ANR 14 CE20 0006 01 and the Labex MME-DII are also gratefully acknowledged by the author.

Appendix

For reader's convenience, we recall in this part important notions involved in the proof of Theorem 1. We start by the definition of the so-called "tail process" introduced recently by Basrak and Segers [3].

Definition 1 (The Tail Process) Let $(Z_i)_{i\in\mathbb{Z}}$ be a stationary process in $\mathbb{R}^+$ and let $\alpha \in (0, \infty)$. If $(Z_i)_{i\in\mathbb{Z}}$ is jointly regularly varying with index $-\alpha$, that is, all vectors of the form $(X_k, \ldots, X_l)$, $k \leq l \in \mathbb{Z}$ are multivariate regularly varying, then there exists a process $(Y_i)_{i\in\mathbb{Z}}$ in $\mathbb{R}^+$, called the tail process such that $\mathbb{P}(Y_0 > y) = y^{-\alpha}$, $y \geq 1$ and for all $(n, m) \in \mathbb{Z}^2$, $n \geq m$

$$\lim_{z\to\infty} \mathbb{P}((z^{-1}Z_n, \cdots, z^{-1}Z_m) \in \cdot \mid Z_0 > z) = \mathbb{P}((Y_n, \cdots, Y_m) \in \cdot).$$

We recall now the strong mixing and anti-clustering conditions.

Definition 2 (Strong Mixing Condition) A stationary sequence $(Z_k)_{k\in\mathbb{Z}}$ is said to be strongly mixing with rate function α_h if

$$\sup |\mathbb{P}(A \cap B) - \mathbb{P}(A)\mathbb{P}(B)| = \alpha_h \to 0, \quad h \to \infty, \tag{18}$$

where the supremum is taken over all sets $A \in \sigma(\cdots, Z_{-1}, Z_0)$ and $B \in \sigma(Z_h, Z_{h+1}, \cdots)$

Definition 3 (Anti-clustering Condition) A positive stationary sequence $(Z_k)_{k\in\mathbb{Z}}$ is said to satisfy the anti-clustering condition if for all $u \in (0, \infty)$,

$$\lim_{k\to\infty} \limsup_{n\to\infty} \mathbb{P}\left(\max_{k\leq|i|\leq r_n} Z_i > a_n u \mid Z_0 > a_n u \right) = 0. \tag{19}$$

"with (a_n) a sequence such that $\lim_{n\to\infty} n\mathbb{P}(|Z_0| > a_n) = 1$" and $r_n \to \infty$ is an integer sequence such that $r_n = o(n)$.

References

1. Asmussen, S. (2003). *Applied probabilities and queues*. New York: Springer.
2. Bartkiewicz, K., Jakubowski, A., Mikosch T., & Wintenberger, O. (2011). Stable limits for sums of dependent infinite variance random variables. *Probability Theory and Related Fields, 150*(3), 337–372.
3. Basrak, B., & Segers, J. (2009). Regularly varying multivariate time series. *Stochastic Processes and Their Applications, 119*(4), 1055–1080.
4. Bertail, P., Ciolek, G., & Tillier, C. (2018). Robust estimation for Markov chains with application to PDMPs. In *Statistical inference for piecewise-deterministic Markov processes*. Hoboken: Wiley.

5. Bertail, P., Clémençon, S., & Tillier, C. (2016) Extreme values statistics for Markov chains with applications to finance and insurance. In F. Longin (Ed.), *Extreme events in finance: a handbook of extreme value theory and its applications*. Hoboken, NJ: Wiley. https://doi.org/10.1002/9781118650318.ch7
6. Bertail, P., Clémençon, S., & Tressou, J. (2008). A storage model with random release rate for modelling exposure to food contaminants. *Mathematical Bioscience and Engineering, 5*, 35–60.
7. Bertail, P., Clémençon, S., & Tressou, J. (2010). Statistical analysis of a dynamic model for food contaminant exposure with applications to dietary methylmercury contamination. *Journal of Biological Dynamics, 4*, 212–234.
8. Bondesson, L. (2006). Shot-noise processes and distributions. In *Encyclopedia of statistical sciences* (Vol. 12). Hoboken: Wiley.
9. Chernick, M. R., Hsing, T., & McCormick, W. P. (1991). Calculating the extremal index for a class of stationary sequences. *Advances in Applied Probability, 25*, 835–850.
10. Cline, D. B. H., & Samorodnitsky, G. (1994). Subexponentiality of the product of independent random variables. *Stochastic Process and Their Applications, 49*, 75–98.
11. Doney, R. A., O'Brien, G. L. (1991). Loud shot noise. *Annals of Applied Probabilities, 1*, 88–103.
12. Embrechts, P., Kluppelberg, C., & Mikosch, T. (1997). *Modelling Extremal Events for Insurance and Finance: Vol. 33. Applications of mathematics*. Berlin, Springer.
13. Homble, P., & William P. (1995). Weak limit results for the extremes of a class of shot noise processes. *Journal of Applied Probability, 32*, 707–726.
14. Hsing, T. L., & Tengels, J. L. (1989). Extremal properties of shot noise processes. *Advances in Applied Probability, 21*, 513–525.
15. Hult, H., & Samorodnitsky, G. (2008). Tail probabilities for infinite series of regularly varying random vectors. *Bernouilli, 14*, 838–864.
16. Leadbetter, M. R. (1983). Extremes and local dependence in stationary sequences. *Zeitschrift für Wahrscheinlichkeitsth, 65*, 291–306.
17. Lefebvre, M., & Guilbault, J. L. (2008). Using filtered Poisson processes to model a river flow. *Applied Mathematical Modelling, 32*(12), 2792–2805.
18. McCormick, W. P. (1997). Extremes for shot noise processes with heavytailed amplitudes. *Journal of Applied Probability, 34*, 643–656.
19. McCormick, W. P., & Seymour, L. (2001). Extreme values for a class of shot-noise processes. In I. V. Basawa, C. C. Heyde, & R. L. Taylor (Eds.) *Selected proceedings of the symposium on inference for stochastic processes. Lecture notes-monograph series* (Vol. 37, pp. 33–46). Beachwood, OH: Institute of Mathematical Statistics. https://doi.org/10.1214/lnms/1215090682
20. Mikosch, T. (2010). *Non life insurance mathematics*. Berlin: Springer.
21. Perfekt, R. (1994). Extremal behaviour of stationary Markov chains with applications. *The Annals of Applied Probability, 4*(2), 529–548.
22. Resnick, S. I. (2007). *Heavy-Tail Phenomena and statistical modeling. Springer series in operations research and financial engineering*. New-York: Springer.
23. Tillier, C., & Wintenberger, O. (2017). Regular variation of a random length sequence of random variables and application to risk assessment. *Extremes, 21*, 27–56. https://doi.org/10.1007/s10687-017-0297-1
24. Weng, C., Zhang, Y., & Tang, K. S. (2013). Tail behavior of Poisson shot noise processes under heavy-tailed shocks and actuarial applications. *Methodology and Computing in Applied Probability, 15*, 665–682.

Subsampling for Big Data: Some Recent Advances

P. Bertail, O. Jelassi, J. Tressou, and M. Zetlaoui

Abstract The goal of this contribution is to develop subsampling methods in the framework of big data and to show their feasibility in a simulation study. We argue that using different subsampling distributions with different subsampling sizes brings a lot of information on the behavior of statistical procedures: subsampling allows to estimate the rate of convergence of different procedures and to construct confidence intervals for general parameters including the generalization error of an algorithm in machine learning.

1 Introduction

Collecting data is becoming faster and faster but standard statistical tools are not adapted to analyze such big datasets. Because optimization methods are too time consuming even for polynomial complexities and most of the time standard methods (for instance, maximum likelihood estimations) require too many access to the data. As consequences, maximum likelihood estimations or general methods based on contrast minimization may be difficult to implement on large scale. Subsampling techniques is a well-known remedy to the apparent intractability of learning from databases of explosive size. Such an approach has been implemented in many applied problems and has been, for instance, developed in [22]. It is also at the core of some recent developments on survey sampling method in the framework of big data (see [6–8]).

P. Bertail (✉) · M. Zetlaoui
Modal'X, UPL, Université Paris Nanterre, Nanterre, France

O. Jelassi
Telecom ParisTech, Paris, France
e-mail: ons.jelassi@telecom-paristech.fr

J. Tressou
MORSE, INRA-MIA, Paris, France

P. Bertail et al. (eds.), *Nonparametric Statistics*, Springer Proceedings in Mathematics & Statistics 250, https://doi.org/10.1007/978-3-319-96941-1_13

One of the main theoretical ideas underlying subsampling related methods is to use the universal validity of the subsampling method as proved in [25] and further by Bertail et al. [9, 11] for general converging or diverging statistics. We recall that these authors proved, under the minimal assumptions that the statistics of interest has a nondegenerate distribution (for some potentially unknown rate of convergence) which is continuous at the point of interest that a subsampling distribution constructed with a much smaller size than the original one is a correct approximation of the distribution of the statistics of interest. This result allows then to extrapolate repeated inferential methods from smaller sizes to bigger size. Such ideas are not new (see [16] or even the first works of Mahalanobis in the 1930s). They have also been developed in some earlier works by Bickel and Yahav [13] about bootstrap and Richardson extrapolation, when the computer capacities were not sufficient to treat even moderate sample size. Such methods are themselves related to well-known numerical methods (see, for instance, [21]). However, most of these methods rely on an adequate standardization of the statistics of interests (see the discussion about interpolations and extrapolations in [4, 10] and Bickel et al. [14]). Such standardization may be hard to obtain for complicated procedure (including statistical learning procedure) and even more difficult to extrapolate to very large sample size.

Indeed to extrapolate the value of statistics from smaller scales to a large one, we need first to be able to determine or estimate τ_n, the rate of convergence of the procedure of interest (a statistic or a statistical learning algorithm) : in many situations, this task is difficult, because this rate depends itself on the true data generating mechanism. We will present a variant of the subsampling rate estimation methodology of [9, 11] to derive a consistent estimator of the rate $\tau_.$ for moderate sizes. It is, then, possible to extrapolate its value to **large** datasets and construct confidence intervals for many difficult to analysis procedures. The underlying idea is that it is possible to construct several subsampling distributions of the statistics/procedure of interest T_n (without standardization). The speed at which it diverges or degenerates to a Dirac measure as $n \to \infty$ is directly related to the adequate standardization τ_{b_n}. As a consequence, constructing several subsampling distributions for different choices of b_n gives valuable information on the shape of τ_n as a function of n which allows to extrapolate to bigger size.

In this chapter, we show that subsampling gives invaluable information on possible **Variability** (or robustness toward the possible **Values**) of the procedure of interest. We will prove the validity of the method and show how it can be practically efficiently implemented. We will also discuss how to implement the method when the size of the dataset evolves in time (a fact linked to the **Velocity** problem).

In Sect. 1, we present the state of the art of the subsampling methods. Then, we demonstrate how we estimate the convergence rate of the samples statistics distribution. In Sect. 3, we give our mathematical results and the subsamples sizes. We show how we integrate the dynamic aspect of the big data environments (specially in case of streaming and IoT) in our method. Section 5 presents our results on simulated data. We implemented subsampling techniques on potentially time-consuming procedures.

2 Subsampling Methods for Big Data

2.1 Definition

In [9, 11, 25], a general subsampling methodology has been put forth for the construction of large-sample confidence regions for a general unknown parameter $\theta = \theta(P) \in \mathbb{R}^q$ under very minimal conditions. Consider $\underline{X}_n = (X_{1,\ldots}, X_n)$ an i.i.d. sample. To construct confidence intervals for θ, we require an approximation to the (generally unknown) sampling distribution (under P) of a standardized statistic $T_n = T_n(\underline{X}_n)$ that is consistent for θ at some *known* rate τ_n. In the statistical learning methodology, θ may be the Bayes Risk and T_n the estimated risk linked to a given algorithm. Or in the framework of prediction, θ may be a value to predict and T_n a predictor.

To fix some notations, assume that there is a nondegenerate asymptotic distribution for the centered "dilated" statistic $\tau_n(T_n - \theta)$, i.e., there is a distribution $K(x, P)$, continuous in x, such that for any real number x,

$$K_n(x, P) \equiv \Pr{}_P\{\tau_n(T_n - \theta) \leq x\} \underset{n\to\infty}{\longrightarrow} K(x, P) \tag{1}$$

then the subsampling distribution with subsampling size b_n is defined by

$$K_{b_n}(x \mid \underline{X}_n, \tau_.) \equiv q^{-1} \sum_{i=1}^{q} 1\{\tau_{b_n}(T_{b_n,i} - T_n) \leq x\}, \tag{2}$$

where $q = \binom{n}{b_n}$ and $T_{b_n,i}$ is a value of the statistic of interest calculated on a subset of size b_n chosen from $\underline{X}_n = \{X_1, \ldots, X_n\}$. Using very simple U-statistics arguments, it was shown in [25] that the subsampling methodology "works", provided that

$$b_n \underset{n\to\infty}{\longrightarrow} \infty \tag{3}$$

and

$$\frac{b_n}{n} \underset{n\to\infty}{\longrightarrow} 0. \tag{4}$$

and

$$\frac{\tau_{b_n}}{\tau_n} \underset{n\to\infty}{\longrightarrow} 0, \tag{5}$$

meaning that, under these assumptions, we have

$$K_{b_n}(x \mid \underline{X}_n, \tau_.) - K_n(x, P) \underset{n\to\infty}{\longrightarrow} 0,$$

uniformly in x over neighborhoods of continuity points of $K(x, P)$. The key point is that when T_n is replaced by θ in (2) we obtain a U-statistics of degree b_n whose variance is of order $\frac{b_n}{n}$, condition (3) ensures that the mean of this U-statitics $K_{b_n}(x, P)$ converges to a limiting distribution, condition (4) ensures that the variance of the U-statistics converges to 0, whereas conditions (1) and (5) ensure that we can replace the re-centering T_n by the true value of the parameter. When choosing an adequate re-centering (for instance, the median of the subsampling distribution) then condition (5) may be completely dropped as discussed below.

This method may be generalized to dependent data in a weakly mixing context but even for long range dependent series by constructing blocks of contiguous observations of length b_n. Since for large databases, computing q values of the statistics $T_{b_n,i}$ may be unfeasible it is recommended to use its Monte-Carlo approximation

$$K_{b_n}^{(B)}(x \mid \underline{X}_n, \tau_.) = B^{-1} \sum_{j=1}^{B} 1\{\tau_{b_n}(T_{b_n,j} - T_n) \leq x\},$$

where now $\{T_{b_n,j}\}_{j=1,\ldots,B}$ are the values of the statistic calculated on B subsamples of size b_n taken without replacement from the original population. It can be easily shown by incomplete U-statitics arguments that if B is large then the error induced by the Monte-Carlo step is only of size

$$K_{b_n}^{(B)}(x \mid \underline{X}_n, \tau_.) - K_{b_n}(x \mid \underline{X}_n, \tau_.) = 0_P\left(\frac{1}{\sqrt{B}}\right),$$

so that if one controls the error induced by $K_{b_n}(x \mid \underline{X}_n, \tau_.)$ on the true distribution, it is always possible to find a value of B (eventually linked to n) so that the Monte-Carlo approximation is negligible.

This approach may also be used to infer on the generalization capability of a given algorithm or an estimation method by estimating some risk θ by some empirical counterpart. Moreover, the centering by T_n may not be adapted for big data, since calculation of T_n itself may be too complicated (either because the exact size is unknown or because the complexity of the algorithm and the cost induced by retrieving all the information are too high).

The main reason for using the centering by T_n (which converges to the true value θ) is simply due to the fact that under (5)

$$\begin{aligned}
\tau_{b_n}(T_{b_n,j} - T_n) &= \tau_{b_n}(T_{b_n,j} - \theta) + \tau_{b_n}(\theta - T_n) \\
&= \tau_{b_n}(T_{b_n,j} - \theta) + O_P\left(\frac{\tau_{b_n}}{\tau_n}\right) \\
&= \tau_{b_n}(T_{b_n,j} - \theta) + o_P(1).
\end{aligned}$$

This suggests to use any centering whose convergence rate is actually faster than τ_{b_n}. This is, for instance, the case if one constructs a subsampling distribution without any centering nor standardization with a subsampling size $m_n >> b_n$ such that $\frac{b_n}{m_n} \to 0$ and $\frac{\tau_{b_n}}{\tau_{m_n}} \to 0$. In this case we have that $\frac{1}{B}\sum_{j=1}^{B} T_{m_n,j}$ which is a proxy of $\frac{1}{q}\sum_{j=1}^{q} T_{m_n,j}$ (with an error of size $1/\sqrt{B}$) converges to θ at a rate as fast as τ_{m_n} (provided that the expectation of these quantities exists). The same results hold if one chooses the median rather than the mean (as considered in [9, 11]), this avoid additional assumption (existence of expectation) which may be difficult to check in practice. In the following, we will denote by $\widehat{\theta}_{m_n}$ any centering based on a subsampling distribution m_n such that

$$\begin{aligned}\tau_{b_n}(T_{b_n,j} - \widehat{\theta}_{m_n}) &= \tau_{b_n}(T_{b_n,j} - \theta) + O_P\left(\frac{\tau_{b_n}}{\tau_{m_n}}\right)\\ &= \tau_{b_n}(T_{b_n,j} - \theta) + o_P(1).\end{aligned}$$

For simplicity, we use the same notation as before and define the subsampling distribution as

$$K_{b_n}(x \mid \underline{X}_n, \tau_.) \equiv q^{-1}\sum_{i=1}^{q} 1\{\tau_{b_n}(T_{b_n,i} - \widehat{\theta}_{m_n}) \le x\},$$

and its Monte-Carlo approximation

$$K_{b_n}^{(B)}(x \mid \underline{X}_n, \tau_.) = B^{-1}\sum_{j=1}^{B} 1\{\tau_{b_n}(T_{b_n,j} - \widehat{\theta}_{m_n}) \le x\}.$$

3 Estimating the Convergence Rate

The main drawback of this approach which is also inherent to the methods proposed in [22] is the knowledge of the standardization (or rate) τ_n. However, this rate may be easily estimated at least when the rate of convergence is of the form $\tau_n = n^{\alpha}L(n)$ where α is unknown and L is a normalized slowly varying function that is such that $L(1) = 1$ and for any $\lambda > 0$, $\lim_{x\to\infty}\frac{L(\lambda x)}{L(x)} = 1$ (see [15]).

For this, we will first construct the subsampling without any standardization. Denote by

$$K_{b_n}(x \mid \underline{X}_n) \equiv K_{b_n}(x \mid \underline{X}_n, 1) = q^{-1}\sum_{i=1}^{q} 1\{T_{b_n,i} - \widehat{\theta}_{m_n} \le x\}$$

the subsampling distribution of the root $(T_n - \theta)$.

Given a distribution F on the real line and a number $t \in (0, 1)$, we will let $F^{-1}(t)$ denote the quantile transformation, i.e., $F^{-1}(t) = \inf\{x \ : F(x) \geq t\}$, which reduces to the regular inverse of the function F if F happens to be continuous non-decreasing. Note that we have

$$K_{b_n}(x\ \tau_{b_n}^{-1} \mid \underline{X}_n) = K_{b_n}(x \mid \underline{X}_n, \tau_.) \tag{6}$$

and thus it is easy to see as in [10] that

$$\begin{aligned} K_{b_n}^{-1}(t \mid \underline{X}_n, \tau_.) &= \tau_{b_n}\ K_{b_n}^{-1}(t \mid \underline{X}_n) && (7)\\ &= K^{-1}(t, P) + o_P(1). && (8) \end{aligned}$$

If $\tau_n = n^{\alpha} L(n)$ where L is a positive normalized slowly varying function , by the Karamata representation theorem, there exists $\epsilon(.)$, $\epsilon(u) \underset{u \to \infty}{\longrightarrow} 0$ such that $L(n) = \exp \int_1^n u^{-1}\epsilon(u)du$, and (8) may be written

$$\log\left(|K_{b_n}^{-1}(t \mid \underline{X}_n)|\right) = \log\left(|K^{-1}(t, P)|\right) - \alpha \log(b_n) + \int_1^{b_n} u^{-1}\epsilon(u)du + o(1)$$

It follows that if we choose two different subsampling sizes satisfying the conditions stated before and such that $b_{n_1}/b_{n2} = e$, then we have

$$\begin{aligned} &\log\left(|K_{b_{n_1}}^{-1}(t \mid \underline{X}_n)|\right) - \log\left(|K_{b_{n2}}^{-1}(t \mid \underline{X}_n)|\right) && (9)\\ &= \alpha + \int_{b_{n_1}}^{b_{n_1}} u^{-1}\epsilon(u)du + o(1)\\ &= \alpha + o(1) \end{aligned}$$

uniformly in t. The last equality relies on the property of the Karamata distribution and slowly varying functions. Indeed, we have $\frac{L(b_{n_1})}{L(b_{n_2})} = \frac{L(eb_{n_2})}{L(b_{n_2})} \to 1$ as $b_{n_2} \to \infty$. This trick based on using sample sizes of the same order avoids the complicated constructions used in [9]. This suggests that the parameter α may be simply estimated by averaging this quantity over several quantiles and/or several subsampling distributions even in presence of a slowly varying functions.

Since computing these three subsampling distributions requires mainly the computation of $B * (b_{n_1}(1 + e) + m_n)$ values of the statistic of interest (whose calculus may be easily parallelized), we will essentially have to choose a resampling size which does not perturb too much the subsampling distributions (which is big enough) but sufficiently small so that the cost in computing these quantities is small in comparison with the global cost of computing a single statistic over the whole database.

Another solution is to consider regression of log range on the log of the subsampling size by remarking that we also have for any $0 < t_1 < 1/2 < t_2 < 1$,

$$\log\left(K_{b_n}^{-1}(t_2 \mid \underline{X}_n) - K_{b_n}^{-1}(t_1 \mid \underline{X}_n)\right)$$
$$= \log\left(K^{-1}(t_2, P) - K^{-1}(t_1, P)\right) - \alpha \log(b_n)$$
$$+ \int_1^{b_n} u^{-1}\epsilon(u)du + o(1)$$

The main interest in this version is that it does not depend on the re-centering of the subsampling distribution. One may choose, for instance, $t_1 = 0.75$ and $t_2 = 0.25$, corresponding to the log of inter-quartiles, that will be used in our simulations. Then we may choose two different sizes $b_{n,1}$ and $b_{n,2} = b_{n,1}/e$ then we have similarly that

$$\log\left(\frac{K_{b_{n,1}}^{-1}(t_2 \mid \underline{X}_n) - K_{b_{n,1}}^{-1}(t_1 \mid \underline{X}_n)}{K_{b_{n,2}}^{-1}(t_2 \mid \underline{X}_n) - K_{b_{n,2}}^{-1}(t_1 \mid \underline{X}_n)}\right) = \alpha + o(1) \tag{10}$$

By looking simply at two subsampling distributions, we are able to estimate the parameter α.

4 Main Results

4.1 A General Subsampling Theorem

For simplicity, we will now assume that $\tau_n = n^\alpha$. The general case $\tau_n = n^\alpha L(n)$ may be treated similarly with a few additional assumptions on the slowly varying function (see [11]). For a given estimator of τ_n, typically $\widehat{\tau}_n = n^{\widehat{\alpha}}$, we will use

$$\widehat{K}_n(x, P) = \Pr{}_P\{\widehat{\tau}_n(T_n - \theta) \leq x\}$$

Theorem 1 *Assume that (1) holds for $\tau_n = n^\alpha$, for some $\alpha > 0$ and some $K(x, P)$ continuous in x; also assume (3) and (4). Let $\hat{\alpha} = \alpha + o_P((\log n)^{-1})$, and put $\widehat{\tau}_n = n^{\hat{\alpha}}$. Then*

$$\sup_x |K_{b_n}(x \mid \underline{X}_n, \widehat{\tau}_.) - \widehat{K}(x, P)| = o_P(1). \tag{11}$$

Let $\beta \in (0,1)$, and let $c_n(1-\beta) = K_{b_n}^{-1}(1-\beta \mid \underline{X}_n, \widehat{\tau}_.)$ be the $1-\beta$th quantile of the subsampling distribution $K_{b_n}(x \mid \underline{X}_n, \widehat{\tau}_.)$. Then

$$\Pr{}_P\{\hat{\tau}_n(T_n - \theta) \geq c_n(1-\beta)\} \underset{n\to\infty}{\longrightarrow} \beta. \tag{12}$$

Thus with an asymptotic coverage probability of $1-\beta$, we have

$$-\hat{\tau}_n^{-1} c_n(1-\beta) \leq \theta - T_n$$

and by symmetry

$$\theta - T_n \leq \hat{\tau}_n^{-1} c_n(\beta).$$

Recall that $K_{b_n}^{-1}(1-\beta \mid \underline{X}_n, \widehat{\tau}_.)$ is the $1-\beta$ quantile of the rescaled subsampling distribution. Assuming that B is such $(B+1)\beta$ is an integer (for instance, for $\beta = 5\%$ $\beta = 1\%$ then $B = 999$ is fine). Then, it is simply given by $\tau_{b_n}(T_{b_n}^{((B+1)(1-\beta))} - \widehat{\theta}_{m_n})$ where $T_{b_n}^{((B+1)(1-\beta))}$ is the $(B+1)(1-\beta)$ largest value over the B sub-sampled values. It then follows that the bound is given by

$$\widehat{B}_n = \frac{\widehat{\tau}_{b_n}}{\hat{\tau}_n}(T_{b_n}^{((B+1)(1-\beta)} - \widehat{\theta}_{m_n}). \tag{13}$$

A straightforward utilization of this result is to compare generalization capability of statistical learning algorithm, when n is so large that most algorithms, even with polynomial complexity, may be hardly used in a reasonable time.

This result also allows to build confidence intervals for θ. For this, assume that $(B+1)\beta/2$ is an integer. In that case, by combining (12) and (13) and choosing $\widehat{\theta}_{m_n} = T_n$, a confidence interval for θ is simply given by

$$\widehat{\theta}_{m_n} - \frac{\widehat{\tau}_{b_n}}{\hat{\tau}_{m_n}}\left(T_{b_n}^{((B+1)(1-\beta/2))} - \widehat{\theta}_{m_n}\right) \leq \theta$$

$$\leq \widehat{\theta}_{m_n} - \frac{\widehat{\tau}_{b_n}}{\hat{\tau}_{m_n}}\left(T_{b_n}^{((B+1)\beta/2)} - \widehat{\theta}_{m_n}\right)$$

which unfortunately is not on the right scale. However, if T_n may be computed on the whole database, a scalable confidence interval is simply given by

$$T_n - \frac{\widehat{\tau}_{b_n}}{\hat{\tau}_n}\left(T_{b_n}^{((B+1)(1-\beta/2))} - T_n\right) \leq \theta$$

$$\leq T_n - \frac{\widehat{\tau}_{b_n}}{\hat{\tau}_n}\left(T_{b_n}^{((B+1)\beta/2)} - T_n\right).$$

In our simulation study, it is seen that the variability of the data may be so high that somehow there is very little difference between confidence interval even if the first one should be larger.

Example 1 (Estimating a Parameter on a Large Database: Logistic Regression) We consider here a very simple parametric model to highlight some inherent difficulties with subsampling. Consider a linear logistic regression model with parameter $\theta = (\beta_0, \beta) \in \mathbb{R} \times \mathbb{R}^d$. Let X be a d-dimensional marginal vector of the input random variables. The linear logistic regression model related to a pair (X, Y) can be written as

$$\mathbb{P}_\theta\{Y = +1 \mid X\} = \exp(\beta_0 + \beta^T X)/(1 + \exp(\beta_0 + \beta^T X)).$$

In high-dimension, i.e. when d is very large and for large n, the computation of the full parametric maximum likelihood estimator (MLE) of θ may be difficult to obtain in a reasonable time. We assume that $d << n$ but also that the subsampling sizes which will be used are such that $d << b_n$. For unbalanced populations (a lot of 1's in comparison with 0's and vice versa), the probability to get a subsample with only unit values (or zeros) may be high and the MLE will not be convergent (a similar problem appears if the labels are fully separated). This is by no means contradictory with the asymptotic validity of subsampling in this case: actually it has been shown in [23] that the true variance of the MLE in a finite population is $+\infty$. Subsampling simply reproduces this fact on a smaller scale. In that case, one should condition on the fact that the ratio of the numbers of 1's to the number of 0's is not too small (or not too close to 1). Else, the subsample should be eliminated.

Even on reasonable sizes, this estimation procedure may be useful. For instance in R, with 1 GB of memory, the usual libraries (sampleSelection, glm) fail to estimate the model with a size of $n = 10^7$ observations (for capacity reasons), whereas it takes only 12 s to get a bound with $B = 999$ replications of the procedure and a subsampling size of the order $b_n = n^{1/3}$. First, we do not estimate the rate of convergence since we know that the rate will be of order $\tau_n = n^{1/2}$. The true extrapolated bound obtained by subsampling is of the same order as the true one, with an error on the variance less than 10^{-5}, for all simulations. If we estimate the rate of convergence with $J = 29$ subsampling distributions based on subsampling sizes equal $n^{1/3+j/(3(J-1))}$, $j = 0, \ldots 28$,, the largest subsampling size is of order $n^{2/3}$, we then get similar results but we need in that case 999×29 simulations : it then takes 6 min to complete these tasks on the same computer.

Example 2 (Pattern Recognition) It is assumed that $((X_1, Y_1), \ldots, (X_N, Y_N)$ is a sample of i.i.d. random pairs taking their values in some measurable product space $\mathscr{X} \times \{-1, +1\}$. In this standard binary classification framework, the r.v. X models some observations are used to predict the binary label Y. The distribution P can also be described by the pair (F, η) where $F(dx)$ denotes the marginal distribution of the input variable X and $\eta(x) = \mathbb{P}\{Y = +1 \mid X = x\}$, $x \in \mathscr{X}$, is the *conditional distribution*. The goal is to build a measurable classifier $\phi : \mathscr{X} \mapsto \{-1, +1\}$ with minimum risk defined by

$$L(Y, \phi(X)) \stackrel{def}{=} \mathbb{I}\{\phi(X) \neq Y\} \tag{14}$$

where $\mathbb{I}\{.\}$ is the indicator function. It is well known that the *Bayes classifier* $\phi^*(x) = 2\mathbb{I}\{\eta(x) > 1/2\} - 1$ is a solution of the risk minimization problem over the collection of all classifiers, $\mathscr{F}$, defined on the input space $\mathscr{X}$. In that case we define the minimizer

$$\widehat{\phi}_n = \arg\min_{\phi \in \mathscr{F}} \frac{1}{n} \sum_{i=1}^{n} \mathbb{I}\{\phi(X_i) \neq Y_i\}. \tag{15}$$

The statistics of interest is now the empirical error on a test set of this optimal classifier. It is now possible to apply the subsampling procedure to different classes of functions (algorithm) to estimate their prediction capability.

4.2 How to Choose the Subsampling Sizes

The choice of the subsampling size is a delicate subject which has been discussed in very few papers including [5, 12, 12, 19]. The main idea underlying most propositions is to construct several subsampling distributions by using two different subsampling sizes, say b_n and $b_{n,2} = qb_n$ for $q \in]0, 1[$. It is easy to see that when the subsampling distribution is a convergent estimator of the true distribution then the distance d between the subsampling distribution and the true one is stochastically equivalent to

$$d(K_{b_n}, K_{qb_n}).$$

The idea is then to find the largest b_n, which minimizes this quantity. Several distances (Kolmogorov distance, Wasserstein metrics, etc.) may be used.

Of course, for large datasets such method is very computationally expensive, so that we recommend only to choose a limited range of values for b_n and to discretize this range so as to compute the distance $d(K_{b_n}, K_{qb_n})$ only on a limited number of points and to select the ones which minimize this quantity.

Another empirical approach has been proposed in [5] based on the bad behavior (high volatility) of subsampling distributions for too large subsampling sizes. Indeed, up to the re-centering which converges quickly to the true value of the parameter, a subsampling distribution may simply be seen as a U-statistic with varying kernel of size b_n. The main tools for studying the behavior of subsampling distribution are Hoeffding decomposition of the U-statistics and empirical process theory as considered in [1] and [20]. A subsampling distribution may be roughly seen as a U-statistic with increasing size b_n so that quantiles may also themselves be seen as U-quantiles. The difficulty for choosing the subsampling size is that in comparison with U-statistics with fixed degree, the linear part of the U-statistic is not always the dominating part in the Hoeffding decomposition. For rather small or moderate b_n, it can be shown that the U-statistic is asymptotically normal with a convergence rate of order $\sqrt{\frac{n}{b_n}}$. However, when b_n becomes too large, the remainder in the Hoeffding decomposition dominates and the U-statistic behaves very erratically (we conjecture that the limiting distribution belongs to linear combination of Wiener Chaos). Nevertheless, this idea gives a very easy and practical way of choosing the optimal subsampling size for the problem of interest. The idea is simply to look at the quantiles of subsampling distributions and to find the largest value such that the quantile remains stable.

5 Subsampling Algorithm Versus Velocity and Variability

5.1 Subsampling in a Growing Environment

The size of some database may evolve quickly in time so that we may wish to implement simple subsampling techniques based on previous observations of subsamples without having to access to the whole database again. How is it possible to use the techniques exposed before when the size of the database is large and increase so fast that taking new subsamples may be too computer expensive? To solve this problem, we present a very simple sequential algorithm.

The idea is as follows: assume that at time t, we have obtained a subsample without replacement of size b_n (uniformly) from the original population n. That is, the probability of a given subsample is $\binom{n}{b_n}^{-1}$. At time $t+1$, the new sample size is $n+1$. Then for this newcomer proceed as follows:

- keep the original subsample with probability $1-b_n/(n+1)$, that is simply draw a Bernoulli rvs B_1 with parameter $1-b_n/(n+1)$ and stay with the same subsample if one gets a 1,
- else with probability $b_n/(n+1)$, choose one element of the current subsample (without replacement, uniformly with probability $1/b_n$) and replace it by this newcomer.

If several newcomers arrive at the same date, then use sequentially the same algorithm by increasing the size of the population. Notice that this algorithm may be easily implemented sequentially to update all the subsamples already obtained at some given time.

The arguments below show that the resulting algorithm is the realization of subsampling without replacement from the total new population.

It may be simply proved by recurrence. Indeed, assume that the probability of the sample is $\binom{n}{b_n}^{-1}$ then

- if $B_1 = 1$, the probability of the new sample is $\binom{n}{b_n}^{-1} * (1 - \frac{b_n}{n+1}) = \binom{n+1}{b_n}^{-1}$
- if $B_1 = 0$, the probability of the new sample is $\binom{n}{b_n}^{-1} * \frac{b_n}{n+1}(\frac{1}{b_n} + (n-b_n)/b_n) = \binom{n+1}{b_n}^{-1}$.

Such result is fully proved in [24]. It follows that the corresponding subsample at any step is actually a subsample obtained without replacement from the total population.

If we want to increase the size of subsample, starting from a subsample of size b_n in a population of size n then we simply draw uniformly in the $n - b_n$ remaining observation an individual (with probability $1/(n - b_n)$). It may be sometimes easier (for instance, using Apache Spark) to use sampling with replacement. It is known in that case that when b_n is small enough such that $\frac{b_n}{\sqrt{n}} \to 0$, then the probability to draw the same individual twice converges to 0, for large n . Indeed when $\frac{b_n}{\sqrt{n}} \to 0$, by Stirling formula we have $\frac{\binom{n}{b_n}}{n^{b_n}} \to 1$, so that with and without replacement methods are asymptotically equivalent under this condition.

5.2 *Subsampling to Assess Variability and Stability (Veracity) of Learning Algorithms (on the Data)*

One question which is also of interest is whether the method of interest is stable over the whole database especially if its size becomes more and more important. This question is of prime importance when the data itself is indexed by time. A first approach hardly applicable with big data is to test for structural changes in the parameter of interest. This problem has been extensively studied in the econometric literature in the case of a single break-point and has been extended to various econometric specification (nonlinear regression model, time series models, nonlinear simultaneous equations models, etc…) and different stability problems (tests of finite multiple structural changes, tests of cross sectional consistency), see, for instance, [17, 18] and the references therein. The intuition behind the proposed tests is that if we split the sample into two subsamples, the set of observations before and after a date t, then the difference between estimations (or monotone transformations) should be equal to 0 if there is no structural change.

A simple generalization of this idea is to base the estimation of the parameter of interest θ or a risk indicator over subsamples of growing sizes. The intuition behind this idea is that, under the hypothesis of global stability, all the estimations over subsets of observations must be close to the true parameter. This approach is closely related to Jackknife techniques, used, for instance, in the detection of outliers (see [3]). Subsampling can actually be seen as the ($N - b_n$ out of n)-jackknife.

The algorithm that we propose is the following :

In a population of size n, consider B subsamples of size b_n , denoted by S_j^{n,b_n}, $j = 1\ldots, B$ and compute the corresponding statistics of interest $T_{b_n,j} = T(S_j^{n,b_n})$. Then to evaluate the impact of a new small set of observation of size l (with $l << b_n$) say s_l, we now consider all the new samples of size b_n+l by including s_l, $S_j^{n+l,b_n+l} = S_j^{n,l} \cup s_l$ and compute the corresponding statistics $T_{b_n+l,j} = T_{b_n+l}(S_j^{n+l,b_n+l})$.

If the observation comes from the same distribution as before (the model is stable) or if the procedure is robust (in terms of quantitative robustness), then the corresponding subsampling distributions defined by (for any re-centering $\widehat{\theta}_{m_n}$)

$$K_{b_n}^{(B)}(x \mid \underline{X}_n, \widehat{\tau}_.) = B^{-1} \sum_{j=1}^{B} 1\{\widehat{\tau}_{b_n}(T_{b_n,j} - \widehat{\theta}_{m_n}) \le x\}$$

$$K_{b_n+l}^{(B)}(x \mid \underline{X}_{n+l}, \widehat{\tau}_.) = B^{-1} \sum_{j=1}^{B} 1\{\widehat{\tau}_{b_n+l}(T_{b_n+l,j} - \widehat{\theta}_{m_n}) \le x\}$$

should be close. In particular, we should have, following the preceding argument that for any t_1 and t_2 , the ratio of the ranges of the two distributions close to 0 that is the ratio

$$r_{b_n,l} = \frac{K_{b_n}^{(B)-1}(t_2 \mid \underline{X}_n, \widehat{\tau}_.) - K_{b_n}^{(B)-1}(t_1 \mid \underline{X}_n, \widehat{\tau}_.)}{K_{b_n+l}^{(B)-1}(t_2 \mid \underline{X}_n, \widehat{\tau}_.) - K_{b_n+l}^{(B)-1}(t_1 \mid \underline{X}_{n+l}, \widehat{\tau}_.)} \to 1$$

Notice that this quantity is independent of the choice of the centering $\widehat{\theta}_{m_n}$. Thus a simple graphical diagnostic for ensuring that the model is stable or the estimator robust is to plot this quantity for some given values of t_1 and t_2, for instance $t_1 = 1 - \beta/2$, $t_2 = \beta/2$, for $\beta = 0.01, 0.02, \ldots .0.25$.

It may be difficult to control the rate of convergence of this quantity to one in the general case with unknown τ_{b_n}. In the following, we will assume for simplicity that the rate of convergence $\tau_n = n^{1/2}$ and give some hint on the optimal choice of b_n, when dealing with simple linear statistics, often encountered when dealing with empirical risks. Indeed from Babu and Singh [2], for any statistics which are a function of moments, and provided that we have sufficient moments (and a absolutely continuous part for this statistics) then we have an Edgeworth expansion (uniform in x)

$$K_{b_n}(x, |\underline{X}_n, \tau_.) = \Phi(\frac{x}{\sigma_\infty}) + O_P(\frac{1}{\sqrt{b_n}}) + O(\frac{b_n}{n})$$

where σ_∞^2 is the asymptotic variance of the quantity of interest. This result holds almost surely up to $O(\frac{\log(b_n)}{\sqrt{b_n}}) + O(\frac{b_n}{n})$. It follows that in that case by simple inversion of this expansion that we have

$$r_{b_n,l} = \frac{\sigma_\infty(\Phi^{-1}(t_2) - \Phi^{-1}(t_1)) + O(\frac{\log(b_n)}{\sqrt{b_n}}) + O(\frac{b_n}{n})}{\sigma_\infty(\Phi^{-1}(t_2) - \Phi^{-1}(t_1)) + O(\frac{\log(b_n+l)}{\sqrt{b_n+l}}) + O(\frac{b_n+l}{n+l})}$$
$$= 1 + O(\frac{\log(b_n)}{\sqrt{b_n}}) + O(\frac{b_n}{n})\ a.s.$$

It follows that the optimal subsampling size in that case, which equilibrates the two errors is of size $b_n = (n\log(n))^{2/3}$ yielding an a.s. approximation of order $\frac{\log(n)}{n^{1/3}}$ a.s. Thus we can construct an as. confidence interval for testing the equality of $r_{b_n,l}$ to 1 and detect variability in the data.

6 Some Empirical Results

In this section, the implementations were executed under R on a standard PC with a 5 GHz Intel processor and 2G of Ram.

6.1 *Maximum Likelihood Estimation for a Simple Probit Model (See Example 1)*

We consider the framework of Example 1. For this we simulate the toy probit model

$$Y_i = \begin{cases} 1 \text{ if } 3X_i + \varepsilon_i > 0 \\ 0 \text{ if } else \end{cases}$$

with X_i and ε_i independent $N(0, 1)$ random variables. We choose, respectively, $n = 10^6$ and $n = 10^7$.

The mean of the estimations of β (and the variances) over the 999 repetitions with the subsampling procedure are given in table for different subsampling sizes $n^{1/3}, n^{1/2}, n^{2/3}$ and on the whole sample with the corresponding execution time (Table 1).

Notice that even with a size of $\mathrm{n}^{1/3}$ we are able to get the correct order for the variance, the bias may be important for small subsampling size but almost vanish for $n = n^{2/3}$. With a subsampling size of order $n^{2/3} = 46{,}415$ even if the model is true, we get the same order as the one on the m.l.e. on the whole database: but in terms of calculus $n^{2/3}$ is too big, since in that case we are able to proceed the m.l.e

Table 1 MLE for a probit model with $n = 10^6, 10^7$ and variance estimations

	Subsample ($B = 999$ replications)				Whole sample
n	b_n	$\hat{\beta}_{b_n}$	$\hat{var}(\beta_{b_n})^{1/2}$	$Time$	$\hat{\beta}_n$ ($\hat{var}(\beta_n)^{1/2}$) $time$
10^6	$n^{1/3}$	3.19	0.0064	13 s	2.992
	$n^{1/2}$	3.022	0.0063	36 s	(0.0061)
	$n^{2/3}$	2.996	0.0060	3.26 min	28.75 s
10^7	$n^{1/3}$	3.10	0.0020	41 s	2.998
	$n^{1/2}$	3.009	0.0020	1.25 min	(0.0019)
	$n^{2/3}$	2.998	0.0019	12 min	4.69 min

on the whole database in less than 5 min (whereas it takes 12 min to replicate 999 times the procedure on the $n^{2/3}$ sample size). But for $n^{1/2}$ we get a gain of 4 for a similar accuracy : of course this strongly depends on the degree of accuracy that one wishes to obtain on the parameter of interest.

6.2 *Estimation of the Out-of-Sample Error with knn (See Example 2)*

Considering the preceding example we now use the subsampling method to estimate the out-of-sample errors of k-nearest neighbor estimators on several subsampling sizes and compare them to the one obtained on the full database. We consider a training set equal to $0.7n$ and a test set of size $0.3n$ (similar results have been obtained for other test sets). The computation times in Table 2 clearly show the computation gains. A striking result is for $n = 10^7$ because it takes almost 5 h to get an estimator of this quantity on the whole sample whereas the subsampling method takes at the worst 15 min with $n^{2/3}$. It seems that even with a size of order $n^{1/3}$ we still get a good approximation in less than 45 s. With the subsampling method by using an extrapolated variance, we are also able to estimate the variance of the

Table 2 Estimation of the out-of-sample error by subsampling and on the whole sample

KNN	Subsample ($B = 999$ replications)			Whole sample	
n	b_n	Out-of-samp. error	Time	Out-s. err	Time
10^6	$n^{1/3}$	0.1177	4.79 s	0.1158	5.252 min
	$n^{1/2}$	0.1165	5.76 s	(0.008)	
	$n^{2/3}$	0.1167	43.5 s		
10^7	$n^{1/3}$	0.1166	44.7 s	0.114082	4 h 57 min
	$n^{1/2}$	0.1163	50.7 s	(0.006)	
	$n^{2/3}$	0.1161	15.35 min		

out-of-sample error (in parenthesis in the table), which shows that the estimation is quite accurate.

Notice that, for this data, the out-of-sample error of the probit model is better (of order 0.050 for both size) : it is because of the way in which the data has been simulated. For real data, these simulations show that it is possible to compare in a reasonable time the out-of-sample errors for several competing methods (with confidence intervals).

7 Technical Proofs

Proof of Theorem 1 Let $\epsilon > 0$; assuming asymptotic convergence of T_n , we have that

$$\Pr{}_P\{|K_{b_n}(x \mid \underline{X}_n, \tau_.) - K(x, P)| \geq \epsilon\} \to 0,$$

uniformly in x. Define the quantile $z = K_{b_n}^{-1}(t - \epsilon \mid \underline{X}_n, \tau_.)$, then we have with probability tending to one that

$$\begin{aligned} K_{b_n}(z \mid \underline{X}_n, \tau_.) \geq t - \epsilon &\Rightarrow K(z, P) \geq t - 2\epsilon \\ &\Rightarrow z \geq K^{-1}(t - 2\epsilon, P). \end{aligned}$$

Similarly, define $y = K^{-1}(t, P)$, thus with probability tending to one, we get

$$\begin{aligned} K(y, P) \geq t &\Rightarrow K_{b_n}(y \mid \underline{X}_n, \tau_.) \geq t - \epsilon \\ &\Rightarrow y \geq K_{b_n}^{-1}(t - \epsilon \mid \underline{X}_n, \tau_.). \end{aligned}$$

Hence, for any t and any $\epsilon > 0$, we have the inequality

$$K^{-1}(t - 2\epsilon, P) \leq K_{b_n}^{-1}(t - \epsilon \mid \underline{X}_n, \tau_.) \leq K^{-1}(t, P).$$

so that by letting $\epsilon \to 0^+$, we obtain that

$$K_{b_n}^{-1}(t \mid \underline{X}_n, \tau_.) = K^{-1}(t, P) + o_P(1).$$

Now, let x be a real number and note that

$$\begin{aligned} K_{b_n}(x \mid \underline{X}_n, \widehat{\tau}_.) &\equiv q^{-1} \sum_{i=1}^{q} 1\{b_n^{\hat{\alpha}}(T_{b_n,i} - \widehat{\theta}_{m_n}) \leq x\} \\ &= q^{-1} \sum_{i=1}^{q} 1\{b_n^{\hat{\alpha}}(T_{b_n,i} - \theta) - b_n^{\hat{\alpha}}(\widehat{\theta}_{m_n} - \theta) \leq x\}. \end{aligned}$$

Now, define $U_n(x) = q^{-1}\sum_{i=1}^{q} 1\{b_n^{\hat{\alpha}}(T_{b_n,i} - \theta) \le x\}$ and the set $E_n = \{b_n^{\hat{\alpha}}|\widehat{\theta}_{m_n} - \theta| \le \epsilon\}$, for some $\epsilon > 0$. Since $\hat{\alpha} = \alpha + o_P((\log n)^{-1})$, it follows that $n^{\hat{\alpha}} = n^{\alpha}(1 + o_P(1))$ and $b_n^{\hat{\alpha}} = b_n^{\alpha}(1 + o_P(1))$.

Equations (3) and (4) imply that $P(E_n) \underset{n\to\infty}{\longrightarrow} 1$; hence, with probability tending to one, we get that

$$U_n(x-\epsilon) \le K_{b_n}(x \mid \underline{X}_n, \widehat{\tau}_{.}) \le U_n(x+\epsilon).$$

We will first show that $U_n(x)$ converges to $K(x, P)$ in probability. For this we introduce the U-statistics with varying kernel defined by

$$V_n(x) = q^{-1}\sum_{i=1}^{q} 1\{b_n^{\alpha}(T_{b_n,i} - \theta) \le x\},$$

that is the equivalent of $U_n(x)$, with the true rate rather than the estimated one. Note that since $V_n(x)$ is a U-statistics of degree b_n, such that $\frac{b_n}{n} \to 0$, by Hoeffding inequality we have $V_n(x) \to K(x, P)$ in probability.

Now, for any $\epsilon_1 > 0$, we have that

$$U_n(x) = q^{-1}\sum_{i=1}^{q} 1\{b_n^{\alpha}(T_{b_n,i} - \theta) \le x\frac{b_n^{\alpha}}{b_n^{\hat{\alpha}}}\} \le V_n(x+\epsilon_1)$$

where the above inequality holds with probability tending to one. A similar argument shows that $U_n(x) \ge V_n(x - \epsilon_1)$ with probability tending to one.

But we have $V_n(x+\epsilon_1) \to K(x+\epsilon_1, P)$ and $V_n(x-\epsilon_1) \to K(x-\epsilon_1, P)$ in probability. Therefore, letting $\epsilon_1 \to 0$, we have that $U_n(x) \to K(x, P)$ in probability as required.

Proving that we have

$$\widehat{K}_n(x, P) - K(x, P) \to 0 \text{ as n} \to \infty$$

follows now by the same arguments as before by recalling that

$$\begin{aligned}\widehat{K}_n(x, P) &= P(\tau_n(T_n - \theta) \le x\frac{\tau_n}{\widehat{\tau}_n}) \\ &= P(\tau_n(T_n - \theta) \le x(1 + o_P(1))\end{aligned}$$

and using the continuity of the limiting distribution.

The second part of the theorem is a straightforward consequence of the uniform convergence of $K_{b_n}(x \mid \underline{X}_n, \widehat{\tau}_{.}) - \widehat{K}_n(x, P)$ to 0, for any point of continuity of the true limiting distribution.

Acknowledgements This research has been conducted as part of the project Labex MME-DII (ANR11-LBX-0023- 01), and the industrial chair "Machine Learning for Big Data," Télécom--ParisTech.

References

1. Arcones, M. A., Giné, E. (1993). Limit theorems for U-processes. *Annals of Probability, 21*(3). 1494–1542.
2. Babu, G., & Singh, K. (1985). Edgeworth expansions for sampling without replacement from finite populations. *Journal of Multivariate Analysis, 17*, 261–278.
3. Belsley, D. A., Kuh, E., & Welsh, R. E. (1980). *Regression diagnostics: Identifying influential data and sources of collinearity.* New York: Wiley.
4. Bertail, P. (1997). Second order properties of an extrapolated bootstrap without replacement under weak assumptions: The i.i.d. and strong mixing case. *Bernoulli, 3*, 149–179.
5. Bertail, P. (2011). Somme comments on Subsampling weakly dependent time series and application to extremes. *TEST, 20*, 487–490.
6. Bertail, P., Chautru, E., & Clémençon, S. (2014). Scaling-up M-estimation via sampling designs: The Horvitz-Thompson stochastic gradient descent. In *Proceedings of the 2014 IEEE International Conference on Big Data, Washington (USA).*
7. Bertail, P., Chautru, E., & Clémençon, S. (2015). Tail index estimation based on survey data. *ESAIM Probability & Statistics, 19*, 28–59.
8. Bertail, P., Chautru, E., & Clémençon, S. (2016). Empirical processes in survey sampling. *Scandinavian Journal of Statistics, 44*(1), 97–111.
9. Bertail, P., Haeffke, C., Politis, D., & White H. (2004). A subsampling approach to estimating the distribution of diverging statistics with applications to assessing financial market risks. *Journal of Econometrics, 120*, 295–326.
10. Bertail, P., & Politis, D. (2001). Extrapolation of subsampling distribution estimators in the i.i.d. strong-mixing cases. *Canadian Journal of Statistics, 29*(4), 667–680.
11. Bertail, P., Politis, D., & Romano, J. (1999). Undersampling with unknown rate of convergence. *Journal of the American Statistical Association, 94*(446), 569–579.
12. Bickel, P. J., & Sakov, A. (2008). On the choice of the m out n bootstrap and confidence bounds for extrema. *Statistica Sinica, 18*, 967–985.
13. Bickel P. J., & Yahav, J. A. (1988). Richardson extrapolation and the bootstrap. *Journal of the American Statistical Association, 83*(402), 387–393.
14. Bickel, P. J., Götze, F., & van Zwet, W. R. (1997). Resampling fewer than n observations, gains, losses and remedies for losses. *Statistica Sinica, 7*, 1–31.
15. Bingham, N. H., Goldie, C. M., & Teugels, J. L. (1987). *Regular variation.* Cambridge: Cambridge University Press.
16. Bretagnolle, J. (1983). Lois limites du bootstrap de certaines fonctionelles. *Annales de l'Institut Henri Poincaré B: Probability and Statistics*, 19, 281–296.
17. Carlstein, E. (1988). Nonparametric change-point estimation. *Annals of Statistics, 16*(1), 188–197.
18. Darkhovshk, B. S. (1976). A non-parametric method for the a posteriori detection of the "disorder" time of a sequence of independent random variables. *Theory of Probability and Its Applications, 21*, 178–83.
19. Götze Rauckauskas, F. A. (1999). Adaptive choice of bootstrap sample sizes. In M. de Gunst, C. Klaassen, & A. van der Vaart (Eds.), *State of the art in probability statistics: Festschrift for Willem R. van Zwet. IMS lecture notes, monograph series* (pp. 286–309). Beachwood, OH: Institute of Mathematical Statistics.
20. Heilig, C., & Nolan, D. (2001). Limit theorems for the infinite degree U-process. *Statistica Sinica, 11*, 289–302.

21. Isaacson, E., & Keller, H. B. (1966). *Analysis of numerical methods*. New York: John Wiley.
22. Kleiner, A., Talwalkar, A., Sarkar, P., & Jordan, M. I. (2014). A scalable bootstrap for massive data. *Journal of the Royal Statistical Society: Series B, 76*(4), 795–816.
23. Le Cam, L. (1990). Maximum likelihood: An introduction. *Revue Internationale de Statistique, 58*(2), 153–171.
24. McLeod, I., & Bellhouse, D. R. (1983). Algorithm for drawing a simple random sample. *Journal of the Royal Statistical Society. Series C (Applied Statistics), 32*(2), 182–184.
25. Politis, D., & Romano, J. P. (1994). Large sample confidence regions based on subsamples under minimal assumptions. *Annals of Statistics, 22*, 2031–2050.

[illegible]

Probability Bounds for Active Learning in the Regression Problem

A.-K. Fermin and C. Ludeña

Abstract In this contribution we consider the problem of active learning in the regression setting. That is, choosing an optimal sampling scheme for the regression problem simultaneously with that of model selection. We consider a batch type approach and an on-line approach adapting algorithms developed for the classification problem. Our main tools are concentration-type inequalities which allow us to bound the supreme of the deviations of the sampling scheme corrected by an appropriate weight function.

1 Introduction

Consider the following regression model

$$y_i = x_0(t_i) + \varepsilon_i,\ i = 1, \ldots, n \tag{1}$$

where the observation noise ε_i are i.i.d. realizations of a random variable ε.

The problem we consider in this chapter is that of estimating the real-valued function x_0 based on $t_1, \ldots, t_n$ and a subsample of size $N < n$ of the observations $y_1, \ldots, y_n$ measured at a well-chosen subsample of $t_1, \ldots, t_n$. This is relevant when, for example, obtaining the values of y_i for each sample point t_i is expensive or time consuming.

A.-K. Fermin (✉)
Université Paris Nanterre, laboratoire Modal'X, Nanterre, France
e-mail: aferminrodriguez@parisnanterre.fr

C. Ludeña
Universidad Jorge Tadeo Lozano, Dpto. de Ciencias Básicas y Modelado, Carrera, Bogotá, Colombia
e-mail: carennec.ludenac@utadeo.edu.co

P. Bertail et al. (eds.), *Nonparametric Statistics*, Springer Proceedings in Mathematics & Statistics 250, https://doi.org/10.1007/978-3-319-96941-1_14

In this work we propose a statistical regularization approach for selecting a good subsample of the data in this regression setting by introducing a weighted sampling scheme (importance weighting) and an appropriate penalty function over the sampling choices.

We begin by establishing basic results for a fixed model, and then the problem of model selection and choosing a good sampling set simultaneously. This is what is known as active learning. We will develop two approaches. The first, a batch approach (see, for example, [7]), assumes the sampling set is chosen all at once, based on the minimization of a certain penalized loss function for the weighted sampling scheme. The second, an iterative approach [1], considers a two-step iterative method choosing alternatively the best new point to be sampled and the best model given the set of points.

The weighted sampling scheme requires each data point t_i to be sampled with a certain probability $p(t_i)$ which is assumed to be inferiorly bounded by a certain constant p_{min}. This constant plays an important role because it controls the expected sample size $\mathbb{E}(N) = \sum_{i=1}^{n} p(t_i) > np_{min}$. However, it also is inversely proportional to the obtained error terms in the batch procedure (see Theorems 2.1 and 2.2), so choosing p_{min} too small will lead to poor bounds. Thus essentially, the batch procedure aims at selecting the best subset of data points (points with high probability) for the user chosen error bound. In the iterative procedure this problem is addressed by considering a sequence of sampling probabilities $\{p_j\}$ where at each step j $p_j(t_i)$ is chosen to be as big as the greatest fluctuation for this data point over the hypothesis model for this step.

Following the active learning literature for the regression problem based on ordinary least squares (OLS) and weighted least squares learning (WLS) (see, for example [5–7] and the references therein) in this chapter we deal mainly with a linear regression setting and a quadratic loss function. This will be done by fixing a spanning family $\{\phi_j\}_{j=1}^{m}$ and considering the best L^2 approximation x_m of x_0 over this family. However, our approach is based on empirical error minimization techniques and can be readily extended to consider other models whenever bounds in probability are available for the error term.

Our results are based on concentration-type inequalities. Although variance minimization techniques for choosing appropriate subsamples are a well-known tool, giving adequate bounds in probability allowing for optimal non-asymptotic rates has been much less studied in the regression setting.

This is also true for the iterative procedure, where our results generalize previous ones obtained only in the classification setting for finite model spaces.

This chapter is organized as follows. In Sect. 2 we formulate the basic problem and study the batch approach for simultaneous sample and model selection. In Sect. 3 we study the iterative approach to sample selection and we discuss effective sample size reduction. All the proofs are available in the extended arXiv version [3].

2 Preliminaries

2.1 Basic Assumptions

We assume that the observations noise ε_i in (1) are i.i.d. realizations of a random variable ε satisfying the moment condition

MC Assume the r.v. ε satisfies $\mathbb{E}\varepsilon = 0$, $\mathbb{E}(|\varepsilon|^r/\sigma^r) \le r!/2$ for all $r > 2$ and $\mathbb{E}(\varepsilon^2) = \sigma^2$.

It is important to stress that the observations depend on a fixed design $t_1, \ldots, t_n$. For this, we need some notation concerning this design. For any vectors u, v r, we define the normalized norm and the normalized scalar product by

$$\|u\|_{n,r}^2 = \frac{1}{n}\sum_{i=1}^{n} r_i (u_i)^2, \quad \text{and} \quad < u, v >_{n,r} = \frac{1}{n}\sum_{i=1}^{n} r_i u_i v_i.$$

We drop the letter r from the notation when $r = 1$. With a slight abuse of notation, we will use the same notation when u, v, or r are functions by identifying each function (e.g. u) with the vector of values evaluates as t_i (e.g. $(u(t_1), \ldots, u(t_n))$. We also require the empirical max-norm $\|u\|_\infty = \max_i |u_i|$.

2.2 Discretization Scheme

To start with we will consider the approximation of function x_0 over a finite-dimensional subspace S_m. This subspace will be assumed to be linearly spanned by the set $\{\phi_j\}_{j\in\mathcal{I}_m} \subset \{\phi_j\}_{j\ge 1}$, with $\mathcal{I}_m$ a certain index set. Moreover, we shall, in general, be interested only in the vector $(x_0(t_i))_{i=1}^n$ which we shall typically denote just by x_0 stretching notation slightly.

We will assume the following properties hold:

AB There exists an increasing sequence c_m such that $\|\phi_j\|_\infty \le c_m$ for $j \le m$.

AQ There exist a certain density q and a positive constant Q such that $q(t_i) \le Q, i = 1, \ldots, n$ and

$$\int \phi_l(t)\phi_k(t)q(t)\,dt = \delta_{k,l},$$

where δ is the Kronecker delta.

We will also require the following discrete approximation assumption. Let $G_m = [\phi_j(t_i)]_{i,j}$ be the associated empirical $n \times m$ Gram matrix. We assume that $G_m^t D_q G_m$ is invertible and moreover that $\frac{1}{n} G_m^t D_q G_m \to I_m$, where D_q is the diagonal matrix with entries $q(t_i)$, for $i = 1, \ldots, n$ and I_m the identity matrix of size m. More precisely, we will assume

AS There exist positive constants α and c, such that

$$\|I_m - \frac{1}{n} G_m^t D_q G_m\| \le c n^{-1-\alpha}.$$

Given [AQ], assumption [AS] is a numerical approximation condition which is satisfied under certain regularity assumptions over q and $\{\phi_j\}$. To illustrate this condition we include the following example.

Example 2.1 Haar Wavelets: let $\phi(t) = \mathbf{1}_{[0,1]}(t)$, $\psi(t) = \phi(2t) - \phi(2t-1)$ (see, for example, [4]), with $q(t) = \mathbf{1}_{[0,1]}(t)$. Define

$$\phi_{j,k}(t) = 2^{j/2}\phi(2^j t - k)\,,\ t \in [0,1]\,,\ j \ge 0 \text{ and } k \in \mathbb{Z}\,;$$
$$\psi_{j,k}(t) = 2^{j/2}\psi(2^j t - k)\,,\ t \in [0,1]\,,\ j \ge 0 \text{ and } k \in \mathbb{Z}\,.$$

For all $m \ge 0$, S_m denotes the linear space spanned by the functions $(\phi_{m,k}, k \in \mathbb{Z})$. In this case $c_m \le 2^{m/2}$ and condition [AS] is satisfied for the discrete sample $t_i = i/2^m, i = 0, \cdots, 2^{m-1}$.

We will denote by $\hat{x}_m \in S_m$ the function that minimizes the weighted norm $\|x - y\|_{n,q}^2$ over S_m evaluated at points $t_1, \ldots, t_n$. This is,

$$\hat{x}_m = \arg\min_{x \in S_m} \frac{1}{n} \sum_{i=1}^{n} q(t_i)(y_i - x(t_i))^2 = R_m y,$$

with $R_m = G_m(G_m^t D_q G_m)^{-1} G_m^t D_q$ the orthogonal projector over S_m in the q-empirical norm $\| \cdot \|_{n,q}$.

Let $x_m := R_m x_0$ be the projection of x_0 over S_m in the q-empirical norm $\| \cdot \|_{n,q}$, evaluated at points $t_1, \ldots, t_n$. Our goal is to choose a good subsample of the data collection such that the estimator of the unobservable vector x_0 in the finite-dimensional subspace S_m, based on this subsample, attains near optimal error bounds. For this we must introduce the notion of subsampling scheme and importance weighted approaches (see [1, 7]), which we discuss below.

2.3 Sampling Scheme and Importance Weighting

In order to sample the data set we will introduce a sampling probability $p(t)$ and a sequence of Bernoulli($p(t_i)$) random variables w_i, $i = 1, \ldots, n$ independent of ε_i

with $p(t_i) > p_{\min}$. Let $D_{w,q,p}$ be the diagonal matrix with entries $q(t_i)w_i/p(t_i)$. So that $\mathbb{E}\left(D_{w,q,p}\right) = D_q$. Sometimes it will be more convenient to rewrite $w_i = \mathbf{1}_{u_i<p(t_i)}$ for $\{u_i\}_i$ an i.i.d. sample of uniform random variables, independent of $\{\varepsilon_i\}_i$ in order to stress the dependence on p of the random variables w_i.

The next step is to construct an estimator for $x_m = R_m x_0$, based on the observation vector y and the sampling scheme p. For this, we consider a modified version of the estimator $\hat{x}_m$.

Consider a uniform random sample $u_1, \ldots, u_n$ and let $w_i = w_i(p) = \mathbf{1}_{u_i<p(t_i)}$ for a given p. For the given realization of $u_1, \ldots, u_n$, $D_{w,q,p}$ will be strictly positive for those $w_i = 1$. Moreover, as follows from the singular value decomposition, the matrix $(G_m^t D_{w,q,p} G_m)$ is invertible as long as at least one $w_i \neq 0$. Set $R_{m,p} = G_m(G_m^t D_{w,q,p} G_m)^{-1} G_m^t D_{w,q,p}$. Then $R_{m,p}$ is the orthogonal projector over S_m in the wq/p-empirical norm $\|\cdot\|_{n,wq/p}$ and it is well defined if at least one $w_i \neq 0$. If all $w_i = 0$, the projection is defined to be 0.

As the approximation of x_m, we then consider (for a fixed m, p and $(u_1, \ldots, u_n)$) the random quantity

$$\hat{x}_{m,p} = \arg\min_{x\in S_m} \|x - y\|^2_{n,\frac{qw}{p}} = \arg\min_{x\in S_m} \frac{1}{n}\sum_{i=1}^{n} \frac{w_i}{p(t_i)} q(t_i)(y_i - x(t_i))^2.$$

Note that

$$\hat{x}_{m,p} = R_{m,p} y, \tag{2}$$

This estimator depends on y_i only if $w_i = 1$. However, as stated above, this depends on $p(t_i)$ for the given probability p.

2.4 Choosing a Good Sampling Scheme

To begin with, given n, we will assume that S_m is fixed with dimension $|\mathscr{I}_m| = d_m$ and $d_m = o(n)$. Remark that the bias $\|x_0 - x_m\|^2_{n,q}$ is independent of p so for our purposes it is only necessary to study the approximation error $\|x_m - \hat{x}_{m,p}\|^2_{n,q}$ which does depend on how p is chosen.

Let $\mathscr{P} := \{p_k, k \geq 1\}$ be a numerable collection of $[0, 1]$ valued functions over $\{t_1, \ldots, t_n\}$. Set $p_{k,min} = \min_i p_k(t_i)$. We will assume that $\min_k p_{k,min} > p_{\min}$. The way the candidate probabilities are ordered is not a major issue, although in practice it is sometimes convenient to incorporate prior knowledge (certain sample points are known to be needed in the sample, for example) letting favourite candidates appear first in the order. To get the idea of what a sampling scheme may be, consider the following toy example:

Example 2.2 Let $\Pi = \{0.1, 0.4, 0.6, 0.9\}$ and set $\mathscr{P} = \{p, p(t_i) = \pi_j \in \Pi, i = 1, \ldots, n\}$ which is a set of $|\Pi|^n$ functions. In this example, any given p will tend to favour the appearance of points t_i with $p(t_i) = 0.9$ and disfavour the appearance of those t_i with $p(t_i) = 0.1$.

A good sampling scheme p, based on the data, should be the minimizer over $\mathscr{P}$ of the non-observable quantity $\|x_m - \hat{x}_{m,p}\|^2_{n,q}$. In order to find a reasonable observable equivalent we start by writing,

$$
\begin{aligned}
[\hat{x}_{m,p} - x_m] &= R_{m,p}[x_0 - x_m] + R_{m,p}\varepsilon \\
&= \mathbb{E}\left(R_{m,p}\right)[x_0 - x_m] + (R_{m,p} - \mathbb{E}\left(R_{m,p}\right))[x_0 - x_m] + R_{m,p}\varepsilon. \quad (3)
\end{aligned}
$$

Consider first the deterministic term $\mathbb{E}\left(R_{m,p}\right)[x_0 - x_m]$ in (3). We have the next lemma which is proved in the extended arXiv version.

Lemma 2.1 *Under condition [AS] if $m = o(n)$, then*

$$\|\mathbb{E}\left(R_{m,p}\right)[x_0 - x_m]\|_{n,q} = O\left(\frac{n^{-1-\alpha}\|x_0 - x_m\|_{n,q}}{p_{min}}\right).$$

From Lemma 2.1, we can derive that the deterministic term is small with respect to the other terms. Thus, it is sufficient for a good sampling scheme to take into account the second and third terms in (3). We propose to use an upper bound with high probability of those two last terms as in a penalized estimation scheme and to base our choice on this bound.

Define

$$\widetilde{B}_1(m, p_k, \delta) = \|x_0 - x_m\|^2_{n,q}(\widetilde{\beta}_{m,k}(1 + \widetilde{\beta}^{1/2}_{m,k}))^2 \quad (4)$$

with

$$\widetilde{\beta}_{m,k} = \frac{c_m(\sqrt{17}+1)}{2}\sqrt{\frac{d_m Q}{n p_{k,\min}}}\sqrt{2\log(2^{7/4} d_m k(k+1)/\delta)}. \quad (5)$$

The second square root appearing in the definition of $\widetilde{\beta}_{m,k}$ is included in order to give uniform bounds over the numerable collection $\mathscr{P}$.

In the following, the expression $\mathrm{tr}(A)$ stands for the trace of the matrix A. Set $T_{m,p_k} = \mathrm{tr}((R_{m,p_k} D_q^{1/2})^t R_{m,p_k} D_q^{1/2})$ and define

$$\widetilde{B}_2(m, p_k, \delta) = \sigma^2 r(1+\theta_k)\frac{T_{m,p_k} + Q}{n} + \sigma^2 Q\frac{log^2(2/\delta)}{dn}, \quad (6)$$

with $r > 1$ and $d = d(r) < 1$ a positive constant that depends on r. The sequence $\theta_k \geq 0$ is such that $\sum_k e^{-\sqrt{dr\theta_k(d_m+1)}} < 1$ holds.

It is thus reasonable to consider the best p as the minimizer

$$\hat{p} = \underset{p_k \in \mathscr{P}}{\operatorname{argmin}} \widetilde{B}(m, p_k, \delta, \gamma, n), \tag{7}$$

where, for a given $0 < \gamma < 1$,

$$\widetilde{B}(m, p_k, \delta, \gamma, n) = \{(1+\gamma)\widetilde{B}_1(m, p_k, \delta) + (1 + 1/\gamma)\widetilde{B}_2(m, p_k, \delta)\}.$$

The different roles of $\widetilde{B}_1$ and $\widetilde{B}_2$ appear in the following lemmas:

Lemma 2.2 *Assume that the conditions [AB], [AS], and [AQ] are satisfied and that there is a constant* $p_{min} > 0$ *such that for all* $i = 1, \ldots, n$, $p(t_i) > p_{k,\min} > p_{\min}$. *Assume* $\widetilde{B}_1$ *to be selected according to (4). Then for all* $\delta > 0$ *we have*

$$P\left[\sup_{\mathscr{P}}\{\|(R_{m,p} - \mathbb{E}\left(R_{m,p}\right))[x_0 - x_m]\|^2_{n,q} - \widetilde{B}_1(m, p, \delta)\} > 0\right] \leq \delta/2$$

Lemma 2.3 *Assume the observation noise in Eq.* (1) *is an i.i.d. collection of random variables satisfying the moment condition [MC]. Assume that the condition [AQ] is satisfied and assume that there is a constant* $p_{min} > 0$ *such that* $p(t_i) > p_{min}$ *for all* $i = 1, \ldots, n$. *Assume* $\widetilde{B}_2$ *to be selected according to (6) with* $r > 1$, $d = d(r)$ *and* $\theta_k \geq 0$, *such that the following Kraft inequality* $\sum_k e^{-\sqrt{dr\theta_k(m+1)}} < 1$ *holds. Then,*

$$P(\sup_{\mathscr{P}}\{\|R_{m,p}\varepsilon\|^2_{n,q} - \widetilde{B}_2(m, p, \delta)\} > 0) < \delta/2.$$

Those two lemmas together with Lemma 2.1 assure that the proposed estimation procedure, based on the minimization of $\widetilde{B}$, is consistent establishing non-asymptotic rates in probability.

We may now state the main result of this section, namely, non-asymptotic consistency rates in probability of the proposed estimation procedure. The proof follows from Lemmas 2.2 and 2.3 and is given in the extended arXiv version along with the proof of the lemmas.

Theorem 2.1 *Assume that the conditions [AB], [AS], and [AQ] are satisfied. Assume* $\hat{p}$ *to be selected according to (7). Then the following inequality holds with probability greater than* $1 - \delta$

$$\|x_m - \hat{x}_{m,\hat{p}}\|^2_{n,q} \leq \inf_{p \in \mathscr{P}} 6\left(\|\mathbb{E}\left(R_{m,p}\right)(x_m - x_0)\|^2_{n,q} + \widetilde{B}(m, p, \delta, \gamma, n)\right).$$

Remark 2.1 In the minimization scheme given above it is not necessary to know the term $\|x_0 - x_m\|^2_{n,q}$ in $\widetilde{B}_1$ as this term is constant with regard to the sampling scheme p. Including this term in the definition of $\widetilde{B}_1$, however, is important because it leads to optimal bounds in the sense that it balances p_{min} with the mean variation, over the sample points, of the best possible solution x_m over the hypothesis model set S_m. This idea shall be pursued in depth in Sect. 3.

Moreover, minimizing $\widetilde{B}_1$ essentially just requires selecting k such that $p_{k,\min}$ is largest and doesn't intervene at all if $p_{k,\min} = p_{min}$ for all k. Minimization based on $p_k(t_i)$ for all sample points is given by the trace T_{m,p_k} which depends on the initial random sample u independent of $\{(t_i, y_i), i = 1, \ldots, n\}$. A reasonable strategy in practice, although we do not have theoretical results for it, is to consider several realizations of u and select sample points which appear more often in the selected sampling scheme $\hat{p}$.

Remark 2.2 Albeit the appearance of weight terms which depend on k both in the definition of $\widetilde{B}_1$ and $\widetilde{B}_2$, actually the ordering of $\mathcal{P}$ does not play a major role. The weights are given in order to assure convergence over the numerable collection $\mathcal{P}$. Thus in the definition of $\widetilde{\beta}_{m,k}$ any sequence of weights θ'_k (instead of $[k(k+1)]^{-1}$) assuring that the series $\sum_k \theta'_k < \infty$ is valid. Of course, in practice $\mathcal{P}$ is finite. Hence for $M = |\mathcal{P}|$ a more reasonable bound is just to consider uniform weights $\theta'_k = 1/M$ instead.

Remark 2.3 Setting $H_{m,p_k} := (G^t_m D_{w,q,p_k} G_m)^{-1} G^t_m D_{w,q,p_k}$ we may write $T_{m,p_k} = \mathrm{tr}(G^t_m D_q G_m H_{m,p_k} H^t_{m,p_k})$ in the definition of $\widetilde{B}_2$. Thus our convergence rates are as in Lemma 1, [5]. Our approach, however, provides non-asymptotic bounds in probability as opposed to asymptotic bounds for the quadratic estimation error.

Remark 2.4 As mentioned at the beginning of this section, the expected "best" sample size given u is $\hat{N} = \sum_i \hat{p}(t_i)$, where u is the initial random sample independent of $\{(t_i, y_i), i = 1, \ldots, n\}$. Of course, a uniform inferior bound for this expected sample size is $\mathbb{E}\left(\hat{N}\right) > np_{min}$, so that the expected size is inversely proportional to the user chosen estimation error. In practice, considering several realizations of the initial random sample provides an empirical estimator of the non-conditional "best" expected sample size.

2.5 Model Selection and Active Learning

Given a model and n observations $(t_1, y_1), \ldots, (t_n, y_n)$ we know how to estimate the best sampling scheme $\hat{p}$ and to obtain the estimator $\hat{x}_{m,\hat{p}}$. The problem is that the model m might not be a good one. Instead of just looking at *fixed m* we would like to consider simultaneous model selection as in [7]. For this we shall pursue a more global approach based on loss functions.

We start by introducing some notation. Set $l(u, v) = (u - v)^2$ the squared loss and let $L_n(x, y, p) = \frac{1}{n} \sum_{i=1}^n q(t_i) \frac{w_i}{p(t_i)} l(x(t_i), y_i)$ be the empirical loss function for the quadratic difference with the given sampling distribution. Set $L(x) := \mathbb{E}\left(L_n(x, y, p)\right)$ with the expectation taken over all the random variables involved.

Let $L_n(x, p) := \mathbb{E}_\varepsilon (L_n(x, y, p))$ where $\mathbb{E}_\varepsilon ()$ stands for the conditional expectation given the initial random sample u, that is the expectation with respect to the random noise ε. It is not hard to see that

$$L(x) = \frac{1}{n} \sum_{i=1}^{n} q(t_i) \mathbb{E} \left(l(x(t_i), y_i) \right),$$

and

$$L_n(x, p) = \frac{1}{n} \sum_{i=1}^{n} q(t_i) \frac{w_i}{p(t_i)} \mathbb{E} \left(l(x(t_i), y_i) \right).$$

Recall that $\hat{x}_{m,p} = R_{m,p} y$ is the minimizer of $L_n(x, y, p)$ over each S_m for given p and that $x_m = R_m x_0$ is the minimizer of $L(x)$ over S_m. Our problem is then to find the best approximation of the target x_0 over the function space $S_0 := \bigcup_{m \in \mathscr{I}} S_m$. In the notation of Sect. 2.2 we assume for each m that S_m is a bounded subset of the linearly spanned space of the collection $\{\phi_j\}_{j \in I_m}$ with $|\mathscr{I}_m| = d_m$.

Unlike the fixed m setting, model selection requires controlling not only the variance term $\|x_m - \hat{x}_{m,p}\|_{n,q}$ but also the unobservable bias term $\|x_0 - x_m\|^2_{n,q}$ for each possible model S_m. If all samples were available this would be possible just by looking at $L_n(x, y, p)$ for all S_m and p, but in the active learning setting labels are expensive.

Set $e_m := \|x_0 - x_m\|_\infty$. In what follows we will assume that there exists a positive constant C such that $\sup_m e_m \le C$. Remark this implies $\sup_m \|x_0 - x_m\|_{n,q} \le QC$, with Q defined in [AQ].

As above $p_k \in \mathscr{P}$ stands for the set of candidate sampling probabilities and $p_{k,\min} = \min_i (p_k(t_i))$.

Define

$$pen_0(m, p_k, \delta) = \frac{QC^2}{p_{k,\min}} \sqrt{\frac{1}{2n} \ln(\frac{6 d_m (d_m + 1)}{\delta})}, \tag{8}$$

$$pen_1(m, p_k, \delta) = QC \beta^2_{m,k} (1 + \beta^{1/2}_{m,k})^2, \tag{9}$$

with

$$\beta_{m,k} = \frac{c_m(\sqrt{17} + 1)}{2} \sqrt{\frac{d_m Q}{n p_{k,\min}}} \sqrt{2 \log(\frac{3 * 2^{7/4} d_m^2 (d_m + 1) k(k+1)}{\delta})},$$

and finally setting $T_{p_k,m} = \operatorname{tr}((R_{m,p_k} D_q^{1/2})^t R_{m,p_k} D_q^{1/2})$, define

$$pen_2(m, p_k, \delta) = \sigma^2 \left\{ r(1 + \theta_{m,k}) \frac{T_{p_k,m} + Q}{n} + \frac{Q \ln^2(6/\delta)}{dn} \right\} \tag{10}$$

where $\theta_{m,k} \geq 0$ is a sequence such that $\sum_{m,k} e^{-\sqrt{dr\theta_{m,k}(d_m+1)}} < 1$ holds.

We remark that the change from δ to $\delta/(d_m(d_m+1))$ in pen_0 and pen_1 is required in order to account for the supremum over the collection of possible model spaces S_m.

Also, we remark that introducing simultaneous model and sample selection results in the inclusion of term $pen_0 \sim C^2/p_{k,\min}\sqrt{1/n}$ which includes an L_∞ type bound instead of an L_2 type norm which may yield non-optimal bounds. Dealing more efficiently with this term would require knowing the (unobservable) bias term $\|x_0-x_m\|_{n,q}$. A reasonable strategy is selecting $p_{k,min} = p_{k,min}(m) \geq \|x_0-x_m\|_{n,q}$ whenever this information is available.

In practice, $p_{k,min}$ can be estimated for each model m using a previously estimated empirical error over a subsample if this is possible. However this yields a conservative choice of the bound. One way to avoid this inconvenience is to consider iterative procedures, which update on the unobservable bias term. This course of action shall be pursued in Sect. 3.

With these definitions, for a given $0 < \gamma < 1$ set

$$pen(m, p, \delta, \gamma, n) = 2p_0(m, p, \delta) + (\frac{1}{p_{\min}} + \frac{1}{\gamma})pen_1(m, p, \delta)$$
$$+(\frac{1}{p_{\min}^2}(\frac{2}{\gamma} + 1) + \frac{1}{\gamma})pen_2(m, p, \delta) + 2((c + 1)\frac{n^{-(1+\alpha)}QC}{p_{min}})^2.$$

and define

$$L_{n,1}(x, y, p) = L_n(x, y, p) + pen(m, p, \delta, \gamma, n).$$

The appropriate choice of an optimal sampling scheme simultaneously with that of model selection is a difficult problem. We would like to choose simultaneously m and p, based on the data in such a way that optimal rates are maintained. We propose for this a penalized version of $\hat{x}_{m,\hat{p}}$, defined as follows.

We start by choosing, for each m, the best sampling scheme

$$\hat{p}(m) = \arg\min_{p} pen(m, p, \delta, \gamma, n), \tag{11}$$

computable before observing the output values $\{y_i\}_{i=1}^n$, and then calculate the estimator $\hat{x}_{m,\hat{p}(m)} = R_{m,\hat{p}(m)}y$ which was defined in (2).

Finally, choose the best model as

$$\hat{m} = \arg\min_{m} L_{n,1}(y, \hat{x}_{m,\hat{p}(m)}, \hat{p}(m)). \tag{12}$$

The penalized estimator is then $\hat{x}_{\hat{m}} := \hat{x}_{\hat{m},\hat{p}(\hat{m})}$. It is important to remark that for each model m, $\hat{p}(m)$ is independent of y and hence of the random observation error structure. The following result assures the consistency of the proposed estimation procedure, although the obtained rates are not optimal as observed at the beginning of this section.

Theorem 2.2 *With probability greater than* $1 - \delta$, *we have*

$$L(\hat{x}_{\hat{m}}) \leq \frac{1+\gamma}{1-4\gamma}[L(x_m) + \min_{m,k}(2p_0(m, p_k, \delta) + \frac{1}{p_{\min}} pen_1(m, p_k, \delta)$$
$$+\frac{1}{p_{\min}^2}(1 + 2/\gamma)pen_2(m, p_k, \delta))]$$
$$\leq \frac{1+\gamma}{1-4\gamma} \min_m[L(x_m) + \min_k pen(m, p_k, \delta, \gamma, n)]$$

Remark 2.5 In practice, a reasonable alternative to the proposed minimization procedure is estimating the overall error by cross-validation or leave one out techniques and then choose m minimizing the error for successive essays of probability $\hat{p}$. Recall that in the original procedure of Sect. 2.5, labels are not required to obtain $\hat{p}$ for a fixed model. Cross-validation or empirical error minimization techniques do, however, require a stock of "extra" labels, which might not be affordable in the active learning setting. Empirical error minimization is specially useful for applications where what is required is a subset of very informative sample points, as for example when deciding what points get extra labels (new laboratory runs, for example) given a first set of complete labels is available. Applications suggest that $\hat{p}$ obtained with this methodology (or a threshold version of $\hat{p}$ which eliminates points with sampling probability $\hat{p}_i \leq \eta$ a certain small constant) is very accurate in finding "good" or informative subsets, over which model selection may be performed.

3 Iterative Procedure: Updating the Sampling Probabilities

A major drawback of the batch procedure is the appearance of $p_{\min}$ in the denominator of error bounds, since typically $p_{\min}$ must be small in order for the estimation procedure to be effective. Indeed, since the expected number of effective samples is given by $\mathbb{E}(N) := \mathbb{E}\left(\sum_i p(t_i)\right)$, small values of $p(t_i)$ are required in order to gain in sample efficiency.

Proofs in Sect. 2.5 depend heavily on bounding expressions such as

$$\frac{1}{n}\sum_{i=1}^{n} q(t_i)\frac{w_i}{p(t_i)}\varepsilon_i(x - x')(t_i)$$

or

$$\frac{1}{n}\sum_{i=1}^{n} q(t_i)(\frac{w_i}{p(t_i)} - 1)(x - x')^2(t_i)$$

where x and x' belong to a given model family S_m. Thus, it seems like a reasonable alternative to consider iterative procedures for which at time j, $p_j(t_i) \sim \max_{x,x' \in S_j} |x(t_i) - x'(t_i)|$ with S_j the current hypothesis space. In what follows we develop this strategy, adapting the results of [1] from the classification to the regression problem. Although we continue to work in the setting of model selection over bounded subsets of linearly spanned spaces, results can be readily extended to other frameworks such as additive models or kernel models. Once again, we will require certain additional restrictions associated to the uniform approximation of x_0 over the target model space.

More precisely, we start with an initial model set $S(= S_{m_0})$ and set x^* to be the overall minimizer of the loss function $L(x)$ over S. Assume additionally

AU $\sup_{x \in S} \max_{t \in \{t_1, \ldots, t_n\}} |x_0(t) - x(t)| \leq B$

Let $L_n(x) = L_n(x, y, p)$ and $L(x)$ be as in Sect. 2.5. For the iterative procedure introduce the notation

$$L_j(x) := \frac{1}{n_j} \sum_{i=1}^{nj} q(t_{j_i}) \frac{w_i}{p(t_{j_i})} (x(t_{j_i}) - y_{j_i})^2, \quad j = 0, \ldots, n$$

with $n_j = n_0 + j$ for $j = 0, \ldots, n - n_0$.

In the setting of Sect. 2 for each $0 \leq j \leq n$, S_j will be the linear space spanned by the collection $\{\phi_\ell\}_{\ell \in \mathcal{I}_j}$ with $|\mathcal{I}_j| = d_j$, $d_j = o(n)$.

In order to bound the fluctuations of the initial step in the iterative procedure we consider the quantities defined in Eqs. (4) and (6) for $r = \gamma = 2$. That is,

$$\Delta_0 = 2\sigma^2 Q \left\{ \frac{2(d_0 + 1)}{n_0} + \frac{\log^2(2/\delta)}{n_0} \right\}$$
$$+ 2(\widetilde{\beta}_{m_0}(1 + \widetilde{\beta}_{m_0}))^2 B^2.$$

with

$$\widetilde{\beta}_{m_0} = \frac{c_{m_0}(\sqrt{17} + 1)}{2} \sqrt{\frac{d_0 Q}{n_0 p_{\min}}} \sqrt{2 \log(2^{7/4} m_0/\delta)}.$$

As discussed in Sect. 2.4, Δ_0 requires some initial guess of $\|x_0 - x_{m_0}\|_{n,q}^2$. Since this is not available, we consider the upper bound B^2. Of course this will possibly slow down the initial convergence as Δ_0 might be too big, but will not affect the overall algorithm. Also remark we do not consider the weighting sequence θ_k of Eq. (6) because the sampling probability is assumed fixed.

Next set $B_j = \sup_{x,x' \in S_{j-1}} \max_{t \in \{t_1,\ldots,t_n\}} |x(t) - x'(t)|$ and define

$$\Delta_j = \sqrt{\sigma^2 Q[(\frac{2(d_j+1)}{n_j}) + \frac{\log^2(4n_j(n_j+1)/\delta)}{n_j}]}$$
$$+\sqrt{\log(4n_j(n_j+1)/\delta)\frac{16B_j^2(2B_j \wedge 1)^2 Q^2}{n_j}} + 4\sqrt{4\frac{(d_j+1)\log n}{n_j}}.$$

The iterative procedure is stated as follows:

1. For $j = 0$:
   - Choose (randomly) an initial sample of size n_0, $M_0 = \{t_{k_1}, \ldots, t_{k_{n_0}}\}$.
   - Let $\hat{x}_0$ be the chosen solution by minimization of $L_0(x)$ (or possibly a weighted version of this loss function).
   - Set $S_0 \subset \{x \in S : L_0(x) < L_0(\hat{x}_0) + \Delta_0\}$
2. At step j:
   - Select (randomly) a sample candidate point t_j, $t_j \notin M_{j-1}$.
     Set $M_j = M_{j-1} \cup \{t_j\}$
   - Set $p(t_j) = (\max_{x,x' \in S_{j-1}} |x(t_j) - x'(t_j)| \wedge 1)$ and generate $w_j \sim Ber(p(t_j))$.
     If $w_j = 0$, set $j = j+1$ and go to (2) to choose a new sample candidate.
     If $w_j = 1$ sample y_j and continue.
   - Let $\hat{x}_j = \arg\min_{x \in S_{j-1}} L_j(x) + \Delta_{j-1}(x)$
   - Set $S_j \subset \{x \in S_{j-1} : L_j(x) < L_j(\hat{x}_j) + \Delta_j\}$
   - Set $j = j+1$ and go to (2) to choose a new sample candidate.

Remark that, such as it is stated, the procedure can continue only up until time n (when there are no more points to sample). If the process is stopped at time $T < n$, the term $\log(n(n+1))$ can be replaced by $\log(T(T+1))$. We have the following result, which generalizes Theorem 2 in [1] to the regression case.

Theorem 3.1 *Let $x^* = \arg\min_{x \in S} L(x)$. Set $\delta > 0$. Then, with probability at least $1-\delta$ for any $j \leq n$*

- $|L(x) - L(x')| \leq 2\Delta_{j-1}$*, for all* $x, x' \in S_j$
- $L(\hat{x}_j) \leq [L(x^*) + 2\Delta_{j-1}]$

Remark 3.1 An important issue is related to the initial choice of m_0 and n_0. As the overall precision of the algorithm is determined by $L(x^*)$, it is important to select a sufficiently complex initial model collection. However, if $d_{m_0} >> n_0$, then Δ_0 can be big and $p_j \sim 1$ for the first samples, which leads to a more inefficient sampling scheme.

3.1 Effective Sample Size

For any sampling scheme the expected number of effective samples is, as already mentioned, $\mathbb{E}\left(\sum_i p(t_i)\right)$. Whenever the sampling policy is fixed, this sum is not random and effective reduction of the sample size will depend on how small sampling probabilities are. However, this will increase the error bounds as a consequence of the factor $1/p_{\min}$. The iterative procedure allows a closer control of both aspects and under suitable conditions will be of order $\sum_j \sqrt{L(x^*) + \Delta_j}$. Recall from the definition of the iterative procedure we have $p_j(t_i) \sim \max_{x,x' \in S_j} |x(t_i) - x'(t_i)|$, whence the expected number of effective samples is of the order of $\sum_j \max_{x,x' \in S_j} |x(t_i) - x'(t_i)|$. It is then necessary to control $\sup_{x,x' \in S_{j-1}} |x(t_i) - x'(t_i)|$ in terms of the (quadratic) empirical loss function L_j. For this we must introduce some notation and results relating the supremum and L_2 norms [2].

Let $S \subset L_2 \cap L_\infty$ be a linear subspace of dimension d, with basis $\Phi := \{\phi_j, j \in m_S\}$, $|m_S| = d$. Set $\bar{r} := \inf_\Lambda r_\Lambda$, where Λ stands for any orthonormal basis of S.

We have the following result

Lemma 3.1 *Let $\hat{x}_j$ be the sequence of iterative approximations to x^* and $p_j(t)$ be the sampling probabilities in each step of the iteration, $j = 1, \ldots, T$. Then, the effective number of samples, that is, the expectation of the required samples $N_e = \mathbb{E}\left(\sum_{j=1}^T p_j(t_j)\right)$ is bounded by*

$$N_e \leq 2\sqrt{2\bar{r}}\left(\sqrt{L(x*)}\sum_{j=1}^{T}\sqrt{d_j} + \sum_{j=1}^{T}\sqrt{d_j \Delta_j}\right).$$

References

1. Beygelzimer, A., Dasgupta, S., & Langford, J. (2009). Importance weighted active learning. In *Proceedings of the 26th Annual International Conference on Machine Learning* (pp. 49–56). New York: ACM.
2. Birgé, L., & Massart, P. (1998). Minimum contrast estimators on sieves: Exponential bounds and rates of convergence. *Bernoulli, 4*, 329–395.
3. Fermin, A. K., & Ludeña, C. (2018). Probability bounds for active learning in the regression problem. arXiv: 1212.4457
4. Härdle, W., Kerkyacharian, G., Picard, D., & Tsybakov, A. (1998). *Wavelets, approximation and statistical applications: Vol. 129. Lecture notes in statistics.* New York: Springer.
5. Sugiyama, M. (2006). Active learning in approximately linear regression based on conditional expectation generalization error with model selection. *Journal of Machine Learning Research, 7*, 141–166.
6. Sugiyama, M., Krauledat, M., & Müller, K. R. (2007). Covariate shift adaptation by importance weighted cross validation. *Journal of Machine Learning Research, 8*, 985–1005.
7. Sugiyama, M., & Rubens, N. (2008). A batch ensemble approach to active learning with model selection. *Neural Networks, 21*(9), 1278–1286.

Elemental Estimates, Influence, and Algorithmic Leveraging

K. Knight

Abstract It is well-known (Subrahmanyam, Sankhya Ser B 34:355–356, 1972; Mayo and Gray, Am Stat 51:122–129, 1997) that the ordinary least squares estimate can be expressed as a weighted sum of so-called elemental estimates based on subsets of p observations where p is the dimension of parameter vector. The weights can be viewed as a probability distribution on subsets of size p of the predictors $\{\boldsymbol{x}_i : i = 1, \cdots, n\}$. In this contribution, we derive the lower dimensional distributions of this p dimensional distribution and define a measure of potential influence for subsets of observations analogous to the diagonal elements of the "hat" matrix for single observations. This theory is then applied to algorithmic leveraging, which is a method for approximating the ordinary least squares estimates using a particular form of biased subsampling.

1 Introduction

Given observations $\{(\boldsymbol{x}_i, y_i) : i = 1, \cdots, n\}$, we define the ordinary least squares (OLS) estimate $\widehat{\boldsymbol{\beta}}$ as the minimizer of

$$\sum_{i=1}^{n} (y_i - \boldsymbol{x}_i^T \boldsymbol{\beta})^2.$$

We are implicitly assuming that $\widehat{\boldsymbol{\beta}}$ estimates a p-dimensional parameter $\boldsymbol{\beta}$ in the model $y_i = \boldsymbol{x}_i^T \boldsymbol{\beta} + \varepsilon_i$ $(i = 1, \cdots, n)$ for some errors $\{\varepsilon_i\}$. However, we will not use this assumption in the sequel.

K. Knight (✉)
University of Toronto, Toronto, ON, Canada
e-mail: keith@utstat.toronto.edu

P. Bertail et al. (eds.), *Nonparametric Statistics*, Springer Proceedings in Mathematics & Statistics 250, https://doi.org/10.1007/978-3-319-96941-1_15

The OLS estimate can be written as a weighted sum of so-called elemental estimates, which are based on subsets of observations of size p. If $s = \{i_1 < \cdots < i_p\}$ is a subset of $\{1, \cdots, n\}$, then we can define the elemental estimate $\widehat{\boldsymbol{\beta}}_s$ satisfying

$$\boldsymbol{x}_{i_j}^T \widehat{\boldsymbol{\beta}}_s = y_{i_j} \quad \text{for } j = 1, \cdots, p$$

provided that the solution $\widehat{\boldsymbol{\beta}}_s$ exists. Subrahmanyam [16] showed that the OLS estimate can be written as

$$\widehat{\boldsymbol{\beta}} = \sum_s \frac{|X_s|^2}{\sum_u |X_u|^2} \widehat{\boldsymbol{\beta}}_s$$

where the summation is over all subsets of size p, $|K|$ denotes the determinant of a square matrix K and

$$X_s = \left(\boldsymbol{x}_{i_1}\ \boldsymbol{x}_{i_2} \cdots \boldsymbol{x}_{i_p}\right). \tag{1}$$

Therefore, we can think of the OLS estimate $\widehat{\boldsymbol{\beta}}$ as an expectation of elemental estimates with respect to a particular probability distribution; that is,

$$\widehat{\boldsymbol{\beta}} = E_{\mathscr{P}}(\widehat{\boldsymbol{\beta}}_S)$$

where the random subset S has a probability distribution

$$\mathscr{P}(s) = P(S = s) = \frac{|X_s|^2}{\sum_u |X_u|^2} \tag{2}$$

where X_s is defined in (1). Hoerl and Kennard [10] note that the OLS estimate can also be expressed as a weighted sum of all OLS estimates based on subsets of $k > p$ observations.

An analogous result holds for weighted least squares (WLS) where we minimize

$$\sum_{i=1}^n w_i (y_i - \boldsymbol{x}_i^T \boldsymbol{\beta})^2$$

for some non-negative weights $\{w_i\}$. Again in this case, the WLS estimate $\widetilde{\boldsymbol{\beta}}$ can be written as $\widetilde{\boldsymbol{\beta}} = E(\widehat{\boldsymbol{\beta}}_S)$ where now S has the probability distribution

$$P(S = s) = \frac{|X_s|^2 \prod_{j \in s} w_j}{\sum_u \left\{|X_u|^2 \prod_{j \in u} w_j\right\}} = \frac{\mathscr{P}(s) \prod_{j \in s} w_j}{\sum_u \left\{\mathscr{P}(u) \prod_{j \in u} w_j\right\}}.$$

Henceforth, we will focus on the distribution $\mathscr{P}(s)$ defined in (2) for the OLS estimate where the results for the WLS estimate will follow *mutatis mutandis*.

The probability $\mathscr{P}(s)$ defined in (2) describes the weight and therefore the potential influence of a subset s (of size p) on the OLS estimate $\widehat{\boldsymbol{\beta}}$. In particular, greater weight is given to subsets where the vectors $\boldsymbol{x}_{i_1}, \cdots, \boldsymbol{x}_{i_p}$ are more dispersed; for example, if $\boldsymbol{x}_i = (1, x_i)^T$, then $|(\boldsymbol{x}_i\ \boldsymbol{x}_j)|^2 = (x_i - x_j)^2$. We can also use the probability distribution $\mathscr{P}$ to define measures of influence of arbitrary subsets of observations.

In Sect. 2, we will derive the lower dimensional distributions of $\mathscr{P}$ defined in (2) while in Sect. 3, we will discuss the potential influence of a subset of observations.

In situations where n and p are large, the OLS estimate $\widehat{\boldsymbol{\beta}}$ may be difficult to compute in which case one can attempt to approximate $\widehat{\boldsymbol{\beta}}$ by sampling $m \ll n$ observations from $\{(\boldsymbol{x}_i, y_i)\}$ leading to a subsampled estimate $\widehat{\boldsymbol{\beta}}_{ss} = E_{\mathscr{Q}}(\widehat{\boldsymbol{\beta}}_S)$ where S has a distribution $\mathscr{Q}$. The goal here is to find a subsampling scheme so that $\mathscr{Q} \approx \mathscr{P}$ in some sense. This will be explored further in Sect. 4.

2 Lower Dimensional Distributions of $\mathscr{P}$

The probability distribution $\mathscr{P}$ defined in (2) describes the weight given to each subset of p observations in defining the OLS estimate. It is also of interest to consider the total weight given to subsets of $k < p$ observations. It turns out that these lower dimensional distributions depend on the elements of the so-called "hat" matrix. (The "hat" matrix is the orthogonal projection onto the column space of the matrix X whose rows are $\boldsymbol{x}_1^T, \cdots, \boldsymbol{x}_n^T$.)

We start by re-expressing $\mathscr{P}(s)$. Since

$$\sum_u |X_u|^2 = \left|\sum_{i=1}^n \boldsymbol{x}_i \boldsymbol{x}_i^T\right|$$

[14], it follows that

$$\mathscr{P}(s) = \left| X_s^T \left(\sum_{i=1}^n \boldsymbol{x}_i \boldsymbol{x}_i^T\right)^{-1} X_s \right| = \left| \begin{pmatrix} h_{i_1 i_1} & h_{i_1 i_2} & \cdots & h_{i_1 i_p} \\ \vdots & \vdots & \ddots & \vdots \\ h_{i_p i_1} & h_{i_p i_2} & \cdots & h_{i_p i_p} \end{pmatrix} \right|$$

where $\{h_{ij} : i, j = 1, \cdots, n\}$ are the elements of the "hat" matrix [9]:

$$h_{ij} = h_{ji} = \boldsymbol{x}_i^T \left(\sum_{i=1}^n \boldsymbol{x}_i \boldsymbol{x}_i^T\right)^{-1} \boldsymbol{x}_j.$$

Henceforth, unless specified otherwise, all probabilities and expected values are based on the probability distribution $\mathscr{P}$.

If S is a random subset of size p drawn from $\{1, \cdots, n\}$ with probability distribution $\mathscr{P}$, it is convenient to describe the distribution of S using the equivalent random vector $\boldsymbol{W} = (W_1, \cdots, W_n)$ where $W_j = I(j \in S)$ and $W_1 + \cdots + W_n = p$. The moment generating function $\varphi(\boldsymbol{t}) = E[\exp(\boldsymbol{t}^T \boldsymbol{W})]$ of $\boldsymbol{W}$ is given by

$$\varphi(\boldsymbol{t}) = \left|\sum_{i=1}^{n} \boldsymbol{x}_i \boldsymbol{x}_i^T\right|^{-1} \sum_{s} |X_s|^2 \left\{\prod_{i_j \in s} \exp(t_{i_j})\right\} = \frac{\left|\sum_{i=1}^{n} \exp(t_i) \boldsymbol{x}_i \boldsymbol{x}_i^T\right|}{\left|\sum_{i=1}^{n} \boldsymbol{x}_i \boldsymbol{x}_i^T\right|}.$$

Thus for $k \le p$,

$$P(\{i_1, \cdots, i_k\} \subset S) = E\left(\prod_{j=1}^{k} W_{i_j}\right) = \left.\frac{\partial^k}{\partial t_{i_1} \cdots \partial t_{i_k}} \varphi(\boldsymbol{t})\right|_{t_1 = t_2 = \cdots = t_n = 0}.$$

The following result gives the lower dimensional distributions of $\mathscr{P}$.

Proposition 1 *Suppose that S has the distribution $\mathscr{P}$ defined in (2). Then for $k \le p$,*

$$P(\{i_1, \cdots, i_k\} \subset S) = \left|\begin{pmatrix} h_{i_1 i_1} & h_{i_1 i_2} & \cdots & h_{i_1 i_k} \\ \vdots & \vdots & \ddots & \vdots \\ h_{i_k i_1} & h_{i_k i_2} & \cdots & h_{i_k i_k} \end{pmatrix}\right|.$$

Proof Define the $k \times k$ matrix

$$H_{i_1 \cdots i_k}(\boldsymbol{t}) = \exp(t_{i_1} + \cdots + t_{i_k}) \begin{pmatrix} \boldsymbol{x}_{i_1}^T \\ \vdots \\ \boldsymbol{x}_{i_k}^T \end{pmatrix} \left(\sum_{i=1}^{n} \exp(t_i) \boldsymbol{x}_i \boldsymbol{x}_i^T\right)^{-1} (\boldsymbol{x}_{i_1} \cdots \boldsymbol{x}_{i_k})$$

and define for $1 \le i, j \le n$,

$$h_{ij}(\boldsymbol{t}) = \boldsymbol{x}_j^T \left(\sum_{i=1}^{n} \exp(t_i) \boldsymbol{x}_i \boldsymbol{x}_i^T\right)^{-1} \boldsymbol{x}_i.$$

It suffices to show that

$$\frac{\partial^k}{\partial t_{i_1} \cdots \partial t_{i_k}} \varphi(\boldsymbol{t}) = \varphi(\boldsymbol{t}) \left|H_{i_1 \cdots i_k}(\boldsymbol{t})\right|. \tag{3}$$

We will prove (3) by induction using Jacobi's formula [8]

$$\frac{d}{dt}|K(t)| = \text{trace}\left(\text{adj}(K(t))\frac{d}{dt}K(t)\right) = |K(t)|\text{trace}\left(K^{-1}(t)\frac{d}{dt}K(t)\right)$$

where $\text{adj}(K(t))$ is adjugate (the transpose of the cofactor matrix) of $K(t)$ as well as the identity

$$\left|\begin{pmatrix} D & \boldsymbol{v} \\ \boldsymbol{v}^T & a \end{pmatrix}\right| = a|D| - \boldsymbol{v}^T \text{adj}(D)\boldsymbol{v} \tag{4}$$

where a is a real number, $\boldsymbol{v}$ a vector of length k, and D a $k \times k$ matrix. For $k = 1$, we have

$$\begin{aligned}
\frac{\partial}{\partial t_{i_1}}\varphi(\boldsymbol{t}) &= \varphi(\boldsymbol{t})\text{trace}\left\{\left(\sum_{i=1}^{n}\exp(t_i)\boldsymbol{x}_i\boldsymbol{x}_i^T\right)^{-1}\exp(t_{i_1})\boldsymbol{x}_{i_1}\boldsymbol{x}_{i_1}^T\right\} \\
&= \varphi(\boldsymbol{t})\exp(t_{i_1})\boldsymbol{x}_{i_1}^T\left(\sum_{i=1}^{n}\exp(t_i)\boldsymbol{x}_i\boldsymbol{x}_i^T\right)^{-1}\boldsymbol{x}_{i_1} \\
&= \varphi(\boldsymbol{t})\exp(t_{i_1})h_{i_1 i_1}(\boldsymbol{t}) \\
&= \varphi(\boldsymbol{t})\left|H_{i_1}(\boldsymbol{t})\right|.
\end{aligned}$$

Now suppose that (3) holds for some $k < p$ and set $\ell = k + 1$. Then

$$\begin{aligned}
\frac{\partial^{\ell}}{\partial t_{i_1}\cdots\partial t_{i_\ell}}\varphi(\boldsymbol{t}) &= \frac{\partial}{\partial t_{i_\ell}}\left\{\varphi(\boldsymbol{t})\left|H_{i_1\cdots i_k}(\boldsymbol{t})\right|\right\} \\
&= \left|H_{i_1\cdots i_k}(\boldsymbol{t})\right|\frac{\partial}{\partial t_{i_\ell}}\varphi(\boldsymbol{t}) + \varphi(\boldsymbol{t})\frac{\partial}{\partial t_{i_\ell}}\left|H_{i_1\cdots i_k}(\boldsymbol{t})\right|.
\end{aligned}$$

First,

$$\frac{\partial}{\partial t_{i_\ell}}\varphi(\boldsymbol{t}) = \varphi(\boldsymbol{t})\left|H_{i_\ell}(\boldsymbol{t})\right| = \varphi(\boldsymbol{t})\exp(t_{i_\ell})h_{i_\ell i_\ell}(\boldsymbol{t}).$$

Second,

$$\varphi(\boldsymbol{t})\frac{\partial}{\partial t_{i_\ell}}\left|H_{i_1\cdots i_k}(\boldsymbol{t})\right| = \varphi(\boldsymbol{t})\left\{\text{trace}\left(\text{adj}(H_{i_1\cdots i_k}(\boldsymbol{t}))\frac{\partial}{\partial t_{i_\ell}}H_{i_1\cdots i_k}(\boldsymbol{t})\right)\right\}$$

with

$$\frac{\partial}{\partial t_{i_\ell}} H_{i_1\cdots i_k}(t) = -\exp(t_{i_1} + \cdots + t_{i_k} + t_{i_\ell}) \begin{pmatrix} h_{i_1 i_\ell}(t) \\ \vdots \\ h_{i_k i_\ell}(t) \end{pmatrix} \left(h_{i_1 i_\ell}(t) \cdots h_{i_k i_\ell}(t)\right).$$

Applying (4) with

$$a = h_{i_\ell i_\ell}(t), \quad D = H_{i_1\cdots i_k}(t), \quad \text{and} \quad \boldsymbol{v} = \begin{pmatrix} h_{i_1 i_\ell}(t) \\ \vdots \\ h_{i_k i_\ell}(t) \end{pmatrix},$$

we get

$$\frac{\partial^\ell}{\partial t_{i_1} \cdots \partial t_{i_\ell}} \varphi(t) = \varphi(t) \left| H_{i_1\cdots i_\ell}(t) \right|$$

and the conclusion follows by setting $\boldsymbol{t} = \boldsymbol{0}$.

3 Measuring Influence for Subsets of Observations

The diagonal elements $\{h_{ii}\}$ of the "hat" matrix are commonly used in regression analysis to measure the potential influence of observations [9]. Similar influence measures for subsets of observations have been proposed; see [2] as well as [15] for surveys of some of these methods.

From Proposition 1, it follows that $P(W_i = 1) = h_{ii} = E(W_i)$, which suggests that an analogous measure of the influence of a subset of observations whose indices are $i_1, \cdots, i_k$ might be based on the distribution of $W_{i_1}, \cdots, W_{i_k}$.

Suppose that A is a subset of $\{1, \cdots, n\}$ and define

$$N(A) = \sum_{j \in A} W_j. \tag{5}$$

Given that $E(W_i) = h_{ii}$ and $P(W_i = 1, W_j = 1) = E(W_i W_j) = h_{ii}h_{jj} - h_{ij}^2$ from Proposition 1, it follows that

$$E[N(A)] = \sum_{j \in A} h_{jj}$$

$$\mathrm{Var}[N(A)] = \sum_{j \in A} h_{jj} - \sum_{i \in A} \sum_{j \in A} h_{ij}^2.$$

More generally, the probability distribution of $N(A)$ in (5) can be determined from the probability generating function

$$E\left[t^{N(A)}\right] = \frac{\left| t \sum_{j \in A} \boldsymbol{x}_j \boldsymbol{x}_j^T + \sum_{j \notin A} \boldsymbol{x}_j \boldsymbol{x}_j^T \right|}{\left| \sum_{i=1}^{n} \boldsymbol{x}_i \boldsymbol{x}_i^T \right|}.$$

This gives, for example, if $A = \{i_1, \cdots, i_k\}$,

$$\begin{aligned} P\left(N(A)=0\right) &= \frac{\left| \sum_{j \notin A} \boldsymbol{x}_j \boldsymbol{x}_j^T \right|}{\left| \sum_{i=1}^{n} \boldsymbol{x}_i \boldsymbol{x}_i^T \right|} \\ &= \left| I - \left(\sum_{i=1}^{n} \boldsymbol{x}_i \boldsymbol{x}_i^T \right)^{-1} \sum_{j \in A} \boldsymbol{x}_j \boldsymbol{x}_j^T \right| \\ &= \left| \begin{pmatrix} 1 - h_{i_1 i_1} & -h_{i_1 i_2} & \cdots & -h_{i_1 i_k} \\ -h_{i_2 i_1} & 1 - h_{i_2 i_2} & \cdots & -h_{i_2 i_k} \\ \vdots & \vdots & \ddots & \vdots \\ -h_{i_k i_1} & -h_{i_k i_2} & \cdots & 1 - h_{i_k i_k} \end{pmatrix} \right|. \end{aligned} \tag{6}$$

In the case where $h_{i_1 i_1}, \cdots, h_{i_k i_k}$ are uniformly small and $k \ll n$ then

$$P\left(N(A)=0\right) \approx \exp\left(-\sum_{j=1}^{k} h_{i_j i_j} - \frac{1}{2} \sum_{j=1}^{k} \sum_{\ell=1}^{k} h_{i_j i_\ell}^2 \right).$$

Also note that (6) can also be computed as

$$\mathscr{P}\left(N(A)=0\right) = \prod_{j=1}^{k} \left\{ 1 - \boldsymbol{x}_{i_j}^T \left(\sum_{i \in A \setminus \{i_1, \cdots, i_{j-1}\}} \boldsymbol{x}_i \boldsymbol{x}_i^T \right)^{-1} \boldsymbol{x}_{i_j} \right\}$$

where the quadratic form

$$\boldsymbol{x}_{i_j}^T \left(\sum_{i \in A \setminus \{i_1, \cdots, i_{j-1}\}} \boldsymbol{x}_i \boldsymbol{x}_i^T \right)^{-1} \boldsymbol{x}_{i_j}$$

is a diagonal of the "hat" matrix with observations $i_1, \cdots, i_{j-1}$ deleted.

Suppose that $\widehat{\boldsymbol{\beta}}_A$ is the OLS estimate of $\boldsymbol{\beta}$ based on $\{(\boldsymbol{x}_i, y_i) : i \notin A\}$ and define $\mathscr{P}_A$ to be the probability distribution on subsets S so that $\widehat{\boldsymbol{\beta}}_A = E_{\mathscr{P}_A}(\widehat{\boldsymbol{\beta}}_S)$. If $\mathscr{P}_A$ is close to $\mathscr{P}$, then we would expect $\widehat{\boldsymbol{\beta}}_A$ to be close to $\widehat{\boldsymbol{\beta}}$—in other words, the influence of the subset A on estimation of $\boldsymbol{\beta}$ is small. More generally, if we delete the observations in A, we may want to define an estimate based on elemental estimates from subsets s with $s \cap A = \emptyset$ using a different probability distribution $\mathscr{Q}$ (with $\mathscr{Q}(s) = 0$ if $s \cap A \neq \emptyset$) so that

$$\widetilde{\boldsymbol{\beta}}_A = \sum_s \widehat{\boldsymbol{\beta}}_s \mathscr{Q}(s).$$

The following result provides a simple formula the total variation (TV) distance between $\mathscr{P}_A$ and $\mathscr{P}$ as well as giving a condition on $\mathscr{Q}$ that minimizes the TV distance between $\mathscr{Q}$ and $\mathscr{P}$.

Proposition 2 *(a) Define $\mathscr{P}_A(s) = \mathscr{P}(s)/P(N(A) = 0)$ for subsets s with $s \cap A = \emptyset$. Then*

$$d_{tv}(\mathscr{P}_A, \mathscr{P}) = \sup_B |\mathscr{P}_A(B) - \mathscr{P}(B)| = P(N(A) \geq 1)$$

where $P(N(A) \geq 1)$ can be evaluated using (6). (b) Suppose that $\mathscr{Q}$ is a probability distribution on subsets s with $\mathscr{Q}(s) = 0$ if $s \cap A \neq \emptyset$. Then $d_{tv}(\mathscr{Q}, \mathscr{P}) \geq P(N(A) \geq 1)$ where the lower bound is attained if $\mathscr{Q}(s) = \lambda(s)\mathscr{P}(s)$ (for $s \cap A = \emptyset$) where $\lambda(s) \geq 1$.

Proof

(a) We can compute the TV distance as

$$d_{tv}(\mathscr{P}_A, \mathscr{P}) = \frac{1}{2} \sum_s |\mathscr{P}_A(s) - \mathscr{P}(s)|.$$

If $s \cap A = \emptyset$, then

$$\mathscr{P}_A(s) = \frac{\mathscr{P}(s)}{P(N(A) = 0)}$$

with $\mathscr{P}_A(s) = 0$ when $s \cap A \neq \emptyset$. Thus

$$d_{tv}\left(\mathscr{P}_A, \mathscr{P}\right) = \frac{1}{2}\sum_s |\mathscr{P}_A(s) - \mathscr{P}(s)|$$

$$= \frac{1}{2}\left\{\sum_{s\cap A=\emptyset} |\mathscr{P}_A(s) - \mathscr{P}(s)| + \sum_{s\cap A\neq\emptyset} |\mathscr{P}_A(s) - \mathscr{P}(s)|\right\}$$

$$= P\left(N(A) \geq 1\right).$$

(b) For probability distributions $\mathscr{Q}$ concentrated on subsets s satisfying $s \cap A = \emptyset$, we have

$$\sum_{s\cap A\neq\emptyset} |\mathscr{Q}(s) - \mathscr{P}(s)| = P\left(N(A) \geq 1\right);$$

thus it suffices to minimize

$$\sum_{s\cap A=\emptyset} |\mathscr{Q}(s) - \mathscr{P}(s)|$$

subject to

$$\sum_{s\cap A=\emptyset} \mathscr{Q}(s) = 1.$$

The first order condition implies that the minimizer $\mathscr{Q}^*$ must satisfy $\mathscr{Q}^*(s) \geq \mathscr{P}(s)$ for all s and so $\mathscr{Q}^*(s) = \lambda(s)\mathscr{P}(s)$ where $\lambda(s) \geq 1$ and

$$\sum_{s\cap A=\emptyset} \lambda(s)\mathscr{P}(s) = 1.$$

Now

$$d_{tv}\left(\mathscr{Q}^*, \mathscr{P}\right) = \frac{1}{2}\sum_s |\mathscr{Q}^*(s) - \mathscr{P}(s)|$$

$$= \frac{1}{2}\left\{\sum_{s\cap A=\emptyset} |\mathscr{Q}^*(s) - \mathscr{P}(s)| + \sum_{s\cap A\neq\emptyset} |\mathscr{Q}^*(s) - \mathscr{P}(s)|\right\}$$

$$= \frac{1}{2}\left\{\sum_{s\cap A=\emptyset} (\lambda(s) - 1)\mathscr{P}(s) + \sum_{s\cap A\neq\emptyset} \mathscr{P}(s)\right\}$$

$$= P\left(N(A) \geq 1\right),$$

which completes the proof.

Part (a) of Proposition 2 suggests that $P(N(A) \geq 1)$ is a natural analogue of the "hat" diagonals for measuring the potential influence of observations with indices in A. More precisely, we can define the leverage $\text{lev}(A)$ of the subset $A = \{i_1, \cdots, i_k\}$ as

$$\text{lev}(A) = P(N(A) \geq 1) = 1 - \left| \begin{pmatrix} 1 - h_{i_1 i_1} & -h_{i_1 i_2} & \cdots & -h_{i_1 i_k} \\ -h_{i_2 i_1} & 1 - h_{i_2 i_2} & \cdots & -h_{i_2 i_k} \\ \vdots & \vdots & \ddots & \vdots \\ -h_{i_k i_1} & -h_{i_k i_2} & \cdots & 1 - h_{i_k i_k} \end{pmatrix} \right| . \quad (7)$$

As before, if $h_{i_1 i_1}, \cdots, h_{i_k i_k}$ are uniformly small and $k \ll n$, then we can approximate $\text{lev}(A)$ in (7) by

$$\text{lev}(A) \approx 1 - \exp\left(-\sum_{j=1}^{k} h_{i_j i_j} - \frac{1}{2} \sum_{j=1}^{k} \sum_{\ell=1}^{k} h_{i_j i_\ell}^2 \right) \approx \sum_{j=1}^{k} h_{i_j i_j} + \frac{1}{2} \sum_{j=1}^{k} \sum_{\ell=1}^{k} h_{i_j i_\ell}^2 .$$

As noted in [4], the matrix in Eq. (7) as well as its determinant (that is, $1 - \text{lev}(A)$) plays a role in a number of diagnostic tests (for example, those of [1] and [3]) for assessing the influence of observations whose indices lie in A; see also [11].

Part (b) of Proposition 2 implies that when $P(N(A) \geq 1) < 1$, any probability distribution of the form $\mathcal{Q}^*(s) = \lambda(s)\mathcal{P}(s)$ where $\lambda(s) \geq 1$ for $s \cap A = \emptyset$ attains the minimum TV distance to $\mathcal{P}$; this condition is always satisfied by $\mathcal{P}_A$. In particular, as $P(N(A) \geq 1)$ decreases, the family of distributions attaining the minimum TV distance becomes richer. (If we replace the TV distance by the Hellinger distance in part (b), then the minimum is attained uniquely at $\mathcal{P}_A$.)

4 Application: Algorithmic Leveraging

In least squares problems where n and p are very large, it is often useful to solve a smaller problem where $m \ll n$ observations are sampled (possibly using some weighting scheme) with $\boldsymbol{\beta}$ estimated using OLS or WLS estimation on the sampled observations. For example, algorithmic leveraging [6, 7, 12, 13] samples observations using biased sampling where the probability that an observation $(\boldsymbol{x}_i, y_i)$ is sampled is proportional to its leverage h_{ii} or an approximation to h_{ii}; efficient methods for approximating $\{h_{ii}\}$ are discussed in [5]. The sampled observations are then used to estimate $\boldsymbol{\beta}$ using OLS or some form of WLS. In addition, the observations may also be "pre-conditioned": If $\boldsymbol{y}$ is the vector of responses and X is the $n \times p$ matrix whose i row is $\boldsymbol{x}_i^T$, then we can transform $\boldsymbol{y} \mapsto V\boldsymbol{y}$ and $X \mapsto VX$ for some $n \times n$ matrix; V is chosen so that the "hat" diagonals of VX are less dispersed than those of X.

Suppose that a given subsample does not include observations with indices in A; in the case of leveraging, these observations more likely have small values of h_{ii} and so $P(N(A) \geq 1)$ will be smaller than if the observations were sampled using simple random sampling. We now estimate $\boldsymbol{\beta}$ by minimizing

$$\sum_{i \notin A} w_i (y_i - \boldsymbol{x}_i^T \boldsymbol{\beta})^2$$

for some weights $\{w_i > 0 : i \notin A\}$. The resulting estimate $\widehat{\boldsymbol{\beta}}_{ss}$ can be written as

$$\widehat{\boldsymbol{\beta}}_{ss} = \sum_{s \cap A = \emptyset} \mathscr{Q}(s) \widehat{\boldsymbol{\beta}}_s$$

where

$$\mathscr{Q}(s) = \frac{\mathscr{P}(s) \prod_{j \in s} w_j}{\sum_{u \cap A = \emptyset} \mathscr{P}(u) \prod_{j \in u} w_j}.$$

From Proposition 2, $\mathscr{Q}$ attains the lower bound on the TV distance to $\mathscr{P}$ if $\mathscr{Q}(s) = \lambda(s)\mathscr{P}(s)$ for some $\lambda(s) \geq 1$ when $s \cap A = \emptyset$; in other words, we require

$$\prod_{j \in s} w_j \geq P(N(A) = 0) \sum_{u \cap A = \emptyset} \left\{ \frac{\mathscr{P}(u)}{P(N(A) = 0)} \prod_{j \in u} w_j \right\} \tag{8}$$

for all s with $s \cap A = \emptyset$. The condition (8) is always satisfied if all the weights $\{w_i\}$ are equal, in which case, $\widehat{\boldsymbol{\beta}}_{ss}$ is an OLS estimate. For non-equal weights, the situation becomes more complicated. For example, if $w_i = 1/h_{ii}$ and the variability of $\{h_{ii} : i \notin A\}$ is relatively large, then (8) may be violated for some subsets s, particularly when $P(N(A) = 0)$ is close to 1 (so that the lower bound for the TV distance is close to 0). This observation is consistent with the results in [12] as well as [13] where unweighted estimation (setting $w_i = 1$) generally outperforms weighted estimation. Proposition 2 also suggests that it may be worthwhile selecting m observations so as to maximize $P(N(A) = 0)$ and thereby minimizing the TV distance. This effectively excludes low-leverage observations from the sample, which may not be desirable from a statistical point of view; moreover, determining the exclusion set A will be computationally expensive for large p and n.

To illustrate, we consider a simple linear regression with $\boldsymbol{x}_i^T = (1\ x_i)$ for $i = 1, \cdots, n = 1000$ where $\{x_i\}$ are drawn from a two-sided Gamma distribution with shape parameter $\alpha = 0.5$; this produces a large number of both large ($h_{ii} > 4p/n = 0.008$) and small ($h_{ii} \approx 1/n = 0.001$) leverage points. We then draw a sample of 200 (unique) observations using leverage sampling and compute the TV distance

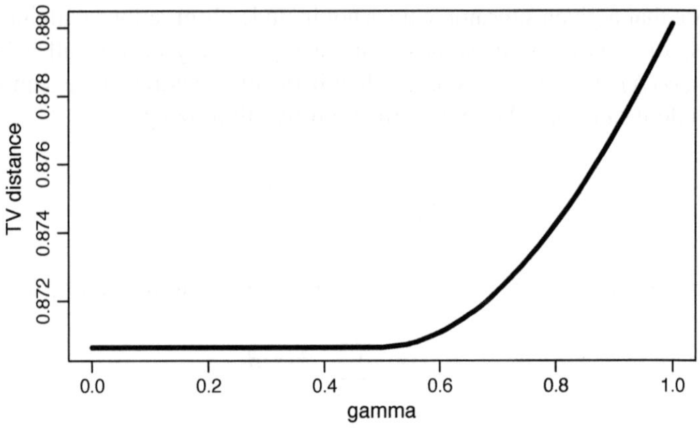

Fig. 1 TV distance as a function of γ for a leverage sample of size $m = 200$

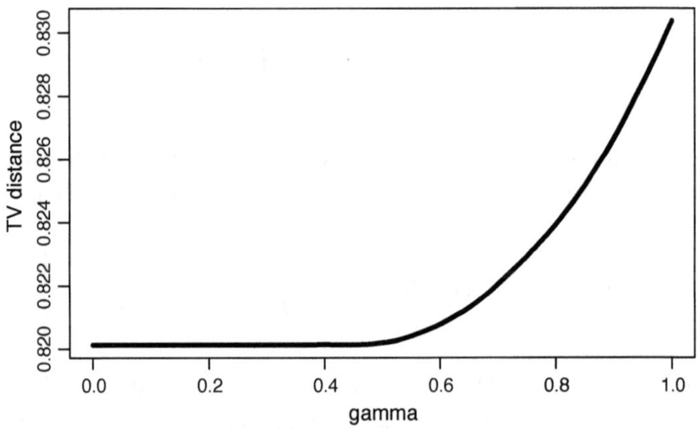

Fig. 2 TV distance as a function of γ for a sample of $m = 200$ where the exclusion set A is chosen to (approximately) maximize $P(N(A) = 0)$

for WLS with $w_i = h_{ii}^{-\gamma}$ for $0 \leq \gamma \leq 1$; a plot of the TV distance as a function of γ is shown in Fig. 1.

A second sample of 200 (unique) observations is obtained by excluding a set A of 800 observations to maximize (approximately) $P(N(A) = 0)$; a plot of the TV distance as a function of γ is shown in Fig. 2. In both cases, the TV distance is minimized (that is, condition (8) is satisfied) for values of γ between 0 and approximately 0.5 with the minimum TV distance being smaller (0.82 versus 0.87) for the second sample.

References

1. Andrews, D. F., & Pregibon, D. (1978). Finding the outliers that matter. *Journal of the Royal Statistical Society. Series B, 40*, 85–93.
2. Chatterjee, S., & Hadi, A. S. (1986). Influential observations, high leverage points, and outliers in linear regression. *Statistical Science, 1*, 379–393.
3. Cook, R. D. (1977). Detection of influential observations in linear regression. *Technometrics, 19*, 15–18.
4. Draper, N. R., & John, J. A. (1981). Influential observations and outliers in regression. *Technometrics, 23*, 21–26.
5. Drineas, P., Magdon-Ismail, M., Mahoney, M. W., & Woodruff, D. P. (2012). Fast approximation of matrix coherence and statistical leverage. *Journal of Machine Learning Research, 13*, 3475–3506.
6. Drineas, P., Mahoney, M. W., Muthukrishnan, S., & Sarlós, T. (2011). Faster least squares approximation. *Numerische Mathematik, 117*, 219–249.
7. Gao, K. (2016). Statistical inference for algorithmic leveraging. Preprint, arXiv:1606.01473.
8. Golberg, M. A. (1972). The derivative of a determinant. *The American Mathematical Monthly, 79*, 1124–1126.
9. Hoaglin, D. C., & Welsch, R. E. (1978). The hat matrix in regression and ANOVA. *The American Statistician, 32*, 17–22.
10. Hoerl, A. E., & Kennard, R. W. (1980). M30. A note on least squares estimates. *Communications in Statistics–Simulation and Computation, 9*, 315–317.
11. Little, J. K. (1985). Influence and a quadratic form in the Andrews-Pregibon statistic. *Technometrics, 27*, 13–15.
12. Ma, P., Mahoney, M. W., & Yu, B. (2015). A statistical perspective on algorithmic leveraging. *Journal of Machine Learning Research, 16*, 861–911.
13. Ma, P., & Sun, X. (2015). Leveraging for big data regression. *Wiley Interdisciplinary Reviews: Computational Statistics, 7*, 70–76.
14. Mayo, M. S., & Gray, J. B. (1997). Elemental subsets: The building blocks of regression. *The American Statistician, 51*, 122–129.
15. Nurunnabi, A. A. M., Hadi, A. S., & Imon, A. H. M. R. (2014). Procedures for the identification of multiple influential observations in linear regression. *Journal of Applied Statistics, 41*, 1315–1331.
16. Subrahmanyam, M. (1972). A property of simple least squares estimates. *Sankhya Series B, 34*, 355–356.

References

Bootstrapping Nonparametric M-Smoothers with Independent Error Terms

Matúš Maciak

Abstract On the one hand, nonparametric regression approaches are flexible modeling tools in modern statistics. On the other hand, the lack of any parameters makes these approaches more challenging when assessing some statistical inference in these models. This is crucial especially in situations when one needs to perform some statistical tests or to construct some confidence sets. In such cases, it is common to use a bootstrap approximation instead. It is an effective alternative to more straightforward but rather slow plug-in techniques. In this contribution, we introduce a proper bootstrap algorithm for a robustified version of the nonparametric estimates, the so-called M-smoothers or M-estimates, respectively. We distinguish situations for homoscedastic and heteroscedastic independent error terms, and we prove the consistency of the bootstrap approximation under both scenarios. Technical proofs are provided and the finite sample properties are investigated via a simulation study.

1 Introduction

Let us consider a simple situation where we have some random sample $\{(X_i, Y_i);\ i = 1, \ldots, n\}$ of size $n \in \mathbb{N}$, drawn from some unknown two dimensional population $(\mathscr{X}, \mathscr{Y})$, where the following association structure is assumed to be valid within the data:

$$Y_i = m(X_i) + \sigma(X_i)\varepsilon_i, \qquad \text{for } i = 1, \ldots, n; \tag{1}$$

M. Maciak (✉)
Department of Probability and Mathematical Statistics, Charles University, Prague, Czech Republic
e-mail: maciak@karlin.mff.cuni.cz; maciak@karlin.mff.cz

P. Bertail et al. (eds.), *Nonparametric Statistics*, Springer Proceedings in Mathematics & Statistics 250, https://doi.org/10.1007/978-3-319-96941-1_16

The random error terms $\{\varepsilon_i\}_{i=1}^n$ are assumed to be independent and identically distributed (*i.i.d.*) with some symmetric distribution function G. The expression in (1) is commonly used for a standard heteroscedastic regression model. If, in addition, we assume that the unknown scale function $\sigma(\cdot)$ is constant over the whole domain, for instance, interval [0, 1], the scenario above reduces to a simple homoscedastic regression case. The unknown regression function $m(\cdot)$ is, in general, used to model the dependence of the mean of some generic random variable Y on some random covariate X. The random sample (X_i, Y_i), for $i = 1, \ldots, n$ is assumed to be drawn from the joint distribution of (X, Y). The main task is to estimate the unknown dependence function $m(\cdot)$. Alternatively, if the model is heteroscedastic, a simultaneous estimation of both, $m(\cdot)$ and $\sigma(\cdot)$, is performed instead.

There is a huge body of literature available on how to estimate the unknown regression function and the scale function, respectively. The simplest approach is to assume some well-defined parametric shape for $m(\cdot)$ (and $\sigma(\cdot)$, respectively), and to use the least squares approach to estimate the parameters defining the shape. The whole inference is later based on the estimated parameters only. More flexible approaches are mostly represented by semi-parametric techniques where there are still parameters involved in the estimation, but these parameters do not directly restrict the shape of the unknown regression function. They rather represent some alternative expression of the estimate using, for instance, fractional polynomials, splines, or wavelets. The estimate of $m(\cdot)$ is then defined as some linear combination of the estimated parameters and some, let's say, basis functions. Any consequent inference needs to target the corresponding linear combination. Finally, the nonparametric estimation is considered to be the most flexible modeling technique, but the resulting estimate usually cannot be expressed explicitly any more. This makes any inference in the nonparametric regression model more challenging and also more difficult to prove. This is also the case that we focus on in this work.

In addition to the overall model flexibility we also introduce some robust flavor in the estimation process: we would like to obtain an estimate of the unknown regression function $m(\cdot)$ which will be robust with respect to the distribution of the random error terms, the distribution G. In fact, beside no parametric assumptions on $m(\cdot)$ and $\sigma(\cdot)$, we also assume no specific distribution family for G. It is only required to be symmetric and continuous, with a unit scale, such that $G(1) - G(-1) = \frac{1}{2}$. Thus, the resulting estimate is robust with respect to outlying observations and heavy-tailed random error distributions. The asymptotic properties of the final estimate, however, depend on some unknown quantities; therefore, for practical utilization, either some plug-in techniques need to be adopted to do the proper inference or one can also try some bootstrap approximation instead.

This chapter is organized as follows: local polynomial M-smoothers are briefly discussed in the next session. Some important theoretical properties summarized as well. In Sect. 3, the bootstrap algorithm is introduced for the M-smoother estimates under both scenarios, the homoscedastic and heteroscedastic random error terms. Finite sample properties of the bootstrap approximation are investigated in Sect. 4, and technical details and proofs are given in the appendix part at the end.

2 Asymptotics for the M-Smoothers

For the purposes of this work we only limit our attention to situations where the unknown regression function is continuous, and moreover, it is considered to be smooth up to some specific order, $p \in \mathbb{N}$. Under this smoothness assumption the local polynomial M-smoother estimates are defined as higher order generalizations of the local linear M-smoother introduced in [15] and also discussed in [8]. Robust approaches in nonparametric regression are also presented in [5] and briefly also in [1]. Considering the given data, the local polynomial M-smoother of $m(\cdot)$, at some given point $x \in (0, 1)$, which is considered to be the domain of interest, is defined as a solution of the minimization problem

$$\widehat{\boldsymbol{\beta}}_x = \underset{(b_0,\ldots,b_p)^\top \in \mathbb{R}^{p+1}}{Argmin} \sum_{i=1}^{n} \rho\left(Y_i - \sum_{j=0}^{p} b_j (X_i - x)^j\right) \cdot K\left(\frac{X_i - x}{h_n}\right), \tag{2}$$

where $\widehat{\boldsymbol{\beta}}_x = (\widehat{\beta}_0, \widehat{\beta}_1, \ldots, \widehat{\beta}_p)^\top$. Function $K(\cdot)$ stands for a classical kernel function common for the nonparametric regression estimation (e.g., [13, 17]), and $h_n > 0$ is some bandwidth parameter. Function ρ stands for a general loss function and it is assumed to be symmetric and convex, such that for its derivative (or one-sided derivatives at least) it holds that $\rho' = \psi$ almost everywhere ($a.e.$).

The M-smoother estimate $\widehat{m}(x)$ of the regression function m at the given point $x \in (0, 1)$ is defined as $\widehat{m}(x) = \widehat{\beta}_0$. In general, it holds that $\widehat{m}^{(\nu)}(x) = \nu! \widehat{\beta}_\nu$, where $\nu = 0, \ldots, p$ stands for the order of the corresponding derivative of $m(\cdot)$. Unlike the classical nonparametric regression where $\rho(\cdot) = (\cdot)^2$ and $m(x) = E[Y|X = x]$, the functional representation of the unknown regression function for some general loss function depends now on a specific choice of $\rho(\cdot)$—the parameter estimates are not given in explicit forms by default and the asymptotic mean squared error (AMSE), which is commonly used to define a right value for the bandwidth parameter, can now lead to a biased bandwidth selection once some outlying observations are present (see [9, 10]). Instead, one needs to find an asymptotic representation for the vector of unknown parameter estimates and to use alternative methods to choose the optimal value of the bandwidth parameter (for instance, robust cross-validation criterion). Similarly, there is also a broad discussion on how to choose the degree of the polynomial approximation $p \in \mathbb{N} \cup \{0\}$, or the kernel function K. We do not discuss these issues here in this chapter. If the reader is interested, more details can be found, for instance, in [3].

The M-smoother estimate of the unknown regression function $m(\cdot)$ at some given point $x \in (0, 1)$ is not explicit; however, under some rather mild conditions, it can be shown that it is asymptotically normal with zero mean and some unknown variance. Before we state the asymptotic properties we provide a set of assumptions which are needed for the theoretical results to hold. For brevity, we only present the assumptions for the heteroscedastic scenario and the assumptions for the homoscedastic case follow as a straightforward simplification.

A.1 The marginal density function $f(\cdot)$ of the *i.i.d.* random variables X_i, for $i = 1, \ldots, n$, is absolutely continuous, positive, and bounded on interval $[0, 1]$, which is considered to be the support of X. In addition, the scale function $\sigma(\cdot)$ is Lipschitz and positive on $[0, 1]$;

A.2 The random error terms $\varepsilon_1, \ldots, \varepsilon_n$ are assumed to be *i.i.d.*, mutually independent of X_i, for $i = 1, \ldots, n$, with a symmetric and continuous distribution function $G(\cdot)$, such that $G(1) - G(-1) = \frac{1}{2}$;

A.3 The regression function $m(\cdot)$ and its derivatives $m^{(1)}(x), \ldots, m^{(p+1)}$ for $p \in \mathbb{N}$ being the degree of the local polynomial approximation are Lipschitz on $(0, 1)$. In addition, the loss function $\rho(\cdot)$ is symmetric, convex, and absolutely continuous. Moreover, it holds that $\rho' = \psi$ almost everywhere (*a.e.*);

A.4 Function $\lambda_G(t, v) = -\int \psi(ve - t)\mathrm{d}G(e)$ is Hölder of the order $\gamma > 0$ in argument $v > 0$. The partial derivative $\lambda'_G(t, v) = \frac{\partial}{\partial t}\lambda_G(t, v)$ exists and it is continuous in t for some neighborhoods of $t = 0$ and $v = \sigma(x)$, for given $x \in (0, 1)$. Moreover, it holds that $\int |\psi(\sigma(x)e)|^2 \mathrm{d}G(e) < \infty$ and $\lambda'_G(0, \sigma(x)) = \frac{\partial}{\partial t}\lambda_G(t, \sigma(x))|_{t=0} \neq 0$. Finally, the following

$$\int_{\mathbb{R}} \Big|\psi(\sigma(x)e - \epsilon_N) - \psi(\sigma(x)e)\Big|^2 \mathrm{d}G(e) < \mathcal{K} \cdot |\epsilon_N|, \tag{3}$$

$$\int_{\mathbb{R}} \Big|\psi(\sigma(x + \epsilon_N)e) - \psi(\sigma(x)e)\Big|^2 \mathrm{d}G(e) < \mathcal{K} \cdot |\epsilon_N|, \tag{4}$$

holds for the given $x \in (0, 1)$, any sequence $\epsilon_N \to 0$, and some $\mathcal{K} > 0$.

A.5 Function $K(\cdot)$ is a kernel function which is assumed to be a symmetric density with its support on $[-1, 1]$, such that $\int_{-1}^{1} K^2(u)\mathrm{d}u < \infty$. The bandwidth parameter h_n satisfies the following: $h_n \to 0$ as $n \to \infty$, such that $h_n \sim n^{-\xi}$, for $\xi \in \left(\frac{1+\delta}{5}, \frac{1}{1+\delta}\right)$, where $\delta > 0$ small enough.

The assumptions stated above are derived as a straightforward combination of the assumptions required for the classical local polynomial regression (see, for instance, [3]) and the robust M-estimates introduced in [16]. Expression (3) and (4) in A.4 can be seen as generalized versions of the Lipschitz condition and they are trivially satisfied, for example, for ψ and σ being Lipschitz. Assumption A.5 can be markedly simplified for the homoscedastic case: function $\lambda_G(ve - t)$ does not depend on $v > 0$ and thus, all statements regarding this argument can be omitted.

Let us also introduce some necessary notation which will be needed for the formulation of the results and the consecutive proofs: for K being some arbitrary density function which satisfies A.5, we define $(p+1)\times(p+1)$ type matrices $S_1 = \left\{\int_{-1}^{1} u^{j+l} K(u)(d)u\right\}_{j,l}$ and $S_2 = \left\{\int_{-1}^{1} u^{j+l} K^2(u)(d)u\right\}_{j,l}$, for $j, l = 0, \ldots, p$.

Theorem 1 (Asymptotic Normality for the Heteroscedastic Model) *Let the model in* (1) *hold and let the assumptions in A.1–A.5 be all satisfied. Then the M-smoother estimate of the unknown regression function $m(\cdot)$, at some given point $x \in (0, 1)$, is consistent and moreover, it holds that*

$$\sqrt{Nh_N}\cdot\left(\widehat{m}(x) - m(x)\right) \xrightarrow[N\to\infty]{D} N\left(0, \frac{E[\psi(\sigma(x)\varepsilon_1)]^2\cdot v_{11}}{[\lambda_G'(0,\sigma(x))]^2\cdot f(x)}\right),$$

with v_{11} being the first diagonal element of matrix $V = S_1^{-1}S_2S_1^{-1} = \{v_{ij}\}_{i,j=1}^{(p+1)}$.

Proof A detailed proof of Theorem 1 is given in [11] where the homoscedastic and heteroscedastic scenarios are considered separately. □

Remark 1 For the homoscedastic scenario the scale function $\sigma(x)$ can be considered to be equal to a constant (e.g., one) over the whole domain [0, 1], therefore, the asymptotic variance reduces to $\frac{E\psi^2(\varepsilon_1)\cdot v_{11}}{[\lambda_G'(0)]^2\cdot f(x)}$, which now only depends on x via the density function $f(x)$. For completeness, we used the notation $\lambda_G'(0) \equiv \lambda_G'(0, 1)$.

Remark 2 The result in Theorem 1 can be generalized for an arbitrary ν-order derivative $m^{(\nu)}(x)$ of $m(x)$, for $\nu \in \{1, \ldots, p\}$ and the given $x \in (0, 1)$. In such case the convergence rate changes to $\sqrt{Nh_N^{1+2\nu}}$ and the asymptotic variance equals $\frac{\nu! E[\psi(\sigma(x)\varepsilon_1)]^2}{[\lambda_G'(0,\sigma(x))]^2 f(x)}\cdot \boldsymbol{e}_\nu^\top V \boldsymbol{e}_\nu$, where $\boldsymbol{e}_\nu \in \mathbb{R}^{p+1}$ denotes a unit vector with value one on the position $(\nu+1)$ (and zeros otherwise).

It is easy to see that under both scenarios the asymptotic variance depends on some unknown quantities—indeed, in many practical applications the design density $f(\cdot)$ is usually left unknown and the same also holds for the distribution $G(\cdot)$ which plays the role in the expectation term in the nominator and also in $\lambda_G(0, \sigma(x))$ in the denominator. In addition, for the heteroscedastic models, the scale function $\sigma(\cdot)$ is usually unknown as well. One can either use some plug-in techniques to consistently estimate the unknown quantities firstly and to plug these estimates into the variance expression to obtain the asymptotic distribution. This distribution can be further used for making statistical tests of constructing confidence intervals. The plug-in techniques, however, are well known for their rather slow convergence, therefore, it is usually recommended to use some bootstrap approximation if possible. The asymptotic normality result stated above is, however, crucial for proving the bootstrap consistency. In the next section we present two algorithms which can be used to mimic the asymptotic distribution of interest under both scenarios—the homoscedastic and heteroscedastic model—and we prove the bootstrap consistency for the proposed algorithms.

3 Smooth Bootstrapping of M-Smoothers

The bootstrap approaches are, in general, used to mimic some distribution of interest—either the true distribution is unknown and the bootstrap simulations are used to estimate it or the distribution of interest is too complicated to be used directly and thus, bootstrapping is employed to get an approximated distribution, which is simpler and more straightforward to be used for the statistical inference procedures.

This is also the situation presented in this contribution. The asymptotic distribution of the M-smoothers presented in Theorem 1 is unknown (in terms that unknown quantities are needed to specify the exact normal distribution) and thus, it cannot be used directly to run any inference on the M-Smoothers estimates. However, the bootstrap approximation can be effectively used to mimic this distribution. In the following we provide two versions of the bootstrapping algorithm (homoscedastic and heteroscedastic cases) and we prove the bootstrap consistency for both.

The bootstrap algorithms proposed below are based on the idea presented in [14]. The notion of the *smooth* bootstrap comes from the step B3 in both algorithms: this step firstly ensures the right centering of the bootstrapped residuals, while the second part—the smoothing element introduced by $a_N Z_i$—ensures a proper convergence of the bootstrapped distribution to an unknown distribution of the true random errors. Another advantage of this approach relies in the fact that no additional over-smoothing is needed (see [14] and also [6]) and, at the same time, the bootstrap distribution is automatically centered and symmetric. Using the smooth version of the bootstrap algorithm one can conveniently handle both, a proper centering of bootstrapped residuals in order to eliminate the systematic bias and also preserving the robust flavor of the whole procedure.

The centering of the bootstrapped residuals is usually done by subtracting their empirical mean, e.g., average $\frac{1}{n}\sum_{i=1}^{n}\tilde{\varepsilon}_i$ from each ε_i, however, bearing in mind the robust flavor of the whole M-smoother estimation framework, where we also allow for outlying observations or heavy-tailed distributions of random errors, such centering would not be computationally stable because of the outliers and heavy-tailed error distributions.

The proposed version of the smooth residual bootstrap can preserve the robust flavor of the M-smoother and, therefore, it nicely suits our model scenario(s).

Smooth residual bootstrap under homoscedasticity

B1 Calculate the set of residuals $\{\widehat{\varepsilon}_i;\ i = 1,\dots,n\}$, where $\widehat{\varepsilon}_i = Y_i - \widehat{m}(X_i)$, for $\widehat{m}(X_i)$ being the M-smoothers estimate of $m(X_i)$ defined by (2);

B2 Resample with replacement from the set of residuals $\{\widehat{\varepsilon}_i;\ i = 1,\dots,n\}$ in order to obtain new residuals $\tilde{\varepsilon}_i$, for $i = 1,\dots,n$;

B3 Define new bootstrapped residuals as $\varepsilon_i^\star = V_i \cdot \tilde{\varepsilon}_i + a_n \cdot Z_i$, where $P[V_i = -1] = P[V_i = 1] = \frac{1}{2}$, and $Z_i \sim N(0,1)$ are i.i.d. standard normal random variables and $a_n = o(1)$ is a bootstrap bandwidth parameter, such that $nh_n a_n^2/\log h_n^{-1} \to \infty$, and $a_n^2/h_n^{1+\delta} = o(1)$ as $n \to \infty$, for some $\delta > 0$ small enough;

B4 Define a new dataset—the bootstrapped sample $\{(X_i, Y_i^\star);\ i = 1,\dots,n\}$, where $Y_i^\star = \widehat{m}(X_i) + \varepsilon_i^\star$;

B5 Re-estimate the unknown regression function $m(x)$ at the given point $x \in (0,1)$ based on the new data sample $\{(X_i, Y_i^\star);\ i = 1,\dots,n\}$ using (2) and obtain $\widehat{m}^\star(x)$;

B6 Repeat the steps in B2–B5 to get multiple copies of the M-smoother estimates $\widehat{m}_b^\star(x)$, for $b = 1,\dots,B$, where $B \in \mathbb{N}$ is sufficiently large, and use these quantities to mimic the asymptotic distribution of interest;

Smooth residual bootstrap under heteroscedasticity

B1 Calculate residuals $\{\widehat{\varepsilon}_i;\ i = 1,\dots,n\}$, where $\widehat{\varepsilon}_i = \frac{Y_i - \widehat{m}(X_i)}{\widehat{\sigma}(X_i)}$, for $\widehat{m}(X_i)$ being the M-smoother estimate of $m(X_i)$ defined by (2) and $\widehat{\sigma}(X_i)$ is the corresponding estimate of the scale function $\sigma(\cdot)$, given at the point X_i;

B2 Resample with replacement from the set of residuals $\{\widehat{\varepsilon}_i;\ i = 1,\dots,n\}$ in order to obtain new residuals $\tilde{\varepsilon}_i$, for $i = 1,\dots,n$;

B3 Define new bootstrapped residuals as $\varepsilon_i^\star = V_i \cdot \tilde{\varepsilon}_i + a_n \cdot Z_i$, where $P[V_i = -1] = P[V_i = 1] = \frac{1}{2}$, and $Z_i \sim N(0,1)$ are i.i.d. standard normal random variables and $a_n = o(1)$ is a bootstrap bandwidth parameter, such that $nh_n a_n^2/\log h_n^{-1} \to \infty$, $Na_N^{2(p+1)} \to 0$, and $a_n^2/h_n^{1+\delta} = o(1)$ as $n \to \infty$, for $\delta > 0$ small enough;

B4 Define a new bootstrapped sample $\{(X_i, Y_i^\star);\ i = 1,\dots,n\}$, where now we have $Y_i^\star = \widehat{m}(X_i) + \widehat{\sigma}(X_i)\varepsilon_i^\star$;

B5 Re-estimate the unknown regression function $m(x)$ at the given point $x \in (0,1)$ based on the new data sample $\{(X_i, Y_i^\star);\ i = 1,\dots,n\}$ using (2) and obtain $\widehat{m}^\star(x)$;

B6 Repeat the steps in B2–B5 to get multiple M-smoother estimates $\widehat{m}_b^\star(x)$, for $b = 1,\dots,B$, where $B \in \mathbb{N}$ is sufficiently large, and use these quantities to mimic the underlying asymptotic distribution;

Comparing the bootstrap algorithm for the heteroscedastic model with the previous version for the homoscedastic model one can see some minor difference: indeed, in the later version one needs to deal with the scale function $\sigma(\cdot)$ in addition. On the other hand, it is easy to see that for the scale function being constant the heteroscedastic algorithm reduces to the homoscedastic version. There were many different methods proposed on how to estimate the scale function in the nonparametric regression models (see, for instance, [4, 12]). However, in order to keep the robust flavor of the whole estimation process, the scale function $\sigma(\cdot)$ should be also obtained in some robust manner. The M-estimator of the scale function which is suitable for this situation was proposed in [2] and it is defined as

$$\widehat{\sigma}(x) = \inf\left\{z > 0;\ \sum_{i=1}^{n-1} w_{ni}(x) \cdot \chi\left(\frac{Y_{i+1} - Y_i}{\alpha_1 \cdot z}\right) \leq \alpha_2\right\}, \tag{5}$$

where $w_{ni}(x)$ are some weights, $\chi(\cdot)$ is some score function, and $\alpha_1, \alpha_2 > 0$ are some constants, such that $E\chi(Z_1) = \alpha_1$ and $E\chi\left(\frac{Z_2 - Z_1}{\alpha_1}\right) = \alpha_2$, for Z_1 and Z_2 being some independent random variables with the distribution which corresponds with the distribution or the random error terms $\{\varepsilon_i\}_{i=1}^{n}$. For the score function $\chi(\cdot)$ it is additionally assumed that it is continuous, bounded, and strictly increasing, such that $\chi(0) = 0$ and $0 < \sup_{x \in \mathbb{R}} \chi(x)$. Under some regularity conditions (see [2] for further details and exact proofs) it was derived that the estimate of the scale function $\sigma(\cdot)$ defined by (5) for some given point $x \in (0, 1)$ yields a strong consistency once the number of the sample size n tends to infinity. In addition, it can be shown that the obtained estimate is also asymptotically normal.

Theorem 2 (Bootstrap Consistency) *Let the model in* (1) *hold with Assumptions A.1–A.5 being satisfied, and let the bootstrap bandwidth parameter* $a_n \to 0$ *satisfy the conditions in* B3*. Then the proposed smooth residual bootstrap algorithm is consistent and it holds that*

$$\sup_{z \in \mathbb{R}} \left\{P^\star\left[\sqrt{nh_n}\left(\widehat{m}^\star(x) - \widehat{m}(x)\right) \leq z\right] - P\left[\sqrt{nh_n}\left(\widehat{m}(x) - m(x)\right) \leq z\right]\right\} \xrightarrow[n \to \infty]{P} 0,$$

where $P^\star[\ \cdot\ ]$ *stands for a conditional probability given data* $\{(X_i, Y_i);\ i = 1, \ldots, n\}$.

Proof The proof of the theorem is given in Appendix. □

Having the consistency result stated in Theorem 2 we have an efficient tool for performing practically any statistical inference in the models being estimated within the M-smoothers regression framework. The bootstrapped distribution can be used to construct confidence intervals for $m(x)$, for some $x \in (0, 1)$, or it can be used to draw critical values to decide about some set of hypothesis, again related to $m(x)$, at some given point from the domain.

Remark 3 Theorem 2 can be only used to make inference about the unknown regression function $m(\cdot)$ at some given point from the domain. If one is interested in providing a confidence bound for the whole regression function $m(\cdot)$, there has to be more advanced methods used to do so—see, for instance, [6].

Remark 4 Similarly as in Theorem 1 and Remark 2 the statement in Theorem 2 can be again generalized for $\nu \in \{0, 1, \ldots, p\}$. In such case $m^{(\nu)}(x)$ stands for the corresponding ν-order derivative of m at the given point $x \in (0, 1)$.

Let us briefly mention that it is also possible to deal with data sequences which are not independent. Under some mild assumptions (for instance, an α-mixing structure of the error terms $\{\varepsilon_i\}$) one can prove the consistency results of the M-smoother estimate of $m(x)$ at some given point $x \in (0, 1)$ (see [7] for details). However, when performing the inference based on the bootstrap approach, it is not possible to rely on the simple version of the smooth residual bootstrap presented in this work. The reason is that the independent resampling in B2 step (of both algorithms) is not capable of preserving the covariance structure in the data. Instead, some block bootstrap modification needs to be employed to obtain a valid approximation of the true distribution function (see [7]).

4 Finite Sample Properties

For the simulation purposes we considered a simple regression function of the form $m(x) = (-8x)\sin(\pi x)$, for $x \in (0, 1)$, and we run the simulations for various settings: three different sample sizes $n \in \{50, 100, 500\}$; three different loss functions (squared loss, absolute loss, Huber loss); three degrees of the polynomial approximation $p \in \{0, 1, 3\}$, and also three different distribution functions for the random error terms were considered:

$\mathscr{D}_1$—standard Gaussian;
$\mathscr{D}_2$—mixture of the standard Gaussian with 5% of $N(0, \sigma^2 = 625)$;
$\mathscr{D}_3$—Cauchy $C(0, 1)$.

For each combination of the sample size, error distribution, loss function, and the approximation degree p, we generated the data for 100 times and we ran the M-smoothers estimation to reconstruct the "unknown" regression function. Out of these 100 independent repetitions the empirical behavior of the M-smoother estimate at some given point was investigated (see histograms in Fig. 1).

Later, for one specific data scenario we employed the bootstrap resampling and we obtained the bootstrapped distribution based on 500 independent bootstrap resamples (according to the algorithms stated in Session 3). The bootstrapped distributions are plotted as solid red lines in Fig. 1. The M-smoother behavior together with the bootstrap performance is summarized in Tables 1, 2, and 3 below. As the bootstrap consistency result stated in Theorem 2 is meant to be used for some given point $x \in (0, 1)$, which is the domain of $m(\cdot)$, we only considered the value of $x = 0.2$ and the results are stated for $m(0.2) = -1.5217$. However, quite analogous results can be obtained for any other choice of $x \in (0, 1)$.

From the simulations results in Tables 1, 2, and 3 it is clear that the robust flavor of the M-smoothers estimates is crucial especially in situations when the random error terms have some distribution with heavy tails (e.g., Cauchy distribution). The classical squared loss based estimation is not able to handle this scenario and the variance of the estimates even increases as the sample size tends to infinity

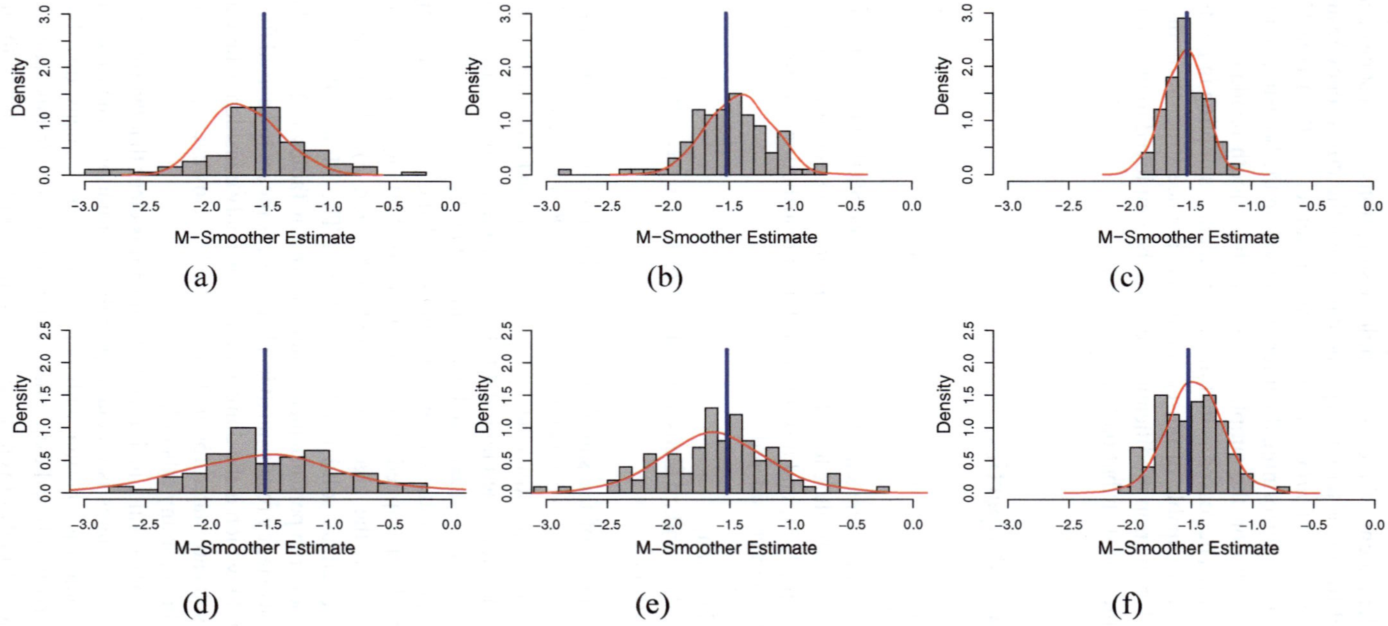

Fig. 1 Bootstrapped distribution (in solid red) compared with the empirical distribution of the M-smoother estimates based on 100 repetitions given in gray histograms. The true value is indicated by the blue vertical line. In the first row the standard normal distribution for error terms is considered together with the squared loss function and below, the heavy-tailed Cauchy distribution with Huber loss function is given instead. (**a**) $n = 50$. (**b**) $n = 100$. (**c**) $n = 500$. (**d**) $n = 50$. (**e**) $n = 100$. (**f**) $n = 500$

Table 1 Local constant M-smoother estimate of the unknown regression function at the given point $x = 0.2$ ($m(x) = -1.522$) calculated as an average out of 100 repetitions with the corresponding standard error (in brackets)

Sample size	Error dist.	Squared loss		Absolute loss		Huber loss	
		M-smoother	Bootstrap	M-smoother	Bootstrap	M-smoother	Bootstrap
n = 50	$\mathscr{D}_1$	−1.587 (*0.473*)	−1.686 (*0.281*)	−1.571 (*0.534*)	−2.249 (*0.281*)	−1.572 (*0.478*)	−1.966 (*0.264*)
	$\mathscr{D}_2$	−1.333 (*1.389*)	−1.378 (*1.440*)	−1.470 (*0.512*)	−1.295 (*0.490*)	−1.471 (*0.460*)	−1.322 (*0.402*)
	$\mathscr{D}_3$	−2.156 (*4.742*)	−1.490 (*1.772*)	−1.503 (*0.772*)	−1.226 (*0.870*)	−1.497 (*0.660*)	−1.467 (*1.070*)
n = 100	$\mathscr{D}_1$	−1.547 (*0.409*)	−1.550 (*0.299*)	−1.554 (*0.460*)	−1.826 (*0.503*)	−1.540 (*0.422*)	−1.714 (*0.363*)
	$\mathscr{D}_2$	−1.355 (*1.510*)	−1.520 (*1.609*)	−1.491 (*0.452*)	−1.462 (*0.498*)	−1.487 (*0.400*)	−1.491 (*0.393*)
	$\mathscr{D}_3$	−2.039 (*3.916*)	−2.567 (*5.532*)	−1.560 (*0.654*)	−1.549 (*0.764*)	−1.547 (*0.572*)	−1.665 (*0.839*)
n = 500	$\mathscr{D}_1$	−1.544 (*0.347*)	−1.633 (*0.288*)	−1.547 (*0.396*)	−1.809 (*0.425*)	−1.541 (*0.358*)	−1.728 (*0.314*)
	$\mathscr{D}_2$	−1.393 (*1.287*)	−1.596 (*1.408*)	−1.491 (*0.385*)	−1.443 (*0.419*)	−1.492 (*0.339*)	−1.477 (*0.335*)
	$\mathscr{D}_3$	−1.576 (*4.773*)	−3.299 (*5.303*)	−1.545 (*0.559*)	−1.604 (*0.642*)	−1.536 (*0.489*)	−1.682 (*0.698*)

The bootstrap counterpart calculated for one specific repetition, based on 500 bootstrap resamples, again with the corresponding bootstrapped standard error (in brackets). Three different sample sizes, three different loss functions, and three different error distributions are considered

Table 2 Local linear M-smoother estimate of the unknown regression function at the given point $x = 0.2$ ($m(x) = -1.522$) calculated as an average out of 100 repetitions with the corresponding standard error (in brackets)

Sample size	Error dist.	Squared loss		Absolute loss		Huber loss	
		M-smoother	Bootstrap	M-smoother	Bootstrap	M-smoother	Bootstrap
n = 50	$\mathscr{D}_1$	−1.598 (*0.494*)	−1.705 (*0.284*)	−1.564 (*0.510*)	−2.104 (*0.242*)	−1.589 (*0.500*)	−1.952 (*0.275*)
	$\mathscr{D}_2$	−1.335 (*1.425*)	−1.191 (*1.555*)	−1.466 (*0.490*)	−1.138 (*0.329*)	−1.462 (*0.473*)	−1.236 (*0.342*)
	$\mathscr{D}_3$	−1.916 (*4.093*)	−1.563 (*1.692*)	−1.445 (*0.781*)	−1.200 (*0.735*)	−1.521 (*0.699*)	−1.493 (*0.888*)
n = 100	$\mathscr{D}_1$	−1.557 (*0.420*)	−1.571 (*0.302*)	−1.553 (*0.447*)	−1.772 (*0.423*)	−1.547 (*0.435*)	−1.687 (*0.377*)
	$\mathscr{D}_2$	−1.362 (*1.536*)	−1.418 (*1.891*)	−1.488 (*0.433*)	−1.371 (*0.440*)	−1.488 (*0.407*)	−1.417 (*0.364*)
	$\mathscr{D}_3$	−1.863 (*3.426*)	−2.727 (*5.732*)	−1.548 (*0.659*)	−1.532 (*0.676*)	−1.571 (*0.589*)	−1.638 (*0.711*)
n = 500	$\mathscr{D}_1$	−1.551 (*0.354*)	−1.646 (*0.286*)	−1.551 (*0.383*)	−1.770 (*0.362*)	−1.545 (*0.367*)	−1.709 (*0.326*)
	$\mathscr{D}_2$	−1.398 (*1.309*)	−1.617 (*1.640*)	−1.492 (*0.370*)	−1.412 (*0.375*)	−1.494 (*0.345*)	−1.440 (*0.315*)
	$\mathscr{D}_3$	−1.506 (*4.000*)	−3.471 (*5.282*)	−1.537 (*0.559*)	−1.582 (*0.570*)	−1.552 (*0.503*)	−1.668 (*0.596*)

The bootstrap counterpart calculated for one specific repetition, based on 500 bootstrap resamples, again with the corresponding bootstrapped standard error (in brackets). Three different sample sizes, three different loss functions, and three different error distributions are considered

Table 3 Local cubic M-smoother estimate of the unknown regression function at the given point $x = 0.2$ ($m(x) = -1.522$) calculated as an average out of 100 repetitions with the corresponding standard error (in brackets)

Sample size	Error dist.	Squared loss		Absolute loss		Huber loss	
		M-smoother	Bootstrap	M-smoother	Bootstrap	M-smoother	Bootstrap
n = 50	$\mathscr{D}_1$	−1.627 (*0.530*)	−1.703 (*0.256*)	−1.576 (*0.533*)	−2.230 (*0.220*)	−1.611 (*0.528*)	−2.017 (*0.245*)
	$\mathscr{D}_2$	−1.276 (*1.622*)	−1.411 (*1.371*)	−1.474 (*0.537*)	−1.243 (*0.367*)	−1.477 (*0.500*)	−1.372 (*0.341*)
	$\mathscr{D}_3$	−1.998 (*4.161*)	−1.932 (*1.648*)	−1.497 (*0.805*)	−1.257 (*0.741*)	−1.571 (*0.755*)	−1.634 (*0.856*)
n = 100	$\mathscr{D}_1$	−1.576 (*0.455*)	−1.556 (*0.282*)	−1.563 (*0.469*)	−1.831 (*0.463*)	−1.569 (*0.465*)	−1.731 (*0.371*)
	$\mathscr{D}_2$	−1.361 (*1.667*)	−1.552 (*1.726*)	−1.496 (*0.469*)	−1.436 (*0.451*)	−1.490 (*0.437*)	−1.506 (*0.338*)
	$\mathscr{D}_3$	−1.947 (*3.503*)	−2.906 (*5.473*)	−1.573 (*0.685*)	−1.583 (*0.674*)	−1.602 (*0.641*)	−1.778 (*0.682*)
n = 500	$\mathscr{D}_1$	−1.567 (*0.385*)	−1.635 (*0.268*)	−1.556 (*0.404*)	−1.813 (*0.390*)	−1.562 (*0.394*)	−1.740 (*0.314*)
	$\mathscr{D}_2$	−1.414 (*1.422*)	−1.682 (*1.495*)	−1.499 (*0.400*)	−1.428 (*0.378*)	−1.496 (*0.372*)	−1.492 (*0.290*)
	$\mathscr{D}_3$	−1.574 (*4.056*)	−3.641 (*5.046*)	−1.559 (*0.585*)	−1.630 (*0.566*)	−1.581 (*0.552*)	−1.765 (*0.569*)

The bootstrap counterpart calculated for one specific repetition, based on 500 bootstrap resamples, again with the corresponding bootstrapped standard error (in brackets). Three different sample sizes, three different loss functions, and three different error distributions are considered

(which is indeed, an expected behavior). On the other hand, the absolute loss and the Huber function can still provide consistent estimates and valid inference.

5 Conclusion

In this contribution, we consider a standard nonparametric regression scenario, however, with a set of very weak assumptions. The error terms are assumed to be independent but the distribution is free of any moment conditions. Thus, the model and the M-smoother estimation approach both allow for some outlying observations in the data or, even, some heavy-tailed random error distribution. The M-smoother estimation approach is stated to be consistent and asymptotically normal but the limiting distribution depends on a few unknown quantities. Instead of using some rather slow plug-in techniques we introduce two versions of the smooth residual bootstrap algorithm which can be used to mimic the underling distribution.

The proposed bootstrap approaches are proved to be consistent and finite sample properties are investigated via an extensive simulation study and the results are shown to correspond with the theory.

The proposed smooth residual bootstrap for nonparametric and robust M-smoother estimates can be considered as an effective alternative to more straightforward, however, less reliable plug-in techniques.

The proposed model scenarios can be further extended in account, for instance, for dependent error terms (such as various mixing sequences) but the bootstrap algorithm needs to be modified accordingly. The smooth residual bootstrap cannot capture the true variance–covariance structure in the data, and therefore, more appropriate block bootstrap version needs to be used instead.

Acknowledgements The author's research was partly supported by the Grant P402/12/G097.

Appendix

In this section we provide some technical details and the proof of the bootstrap consistency result stated in Theorem 2. Let $\{(X_i, Y_i^\star);\ i = 1, \ldots, n\}$ be the bootstrapped data where $Y_i^\star = \widehat{m}(X_i) + \widehat{\sigma}(X_i)\varepsilon_i^\star$, for $\widehat{m}(X_i)$ being the M-smoother estimate of $m(X_i)$, $\widehat{\sigma}(X_i)$ the estimate of $\sigma(X_i)$ in sense of (5), and the random error terms $\{\varepsilon_i^\star\}_{i=1}^n$ are defined in B3 step of the bootstrap algorithm in Sect. 3. Then, we can obtain the bootstrapped version of $\widehat{m}(x)$, for some $x \in (0, 1)$, given as a solution of the minimization problem

$$\widehat{\boldsymbol{\beta}}_x^\star = \underset{(b_0,\ldots,b_p)^\top \in \mathbb{R}^{p+1}}{Argmin} \sum_{i=1}^{n} \rho\left(Y_i^\star - \sum_{j=0}^{p} b_j (X_i - x)^j\right) \cdot K\left(\frac{X_i - x}{h_n}\right), \tag{6}$$

where $\widehat{\boldsymbol{\beta}}_x^\star = (\widehat{\beta}_0^\star, \ldots, \widehat{\beta}_p^\star)^\top$, and $\widehat{m}^\star(x) = \widehat{\beta}_0^\star$. Using the smoothness property of $m(\cdot)$ we can apply the Taylor expansion of the order p and given the model definition in (1) we can rewrite the minimization problem as an equivalent problem given by the following set of equations:

$$\frac{1}{\sqrt{nh_n}} \sum_{i=1}^{n} \psi\left(\widehat{\sigma}(X_i)\varepsilon_i^\star - \sum_{j=0}^{p} b_j \left(\frac{X_i - x}{h_n}\right)^j\right) \cdot \left(\frac{X_i - x}{h_n}\right)^\ell K\left(\frac{X_i - x}{h_n}\right) = 0,$$

for $\ell = 0, \ldots, p$, where $\psi = \rho'$. Next, for any $\ell \in \{0, \ldots, p\}$ and $\boldsymbol{b} \in \mathbb{R}^{p+1}$ let us define an empirical process $M_n(\boldsymbol{b}, \ell)$ and its bootstrap counterpart $M_n^\star(\boldsymbol{b}, \ell)$ as follows:

$$M_n(\boldsymbol{b}, \ell) = \frac{1}{\sqrt{nh_n}} \sum_{i=1}^{n} \left[\psi\left(\sigma(X_i)\varepsilon_i - \sum_{j=0}^{p} b_j \xi_i^j(x)\right) - \psi\left(\sigma(X_i)\varepsilon_i\right)\right] K(\xi_i^1(x))\xi_i^\ell(x), \tag{7}$$

and

$$M_n^\star(\boldsymbol{b}, \ell) = \frac{1}{\sqrt{nh_n}} \sum_{i=1}^{n} \left[\psi\left(\widehat{\sigma}(X_i)\varepsilon_i^\star - \sum_{j=0}^{p} b_j \xi_i^j(x)\right) - \psi\left(\widehat{\sigma}(X_i)\varepsilon_i^\star\right)\right] K(\xi_i^1(x))\xi_i^\ell(x), \tag{8}$$

where for brevity we used the notation $\xi_i^\ell(x) = \left(\frac{X_i - x}{h_n}\right)^\ell$. We need to investigate the behavior of $M_n^\star(\boldsymbol{b}, \ell)$, conditionally on the sample $\{(X_i, Y_i);\ i = 1, \ldots, n)\}$, and we will compare it with the behavior of $M_n(\boldsymbol{b}, \ell)$.

Let $G^\star(\cdot)$ be the distribution function of the bootstrap residuals $\{\varepsilon_i^\star\}_{i=1}^n$ defined in B3. It follows from the definition that

$$\begin{aligned}
G^\star(e) &= P^\star[\varepsilon_i^\star \leq e] = P^\star[V_i \cdot \tilde{\varepsilon}_i + a_n \cdot Z_i \leq e] \\
&= \frac{1}{2n}\left[\int_{\mathbb{R}} \sum_{i=1}^{n} \mathbb{I}_{\{\widehat{\varepsilon}_i \leq e - a_n u\}} \phi(u) \mathrm{d}u + \int_{\mathbb{R}} \sum_{i=1}^{n} \mathbb{I}_{\{\widehat{\varepsilon}_i \geq a_n u - e\}} \phi(u) \mathrm{d}u\right] \\
&= \frac{1}{2n} \sum_{i=1}^{n} \left[\Phi\left(\frac{e - \widehat{\varepsilon}_i}{a_n}\right) + \Phi\left(\frac{e + \widehat{\varepsilon}_i}{a_n}\right)\right],
\end{aligned}$$

where $\phi(\cdot)$ and $\Phi(\cdot)$ stand for the density and the distribution function of Z_i's, which are assumed to be normally distributed with zero mean and unit variance. It is easy to verify that $G^\star(\cdot)$ is continuous, symmetric, and moreover, it satisfies Assumption A.2. Thus, for $E^\star$ being the conditional expectation operator when conditioning by the initial data sample, we obtain the following:

$$E^\star M_n^\star(\boldsymbol{b},\ell) = \frac{1}{\sqrt{nh_n}} \sum_{i=1}^n E^\star\Big[\psi\big(\widehat{\sigma}(X_i)\varepsilon_i - \sum_{j=0}^p b_j \xi_i^j(x)\big)\Big] \cdot K(\xi_i^1(x))\xi_i^\ell(x)$$

$$= \frac{-1}{\sqrt{nh_n}} \sum_{i=1}^n \lambda_{G^\star}\Big(\sum_{j=0}^p b_j \xi_i^j(x), \widehat{\sigma}(X_i)\Big) \cdot K(\xi_i^1(x))\xi_i^\ell(x),$$

where we used the symmetric property of the distribution function $G^\star$. Next, we obtain

$$\lambda_{G^\star}\Big(\sum_{j=0}^p b_j \xi_i^j(x), \widehat{\sigma}(X_i)\Big) = -\int_{\mathbb{R}} \psi\Big(\widehat{\sigma}(X_i)e - \sum_{j=0}^p b_j \xi_i^j(x)\Big) \mathrm{d}G^\star(e)$$

$$= \lambda_G\Big(\sum_{j=0}^p b_j \xi_i^j(x), \widehat{\sigma}(X_i)\Big) \tag{9}$$

$$- \int_{\mathbb{R}} \psi\Big(\widehat{\sigma}(X_i)e - \sum_{j=0}^p b_j \xi_i^j(x)\Big) \mathrm{d}(G^\star - G)(e),$$

where the last term can be shown to be asymptotically negligible due to the properties of $\psi(\cdot)$ and the fact that $\sup_{x\in\mathbb{R}} |G^\star(x) - G(x)| \to 0$ in probability (see Lemma 2.19 in [14]). For (9) we can use the Hölder property of $\lambda_G(\cdot)$ (Assumption A.4) and we get that

$$\left|\lambda_G\Big(\sum_{j=0}^p b_j \xi_i^j(x), \widehat{\sigma}(X_i)\Big) - \lambda_G\Big(\sum_{j=0}^p b_j \xi_i^j(x), \sigma(X_i)\Big)\right| = o(1),$$

and

$$\left|\lambda_G\Big(\sum_{j=0}^p b_j \xi_i^j(x), \sigma(X_i)\Big) - \lambda_G\Big(\sum_{j=0}^p b_j \xi_i^j(x), \sigma(x)\Big)\right| = o(1),$$

where the first equation follows from the fact that $\widehat{\sigma}(X_i)$ is a consistent estimate of $\sigma(X_i)$ and the second follows from the fact that $|X_i - x| \leq h_n$. Both equations hold almost surely.

Putting everything together we obtain that

$$E^\star M_n^\star(\boldsymbol{b},\ell) = EM_n(\boldsymbol{b},\ell) + o_P(1),$$

and, moreover, repeating the same steps also for the second moment $E^{\star}\big[M_n^{\star}(\boldsymbol{b}, \ell)\big]^2$ and applying (3) and (4) we also obtain that $E^{\star}\big[M_n^{\star}(\boldsymbol{b}, \ell)\big]^2 \to 0$ in probability.

To finish the proof we need the following lemma.

Lemma 1 *Let the model in* (1) *hold and let Assumptions A.1–A.5 be all satisfied. Then the following holds:*

$$\sup_{\|\boldsymbol{b}\|\leq C}\left|M_n^{\star}(\delta_n \boldsymbol{b}, \ell) + \frac{\delta_n}{\sqrt{nh_n}}\,\lambda_G'(0, \sigma(x))\sum_{i=1}^{n}\left(\frac{X_i - x}{h_n}\right)^{\ell} \times \right.$$
$$\left. \times \left(\sum_{j=0}^{p} b_j\left(\frac{X_i - x}{h_n}\right)^{j}\right)\cdot K\left(\frac{X_i - x}{h_n}\right)\right| = o_P\left(1\right),$$

where $\ell = 0, \ldots, p$, $C > 0$, $\|\boldsymbol{b}\| = |b_0| + \cdots + |b_p|$, $\delta_n = (nh_n)^{-\gamma/2}$, *and* $\gamma \in (\gamma_0, 1]$, *for some* $0 < \gamma_0 \leq 1$.

Proof Lemma 1 is a bootstrap version of Lemma 4 in [11] or a more general Lemma A.3 in [7]. The proof follows the same lines using the moment properties derived for $M_n^{\star}(\delta_n \boldsymbol{b}, \ell)$. □

Lemma 4 in [11] allows us to express the classical M-smoother estimates $\widehat{\beta}_x$ in terms of the asymptotic Bahadur representations as

$$\frac{1}{(nh_n)^{1/2}}\widehat{\beta}_x = \frac{(nh_n)^{-1/2}}{\lambda_G'(0, \sigma(x))}\cdot\left(X_n^{\top} W_n X_n\right)^{-1}\cdot X_n^{\top} W_n \begin{pmatrix}\psi(\sigma(X_1)\varepsilon_1)\\ \vdots \\ \psi(\sigma(X_n)\varepsilon_n)\end{pmatrix} + o_P(1),$$

while Lemma 1 allows us to express the bootstrapped counterparts $\widehat{\beta}_x^{\star}$ in a similar manner as

$$\frac{1}{(nh_n)^{1/2}}\widehat{\beta}_x^{\star} = \frac{(nh_n)^{-1/2}}{\lambda_G'(0, \sigma(x))}\cdot\left(X_n^{\top} W_n X_n\right)^{-1}\cdot X_n^{\top} W_n \begin{pmatrix}\psi(\widehat{\sigma}(X_1)\varepsilon_1^{\star})\\ \vdots \\ \psi(\widehat{\sigma}(X_n)\varepsilon_n^{\star})\end{pmatrix} + o_P(1),$$

where $W_n = Diag\left\{K\left(\frac{X_1 - x}{h_n}\right), \ldots, K\left(\frac{X_n - x}{h_n}\right)\right\}$, and $X_n = \left(\left(\frac{X_i - x}{h_n}\right)^{j}\right)_{i=1, j=0}^{n, p}$.

To finish the proof one just needs to realize that the sequences of random variables $\{\xi_{ni}\}_{i=1}^{n}$ and $\{\xi_{ni}^{\star}\}_{i=1}^{n}$ for $\xi_{ni} = \frac{1}{\sqrt{nh_n}}\psi(\sigma(X_i)\varepsilon_i)\big(\frac{X_i - x}{h_n}\big)^{\ell} K\big(\frac{X_i - x}{h_n}\big)$ and $\xi_{ni}^{\star} = \frac{1}{\sqrt{nh_n}}\psi(\widehat{\sigma}(X_i)\varepsilon_i^{\star})\big(\frac{X_i - x}{h_n}\big)^{\ell} K\big(\frac{X_i - x}{h_n}\big)$ both comply with the assumptions of the central limit theorem for triangular schemes and thus, random quantities $\sum_{i=1}^{n}\xi_{ni}$ and $\sum_{i=1}^{n}\xi_{ni}^{\star}$ both converge in distribution, conditionally on X_i's and the original data $\{(X_i, Y_i);\ i = 1, \ldots, n\}$, respectively, to the normal distribution with zero mean and the same variance parameter. □

References

1. Antoch, J., & Janssen P. (1989). Nonparametric regression m-quantiles. *Statistics & Probability Letters, 8*, 355–362.
2. Boente, G., Ruiz, M., & Zamar, R. (2010). On a robust local estimator for the scale function in heteroscedastic nonparametric regression. *Statistics & Probability Letters, 80*, 1185–1195.
3. Fan, J., & Gijbels, I. (1995). *Local polynomial modelling and its applications* (1st ed.). Boca Raton, FL: Chapman & Hall
4. Hall, P., Kay, J., & Titterington, D. (1990). Asymptotically optimal difference-based estimation of variance in nonparametric regression. *Biometrica, 77*, 521–528.
5. Härdle, W., & Gasser, T. (1984). Robust nonparametric function fitting. *Journal of the Royal Statistical Society, Series B, 46*, 42–51.
6. Härdle, W. K., & Marron, J. S. (1991). MBootstrap simultaneous error bars for nonparametric regression. *The Annals of Statistics, 19*(2), 778–796.
7. Hušková, M., & Maciak, M. (2017). Discontinuities in robust nonparametric regression with α-mixing dependence. *Journal Nonparametric Statistics, 29*(2), 447–475.
8. Hwang, R. C. (2002). A new version of the local constant M-smoother. *Communications in Statistics: Theory and Methods, 31*, 833–848
9. Leung, D. (2005). Cross-validation in nonparametric regression with outliers. *Annals of Statistics, 33*, 2291–2310.
10. Leung, D., Marriott, F., & Wu, E. (1993). Bandwidth selection in robust smoothing. *Journal of Nonparametric Statistics, 2*, 333–339.
11. Maciak, M. (2011). *Flexibility, Robustness, and Discontinuity in Nonparametric Regression Approaches*. Ph.D. thesis, Charles University, Prague.
12. Müller, H., & Stadtmüller, U. (1987). Estimation of heteroscedasticity in regression analysis. *Annals of Statistics, 15*, 610–625.
13. Nadaraya, E. (1964). On estimating regression. *Theory Probability Applications, 9*, 141–142.
14. Neumeyer, N. (2006). *Bootstrap Procedures for Empirical Processes of Nonparametric Residuals*. Habilitationsschrift, Ruhr-University Bochum, Germany.
15. Rue, H., Chu, C., Godtliebsen, F., & Marron, J. (1998). M-smoother with local linear fit. *Journal of Nonparametric Statistics, 14*, 155–168.
16. Serfling, R. (1980). *Approximation theorems of mathematical statistics*. New York: Wiley.
17. Watson, G. (1964). Smooth regression analysis. *The Indian Journal of Statistics, Series A, 26*, 359–372.

Extension Sampling Designs for Big Networks: Application to Twitter

A. Rebecq

Abstract With the rise of big data, more and more attention is paid to statistical network analysis. However, exact computation of many statistics of interest is of prohibitive cost for big graphs. Statistical estimators can thus be preferable. Model-based estimators for networks have some drawbacks. We study design-based estimates relying on sampling methods that were developed specifically for use on graph populations. In this contribution, we test some sampling designs that can be described as "extension" sampling designs. Unit selection happens in two phases: in the first phase, simple designs such as Bernoulli sampling are used, and in the second phase, some units are selected among those that are somehow linked to the units in the first-phase sample. We test these methods on Twitter data, because the size and structure of the Twitter graph is typical of big social networks for which such methods would be very useful.

1 Introduction

1.1 Problem

With more and more businesses and public administrations producing larger raw datasets every day, statistical analysis of the so-called big data has risen. Consequently, more research in computer science and statistics have focused on methods to tackle such problems. However, a significant part of datasets that fall under the general "big data" framework are actually graphs. Graph-specific data analysis has applications in domains as diverse as social networks, biology, finance, etc. Since the rise of the web, statistical literature for networks has been growing rapidly, especially in the field of model-based estimation. In the past 20 years, models such as Barabási–Albert [1], Watts–Strogatz [33], stochastic block models [22],

A. Rebecq (✉)
Modal'X, UPL, University Paris Nanterre, Nanterre, France
e-mail: antoine.rebecq@polytechnique.org

P. Bertail et al. (eds.), *Nonparametric Statistics*, Springer Proceedings
in Mathematics & Statistics 250, https://doi.org/10.1007/978-3-319-96941-1_17

and many others have induced huge progress in understanding probabilities and statistics for various cases of networks. Yet, model-based estimation is sometimes inconvenient. First, models cannot possibly perform well on all statistics of real-life graphs. Second, model-based estimation obviously requires a fine tuned choice of model before being able to produce statistics. Finally, even when a specific model exists and fits the analysis of a specific graph, computation can be cumbersome. This is a motivation for developing design-based estimates, which have received little attention from a purely statistical point of view. One of the most efficient tools for statistical graph analysis, snap—which has a very convenient Python interface [17]—uses sampling methods to compute various statistics of interest [16].

There are some domains in computer science and social sciences that focus exclusively on social network analysis. A large part of this literature analyzes published content rather than estimates regarding quantification or qualification of accounts engaged in the analyzed content. On Twitter, this means the focus is more on tweets than on who tweets. Most recent analysis and inference based on Twitter data used the Streaming API[1] [3]. Tweets matching a research criterion were collected in real time for a few days, and then analyzed. Many studies perform sentiment analysis on the harvested tweets. Some studies try to unravel political sentiment based on their Twitter data [4, 32], for example, to try and predict election outcomes. Very recently, election prediction based on Twitter data was proven not more accurate than traditional quota sampling, when trying to predict the outcome of the UK 2015 general election [3]. Other topics are very diverse and include, for example, stock market prediction [2]. Non-academic studies based on tweets data are also numerous. Most are made by market research companies, often to measure "engagement" by users to a brand.

However, such analyses are based on biased estimates. First, the Streaming API is often preferred by researchers over the Rest API.[2] The Streaming API provides tweets in real-time matching a certain query, thus allowing collection of a huge amount of data. However, when the data size exceeds a certain threshold, only a fraction of tweets are selected, but Twitter does not disclose the sampling design used to select these tweets. The Rest API allows the collection of a much more limited number of tweets, but the selection is made by account and not by a query on tweets. A statistical analysis using the Rest API is thus much closer to the survey sampling paradigm. In fact, it is possible to derive unbiased estimators using the Rest API. Using the Streaming API to perform a global analysis on the scope of the Twitter users (or *a fortiori* if estimates are to be extrapolated to a larger population, such as in the case of the prediction of election outcomes) can lead to unbalanced profiles of users. When the selection is made by tweets, users tweeting much less about a certain subject have a much lower probability of being selected than users who tweet a lot about the same subject. This is even more telling when the so-called bots (automated accounts set up to regularly write tweets about a pre-defined

[1] https://dev.twitter.com/streaming/overview.

[2] https://dev.twitter.com/rest/public.

subject) account for a non-negligible fraction of the total tweets [8]. When selection is made using a search query, it is very difficult to assess the selection probability, and when not all the tweets are output a large number of probabilities can be equal to 0. As [19] noticed, in many cases the users that precisely have a low selection probability account for the greater variability in estimates. Sloan et al. [26] study precisely these difference in user profiles used in Twitter-based estimations.

The goal of our work is to perform an estimate of the number of accounts tweeting on a specific subject, focusing on the accounts behind the tweets instead of the tweets themselves. We chose to work on the release of the trailer of the movie "Star Wars: the Force awakens" that occurred on October 19, 2015. According to "Twitter Reverb" data, this event generated 390,000 tweets in less than 3 h.[3]

This study is meant to be a proof of concept for further study. Regarding specific analysis on social networks, we think our analysis could be used jointly with analyses based on tweets to shed a new light on data structures. For example, in the case of election prediction, it could be used to balance selection probabilities that are naturally skewed away from people who're less likely to participate in Internet debates (which, unfortunately, is correlated to political opinions). In any other marketing context, this method could be used to weight tweets according to methods extending the "generalized weights share method" such as proposed by Rivers [14, 23]. More generally, understanding how different sampling schemes behave on a huge graph that exhibits many of the particular properties of web and social graphs (see, for example, Sect. 1.2) could help us understand the benefits and disadvantages of these schemes. Estimation by sampling could be used to improve computation time and precision for descriptive statistics or even machine learning algorithms on graphs.

Here, we test two different methods of sampling adapted to graph populations. These two methods are in fact extensions of simple sampling designs. Each of these methods yields unbiased estimates. Theoretically, the precision of the estimates can be spectacularly improved with these extension designs. However, their practical use can be limited by our ability to collect the data on the graph. In this chapter, no inference is made on any population outside the scope of Twitter users (i.e., the vertices of the Twitter graph).

1.2 The Twitter Graph

In computer science, the Twitter graph is said to show both "information network" and "social network" properties (see, for instance, [20]). The degree distributions (both inbound and outbound) are heavy-tailed, resembling a power law distribution and the average path length between two users is short. Thus, in the context of model-based estimation, the Twitter graph should be modelized by both a scale-free network (Barabási-Albert [1]) and a small-world network (Watts-Strogatz [33])

[3]http://reverb.guru/view/597295533668271595.

Let's denote :

$$\begin{aligned}
\mathcal{U} &= \text{population of interest} = \text{vertices of the graph} \\
N &= \#\mathcal{U} = \text{population size} \\
n &= \text{sample size} \\
y_k &= \text{number of times user k tweeted about the Star Wars trailer} \\
z_k &= \mathbb{1}\{k \in \mathcal{U}, y_k \geq 1\} \\
N_C &= \#\{k \in \mathcal{U}, y_k \geq 1\} \\
n_C &= \#\{k \in s, y_k \geq 1\} = \text{number of people in } s \text{ who tweeted about Star Wars} \\
T(Y) &= \sum_{k \in \mathcal{U}} y_k
\end{aligned}$$

We use graph-specific notations to make some formulas clearer: V = set of vertices (= $\mathcal{U}$), E_{ij} = set of edges linking vertices i and j. The goal is to estimate $N_C = T(Z)$, the number of users who have tweeted about the Star Wars trailer during the time span of the study.

For the sake of simplicity, we chose a very simple definition of our sub-population of interest, the Twitter users that tweeted about the Star Wars trailer. In our study, the "Star Wars fans" are detected because the tweets they wrote during the selected time span match a pre-defined list of words, mentions to other accounts, or hashtags. However, a more realistic setting would be to imagine that we have at our disposal a complex classifier f able to identify a broader definition of a "Star Wars fan" (for example, a machine learning classifier based on natural language processing or a community detection algorithm). Therefore our goal would be to estimate $N_C = \mathbb{1}\{k, f(k) = 1\}$, which is statistically speaking identical to the problem we deal with here. It is also worth noting that exhaustive computation of such a classifier would be very computationally intensive, even with a full access to the Twitter graph. Thus the use of a sample also makes practical computation of such statistical objects easier.

2 Sampling Designs

2.1 Notations

Notations for sets and networks :

$$\begin{aligned}
s_0 &= \text{initial sample} \\
C &= \bigcup_i r_i = \text{sub-population of units bearing the characteristic of interest}
\end{aligned}$$

$s_C = s \cap C$

$s =$ total sample, i.e., s_0 inflated with units of C that can be reached

$s_{ri} = s \cap r_i$

$C_k = \forall k \in s, C_k = \mathbb{1}\{k \in C\}$

$\delta =$ Graph geodesic distance between two units in the network

$r_i^0 = \{k, \delta(k, r_i)\}$, "side" of r_i

$$s^0 = \bigcup_i r_i^0 = \{k \in s, \delta(k, s_C) = 1\}$$

$s_{ex} =$ elements which are neither in a network nor a side $= \{k \in s, \delta(k, s_C) \geq 2\}$

In this work, s_0 is selected using stratified Bernoulli sampling.

2.2 Bernoulli Sampling

Poisson sampling consists in selecting in the sample s each unit $k \in \mathcal{U}$ with a Bernoulli experiment of parameter π_k, the first-order inclusion probability of unit k. Second-order inclusion probabilities thus have a very simple expression $\forall k \neq l \in \mathcal{U}, \pi_{kl} = \mathbb{P}(k, l \in s) = \pi_k \pi_l$. Bernoulli sampling is Poisson sampling with equal probabilities ($\forall k \in \mathcal{U}, \pi_k = p$). Bernoulli sampling design is the most simple design that can be used and thus gives very simple formulae for estimators and variance estimators. One can refer to Särndal [24] for a more detailed presentation of Poisson and Bernoulli sampling. For the remainder of this work, we will use Bernoulli sampling as the primary sampling design on which we'll base other more complicated (and more efficient in terms of precision) designs: a stratified design, an adaptive design, and a one-stage snowball design. We write p the selection probability for each unit in the frame (each Twitter user) and $q = 1-p$.

Poisson sampling is seldom used in survey sampling (at least in national statistics institutes), because it yields samples of variable size ($\#s$ is a random variable). A variable size can be problematic when data collection is costly (in most surveys by official statistics institutes, interviewers set up meetings with selected individuals either by phone or face-to-face). In our present case, it doesn't matter because the final goal is to use adaptive or snowball sampling, which will eventually yield a random final sample size. More, the cost of data collection is uniform across all units. We thus chose Poisson designs over simple random sampling, because the expressions for adaptive estimation and Rao–Blackwell estimates are much simpler.

On Twitter, users are assigned an id, ranging from 1 to $N \approx 3.3 \cdot 10^9$ (as of December, 2015). Some of the ids in this range are not assigned, but every new user is given an id greater than the last. In result, when selecting a Poisson sample using the ids in $[\![1, N]\!]$, part of the ids selected ($\approx 30\%$) will not be assigned to a Twitter user. Our sample is thus selected in a set greater than the population. Moreover, we are interested in only people who are *active* on Twitter, i.e., users who tweeted in the last month. There are two reasons for which we consider these units: first we have a population margin to calibrate on (see Sect. 3.4) and second, it is the scope that is generally considered when stats on Twitter users are discussed for business purposes. Consequently, our frame over-covers our scope. But as out-of-scope units are perfectly identified, the over-coverage can be treated very simply (see [25]) by using the restriction to a domain of the usual estimators. Two strategies can be used to produce estimates. If the total population on the domain $\mathscr{U}_d$ is unknown, we use Hájek estimators:

$$\hat{T}(Y)_d = \sum_{k \in \mathscr{U}} y_{kd}, \ \hat{\bar{y}}_d = \frac{\sum_{k \in \mathscr{U}} y_{kd}}{\sum_{k \in \mathscr{U}} z_{kd}}$$

$$\text{with: } y_{kd} = y_k \cdot \mathbb{1}\{k \in \mathscr{U}_d\}, \ z_{kd} = \mathbb{1}\{k \in \mathscr{U}_d\}$$

In our case, we know the total of the population on the domain N_d (see Sect. 3.4), so it is preferable to use $\hat{T}(Y)_d = \sum_{k \in \mathscr{U}_d} y_k$, which means all the estimators will work just like a restriction of the sample data frame on the scope. From now on, all the estimators are implicitly restricted to the scope domain unless stated otherwise.

The fact that some of the units sampled are finally out of scope leads to a final true sample size that can vary, which is another argument for not using fixed-size sampling design. For the simple Bernoulli design, our initial expected sample size (i.e., not taking the scope restriction into account) is **20000**: $p = 20000/3300000000 \approx 6.1 \cdot 10^{-6}$

2.3 *Stratified Bernoulli Sampling*

A usual way to improve the precision of the estimates is to stratify the population. We write $\mathscr{U} = \mathscr{U}_1 \oplus \mathscr{U}_2$ and draw two independent samples in $\mathscr{U}_h, \ h = 1, 2$ (in our case, two Bernoulli samples). With this design, the variance of the estimators writes $\text{Var}(\hat{T}) = \sum_h f_h(S_h^2)$. If strata are wisely selected, they should be homogeneous so that the S_h^2 are much reduced in comparison to the S^2 which determines the variance of the Horvitz–Thompson estimator under the simple Bernoulli design (see Sect. 4.1). Here, we try to estimate a proportion (alternatively a sub-population size), so we have:

$$f_h(S_h^2) = (1 - \frac{n_h}{N_h})\frac{N_h}{N_h - 1}\frac{p_h(1 - p_h)}{n_h}$$

The goal is thus to constitute strata where the probability of tweeting about the "Star Wars : The force awakens" trailer is more or less equal among users.

We divide the population of Twitter users into two strata: $\mathcal{U}_1$, the users who follow the official Star Wars account ("@starwars") (1654874 followers as of October, 21, 2015) and $\mathcal{U}_2$, the users who don't. We use Bernoulli sampling on each stratum independently, but with different inclusion probabilities, so that users who follow the official Star Wars account are "over-represented" in the final sample. By doing this, we hope to increase n_C the number of units who tweeted about Star Wars in s, because users who follow the official Star Wars account are more likely to be interested in tweeting about the new trailer.

In order to be able to compare the precision of the stratified design with the simple Bernoulli, we keep an expected final sample size of $n = n_1 + n_2 = 20000$. But we have to think about how to allocate these 20000 units between the two strata. We choose (again, in expected sample sizes) 9700 units in stratum 1 and 10300 units in stratum 2, which corresponds to a Neyman allocation [21] supposing that approximately half of the people who tweeted were following the official @starwars account and that each person who tweeted about "The force awakens" did it three times. This gives $n_{sw1} = n_{sw2} = 50000$ and thus $p_1 = 3.3$, $p_2 = 0.15$. Then, the Neyman allocation writes $n_1 = n\frac{N_1S_1}{N_1S_1 + N_2S_2} \approx 9700$ and $n_2 = n\frac{N_2S_2}{N_1S_1 + N_2S_2} \approx 10300$. The quantities in these equations determining the allocations are approximations, but this does not matter much, as it is well known that the Neyman optimum is flat (see, for example, [18] for an illustration of this property). Thus, even if the approximated allocation is slightly shifted from the optimal one, the variance of the Horvitz–Thompson estimator will still be very close from the optimal variance.

2.4 One-Stage Snowball Sampling

For a vertex i, let's denote:

$$B_i = \{i\} \cup \{j \in V, E_{ji} \neq \emptyset\}$$

$$A_i = \{i\} \cup \{j \in V, E_{ij} \neq \emptyset\}$$

B_i is called the set of vertices adjacent before i, and A_i the set of vertices adjacent after i. In the case of the Twitter graph, B_i is the set of the users *following user i* and A_i is the set of users *followed by user i*.

The snowball design consists in selecting $s = A(s_0)$. Contrary to adaptive sampling (see Sect. 2.5), the sample is enhanced with the friends of every user $k \in s_0$, even if $k \notin C$. In this design, we also do not care about symmetric relationships. Another difference with the adaptive design is that we cannot remove

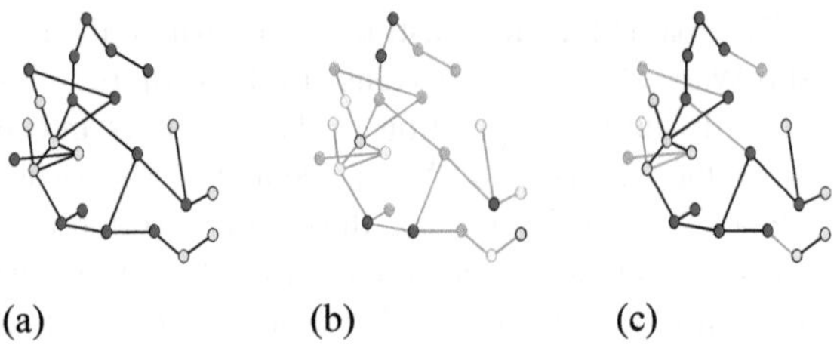

Fig. 1 (**a**) Graph population $\mathscr{U}$. Elements of community C are depicted in yellow. (**b**) Sample s_0. (**c**) Addition of 1-stage snowball $A(s_0)$

out-of-scope units prior to the sample enhancement, as out-of-scope units may have in-scope units in their friends. Figure 1a–c describe the snowball sampling design.

As described in Sect. 1.2 and in [20], the Twitter graph is highly central and clustered. Thus, the number of reachable units *via* the edges of the units in s_0 is huge. Mean number of friends for every unit is estimated from the stratified sample to be approximately 75 for units in stratum 2 and a little more than 400 for units in stratum 1. In order to obtain final samples s of comparable sizes, we base the snowball design on a smaller stratified Bernoulli sample of expected size **1000 units**, with an allocation proportional to the Neyman described in Sect. 2.3: $p_{S1} = 485/1654874 \approx 2.9 \cdot 10^{-4}$ and $p_{S2} = 515/33000000 \approx 1.6 \cdot 10^{-8}$. The final sample size is **159957**, which is much higher than we expected, probably due to the large tail distribution of the degree for units of stratum 1. With such a degree distribution, mean number of contacts is not very informative about the distribution of the sample size. Also, sample size has a very high variability.

This definition of one-stage snowball sampling can easily be generalized to n-stage snowball sampling by including in the sample all units that have a shorter path than n to any unit in s_0. n-stage snowball sampling, although fairly simple to implement, can lead to very complex estimators for $n \geq 2$ (see [13] for a more general discussion). But even more problematic in our case is that we saw in Sect. 1.2 that the average path length for the Twitter is very small (≈ 4.5). A precise analysis of characteristics of the Twitter graph [20] even shows that the distribution of path length is somewhat platykurtic. Thus, selecting an n-snowball sample could lead to huge sample sizes, reaching almost every units for very small values of n. For these reasons, we limit ourselves to one-stage snowball sampling (see Sect. 3.2).

2.5 *Stratified Adaptive Cluster Sampling*

In our problem, the population is made of the vertices of a graph bearing a dummy characteristic. One modality of the characteristic is supposed to be rather rare in the population, and the units bearing it should be more likely to be linked to one another. In other terms, we suppose that users who follow people who tweet about Star Wars have a higher propensity of also tweeting about Star Wars. Adaptive cluster sampling (first described in [27]) consists in enhancing the initial sample s_0 with all units for which $y > 0$ (i.e., who tweeted about the Star Wars Trailer) among the

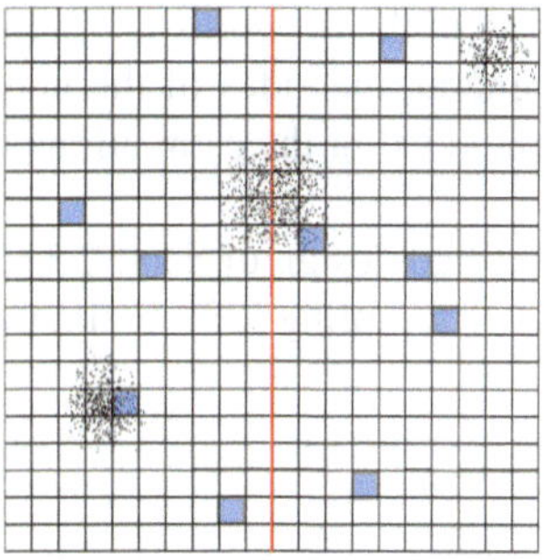

(a) First sampling phase for the stratified adaptive design

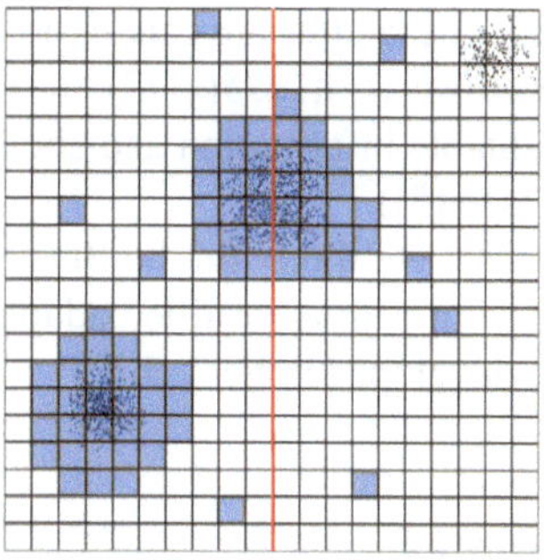

(b) Extension sampling phase for the stratified adaptive design

Fig. 2 Illustration of stratified adaptive sampling. The population is described as a 20 × 20 grid (and two strata partitioning the population in two, marked with the vertical red line). The networks of units are represented by aggregated little dots. In the first sampling phase (**a**), five squares are selected in each stratum (marked in light blue). In the second phase (**b**), the sample is extended with the networks (and their sides) that could be reached using the first-phase sample

units who are connected to the units of s_0. Adaptive sampling is a particular case of cluster sampling, with the clusters being the networks of units having tweeted about the Star Wars trailer (each unit of interest that is friends with a unit in s_0 will be added in the final sample). Units of s_0 who didn't tweet about Star Wars won't have any other unit from their network added to the sample, but they can be seen as a 1-unit cluster. Once a person who tweeted about Star Wars is found in the initial sample s_0, many more can be discovered, which resembles the gameplay of the famous "minesweeper" video game, and is often depicted as such in the literature (see Fig. 2).

The Twitter network is a directed graph. Let us consider a unit $i \in s_0, y_i = 1$. Following the logic of adaptive sampling explained in the previous paragraph, we should look for units who also tweeted about "The force awakens" by searching the friends and followers of i, and if such units were found, look for other units among the friends and followers of these units and so on till the entire network is discovered. But if we did this, the inclusion probabilities for any unit $k \in s$ would not only depend on the units of k's network but also on other units with edges leading to k. This typically cannot be estimated from sample data [29], as there is no reason that any of these units be included in the sample. Thus, we only select units that show symmetric relationships with units in s_0 or in their networks.

For huge networks, we could expect the final sample size to be much greater than the initial sample size, especially if the network is highly clustered (which is the case of the Twitter graph, see Sect. 1.2). However, the fact that the Twitter graph is directed imposes us to only look for symmetric relationship, which will limit the size of the networks s_{ri} added to s_0. Finally, in addition to the s_{ri}, we also include in s units who have symmetric relationships to units in s_C, but who are not in C

themselves. These units are called the "sides" s^0 of the networks s_r. These units can be described as: $\{k \in s, \delta(k, C) = 1\}$. In general, the sides of the networks also contain valuable information and should be included in the estimations to improve precision (see [5]).

In our case, we'll rely on the Bernoulli stratified design of Sect. 2.3 to select s_0: the final design is thus stratified adaptive cluster sampling [28]. Estimators for this design are developed in Sect. 3.3.

3 Estimates

3.1 *Horvitz–Thompson Estimator*

For the simple designs, the privileged estimator is Horvitz and Thompson's [11], which weighs the observations with the inverse of the inclusion probabilities:

$$\hat{T}(Y)_{HT} = \sum_{k \in s} \frac{y_k}{\pi_k}, \ \hat{\bar{y}}_{HT} = \frac{1}{N} \sum_{k \in s} \frac{y_k}{\pi_k}$$

For the Bernoulli design, it simply writes $\hat{T}(Y)_1 = \frac{1}{p} \sum_{k \in s} y_k$ and $\hat{N}_{C1} = \frac{n_C}{p}$.

For the stratified Bernoulli, we get $\hat{T}(Y)_2 = \sum_{s \cap \mathscr{U}_{d1}} \frac{y_k}{p_1} + \sum_{s \cap \mathscr{U}_{d2}} \frac{y_k}{p_2}$ and $\hat{N}_{C2} = \frac{N_1}{n_1} n_{C1} + \frac{N - N_1}{n_2} n_{C2}$

3.2 *One-Stage Snowball Sampling*

3.2.1 Horvitz–Thompson Estimator

Estimation is developed in Frank [9]. The Horvitz–Thompson estimator writes:

$$\hat{T}(Y)_3 = \sum_{k \in s} \frac{y_i}{1 - \bar{\pi}(B_i)}, \ \hat{N}_{C3} = \sum_{k \in s} \frac{z_i}{1 - \bar{\pi}(B_i)}$$

where $\bar{\pi}(B_i) = \mathbb{P}(B_i \subset \bar{s})$, the probability that no unit of B_i is included in s. As the sampling design for s_0 is stratified Bernoulli (in particular, the event of belonging or not to s is independent for each unit of $\mathscr{U}$):

$$\bar{\pi}(B_i) = \prod_{k \in B_i} (1 - \mathbb{P}(k \in s)) \tag{1}$$

$$= q_{S1}^{\#(B_i \cap \mathscr{U}_1)} \cdot q_{S2}^{\#(B_i \cap \mathscr{U}_2)}$$

The computation of this probability does not use any particular knowledge of the shape of the graph.

Just like the classic Horvitz–Thompson estimator, the estimator for the one-stage snowball sampling is linear homogeneous, with weights $d_i = \dfrac{1}{1-\bar{\pi}(B_i)}$. This property will be particularly useful for calibration on margins (Sect. 3.4).

3.2.2 Rao–Blackwell Estimator

In the case of one-stage snowball sampling, [9] shows that a Rao–Blackwell type estimator can be written in closed form:

$$\hat{N}^*_{C3} = \sum_{k\in s} \frac{y_k}{\pi_k}\left[1 - \frac{\sum_{L\subset s}(-1)^{\#L}\bar{\pi}(\{k\}\cup B(L\cup\bar{s}))}{\sum_{L\subset s}(-1)^{\#L}\bar{\pi}(B(L\cup\bar{s})}\right] \tag{2}$$

where the summation is done over all the subsamples L of s.

This Rao–Blackwell estimator is unbiased and as long as selection probabilities are not ill-defined (meaning that $\forall k \in \mathcal{U}, \bar{\pi}(B_k) < 1$), it performs better than the Horvitz–Thompson (1). However, there are $2^{\#s} - 1$ terms in each of two sums in Eq. (2). Computing each term of the sums implies computing the exclusion probabilities for as many B_k, which means at least knowing their sizes. This is extremely costly in time, and in our particular case in number of calls to the Twitter API. Thus, we prefer using $\hat{N}_{C3}$ over $\hat{N}^*_{C3}$.

3.3 *Estimators for Adaptive Sampling*

As explained in Sect. 2.5, the adaptive sample is selected using only units that have symmetric relationships (so that the graph induced by a unit $k \in s_0 \cap C$ can be considered undirected). We have $\hat{T}(Y)_4 = \sum_{k=1}^{K} \frac{y_k^* J_k}{\pi_{gk}}$ and $\hat{N}_{C4} = \sum_{k=1}^{K} \frac{n^*_{Ck} J_k}{\pi_{gk}}$, where k = networks in the population , y_k^* is the total of Y in the network k, n^*_{Ck} the number of people with $y_k \geq 1$ in the network k, $J_k = \mathbb{1}\{k \in C\}$ (i.e., intersection with sample), and π_{gk} is the probability that the initial sample intersects network k: $\pi_{gk} = 1 - (1 - p_k)^{n_g}$, where $n_g = \#\{k \in g\}$.

Other design-unbiased estimators can be used. For example, Hansen–Hurvitz-like [10] estimators in this case would write: $\hat{T}(Y)_{HH} = \sum_{h=1,2} \frac{1}{p_h} \sum_{n=1}^{N_h} \frac{y_i f_i}{m_i}$ where m_i is the number of units in the network that includes unit i and f_i is the number of units from that network included in the initial sample. This rewrites: $\hat{T}(Y)_{HH} = \sum_{h=1,2}\sum_{i=1}^{n_h}\frac{w_i}{p_h}$, where the sum is over the elements of $s_0 \cap \mathscr{U}_h$ and with w_i the average of y in the network r_i, which would give: $\hat{N}_{HH} = \sum_{h=1,2}\frac{1}{p_h}\sum_{i=1}^{n_h} n_{Ci}$.

3.3.1 Rao–Blackwell

In the case of adaptive sampling, it is very common to use the Rao–Blackwell estimator. This means using the conditional design $\mathbb{P}(s|s_0)$ (see, for example, [5] or [30] for more details on exhaustivity in sampling). In most cases, computing the Rao–Blackwell is computationally intensive and is achieved using Markov-Chain Monte-Carlo [30]. However, with Bernoulli stratified sampling it is possible to derive a closed form of the Rao–Blackwell [5]. Following [29], we write:

$$\hat{T}(Y)_{RB} = \sum_{k\in s_{ex}}\frac{y_k}{\pi_k} + \sum_{k\in s^0} y_k + \sum_{k\in s_r}\frac{Y_r}{\pi_r}$$

where $Y_r = \sum_{k\in r} y_k$ denotes the total of the variable of interest y in the network r. In our case, this leads to:

$$\hat{N}_{C5} = \sum_{s^0}\mathbb{1}\{yk \geq 1\} + \sum_{s_r}\frac{n_r}{\pi_r} = n^0 + \sum_{k=1}^{K}\frac{n_r}{1-(1-p)^{n_r}}$$

with $n^0 = \#s^0$.

The Rao–Blackwell for the Hansen–Hurvitz estimator also has a simple closed form that can be found in [5].

3.4 *Calibration on Margins*

3.4.1 The Calibration Estimator

Let us denote X a matrix of J auxiliary variables:

$$\forall j \in [\![1, J]\!], X_j = (X_{j1}\ldots X_{jn})$$

whose values can be computed at least for all units of s. Let's suppose we also know the totals of these auxiliary variables on the population: $T(X_j) = \sum_{k \in \mathcal{U}} X_{jk}$. We write $T(X_j)$ the total of the j-th auxiliary variable and $T(X)$ the column vector of all the J totals $T(X) = (T(X_1) \ldots T(X_J))$.

In the general case, there is no reason that the estimated totals $T(\hat{X})_{j\pi}$ using the Horvitz–Thompson weights match the actual totals $T(X_j)$. For two main reasons, one may want to use a linear homogeneous estimator that is calibrated on some auxiliary variables. First, often in official statistics, a single set of weights is used to compute estimated totals for many variables. Thus, it is often required that these weights are calibrated on some margins for the sake of consistency. For example, $X = 1$ is often chosen as a calibration variable so that the size of the population N is estimated with no variance. Other typical calibration variables include sizes of sub-populations divided into sex and age groups so that the demographic structure of estimates is similar to the demographic structure of the population or quantitative variables linked to revenue—which is a main parameter of interest in most sociological studies, etc. Second, when estimating a variable Y using the calibrated weights w_k, the precision of the estimator will be increased if Y is correlated to one or more of the auxiliary variables X_j (see Sect. 4.4). This is the most interesting feature of calibration in the context of this study.

Let us write: X_s the $n \times j$ matrix of the values of the auxiliary variables and w a set of weights of a linear homogeneous estimator. Calibration on margins $X_1 \ldots X_j$ consists in searching a linear homogeneous estimator such that :

$$X_s' w = T(X) \tag{3}$$

supposing, of course, that this system has at least one solution. In general, when this linear systems has solutions, their number is infinite. To choose among these solutions, we look for weights w_k that are the closest to the Horvitz–Thompson weights d_k with respect to some distance G. Finally, finding the calibrated weights is equivalent to solving a linear optimization problem [6]:

$$\begin{cases} \min\limits_{w_k} \sum\limits_{k \in s} d_k G(\frac{w_k}{d_k}) \\ \text{under constraint: } X_s' w = T(X) \end{cases}$$

Proposition 3.1 (Approximately Design Unbiasedness (ADU)) *The calibration estimator is approximately design-unbiased. Its bias tends to 0 as $N \to +\infty$*

The Horvitz–Thompson estimator is the most used estimator in survey sampling. One of its key properties is that it is design-unbiased. The calibration estimator is the closest estimator to the Horvitz–Thompson estimator (w.r.t a certain distance) that ensures the calibration equation (3). We can thus hope that it shares some properties with the Horvitz–Thompson estimator, in particular its unbiasedness. In fact, the unbiasedness of the calibration estimator holds asymptotically, i.e., when N goes with n to infinity (precise definition of the superpopulation model used to prove this asymptotic property is detailed in Isaki–Fuller, [12]).

3.4.2 Margins Used for Twitter

For the calibration of our estimators $\hat{N}_C$, we use the following margins (i.e., the columns of matrix X):

- N = Total number of accounts who tweeted in the last month (quantitative):
- $T(Y)$ =Total number of tweets about "Star Wars : The Force Awakens" between 10/25, 7:48 PM EST and 10/25, 10:48 PM EST (quantitative): 390000
- Number of verified accounts[4]
- Structure of users in sample in terms of number of followers

Many other margins could be added to the calibration process. Improvement in terms of variance of $\hat{N}_C$ is achieved as long as the margin X_j correlates with N_C and the calibration algorithm has a solution. Other calibration margins could include the number of (active) verified users, the number of tweets about "Star Wars: The Force awakens" per hour, etc. The geographical origin of the tweets is another variable that might strongly correlate to N_C. Twitter does dispose of such a variable, but it is seldom available. Imputation is not worth considering for this particular variable, because the number of accounts for which the variable can be used is very low. However, we could use Time Zones as a proxy for geographical origin, as the Time Zones are disclosed for every account. In general, in order to facilitate sampling estimates, Twitter could release calibration margins on a few characteristics.

Finally, we could calibrate on variables accounting for the graph structure, which probably correlate highly with N_C as well as a lot of other characteristics of interest of the Twitter graph. This could be achieved by finding a method for calibration on non-linear totals (see [15]).

4 Variance and Precision Estimation

4.1 Simple Designs

For simple designs, we have $Var(\hat{T}(Y)_1) = \frac{1}{p}(\frac{1}{p} - 1)\sum_{k \in \mathcal{U}} y_k^2$, which can be easily estimated by $\hat{\text{Var}}_1(\hat{T}(Y)_1) = \frac{1}{p}(\frac{1}{p} - 1)\sum_{k \in s} y_k^2$. Once n (which is random) has been

[4] https://support.twitter.com/articles/119135.

drawn, it is common to work with the variance conditional to the sample size [31]. This gives another variance estimator:

$$\hat{\text{Var}}_2(\hat{T}(Y)_1) = \hat{\text{Var}}_1(\hat{T}(Y) \mid \#s = n) = \frac{1 - \frac{n}{N}}{n}\hat{\sigma}(Y)$$

where $\hat{\sigma}(Y) = \frac{1}{n-1}\sum_{k\in s}(y_k - \bar{y})^2$.

For stratified Bernoulli, the total variance is the sum of the variance of the two independent designs in the strata. If the strata are built correctly, the dispersions $\hat{\Sigma}(Y)$ will be lower among each strata than in the total population, yielding a lower variance for the Horvitz–Thompson estimator. For example, using the simple plugin variance estimator, this writes:

$$\hat{\text{Var}}(\hat{T}(Y)_2) = \sum_{h=1}^{2} \frac{1}{p_h}(\frac{1}{p_h} - 1)\sum_{k\in s_h} y_k^2$$

In our study we chose the variance estimators $\hat{\text{Var}}_2$, which finally write:

$$\hat{\text{Var}}_2(\hat{N}_{C1}) = \frac{N(N-n)(n-n_C)}{n^3(n-1)}$$

$$\hat{\text{Var}}_2(\hat{N}_{C2}) = N^2\left[\frac{(N_1-n_1)(n_1-n_{C1})}{n_1^3(n_1-1)N_1} + \frac{(N_2-n_2)(n_2-n_{C2})}{n_2^3(n_2-1)N_2}\right]$$

4.2 *Snowball Sampling*

An unbiased estimator of the variance of the Horvitz–Thompson estimator is given in [11]. One can easily rewrite the general case formula in the case of snowball sampling:

$$\hat{\text{Var}}(\hat{N}_{C3}) = \sum_{i\in s}\sum_{j\in s} \frac{z_i z_j}{\bar{\pi}(B_i \cup B_j)}\gamma'_{ij}$$

where:

$$\gamma'_{ij} = \frac{\bar{\pi}(B_i \cup B_j) - \bar{\pi}(B_i)\bar{\pi}(B_j)}{[1-\bar{\pi}(B_i)][1-\bar{\pi}(B_j)]} \tag{4}$$

The probabilities in (4) are theoretically easily computed using the formula (1). However, practically, it means browsing the whole set of vertices $B(s)$, which can be huge for highly central and/or clustered graphs (see Sect. 5). This variance estimator is thus potentially highly intensive in time and computation power. In the case of the Twitter graph, this means a high number of calls to the API, which is unfortunately limited in number of calls. For too big sample sizes, we might have trouble computing the estimation.

4.3 *Adaptive Designs*

4.3.1 Horvitz–Thompson

We use the variance estimator proposed by Särndal [24]:

$$\hat{\text{Var}}(\hat{N}_{C4}) = \sum_{k=1}^{K}\sum_{k'=1}^{K} \frac{y_k y_{k'}}{\pi_{gkk'}} \left(\frac{\pi_{gkk'}}{\pi_{gk}\pi_{gk'}} - 1 \right)$$

where:

$$\pi_{gkk'} = 1 - \pi_{gk} - \pi_{gk'} + (1-p)^{n_{gk}+n_{gk'}}$$

Computing this variance estimator only requires the sizes of each network, which can be stored during the data collection process. In terms of number of calls, the variance estimator is thus much less demanding than the variance estimator in the case of the one-stage snowball (Sect. 4.2).

4.3.2 Rao–Blackwell

Although the Bernoulli scheme conveniently yields a closed-form Rao–Blackwell estimation, it's not the case for the variance estimator. We have:

$$\text{Var}(\hat{N}_{C4}) = \text{Var}(\hat{N}_{C5}) + \mathbb{E}(\hat{N}_{C4} - \hat{N}_{C5})^2$$

The second term can be estimated without bias by selecting m samples. An unbiased estimator for $\text{Var}(\hat{N}_{C5})$ then writes:

$$\hat{\text{Var}}(\hat{N}_{C5}) = \hat{\text{Var}}(\hat{N}_{C4}) - \frac{1}{m-1}\sum_{i=1}^{m}(\hat{N}_{C5i} - \overline{\hat{N}}_{C4})$$

Of course, this method increases the number of calls needed to the graph API. In the case of Star Wars, we were already unable to get to the end of s^0 (see Sect. 5). If we'd had to generate several samples to estimate the variance of the estimation, it would obviously have been even harder.

4.4 Adjustment for Calibration

Following [6], the variance for a calibrated Horvitz–Thompson for total Y writes:

$$\hat{\text{Var}}_c(\hat{T}(Y)) = \sum_{k \in s} \sum_{l \in s} \frac{\Delta_{kl}}{\pi_{kl}} (g_k d_k e_k)(g_l d_l e_l)$$

where w_k are the weights of the calibrated estimator, d_k the weights of the non-calibrated estimator, $g_k = \dfrac{w_k}{d_k}$ and $e_k = y_k - \hat{b}' x_k$ residuals of the weighted regression (weights: d_k) of Y on $X_1 \dots X_j \dots X_J$ in s.

We recognize the general form of the Horvitz–Thompson estimator of variance (see [11]) which is used to construct variance estimators of Sects. 4.1–4.3, applied to the linear variable $g_k e_k$. This means that $\hat{\text{Var}}(\hat{N}_{Ci}), i = 1 \dots 5$ can be easily computed by re-using the formulae from Sects. 4.1–4.3, and replacing the z_k by $g_k \cdot e_k$.

5 Results

The variance of all estimators used in this work is bounded by $\mathcal{O}(\frac{1}{n})$. It is thus always possible to reduce variance by increasing the sample size. In order to compare the sampling designs for their respective merits in reducing the variance of the estimation, we define the **design effect** Deff as:

$$Deff = \frac{\text{Var}(\hat{Y}_{design})}{\text{Var}(\hat{Y}_{design})} = \frac{\text{Var}(\hat{Y}_{design})}{(1 - \frac{n}{N} \Sigma(Y)^2)}$$

$$\text{with: } \Sigma(Y) = \frac{1}{N-1} \sum_{k \in \mathscr{U}} (y_k - \bar{y})^2$$

which compares the variance of the estimator to the variance of the estimator under a simple random sampling design of same size. The Deff is greater than 1 for designs that yield worse precision than the simple design (typically cluster or two-degree sampling), and lesser than 1 for designs that are more precise (typically stratified designs). Of course, real variance is impossible to compute, so the Deff can be estimated by:

$$\hat{Deff} = \frac{\hat{\text{Var}}(\hat{Y}_{design})}{\hat{\text{Var}}(\hat{Y}_{SAS})} = \frac{\hat{\text{Var}}(\hat{Y}_{design})}{(1 - \frac{n}{N} \sigma(Y)^2)}$$

$$\text{with: } \sigma(Y) = \frac{1}{n-1} \sum_{k \in s} (y_k - \bar{y})^2$$

Table 1 Estimates and characteristics for three different sampling designs

Design	n	n_{scope}	n_0	$\hat{N}_C$	$\hat{CV}$	$\hat{Deff}$
Bernoulli	20,013	3946		354,121	0.231	1.04
Stratified	20,094	9832		316,889	0.097	0.68
1-snowball	159,957	73,570	1000	331,097	0.031	0.60

The results of the estimators are presented in Table 1:

We can also note that the mean number of tweets about Star Wars according to the one-stage snowball sampling design is 1.18 ± 0.07. This low number suggests that automatic accounts are responsible for a very small amount, if any, of the total number of tweets on this subject [7].

It is important to note that the final sample size is random, just like for any clustered sampling design with non-constant sizes of clusters. When a unit is selected in s_0, there is no way to know in advance what the size of the clusters (networks) that includes it will be. Therefore, the final sample size $\#s$ can be highly variable. In the case of adaptive sampling, the subject studied featured highly clustered networks of Star Wars fans. This led to a very large number of users reached by the adaptive procedure. Due to the limits in number of calls imposed by the Twitter API, we were not able to finish collecting the stratified adaptive sample in less than a month. In order to prevent such issues, we could imagine sampling designs adjusting the selection probabilities as the collection of the units of s^0 goes along. Sampling design and estimation with such sampling designs is one of the future developments of our research we consider on this subject.

Snowball sampling is unlikely prone to the same flaw, as number of units in final sample depends on the degree distribution. This distribution is often known *a priori*, and well modeled by a power law for all web and social network graphs. Contrary to adaptive sampling, we are guaranteed that the extension will be complete after only one browse of the list of users followed by the units in s_0.

6 Conclusion

This chapter describes a design-based statistical method to estimate linear quantities on the Twitter graph based on users rather than queries. The method relies on sampling theory and in particular on developments of sampling theory for graphs. We tried two so-called extension designs: snowball sampling and adaptive sampling, which were first designed in official statistics to measure rare characteristics on a given population. We use them to try and measure the number of accounts having tweeted about the trailer of the Star Wars movie on the day of its release. Despite the event generating 390,000 tweets in approximately 3 h, the users responsible for these tweets are rare among the 1 billion Twitter users. Despite being unable to get through the whole adaptive sample because of the number of calls allowed by the

Twitter API, the stratified snowball proved rather precise. Variance estimators were also computed, although they required a much higher number of calls.

References

1. Barabási, A. L., & Albert, R. (1999). Emergence of scaling in random networks. *Science, 286*(5439), 509–512.
2. Bollen, J., Mao, H., & Zeng, X. (2011). Twitter mood predicts the stock market. *Journal of Computational Science, 2*(1), 1–8.
3. Burnap, P., Gibson, R., Sloan, L., Southern, R., & Williams, M. (2015). 140 characters to victory? Using twitter to predict the UK 2015 general election. arXiv:1505.01511.
4. Conover, M., Ratkiewicz, J., Francisco, M., Gonçalves, B., Menczer, F., & Flammini, A. (2011). Political polarization on twitter. In *ICWSM*.
5. Deville, J. C. (2012). échantillonnage de réseaux, une relecture de s.k thompson avec une nouvelle présentation et quelques nouveautés. Accessed April 7, 2017 from http://jms.insee.fr/files/documents/2012/930_2-JMS2012_S21-4_DEVILLE-ACTE.PDF
6. Deville, J. C., & Särndal, C. E. (1992). Calibration estimators in survey sampling. *Journal of the American statistical Association, 87*(418), 376–382.
7. Ferrara, E. (2015). Manipulation and abuse on social media. *SIGWEB Newsletter, (Spring)*, 4:1–4:9. https://doi.org/10.1145/2749279.2749283
8. Ferrara, E., Varol, O., Davis, C., Menczer, F., & Flammini, A. (2014). The rise of social bots. arXiv:1407.5225.
9. Frank, O. (1977). Survey sampling in graphs. *Journal of Statistical Planning and Inference, 1*(3), 235–264.
10. Hansen, M. H., & Hurwitz, W. N. (1943). On the theory of sampling from finite populations. *The Annals of Mathematical Statistics, 14*(4), 333–362.
11. Horvitz, D. G., & Thompson, D. J. (1952). A generalization of sampling without replacement from a finite universe. *Journal of the American Statistical Association, 47*(260), 663–685.
12. Isaki, C. T., & Fuller, W. A. (1982). Survey design under the regression superpopulation model. *Journal of the American Statistical Association, 77*(377), 89–96.
13. Kolaczyk, E. D. (2009). *Statistical analysis of network data*. Berlin: Springer.
14. Lavallée, P., & Caron, P. (2001). Estimation par la méthode généralisée du partage des poids: Le cas du couplage d'enregistrements. *Survey Methodology, 27*(2), 171–188.
15. Lesage, E. (2009). Calage non linéaire. Accessed April 7, 2017, from http://jms.insee.fr/files/documents/2009/85_2-JMS2009_S11-3_LESAGE-ACTE.PDF
16. Leskovec, J., & Faloutsos, C. (2006). Sampling from large graphs. In *Proceedings of the 12th ACM SIGKDD International Conference on Knowledge Discovery and Data Mining* (pp. 631–636). New York: ACM.
17. Leskovec, J., & Sosič, R. (2014). Snap.py: SNAP for Python, a general purpose network analysis and graph mining tool in Python. http://snap.stanford.edu/snappy
18. Merly-Alpa, T., & Rebecq, A. (2017). L'algorithme CURIOS pour l'optimisation du plan de sondage en fonction de la non-réponse. Accessed April 7, 2017, from http://papersjds15.sfds.asso.fr/submission_29.pdf
19. Mustafaraj, E., Finn, S., Whitlock, C., & Metaxas, P. T. (2011). Vocal minority versus silent majority: Discovering the opinions of the long tail. In *Privacy, Security, Risk and Trust (PASSAT) and 2011 IEEE Third International Conference on Social Computing (SocialCom)* (pp. 103–110).
20. Myers, S. A., Sharma, A., Gupta, P., & Lin, J. (2014). Information network or social network? The structure of the twitter follow graph. In *Proceedings of the Companion Publication of the 23rd International Conference on World Wide Web Companion* (pp. 493–498). International World Wide Web Conferences Steering Committee.

21. Neyman, J. (1934). On the two different aspects of the representative method: The method of stratified sampling and the method of purposive selection. *Journal of the Royal Statistical Society, 97*, 558–625.
22. Nowicki, K., & Snijders, T. (2001). Estimation and prediction for stochastic blockstructures. *Journal of the American Statistical Association, 96*(455), 1077–1087.
23. Rivers, D., & Bailey, D. (2009). Inference from matched samples in the 2008 US National elections. In *Proceedings of the Joint Statistical Meetings* (pp. 627–639)
24. Särndal, C. E., Swensson, B., & Wretman, J. (2003). *Model assisted survey sampling*. New York: Springer Science & Business Media.
25. Sautory, O. (2012). Les enjeux méthodologiques liés à l'usage de bases de sondage imparfaites. Conference Report.
26. Sloan, L., Morgan, J., Housley, W., Williams, M., Edwards, A., Burnap, et al. (2013). Knowing the tweeters: Deriving sociologically relevant demographics from twitter. *Sociological Research Online, 18*(3), 7.
27. Thompson, S. K. (1990). Adaptive cluster sampling. *Journal of the American Statistical Association, 85*(412), 1050–1059
28. Thompson, S. K. (1991). Stratified adaptive cluster sampling. *Biometrika, 78*, 389–397.
29. Thompson, S. K. (1998). Adaptive sampling in graphs. In *Proceedings of the Section on Survey Methods Research, American Statistical Association* (pp. 13–22).
30. Thompson, S. K. (2006). Adaptive web sampling. *Biometrics, 62*(4), 1224–1234.
31. Tillé, Y. (2001). Théorie des sondages. Paris: Dunod.
32. Tumasjan, A., Sprenger, T. O., Sandner, P. G., & Welpe, I. M. (2010). Predicting elections with twitter: What 140 characters reveal about political sentiment. *International AAAI Conference on Web and Social Media, 10*, 178–185.
33. Watts, D. J., & Strogatz, S. H. (1998). Collective dynamics of 'small-world' networks. *Nature, 393*(6684), 440–442.

Wavelet Whittle Estimation in Multivariate Time Series Models: Application to fMRI Data

S. Achard and I. Gannaz

Abstract In many applications such as finance, geophysics or neuroscience, data are multivariate time series. Identification of correlation between time series is an important feature. Yet differences in memory properties of the processes can induce phase-shifts in estimation of correlations. A semiparametric model for multivariate long-range dependent time series is considered. The coupling between time series is characterized by the long-run covariance matrix. The multivariate wavelet-based Whittle estimation is consistent for the estimation of both the long-range dependence and the covariance matrix. Finally an application to the estimation of a human brain functional network based on fMRI test-retest data sets is described. Our study highlights the benefit of the multivariate analysis, namely improved efficiency of estimation of dependence parameters.

1 Introduction

In many areas such as finance, geophysics or neurosciences, data are multivariate time series with long-range dependence properties. When analysing multivariate time series, a challenge is to characterize the coupling between the time series as well as the long-term dependence properties of the recordings. Many statistical developments have been proposed to deal with univariate long-memory processes. Parametric estimates of the long-range memory [4, 5, 7] present the drawback of being sensitive to short-range dependence modelling. Semiparametric approaches were proposed to be robust to model misspecification [11]. Semiparametric Fourier-based procedures were developed such as Geweke–Porter-Hudak [6] and local

S. Achard
CNRS, University of Grenoble Alpes, GIPSA-Lab, Grenoble, France
e-mail: sophie.achard@gipsa-lab.grenoble-inp.fr

I. Gannaz (✉)
Univ Lyon, INSA de Lyon, CNRS UMR 5208, Institut Camille Jordan, Villeurbanne, France
e-mail: irene.gannaz@insa-lyon.fr

P. Bertail et al. (eds.), *Nonparametric Statistics*, Springer Proceedings in Mathematics & Statistics 250, https://doi.org/10.1007/978-3-319-96941-1_18

Whittle estimator [11]. Some wavelet-based estimators were also introduced, e.g. by log-regression on the wavelet coefficients [1] or Whittle procedure [10].

Recently, multivariate estimations of the long-memory parameters were provided. In [9] and [13], multivariate Fourier-based Whittle estimation has been studied. Additionally to the estimation of long-range dependence parameters, the procedure gives an estimation of the long-run covariance, measuring the coupling between the time series. Long-run covariance can be defined as the limit of the cross-spectral density at the zero-frequency. Based on the wavelet representation of the data rather than Fourier, [2] proposes a similar procedure.

Our objective in this manuscript is to give an overview of a phase phenomenon as stated in [2, 13]. Our contribution is to give practical guidelines for the wavelet-based Whittle estimation provided in [2]. Our study supports the benefit of the multivariate approach with respect to a univariate one. Our statement is that multivariate procedure not only provides an estimation of the long-run covariance but also improves the quality of estimation of the long-memory parameters. We finally show the robustness of multivariate estimation on a real data set.

Let $\mathbf{X} = \{X_\ell(k), k \in \mathbb{Z}, \ell = 1, \ldots, p\}$ be a multivariate process with long-memory properties. Suppose that the generalized cross-spectral density of processes X_ℓ and X_m can be written as

$$f_{\ell,m}(\lambda) = \frac{1}{2\pi}\Omega_{\ell,m}(1-e^{-i\lambda})^{-d_\ell}(1-e^{i\lambda})^{-d_m} f^S_{\ell,m}(\lambda), \qquad \lambda \in [-\pi, \pi]. \tag{1}$$

The parameters $(d_\ell)_{\ell=1,\ldots,p}$ quantify the long-range dependence of the time series components $(X_\ell)_{\ell=1,\ldots,p}$. For given $\ell = 1, \ldots, p$, parameter d_ℓ belongs to $(-0.5; \infty)$. When $d_\ell \geqslant 0.5$, the construction is based upon increments of X_ℓ, [2]. The matrix $\boldsymbol{\Omega}$ corresponds to the long-run covariance between time series. It measures the connections between the components of $\mathbf{X}$ at low-range frequencies. The function $f^S_{\ell,m}(\cdot)$ corresponds to the short memory behaviour of the bivariate process (X_ℓ, X_m). We assume that $\boldsymbol{f}^S_{\ell,m}(0) = 1$ for all ℓ, m and that $\boldsymbol{f}^S \in \mathscr{H}(\beta, L, \epsilon)$ with $0 < \beta \leq 2$, $O < L$ and $0 < \epsilon \leq \pi$. The space $\mathscr{H}(\beta, L, \epsilon)$ is defined as the class of non-negative symmetric functions $\mathbf{g}$ on $[\pi, \pi]$ such that for all $\lambda \in (-\epsilon, \epsilon)$, $\|\mathbf{g}(\lambda) - \mathbf{g}(0)\|_\infty \leq L\|g(0)\|_\infty|\lambda|^\beta$.

This model is semiparametric since the short-range behaviour has a nonparametric form. As stated by Robinson [11], procedures are hence more robust to model misspecification. Indeed, it is not necessary to define precisely the short-range dependence. As it is explained in this work, highest frequencies may be influenced by short-range dependence. We propose a procedure which detects them by looking at the behaviour of the wavelet correlations. This enables to discard them in the estimation procedure. Thus, the nonparametric form of $f^S(\cdot)$ adds flexibility to the model and offers a larger scope of applications.

Section 2 illustrates the influence of memory properties on the correlation between time series, through the behaviour of wavelet coefficients. The next section recalls the estimation procedure of [2], for both the long-run covariance $\boldsymbol{\Omega}$ and the memory parameters $\mathbf{d}$. In Sect. 4 this procedure is applied on simulated data, and we highlight the improvement of multivariate estimation for long-memory parameters. The last section deals with a real data example, from neuroscience. We estimate the long-memory parameters on two consecutive recordings of functional magnetic resonance imaging (fMRI). The result shows that the multivariate estimation is more robust than the univariate one, since estimations are more reproducible.

All simulations and calculations were done using `multiwave`[1] package.

2 Influence of Long- and Short-Range Memory on Correlations

The first objective of this contribution is to stress the inherent influence of time dependencies on long-run correlations between time series. We focus on the behaviour of the correlation between the wavelet coefficients. Our simulation study assesses a phase-shift phenomenon caused by long-range dependence properties which may introduce a bias when looking at the wavelet correlations. This observation is related with theoretical results. The study also highlights that some scales should be removed from estimation since their behaviour depends on the short-range dependence.

Simulations were done using FIVARMA (Fractionally Integrated Vector Auto Regressive Moving Average) process models. FIVARMA process is an example of linear processes whose spectral density satisfies (1). FIVARMA corresponds to Model A of [8] and is also defined in [12] or [2].

2.1 *Wavelet Transform*

We are interested in the behaviour of the correlation of the wavelet coefficients in the presence of long- or short-range dependence. Let $(\phi(\cdot), \psi(\cdot))$ be respectively a father and a mother wavelets. Their Fourier transforms are given by $\hat{\phi}(\lambda) = \int_{-\infty}^{\infty} \phi(t)e^{-i\lambda t}dt$ and $\hat{\psi}(\lambda) = \int_{-\infty}^{\infty} \psi(t)e^{-i\lambda t}dt$.

At a given resolution $j \geq 0$, for $k \in \mathbb{Z}$, we define the dilated and translated functions $\phi_{j,k}(\cdot) = 2^{-j/2}\phi(2^{-j}\cdot - k)$ and $\psi_{j,k}(\cdot) = 2^{-j/2}\psi(2^{-j}\cdot - k)$. The wavelet coefficients of the processes X_ℓ, $\ell = 1, \ldots, p$, are defined by

$$W_{j,k}(\ell) = \int_{\mathbb{R}} \tilde{X}_\ell(t)\psi_{j,k}(t)dt \quad j \geq 0, k \in \mathbb{Z},$$

[1] https://cran.r-project.org/web/packages/multiwave/index.html.

where $\tilde{X}_\ell(t) = \sum_{k\in\mathbb{Z}} X_\ell(k)\phi(t-k)$. We denote by $\theta_j(\ell, m)$,

$$\theta_j(\ell, m) = \mathrm{Cor}\left(\{W_{j,k}(\ell),\ k \in \mathbb{Z}\},\ \{W_{j,k}(m),\ k \in \mathbb{Z}\}\right)$$

the empirical correlation between wavelet coefficients at scale $j \geq 0$ between components X_ℓ and X_m.

The regularity assumptions on the wavelet transform are the following:

(W1) The functions $\phi(\cdot)$ and $\psi(\cdot)$ are integrable, have compact supports, $\int_{\mathbb{R}} \phi(t)dt = 1$ and $\int \psi^2(t)dt = 1$;
(W2) There exists $\alpha > 1$ such that $\sup_{\lambda\in\mathbb{R}} |\hat{\psi}(\lambda)|(1+|\lambda|)^\alpha < \infty$, i.e. the wavelet is α-regular;
(W3) The mother wavelet $\psi(\cdot)$ has $M > 1$ vanishing moments.
(W4) The function $\sum_{k\in\mathbb{Z}} k^\ell \phi(\cdot - k)$ is polynomial with degree ℓ for all $\ell = 1, \ldots, M-1$.
(W5) For all $i = 1, \ldots, p$, $(1+\beta)/2 - \alpha < d_i \leq M$.

These conditions are not restrictive, and many standard wavelet bases satisfy them. They hold in particular for Daubechies wavelet basis with sufficiently large M. Implementations below were done using Daubechies wavelet with $M = 4$ vanishing moments.

2.2 Phase-Shift Phenomenon

Following [3] or [17], the coupling between time series at a given scale can be measured using the correlation between the wavelet coefficients at this scale. Let $\{X_1(k), k \in \mathbb{Z}\}$ and $\{X_2(k), k \in \mathbb{Z}\}$ be two time series, with long-memory parameters d_1 and d_2. Denote $\{W_{j,k}(1), k \in \mathbb{Z}\}$ and $\{W_{j,k}(2), k \in \mathbb{Z}\}$ their wavelets coefficients at a given scale j.

We first consider a bivariate $FIVARMA(0, (d_1, d_2), 0)$ with a covariance matrix $\boldsymbol{\Omega} = \begin{pmatrix} 1 & \rho \\ \rho & 1 \end{pmatrix}$ and $\rho = 0.4$. That is, $\mathbf{X}$ is defined by

$$(\mathbb{1} - \mathbb{L})^{d_\ell} X_\ell(k) = u_\ell(k),\ \ell = 1, 2,\ k \in \mathbb{Z} \tag{2}$$

with $\mathbb{L}$ lag-operator. The process $\mathbf{u}$ is a bivariate white noise $\mathbb{E}[\mathbf{u}(t) \mid \mathscr{F}_{t-1}] = 0$ and $\mathbb{E}[\mathbf{u}(t)\mathbf{u}(t)^T \mid \mathscr{F}_{t-1}] = \boldsymbol{\Omega}$, where $\mathscr{F}_{t-1}$ is the σ-field generated by $\{\mathbf{u}(s),\ s < t\}$ and the superscript T denotes the transpose operator. The cross-spectral density of the process (X_1, X_2) is given by

$$f(\lambda) = \boldsymbol{\Lambda}(\mathbf{d})\boldsymbol{\Omega}\boldsymbol{\Lambda}(\mathbf{d})^* \ \text{ with } \boldsymbol{\Lambda}(\mathbf{d}) = \mathrm{diag}((1 - e^{-i\lambda})^{-\mathbf{d}}).$$

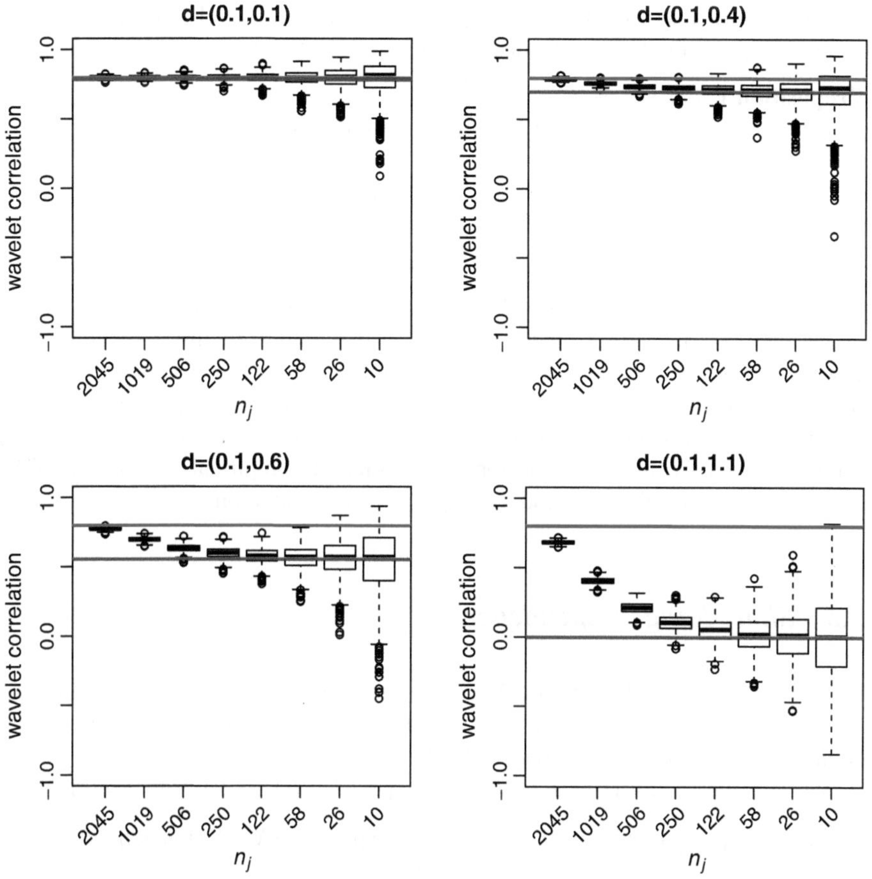

Fig. 1 Boxplots of $\theta_j(1,2)$ at different scales for an FIVARMA(0,**d**,0), with different values of **d**. Number of coefficients available at each scale are given in indexes. The horizontal red line corresponds to the correlation ρ between the two innovations processes. The horizontal blue line corresponds to the first order approximation $\rho\cos(\pi(d_1 - d_2)/2)C_K$

When $\lambda \to 0^+$, this density satisfies the first order approximation $\boldsymbol{f}(\lambda) \sim \mathbf{G}$ where

$$\mathbf{G} = \tilde{\boldsymbol{\Lambda}}(\mathbf{d})^* \boldsymbol{\Omega} \tilde{\boldsymbol{\Lambda}}(\mathbf{d}), \text{ with } \tilde{\boldsymbol{\Lambda}}(\mathbf{d}) = \mathrm{diag}(\lambda^{-\mathbf{d}} e^{-i\pi\mathbf{d}/2}). \quad (3)$$

Figure 1 displays the empirical correlations $\theta_j(1,2)$ at different scales j of the wavelet coefficients of such a process with various (d_1, d_2) and $\rho = 0.8$. Simulations were done with 2^{12} time points and 1000 repetitions. It highlights that the behaviour of $\{\theta_j(1,2),\ j \geq 0\}$ depends on the long-memory parameters. More precisely, when the difference between the values of d_1 and d_2 increases (up to 2), then the bias of $\{\theta_j(1,2),\ j \geq 0\}$ with respect to ρ increases. This phenomenon is observed for all scales j, even if it is more important for low frequencies.

This empirical statement has been quantified theoretically in Proposition 2 of [2]. A similar approximation can be deduced for the correlation :

Proposition 1 *Let* $\rho_{\ell,m} = \Omega_{\ell,m}/\sqrt{\Omega_{\ell,\ell}\Omega_{m,m}}$.
Under assumptions (W1)–(W5), when j goes to infinity,

$$\theta_j(\ell, m) \to \rho_{\ell,m} \cos(\pi(d_\ell - d_m)/2)\, C_K, \tag{4}$$

where $C_K = K(d_\ell + d_m)/\sqrt{K(2d_\ell)K(2d_m)}$ *with* $K(\delta) = \int_{-\infty}^{\infty} |\lambda|^{-\delta}|\hat{\psi}(\lambda)|^2\, d\lambda$.

This approximation corresponds to a first order approximation. A second-order approximation, depending on the scale, is also possible. Figure 1 illustrates the quality of this result. The estimation of correlation between two processes with long-range memory must hence take into account this phase-shift phenomenon. The phase in our setting is equal to $\pi(d_\ell - d_m)/2$ (due to approximation (5)). A multiplicative term appears in the approximation of the empirical correlations of the wavelet coefficients that can lead to a highly biased estimation.

2.3 Short-Range Dependence

In the cross-spectral definition (1), the short-range dependence is nonparametric. This choice leads to a larger scope of applications and enables procedures more robust to model misspecification, see [11].

We consider a bivariate $FIVARMA(q_{AR}, (d_1, d_2), q_{MA})$ with a correlation matrix $\boldsymbol{\Omega} = \begin{pmatrix} 1 & \rho \\ \rho & 1 \end{pmatrix}$ and $\rho = 0.4$. The bivariate process (X_1, X_2) is defined by

$$\mathbf{A}(\mathbb{L})\, \mathrm{diag}(\mathbb{1} - \mathbb{L})^{\mathbf{d}}\, \mathbf{X}(t) = \mathbf{B}(\mathbb{L})\mathbf{u}(t).$$

The sequence $\{\mathbf{A}_k,\ k = 0, 1, \ldots, q_{AR}\}$ is $\mathbb{R}^{p\times p}$-valued matrices with $\mathbf{A}_0$ the identity matrix and $\sum_{k=0}^{q_{AR}} \|\mathbf{A}_k\|^2 < \infty$. Let also $\{\mathbf{B}_k,\ k = 0, 1, \ldots, q_{MA}\}$ be a sequence in $\mathbb{R}^{p\times p}$ with $\mathbf{B}_0$ the identity matrix and $\sum_{k=0}^{q_{MA}} \|\mathbf{B}_k\|^2 < \infty$. Let $\mathbf{A}(\cdot)$ (respectively $\mathbf{B}(\cdot)$) be the discrete Fourier transform of the sequence, that is, $\mathbf{A}(\lambda) = \sum_{k=0}^{q_{AR}} \mathbf{A}_k e^{ik\lambda}$. We assume that all the roots of $|\mathbf{A}(\mathbb{L})|$ are outside the closed unit circle. Let $\mathbf{u}$ be a p-dimensional white noise with $\mathbb{E}[\mathbf{u}(t) \mid \mathscr{F}_{t-1}] = 0$ and $\mathbb{E}[\mathbf{u}(t)\mathbf{u}(t)^T \mid \mathscr{F}_{t-1}] = \boldsymbol{\Sigma}$, where $\mathscr{F}_{t-1}$ is the σ-field generated by $\{\mathbf{u}(s),\ s < t\}$, and $\boldsymbol{\Sigma}$ is a positive definite matrix.

The cross-spectral density satisfies

$$f_{\ell,m}(\lambda) \sim_{\lambda\to 0^+} \frac{1}{2\pi}\Omega_{\ell,m} e^{-i\pi/2(d_\ell - d_m)} \lambda^{-(d_\ell + d_m)}, \quad \ell, m = 1, 2, \tag{5}$$

with $\boldsymbol{\Omega} = \mathbf{A}(1)^{-1}\mathbf{B}(1)\boldsymbol{\Sigma}\mathbf{B}(1)^T\mathbf{A}(1)^{T^{-1}}$. We will denote $r_{12} = \Omega_{12}/\sqrt{\Omega_{11}\Omega_{22}}$ the long-run correlation term obtained by normalizing $\boldsymbol{\Omega}$.

Four cases were simulated :

- FIVARMA(1,(0.1, 0.1),0) with $A_1 = \begin{pmatrix} 0.8 & 0 \\ 0 & 0.6 \end{pmatrix}$,
- FIVARMA(2,(0.1, 0.1),0) with A_1 and with $A_2 = \begin{pmatrix} 0.4 & 0 \\ 0.2 & 0.7 \end{pmatrix}$,
- FIVARMA(0,(0.1, 0.1),1) with B_1 defined equal to matrix A_2,
- FIVARMA(1,(0.1, 0.1),1) with A_1 and B_1.

Numerical results were obtained on 1000 simulations, with 2^{12} time points. Values of d_1 and d_2 were taken equal to 0.1 and parameter ρ was fixed to 0.8. Figure 2 represents the boxplots of $\theta_j(1, 2)$ with respect to scales j.

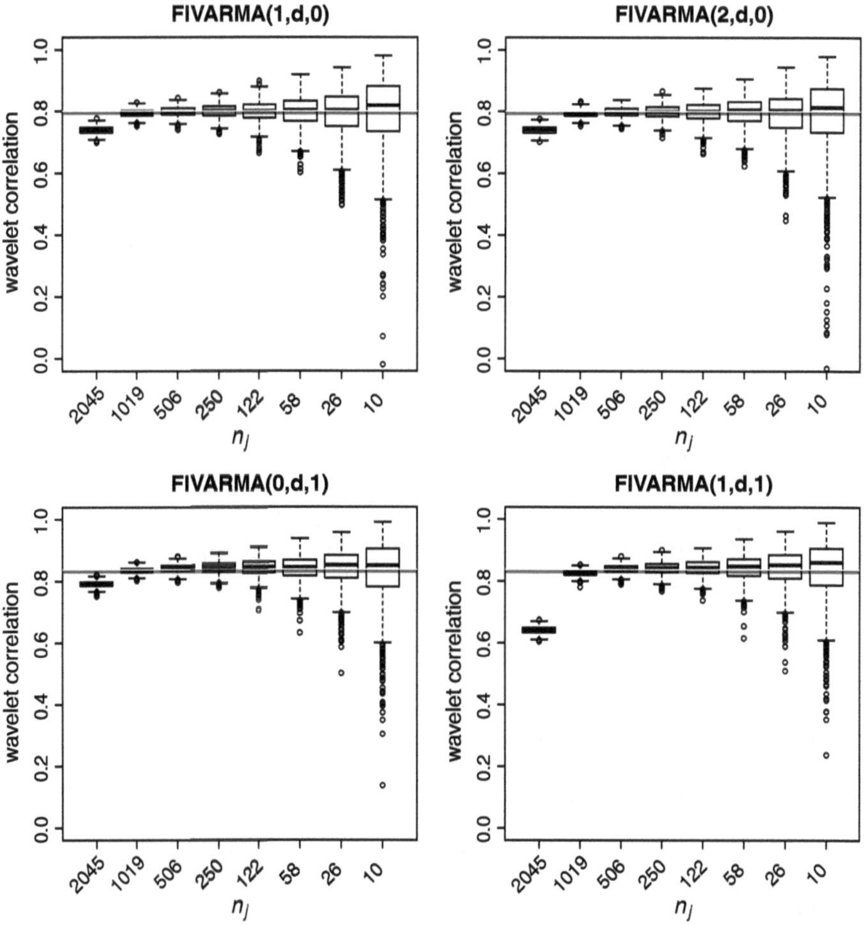

Fig. 2 Boxplots of $\theta_j(1, 2)$ at different scales for an FIVARMA(p,(0.1, 0.1),q), with different values of p and q. Number of coefficients available at each scale are given in indexes. The horizontal red line corresponds to the correlation r_{12}

As illustrated in Fig. 2, the wavelet correlations $\theta_j(1,2)$ are a good approximation of r_{12} at low frequencies, i.e. for high j. Indeed, no phase phenomenon appears since d_1 equals d_2. Yet the presence of short-range dependence disrupts the behaviour of $\theta_j(1,2)$ for the highest frequency. The mean of $\theta_j(1,2)$ is no longer close to r_{12}. This highlights that removing the first scales will improve the quality of estimation of r_{12}. Depending on the short-range dependence, in some simulations, more than one scale can be impacted. The number of scales to remove in estimation will thus depend on the short-range behaviour. But this can be evaluated by visual inspection on the plot of wavelet correlations against scales.

3 Estimation Procedure

As illustrated in Sect. 2, measurements of correlation are influenced by parameters $\mathbf{d}$. A joint estimation of $\boldsymbol{\Omega}$ and $\mathbf{d}$ thus seems more adapted. We recall briefly the estimation procedure described in [2]. The wavelet Whittle criterion is defined as

$$\mathscr{L}(\mathbf{G}(\mathbf{d}),\mathbf{d}) = \frac{1}{n}\sum_{j=j_0}^{j_1}\Big[n_j \log\det\left(\boldsymbol{\Lambda}_j(\mathbf{d})\mathbf{G}(\mathbf{d})\boldsymbol{\Lambda}_j(\mathbf{d})\right) +$$

$$\sum_{k=0}^{n_j}\left(W_{j,k}(1)\,\ldots\,W_{j,k}(p)\right)^T\left(\boldsymbol{\Lambda}_j(\mathbf{d})\mathbf{G}(\mathbf{d})\boldsymbol{\Lambda}_j(\mathbf{d})\right)^{-1}\left(W_{j,k}(1)\,\ldots\,W_{j,k}(p)\right)\Big],$$

where n_j denotes the number of non zero wavelet coefficients at scale j and $j_0 \le j_1$ are two given scales.

The estimators minimizing this criterion satisfy:

$$\hat{\mathbf{d}} = \underset{\mathbf{d}}{\text{argmin}}\ \log\det(\hat{\mathbf{G}}(\mathbf{d})) + 2\log(2)\left(\frac{1}{n}\sum_{j=j_0}^{j_1} j n_j\right)\left(\sum_{\ell=1}^{p} d_\ell\right),$$

$$\hat{\mathbf{G}}(\mathbf{d}) = \frac{1}{n}\sum_{j=j_0}^{j_1}\boldsymbol{\Lambda}_j(\mathbf{d})^{-1}\mathbf{I}(j)\boldsymbol{\Lambda}_j(\mathbf{d})^{-1}.$$

The long-run covariance matrix can then be estimated by

$$\hat{\Omega}_{\ell,m} = \hat{G}_{\ell,m}(\hat{\mathbf{d}})/(\cos(\pi(\hat{d}_\ell - \hat{d}_m)/2)K(\hat{d}_\ell + \hat{d}_m)). \tag{6}$$

These estimators are consistent under an additional assumption. We introduce:

Condition (C)

$$\text{For all } \ell, m = 1,\ldots,p,\quad \sup_n \sup_{j\ge 0} \frac{1}{n_j 2^{2j(d_\ell+d_m)}} Var\left(\sum_k W_{j,k}(\ell)W_{j,k}(m)\right) < \infty.$$

The convergence theorem is the following:

Theorem 1 ([2]) *Assume that (W1)–(W5) and Condition (C) hold. If j_0 and j_1 are chosen such that* $\log(N)^2(2^{-j_0\beta} + N^{-1/2}2^{j_0/2}) \to 0$ *and* $j_0 < j_1 \leq j_N$ *then*

$$\hat{\mathbf{d}} - \mathbf{d}^0 = O_{\mathbb{P}}(2^{-j_0\beta} + N^{-1/2}2^{j_0/2}),$$

$$\forall(\ell, m) \in \{1, \ldots, p\}^2,\ \hat{\Omega}_{\ell,m} - \Omega_{\ell,m} = O_{\mathbb{P}}(\log(N)(2^{-j_0\beta} + N^{-1/2}2^{j_0/2})).$$

Taking $2^{j_0} = N^{1/(1+2\beta)}$, $\hat{\mathbf{d}}$ *achieves the minimax rate since* $\hat{\mathbf{d}} - \mathbf{d}^0 = O_{\mathbb{P}}(N^{-\beta/(1+2\beta)})$.

This result states that the finest frequencies $j < j_0$ in the wavelet procedure must not be taken into account in the procedure. It is in line with Sect. 2.3. The optimal choice of j_0 depends on the strength of the short-range dependence and it is thus fixed by the regularity β of the density $\boldsymbol{f}^S(\cdot)$. A perspective is to develop an adaptive estimation, where the value of β would not be used to calibrate parameters of estimation. Yet, as it can be seen in Fig. 2 it is possible to handle empirically the optimal choice of j_0: plotting boxplots of wavelet correlations at different scales can enable to extract scales where the behaviour is perturbed by short-range behaviour. We illustrate the practical choice of j_0 in Sect. 5.

4 Numerical Results

We simulate bivariate FIVARMA processes and assess the quality of estimation on simulation. First multivariate procedure enables to estimate the long-run covariance matrix $\boldsymbol{\Omega}$ which brings an important information on the structure of multivariate time series. Second, our aim is also to illustrate that multivariate methods improve the quality of estimation of $\mathbf{d}$.

We consider 1000 replications of bivariate FIVARMA(0,$\mathbf{d}$,0) processes, that is, defined by (2), with a covariance matrix $\boldsymbol{\Omega} = \begin{pmatrix} 1 & \rho \\ \rho & 1 \end{pmatrix}$ between innovations $\mathbf{u}$. Various values of $\mathbf{d}$ and ρ are considered. Results of estimations of $\boldsymbol{\Omega}$ and $\mathbf{d}$ are displayed respectively in Table 1 and in Table 2. The quality is quantified by `bias` which is the bias of $\hat{d}_\ell$, $\ell = 1, 2$ and `std` which denotes the standard deviation of $\hat{d}_\ell$, $\ell = 1, 2$. The RMSE is the root mean squared error, equal to $(\texttt{bias}^2 + \texttt{std}^2)^{1/2}$. Table 2 also gives the values of `ratio M/U`. For a given $\ell = 1, 2$, this quantity corresponds to the ratio of the RMSE of $\hat{d}_\ell$ when estimated by multivariate Whittle procedure using both components and the RMSE of $\hat{d}_\ell$ when estimated by Whittle procedure using univariate Whittle procedure only on time series X_ℓ.

A good quality of estimation of the long-run covariance matrix $\boldsymbol{\Omega}$ and of the long-run correlation is observed. The quality slightly decreases when the phase phenomenon occurs, due to correction (6).

Table 1 Wavelet Whittle estimation of $\boldsymbol{\Omega}$ for a bivariate $ARFIMA(0, \mathbf{d}, 0)$ with $\rho = 0.4$, $N = 512$ with 1000 repetitions

d_1	d_2		Bias	Std	RMSE
$\rho = 0.4$					
0.2	0.0	$\Omega_{1,1}$	0.0309	0.0697	0.0762
		$\Omega_{1,2}$	0.0176	0.0540	0.0568
		$\Omega_{2,2}$	−0.0012	0.0732	0.0733
		Correlation	0.0113	0.0417	0.0432
0.2	0.2	$\Omega_{1,1}$	0.0297	0.0733	0.0790
		$\Omega_{1,2}$	0.0116	0.0518	0.0530
		$\Omega_{2,2}$	0.0282	0.0725	0.0778
		Correlation	−0.0003	0.0386	0.0386
0.2	0.4	$\Omega_{1,1}$	0.0356	0.0703	0.0788
		$\Omega_{1,2}$	0.0328	0.0568	0.0655
		$\Omega_{2,2}$	0.0707	0.0728	0.1015
		Correlation	0.0106	0.0422	0.0435

d_1	d_2		Bias	Std	RMSE
$\rho = 0.8$					
0.2	0.0	$\Omega_{1,1}$	0.0309	0.0697	0.0762
		$\Omega_{1,2}$	−0.0284	0.0641	0.0701
		$\Omega_{2,2}$	−0.0062	0.0667	0.0670
		Correlation	−0.0182	0.0236	0.0298
0.2	0.2	$\Omega_{1,1}$	0.0297	0.0733	0.0790
		$\Omega_{1,2}$	−0.0263	0.0656	0.0706
		$\Omega_{2,2}$	0.0318	0.0731	0.0798
		Correlation	−0.0013	0.0172	0.0173
0.2	0.4	$\Omega_{1,1}$	0.0356	0.0703	0.0788
		$\Omega_{1,2}$	−0.0632	0.0659	0.0913
		$\Omega_{2,2}$	0.0708	0.0706	0.1000
		Correlation	−0.0195	0.0232	0.0303

Table 2 Multivariate Whittle wavelet estimation of $\mathbf{d}$ for a bivariate $ARFIMA(0, \mathbf{d}, 0)$, $N = 512$ with 1000 repetitions

d_1	d_2	Bias	Std	RMSE	ratio M/U
$\rho = 0.4$					
0.2	0.2	−0.0298	0.0428	0.0522	0.9631
	0.0	−0.0002	0.0438	0.0438	0.9504
	0.2	−0.0330	0.0456	0.0563	0.9713
	0.2	−0.0333	0.0443	0.0554	0.9831
	0.2	−0.0304	0.0429	0.0526	0.9583
	0.4	−0.0571	0.0461	0.0734	0.9701

d_1	d_2	Bias	std	RMSE	ratio M/U
$\rho = -0.8$					
0.2	0.2	−0.0161	0.0380	0.0413	0.7625
	0.0	0.0129	0.0371	0.0393	0.8980
	0.2	−0.0334	0.0384	0.0509	0.8780
	0.2	−0.0331	0.0391	0.0512	0.8966
	0.2	−0.0164	0.0392	0.0425	0.7742
	0.4	−0.0439	0.0387	0.0585	0.7836

The procedure also gives good quality estimates of $\mathbf{d}$. The rate of convergence of $\hat{\mathbf{d}}$ in Theorem 1 does not depend on the values of the long-run covariance matrix $\boldsymbol{\Omega}$. Table 2 stresses that it does influence the quality of estimation. Since all data are used in the estimation procedure, it is straightforward that the quality of estimation for $\mathbf{d}$ will be improved compared with a univariate estimation. This is confirmed by numerical results, since the `ratio M/U` is always smaller than 1. Additionally, when the correlation between the time series components increases (in absolute value), the `ratio M/U` decreases, meaning that the quality of estimation of $\mathbf{d}$ increases. These results illustrate that multivariate estimation improves the quality of estimation of the long-memory parameters $\mathbf{d}$.

5 Application in Neurosciences

Noninvasive data recorded from the brain are an example where the proposed methodology is efficient. The data consist of time series recording signals from fMRI. These data are intrinsically correlated because of the known interactions of the brain areas (also called regions of interest). These time series present long-memory features. Other data sets presenting similar features are coming from finance, e.g. [15], where time series present long-memory characteristics and are also correlated because of links between companies, for example. In this section, we observed time series extracted using fMRI facilities. The whole description of this data set is detailed in [16]. 100 subjects were scanned twice and we extracted 89 regions of interest for each scan with time series of length 1200 time points. Figure 3 displays 6 arbitrary signals from a subject in this data set.

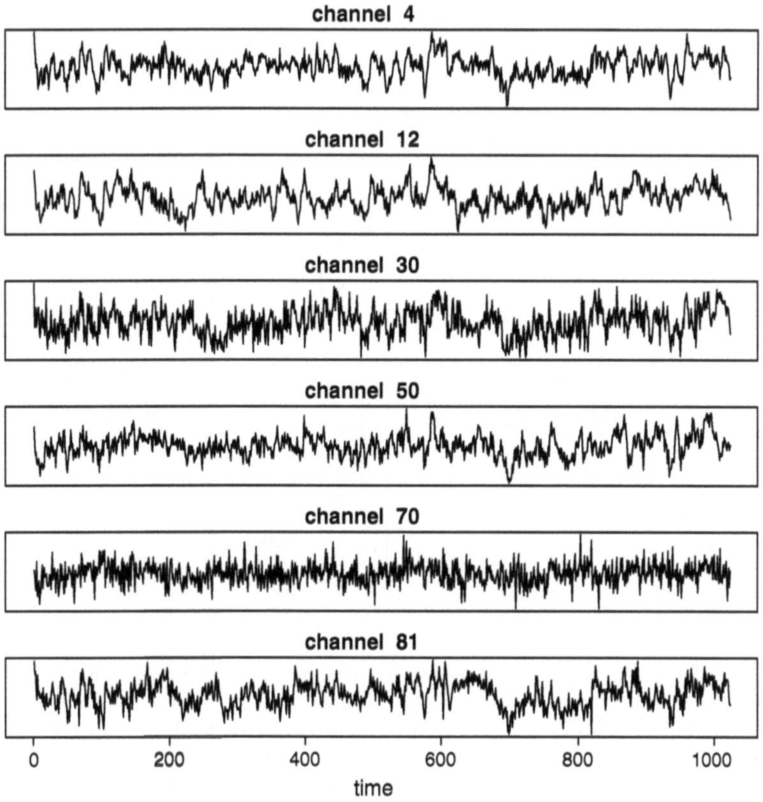

Fig. 3 Plot of 6 arbitrary signals from a subject of fMRI data set

5.1 Estimation of $\mathbf{d}$ and Ω

As stated by Theorem 1, the highest frequencies should not be taken into account, in order to decrease the bias caused by short-range dependence. As illustrated in Sect. 2.3, the advantage of representing the wavelet correlation in terms of scale is to qualitatively assess the scales necessary to estimate the long-memory parameters and long-range covariance matrix. When dealing with real data, bootstrap is providing a way to assess the behaviour of the wavelet correlations. Sliding overlapping windows of the time series containing 512 points were extracted and we repeated the estimation until reaching the final point of the time series. This is illustrated in Fig. 4, where an example of four pairs of fMRI data from one subject is presented. Boxplots are constructed using the sliding window extractions. From these plots, and taking into account neuroscientific hypothesis stating that the signal of interest for resting state is occurring for frequency below 1 Hz, we chose to compute the long-memory parameters between scales 3 and 6.

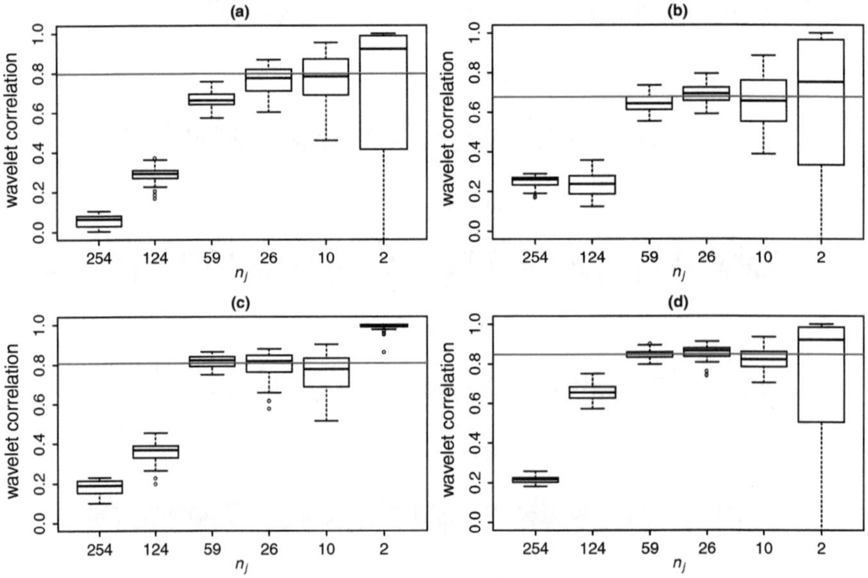

Fig. 4 Boxplots of the empirical correlations between the wavelet coefficients at different scales for real time series from a single subject of fMRI data sets: (**a**) Time series 1 and 2; (**b**) Time series 13 and 14; (**c**) Time series 31 and 32; (**d**) Time series 47 and 48. Boxplots were obtained using sub-time series with N points, extracted from two fMRI time series with length equal to 1200 points, from a single subject. Estimated long parameters d of the two time series are equal with a two digits precision. The index of the horizontal axis displays the number of coefficients available. The horizontal red lines represent the estimated long-run correlation. Calculation was done on 100 sliding windows (with overlap), each of them containing $N = 512$ observations

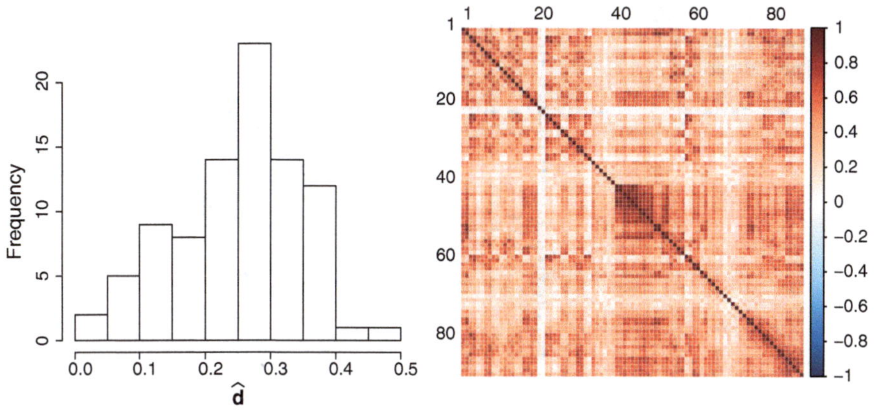

Fig. 5 Histogram of $\hat{\mathbf{d}}$ and estimation of $\boldsymbol{\Omega}$ from a subject of fMRI data set

For each subject, the vector of long-memory parameter $\mathbf{d}$ is estimated using both univariate and multivariate methods. Figure 5 displays an example of long-memory parameters estimated for one subject taken at random among the 100 subjects.

5.2 *Comparison with Univariate Estimates of* $\mathbf{d}$

In addition, as the data sets consist of a test-retest paradigm with two recordings for each subject, a way to evaluate the accuracy of the estimator is to evaluate the reproducibility using intra-class correlation. Following [14], intra-class correlation (ICC) was computed. ICC is a coefficient smaller than 1 that takes into account the variance within subject in comparison to the variance between subject, defined as,

$$ICC = \frac{s_b - s_w}{s_b + (k-1)s_w} \tag{7}$$

where s_b is the variance between subjects, s_w is the variance within subjects and k is the number of sessions per subject. ICC is close to 0 when the reliability is low, and close to 1 when the reliability is high. ICC defined as (7) can have negative values but this reflects a wrong behaviour of the data set.

Figure 6 shows the results obtained using the univariate estimator and multivariate estimator. This result suggests that multivariate estimations are more reproducible in a test-retest paradigm than univariate estimations.

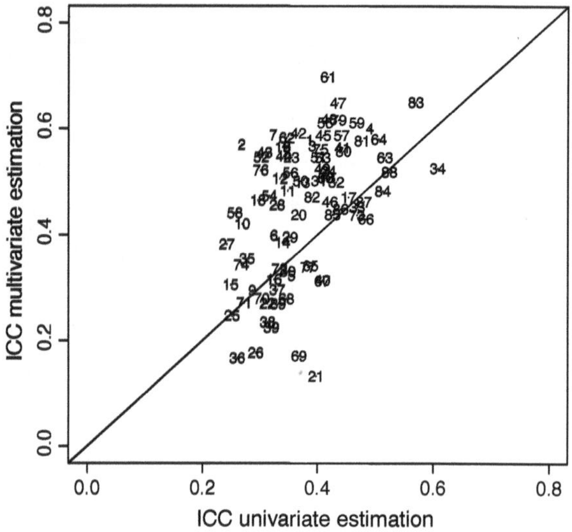

Fig. 6 Estimation of ICC using multivariate or univariate estimations

6 Conclusion

Statistical analysis of multivariate time series with long memory is challenging. Based on the results of [2], our study highlights the influence of the memory properties on the long-run coupling between time series. In particular, due to differences in long-memory parameters a phase phenomenon occurs, that is illustrated by the behaviour of the wavelet coefficients. Next, we recall the wavelet-based Whittle estimation of [2]. The main contribution is to illustrate how multivariate estimation can improve univariate procedure. This is noticeable on numerical results on simulations. We also consider a test-retest fMRI data set. The analysis shows the robustness of multivariate estimation on this real application, compared with univariate estimation.

Acknowledgements This work was partly supported by the project *Graphsip* from Agence Nationale de la Recherche (ANR-14-CE27-0001). S. Achard was partly funded by a grant from la Région Rhône-Alpes and a grant from AGIR-PEPS, Université Grenoble Alpes-CNRS.

References

1. Abry, P., & Veitch, D. (1998). Wavelet analysis of long-range-dependent traffic. *IEEE Transactions on Information Theory, 44*(1), 2–15.
2. Achard, S., & Gannaz, I. (2016). Multivariate wavelet whittle estimation in long-range dependence. *Journal of Time Series Analysis, 37*(4), 476–512.
3. Achard, S., Salvador, R., Whitcher, B., Suckling, J., & Bullmore, E. (2006). A resilient, low-frequency, small-world human brain functional network with highly connected association cortical hubs. *Journal of Neuroscience, 26*(1), 63–72.

4. Dahlhaus, R. (1989). Efficient parameter estimation for self-similar processes. *The Annals of Statistics, 17*(4), 1749–1766.
5. Fox, R., & Taqqu, M. S. (1986). Large-sample properties of parameter estimates for strongly dependent stationary gaussian time series. *The Annals of Statistics, 14*(2), 517–532.
6. Geweke, J., & Porter-Hudak, S. (1983). The estimation and application of long memory time series models. *Journal of Time Series Analysis, 4*(4), 221–238.
7. Giraitis, L., Robinson, P. M., & Samarov, A. (1997). Rate optimal semiparametric estimation of the memory parameter of the gaussian time series with long-range dependence. *Journal of Time Series Analysis, 18*(1), 49–60.
8. Lobato, I. N. (1997). Consistency of the averaged cross-periodogram in long memory series. *Journal of Time Series Analysis, 18*(2), 137–155.
9. Lobato, I. N. (1999). A semiparametric two-step estimator in a multivariate long memory model. *Journal of Econometrics, 90*(1), 129–153.
10. Moulines, E., Roueff, F., & Taqqu, M. S. (2008). A wavelet Whittle estimator of the memory parameter of a nonstationary gaussian time series. *The Annals of Statistics, 36*(4), 1925–1956.
11. Robinson, P. M. (1995). Gaussian semiparametric estimation of long range dependence. *The Annals of Statistics, 23*(5), 1630–1661.
12. Sela, R. J., & Hurvich, C. M. (2008). Computationally efficient methods for two multivariate fractionally integrated models. *Journal of Time Series Analysis, 30*(6), 631–651.
13. Shimotsu, K. (2007). Gaussian semiparametric estimation of multivariate fractionally integrated processes. *Journal of Econometrics, 137*(2), 277–310.
14. Shrout, P. E., & Fleiss, J. L. (March 1979). Intraclass correlations: Uses in assessing rater reliability. *Psychological Bulletin, 86*(2), 420–428.
15. Songsiri, J., & Vandenberghe, L. (2010). Topology selection in graphical models of autoregressive processes. *The Journal of Machine Learning Research, 11*, 2671–2705.
16. Termenon, M., Jaillard, A., Delon-Martin, C., & Achard, S. (2016). Reliability of graph analysis of resting state fMRI using test-retest dataset from the human connectome project. *NeuroImage, 142*, 172–187.
17. Whitcher, B., Guttorp, P., & Percival, D. B. (2000). Wavelet analysis of covariance with application to atmospheric time series. *Journal of Geophysical Research 105(D11)*(14), 941–962.

On Kernel Smoothing with Gaussian Subordinated Spatial Data

S. Ghosh

Abstract We address estimation of a deterministic function μ, that is the mean of a spatial process $y(\mathbf{s})$ in a nonparametric regression context. Here $\mathbf{s}$ denotes a spatial coordinate in R_+^2. Given $k = n^2$ observations, the aim is to estimate μ assuming that y has finite variance, and that the regression errors $\epsilon(\mathbf{s}) = y(\mathbf{s}) - E\{y(\mathbf{s})\}$ are Gaussian subordinated.

1 Introduction

We consider a nonparametric regression setting where our observations are

$$y(\mathbf{s}_r), \; r = 1, 2, \ldots, k \tag{1}$$

on a continuous index random field $y(\mathbf{s})$. Here $\mathbf{s}_r, \mathbf{s} \in R_+^2$ denote two-dimensional spatial locations although these ideas easily generalize to other cases, such as higher dimension or inclusion of predictors or explanatory variables as in regression. Let $k = n^2$, and let us denote

$$\mathbf{s_r} = (s_{1r}, s_{2r}). \tag{2}$$

Suppose also that there is a zero mean, unit variance, stationary latent Gaussian random field $Z(\mathbf{s})$ such that the centered process is Gaussian subordinated, i.e.,

$$\epsilon(\mathbf{s}) = y(\mathbf{s}) - E\{y(\mathbf{s})\} = G(Z(\mathbf{s}), \mathbf{u}) \tag{3}$$

where $\mathbf{u} \in [0, 1]^2$ is obtained by rescaling the spatial coordinate $\mathbf{s}$ and

$$G : R \times [0, 1]^2 \to R \tag{4}$$

S. Ghosh (✉)
Statistics Lab, Swiss Federal Research Institute WSL, Birmensdorf, Switzerland
e-mail: rita.ghosh@wsl.ch

P. Bertail et al. (eds.), *Nonparametric Statistics*, Springer Proceedings
in Mathematics & Statistics 250, https://doi.org/10.1007/978-3-319-96941-1_19

is an unknown function, square integrable with respect to the standard normal density, i.e., for $\mathbf{u} \in [0, 1]^2$,

$$\int_{-\infty}^{\infty} G^2(z, \mathbf{u})\phi(z)dz < \infty \tag{5}$$

$$\int_{-\infty}^{\infty} G(z, \mathbf{u})\phi(z)dz = 0 \tag{6}$$

where ϕ denotes the standard normal density function.

The consequence of the transformation in (3) is that the marginal distribution of the regression errors ϵ may be location dependent. In various applications, this is advantageous. When very large scales are involved, using the above transformation as a model, one may accommodate more flexibility. For instance, one may estimate the marginal error distribution and test if various types of exceedance or non-exceedance probabilities are location dependent, if the spatial quantiles of ϵ are horizontal surfaces, or if there are local modes etc.

There is an extensive literature on non-linear models of this kind; some references are [4–6, 9, 16, 21, 22] and others, whereas [1] provide a review.

Often the spatial observations are available at a discrete set of locations. Let the set of these coordinates be

$$A_n = \{(i, j) \mid i, j = 1, 2, \ldots, n\}. \tag{7}$$

In this case, the rescaled coordinates are $\mathbf{u} = \mathbf{s}/n \in [0, 1]^2$ and the discrete lags are $\mathbf{h} \in \{0, \pm 1, \pm 2, ..\}^2$, $|\mathbf{h}|$ being the Euclidean norm of $\mathbf{h}$.

Written in terms of the spatial coordinates in A_n, our nonparametric regression model is:

$$y(\mathbf{s}_r) = \mu(\mathbf{u}_r) + \epsilon(\mathbf{s}_r), \ \mathbf{s}_r \in A_n, \mathbf{u}_r = \mathbf{s}_r/n, \tag{8}$$

$r = 1, 2, \ldots, k, k = n^2$. Our aim is to estimate $\mu(\mathbf{u})$, ϵ being a zero-mean Gaussian subordinated process as described above. In a similar manner, the marginal distribution of y at location $\mathbf{u}$ may be estimated, e.g., by smoothing an appropriately defined indicator function; see, for instance, Ghosh and Draghicescu [15] for an example in the time series case. Before proceeding, we formulate the correlation structure in the latent Gaussian process Z.

1.1 Latent Gaussian Process $Z(\mathbf{s})$, $\mathbf{s} \in R_+^2$

We assume that $E(Z(\mathbf{s})) = 0$, $Var(Z(\mathbf{s})) = 1$. Also let Z be isotropic, its covariance function being,

$$Cov(Z(\mathbf{s}_1), Z(\mathbf{s}_2)) = \gamma_Z(|\mathbf{s}_1 - \mathbf{s}_2|), \ \mathbf{s}_1, \mathbf{s}_2 \in R_+^2 \tag{9}$$

where $|\cdot|$ denotes the Euclidean norm. When Z has short-memory or long-memory correlations, γ_Z will satisfy:

Short-Memory For any integer $m_0 \geq 1$,

$$\sum_{l=m_0}^{\infty} \int_{[0,1]^2} \int_{[0,1]^2} \left|\gamma_Z(\sqrt{k}|\mathbf{u_1} - \mathbf{u_2}|)\right|^l d\mathbf{u_1} d\mathbf{u_2} = 0\left(\frac{1}{k}\right), \; k \to \infty, \tag{10}$$

or, when the data are on a lattice,

$$\sum_{\mathbf{h}} |\gamma_Z(\mathbf{h})|^{m_0} < \infty; \tag{11}$$

Long-Memory Let $0.5 < H < 1$ be the Hurst coefficient. Then

$$\int_{[0,1]^2} \int_{[0,1]^2} \left|\gamma_Z(\sqrt{k}|\mathbf{u_1} - \mathbf{u_2}|)\right|^{m_0} d\mathbf{u_1} d\mathbf{u_2} = 0\left(k^{m_0(2H-2)}\right), \; k \to \infty, \tag{12}$$

or, in case of lattice data (see [2]), in terms of the discrete lags,

$$\gamma_Z(\mathbf{h}) \sim C_Z |\mathbf{h}|^{-2\alpha} f\left(\frac{\mathbf{h}}{|\mathbf{h}|}\right), \; \text{as } |\mathbf{h}| \to \infty \tag{13}$$

where $0 < \alpha < 1/m_0$, $H = 1 - \alpha/2$, $m_0 \geq 1$ is a positive integer, $C_Z > 0$ and f is a continuous function on the unit circle on R^2; also see [9].

Let m be the Hermite rank of G and assume that $m_0 = m$, so that the regression errors ϵ will also have long-memory when Z is long-range dependent.

1.2 Regression Errors and Its Second Moments

As mentioned above, the errors ϵ have zero mean, finite variance and are assumed to be Gaussian subordinated. Consider the following Hermite polynomial expansion:

$$\epsilon(\mathbf{s}) = G(Z(\mathbf{s}), \mathbf{u}) = \sum_{l=m}^{\infty} \frac{c_l(\mathbf{u})}{l!} H_l\left(Z(\mathbf{s})\right) \tag{14}$$

where as before, $\mathbf{s} \in R_+^2$ is a spatial coordinate, $\mathbf{u} \in [0, 1]^2$ is rescaled $\mathbf{s}$, and G is the unknown Lebesgue-measurable L^2 function mentioned earlier. Moreover, c_l are Hermite coefficients, m is Hermite rank of G, and H_l are Hermite polynomials. Let the Hermite coefficients be in $C^3\left([0, 1]^2\right)$, $l = m, m+1, m+2, \ldots$.

In particular,

$$Cov\left(H_l(Z(\mathbf{s}_1)), H_{l'}(Z(\mathbf{s}_2))\right) = 0, \; \text{if } l \neq l', \tag{15}$$

whereas

$$Cov\,(H_l(Z(\mathbf{s}_1)), H_l(Z(\mathbf{s}_2))) \\ = l!\,\{\gamma_Z\,(|\mathbf{s}_1 - \mathbf{s}_2|)\}^l\,,\ \text{if } l = l', \tag{16}$$

and

$$Var\,(H_l(Z(i, j)) = l!. \tag{17}$$

Due to the transformation (3), the variance of ϵ need not be a constant and from the above Hermite expansion, it follows that we have

$$\sigma^2(\mathbf{u}) = Var\,(\epsilon(\mathbf{s})) = \sum_{l=m}^{\infty} \frac{c_l^2(\mathbf{u})}{l!}, \tag{18}$$

We assume that $\sigma(\mathbf{u}),\ \mathbf{u} \in [0, 1]^2$ is smooth, so that consistent estimation can be facilitated.

2 Estimation

For estimation, we assume that $\mu(\mathbf{u})$ is in $C^3\left([0, 1]^2\right)$. We consider the following regression estimator due to [18]. For $\mathbf{u} = (u_1, u_2)$, and $\mathbf{u}_r = (u_{1r}, u_{2r}),\ r = 1, 2, \ldots, k,\ k = n^2$

$$\widehat{\mu}(\mathbf{u}) = \frac{1}{kb_1b_2} \sum_{r=1}^{k} K\left(\frac{u_{1r} - u_1}{b_1}\right) K\left(\frac{u_{2r} - u_2}{b_2}\right) y(\mathbf{s}_r). \tag{19}$$

Here $b_1 > 0$ and $b_2 > 0$ are bandwidths such that as $n \to \infty$, $b_1, b_2 \to 0$, and $nb_1, nb_2 \to \infty$. Moreover, the kernel K is a continuous symmetric density function on $[-1, 1]$.

Theorem 1 along with its proof and related information can be found in [10, 11] and [12].

Theorem 1 *As $n \to \infty$, for fixed $\mathbf{u} \in (0, 1)^2$,*
Bias:

$$E\,\{\widehat{\mu_n}(\mathbf{u})\} - \mu(\mathbf{u}) = \frac{1}{2}\int_{-1}^{1} v^2 K(v)dv \left\{b_1^2 \frac{\partial^2}{\partial u_1^2}\{\mu(\mathbf{u})\} + b_2^2 \frac{\partial^2}{\partial u_2^2}\{\mu(\mathbf{u})\}\right\} \\ + o(max(b_1^2, b_2^2)). \tag{20}$$

Variance: if $b_1 = b_2 = b$,

$$\textit{short memory:}\quad Var\{\widehat{\mu_n}(\mathbf{u})\} = O\left((nb)^{-2}\right) \tag{21}$$

$$\textit{long memory:}\quad Var\{\widehat{\mu_n}(\mathbf{u})\} = O\left((nb)^{-2m\alpha}\right) \tag{22}$$

where $0 < \alpha < 1/m$, *m being the Hermite rank of* G.

Furthermore, weak uniform consistency can be proved assuming that K has a characteristic function that is absolutely integrable on R.

For $\mathbf{u} = (u_1, u_2) \in (0, 1)^2$, consider a similar kernel smoothing operation applied to the regression errors. Specifically, let

$$S_n(\mathbf{u}) = \frac{1}{kb_1b_2}\sum_{r=1}^{k} K\left(\frac{u_{1r} - u_1}{b_1}\right) K\left(\frac{u_{2r} - u_2}{b_2}\right) \epsilon(\mathbf{s}_r) \tag{23}$$

where let ψ be the characteristic function of K, i.e.,

$$\psi(t) = \int e^{\iota t w} K(w)dw \tag{24}$$

where $\iota = \sqrt{-1}$ and $t \in R$. Let ψ satisfy

$$\int_{-\infty}^{\infty} |\psi(t)|dt < \infty \tag{25}$$

Then the following holds.

Theorem 2 S_n *converges to zero uniformly and in probability as* $n \to \infty$.

For the proof of Theorem 2 see [12], where the arguments are generalizations of [17] to non-linear transformations of Gaussian random fields; also see [3]. For a general background on kernel smoothing see [13, 20, 23] as well as [19].

A consequence of the above theorems is that the nonparametric estimator $\widehat{\mu}$ is also uniformly consistent in probability. This result can be immediately exploited to propose a possible bandwidth selection algorithm by noting that if

$$\widehat{\epsilon}(\mathbf{s}) = y(\mathbf{s}) - \widehat{\mu}(\mathbf{u}) \tag{26}$$

where $\mathbf{u}$ is rescaled $\mathbf{s}$, then, as $n \to \infty$,

Theorem 3 $|\widehat{\epsilon}(\mathbf{s}) - \epsilon(\mathbf{s})|$ *converges to zero uniformly and in probability.*

Moreover, although $\epsilon(\mathbf{s})$ is not covariance stationary, in small neighborhoods, they are locally stationary; also see [7]. We have

Theorem 4 *If* $\mathbf{u}_1, \mathbf{u}_2 \to \mathbf{u}$ *then,*

$$Cov(\epsilon(\mathbf{s}_1), \epsilon(\mathbf{s}_2)) \sim g(|\mathbf{s}_1 - \mathbf{s}_2|, \mathbf{u}) \tag{27}$$

where g is a covariance function.

Due to Theorem 3, g may now be estimated using a local variogram using the regression residuals $\widehat{\epsilon}(\mathbf{s})$ (also see, e.g., [8]). Similarly, σ^2 may be estimated by smoothing the squared residuals. These estimates may then be combined to obtain a direct estimate of the variance of $\widehat{\mu}$. For further information see [12]. A proposal for a bandwidth selection procedure along with these lines can be found in [24]; also see [14] who consider a data-driven bandwidth selection procedure for trend estimation for Gaussian subordinated time series data.

References

1. Beran, J., Feng, Y., Ghosh, S., & Kulik, R. (2013). *Long memory processes - probabilistic properties and statistical models*. Heidelberg: Springer.
2. Beran, J., Ghosh, S., & Schell, D. (2009). Least square estimation for stationary lattice processes with long-memory. *Journal of Multivariate Analysis, 100*, 2178–2194.
3. Bierens, H. J. (1983). Uniform consistency of kernel estimators of a regression function under generalized conditions. *Journal of the American Statistical Association, 77*, 699–707.
4. Breuer, P., & Major, P. (1983). Central limit theorems for nonlinear functionals of Gaussian fields. *Journal of Multivariate Analysis, 13*, 425–441.
5. Csörgö, S., & Mielniczuk, J. (1995). Nonparametric regression under long-range dependent normal errors. *The Annals of Statistics, 23*, 1000–1014.
6. Csörgö, S., & Mielniczuk, J. (1996). The empirical process of a short-range dependent stationary sequence under Gaussian subordination. *Probability Theory and Related Fields, 104*, 15–25.
7. Dahlhaus, R. (1997). Fitting time series models to nonstationary processes. *The Annals of Statistics, 25*, 1–37.
8. Diggle, P. J. (1990). *Time series: A biostatistical introduction*. Oxford: Oxford University Press.
9. Dobrushin, R. L., & Major, P. (1979). Non-central limit theorems for non-linear functional of Gaussian fields. *Probability Theory and Related Fields, 50*, 27–52.
10. Ghosh, S. (2009). The unseen species number revisited. *Sankhya-B, 71*, 137–150.
11. Ghosh, S. (2015). Computation of spatial Gini coefficients. *Communications in Statistics - Theory and Methods, 44*, 4709–4720.
12. Ghosh, S. (2015). Surface estimation under local stationarity. *Journal of Nonparametric Statistics, 27*, 229–240.
13. Ghosh, S. (2018). *Kernel smoothing. Principles, methods and applications*. Hoboken: Wiley.
14. Ghosh, S., & Draghicescu, D. (2002). An algorithm for optimal bandwidth selection for smooth nonparametric quantiles and distribution functions. In Y. Dodge (Ed.), *Statistics in industry and technology: Statistical data analysis based on the L_1-norm and related methods* (pp. 161–168). Basel: Birkhäuser Verlag.
15. Ghosh, S., & Draghicescu, D. (2002). Predicting the distribution function for long-memory processes. *International Journal of Forecasting, 18*, 283–290.
16. Major, P. (1981). Limit theorems for non-linear functionals of Gaussian sequences. *Probability Theory and Related Fields, 57*, 129–158.

17. Parzen, E. (1962). On estimation of a probability density function and mode. *The Annals of Mathematical Statistics, 33*, 1065–1076.
18. Priestley, M. B., & Chao, M. T. (1972). Non-parametric function fitting. *Journal of the Royal Statistical Society. Series B, 34*, 385–392.
19. Robinson, P. M. (1997). Large-sample inference for nonparametric regression with dependent errors. *The Annals of Statistics, 25*, 2054–2083.
20. Silverman, B. W. (1986). *Density estimation*. New York: Chapman and Hall.
21. Taqqu, M. S. (1975). Weak convergence to fractional Brownian motion and to the Rosenblatt process. *Probability Theory and Related Fields, 31*, 287–302.
22. Taqqu, M. S. (1979). Convergence of integrated processes of arbitrary Hermite rank. *Probability Theory and Related Fields, 50*, 53–83.
23. Wand, M. P., & Jones, M. C. (1995). Kernel smoothing. London: Chapman and Hall.
24. Wu, F. (2016). *On Optimal Surface Estimation under Local Stationarity - An Application to Swiss National Forest Inventory Data*. Master thesis 2016, ETH, Zürich.

Strong Separability in Circulant SSA

J. Bógalo, P. Poncela, and E. Senra

Abstract Circulant singular spectrum analysis (CSSA) is an automated variant of singular spectrum analysis (SSA) developed for signal extraction. CSSA allows to identify the association between the extracted component and the frequencies they represent without the intervention of the analyst. Another relevant characteristic is that CSSA produces strongly separable components, meaning that the resulting estimated signals are uncorrelated. In this contribution we deepen in the strong separability of CSSA and compare it to SSA by means of a detailed example. Finally, we apply CSSA to UK and US quarterly GDP to check that it produces reliable cycle estimators and strong separable components. We also test the absence of any seasonality in the seasonally adjusted time series estimated by CSSA.

1 Introduction

Signal extraction time series methods seek to estimate a set of components that isolate each of the main characteristics of a dynamic indicator. The decomposition is usually based on the periodicity of associated fluctuations. In the classic setting in economics, the trend component accounts for the long-run fluctuations, the cycle for fluctuations between 1.5 and 8 years, seasonality for 1 year fluctuations, and the remaining short run deviations are assigned to the irregular component. Additionally, statistical offices regularly produce and publish seasonal adjusted time series that remove the estimated seasonality from the original data. Seasonal

J. Bógalo · E. Senra
Universidad de Alcalá, Madrid, Spain
e-mail: juan.bogalo@edu.uah.es; eva.senra@uah.es

P. Poncela (✉)
European Commission, Joint Research Centre (JRC), Ispra, Italy
Universidad Autónoma de Madrid, Madrid, Spain
e-mail: pilar.poncela@ec.europa.eu

P. Bertail et al. (eds.), *Nonparametric Statistics*, Springer Proceedings in Mathematics & Statistics 250, https://doi.org/10.1007/978-3-319-96941-1_20

adjusted time series are a useful tool for economic policy makers and analysts as they are much easier to interpret than the original data.

One desirable property of the signal extraction method is that the resulting components are orthogonal, however, they usually exhibit cross-correlation. Residual seasonality in seasonal adjusted time series is a concern in any signal extraction method since the very early papers, see, for instance, [3] or [2]. And it is today still a relevant issue. Findley et al. [4] point out that "The most fundamental seasonal adjustment deficiency is detectable seasonality after adjustment." This is also a concern for policy makers as seen in [9].

Circulant singular spectrum analysis (CSSA) is a signal extraction non-parametric automated procedure based on singular spectrum analysis (SSA) developed in [1]. These authors also showed that the components estimated by CSSA were strongly separable and checked that this property was fulfilled by the estimation obtained in the analysis of Industrial Production in 6 countries. In this chapter, we focus more in detail on the property of separability and illustrate by means of an example the main differences between Basic SSA and CSSA in relation to separability.

Finally, we apply CSSA to UK and US quarterly GDP, check the reliability of the estimated components and their separability, and test for any remaining seasonality in the seasonally adjusted time series.

This chapter is organized as follows. Section 2 reviews the classical SSA methodology and introduces the new CSSA. Section 3 deals with separability and the differences between the results obtained by CSSA and Basic CSSA. Section 4 applies the technique to UK and US GDP, and Sect. 5 concludes.

2 SSA Methodology and CSSA

2.1 Classical SSA

In this section we briefly review SSA and then point out the differences with our new approach CSSA. The goal is to decompose a time series into its unobserved components (trend, cycle, etc.). SSA, see [5], is a technique in two stages: decomposition and reconstruction. In the first stage, decomposition, we transform the original vector of data into a related trajectory matrix and perform its singular value decomposition to obtain the so-called elementary matrices. This corresponds to steps 1 and 2 in the algorithm. In the second stage, reconstruction, steps 3 and 4 of the algorithm, we provide estimates of the unobserved components. In the third step, we classify the elementary matrices into groups associating each group to an unobserved component (trend, cycle, etc.). In the final step, we transform every group into an unobserved component of the same size of the original time series by diagonal averaging. To proceed with the algorithm, let $\{x_t\}$ be a real valued zero mean time series of size T, $\mathbf{x} = (x_1, \ldots, x_T)'$, and L a positive integer, called the window length, such that $1 < L < T/2$. The SSA procedure involves the following 4 steps:

1st Step: Embedding

From the original time series we obtain an $L \times N$ trajectory matrix $\mathbf{X}$ given by L dimensional time series of length $N = T - L + 1$ as

$$\mathbf{X} = (\mathbf{x}_1|\ldots|\mathbf{x}_N) = \begin{pmatrix} x_1 & x_2 & x_3 & \ldots & x_N \\ x_2 & x_3 & x_4 & \ldots & x_{N+1} \\ \vdots & \vdots & \vdots & \vdots & \vdots \\ x_L & x_{L+1} & x_{L+2} & \ldots & x_T \end{pmatrix}$$

where $\mathbf{x}_j = (x_j, \ldots, x_{j+L-1})'$ indicates the vector of dimension L and origin at time j. Notice that the trajectory matrix $\mathbf{X}$ is Hankel and both, by columns and rows, we obtain subseries of the original one.

2nd Step: Decomposition

In this step, we perform the singular value decomposition (SVD) of the trajectory matrix $\mathbf{X} = \mathbf{U}\mathbf{D}^{1/2}\mathbf{V}'$. In Basic SSA $\mathbf{U}$ is the $L \times L$ matrix whose columns $\mathbf{u}_k$ are the eigenvectors of the second moment matrix $\mathbf{S} = \mathbf{X}\mathbf{X}'$, $\mathbf{D} = diag(\tau_1, \ldots, \tau_L)$, $\tau_1 \geq \ldots \geq \tau_L \geq 0$, are the eigenvalues of $\mathbf{S}$ and $\mathbf{V}$ is the $N \times L$ matrix whose columns $\mathbf{v}_k$ are the L eigenvectors of $\mathbf{X}'\mathbf{X}$ associated with nonzero eigenvalues. This decomposition allows to write $\mathbf{X}$ as the sum of the so-called elementary matrices $\mathbf{X}_k$ of rank 1,

$$\mathbf{X} = \sum_{k=1}^{r} \mathbf{X}_k = \sum_{k=1}^{r} \mathbf{u}_k \mathbf{w}_k',$$

where $\mathbf{w}_k = \mathbf{X}'\mathbf{u}_k = \sqrt{\tau_k}\mathbf{v}_k$ and $r = \max_{\tau_k > 0}\{k\}$=rank($\mathbf{X}$).

The original and alternative versions of SSA base the decomposition of the trajectory matrix on the second order moment of the series.

3rd Step: Grouping

In this step we group the elementary matrices $\mathbf{X}_k$ into m disjoint groups summing up the matrices within each group. Let $I_j = \{j_1, \ldots, j_p\}$, $j = 1, \ldots, m$ each disjoint group of indexes associated with the corresponding eigenvectors. The matrix $\mathbf{X}_{I_j} = \mathbf{X}_{j_1} + \ldots + \mathbf{X}_{j_p}$ is associated with the I_j group. The decomposition of the trajectory matrix into these groups is given by $\mathbf{X} = \mathbf{X}_{I_1} + \ldots + \mathbf{X}_{I_m}$. The contribution of the component coming from matrix $\mathbf{X}_{I_j}$ is given by $\sum_{k \in I_j} \widehat{\lambda}_k / \sum_{k=1}^{r} \widehat{\lambda}_k$.

4th Step: Reconstruction

Let $\mathbf{X}_{I_j} = (\widetilde{x}_{ij})$. In this step, each matrix $\mathbf{X}_{I_j}$ is transformed into a new time series of the same length T as the original one, denoted as $\widetilde{\mathbf{x}}^{(j)} = (\widetilde{x}_1^{(j)}, \ldots, \widetilde{x}_T^{(j)})'$

by diagonal averaging of the elements of $\mathbf{X}_{I_j}$ over its antidiagonals as follows:

$$\widetilde{x}_t^{(j)} = \begin{cases} \frac{1}{t}\sum_{i=1}^{t} \widetilde{x}_{i,t-i+1}, & 1 \leq t < L \\ \frac{1}{L}\sum_{i=1}^{L} \widetilde{x}_{i,t-i+1}, & L \leq t \leq N \\ \frac{1}{T-t+1}\sum_{i=L-N+1}^{T-N+1} \widetilde{x}_{i,t-i+1}, & N < t \leq T. \end{cases}$$

2.2 Circulant SSA

Circulant SSA modifies steps 2 and 3 of the previous algorithm. It has been introduced in [1]. The decomposition step is made through an alternative second moments matrix that has the property of being Circulant. Let S_C a matrix of second moments where each element of the first row is given by

$$\widehat{c}_m = \frac{L-m}{L} s_m + \frac{m}{L} s_{L-m}, \qquad m = 0, 1, \ldots, L-1;$$

with

$$s_m = \frac{1}{T-m} \sum_{t=1}^{T-m} x_t x_{t+m}.$$

From this first row we construct a Circulant matrix so, for instance, in the second row we will shift one position to the right each element $\widehat{c}_m$ and place the last one in the first position of the row, and so on. Circulant matrices are related to the spectral density of stationary time series. In particular, the elements of the Circulant matrix of population second moments are also given by

$$c_m = \frac{1}{L} \sum_{k=0}^{i-1} f\left(\frac{k}{L}\right) \exp\left(i 2\pi m \frac{k}{L}\right)$$

for $m = 0, 1, \ldots, L-1$ where the spectral density is given by

$$f(w) = \sum_{m=-\infty}^{\infty} \gamma_m \exp\left(i 2\pi m w\right)$$

for $w \in [0, 1]$ and γ_m is the autocovariance of order m-th, that is estimated through s_m. Then, the spectral density function can be estimated at particular frequencies in the sample by

$$\widehat{f}\left(\frac{k-1}{L}\right) = \sum_{m=-\infty}^{\infty} s_m \exp\left(i 2\pi m \frac{k-1}{L}\right).$$

The decomposition step in CSSA proposes to perform an orthogonal diagonalization over the Circulant matrix of second moments S_C. Bógalo et al. [1] show that the eigenvalues $\widehat{\lambda}_k$ and the eigenvectors $\mathbf{u}_k$ of S_C are given by

$$\widehat{\lambda}_k \approx \widehat{f}\left(\frac{k-1}{L}\right), \tag{1}$$

and

$$\mathbf{u}_k = L^{-1/2}\left(u_{k,1}, \ldots, u_{k,L}\right)', u_{k,j} = \exp\left(-i2\pi\,(j-1)\,\frac{k-1}{L}\right). \tag{2}$$

Notice that (1) allows a direct association between the k-th eigenvalue and the frequency $w_k = \frac{k-1}{L}, k = 1, \ldots, L$.

The second variant that CSSA proposes is to make the grouping according to the desired frequencies.

Given the symmetry of the spectral density, we have that $\widehat{\lambda}_k = \widehat{\lambda}_{L+2-k}$. Therefore, to generate the elementary matrices we first form the groups of 2 elements $B_k = \{k, L+2-k\}$ for $k = 2, \ldots, M$ with $B_1 = \{1\}$ and $B_{\frac{L}{2}+1} = \left\{\frac{L}{2}+1\right\}$ if L is even. Second, notice that their corresponding eigenvectors $\mathbf{u}_k$, given by (2), are the k columns of the $\mathbf{U}$ Fourier matrix of dimension L, therefore, they are conjugated complex by pairs, $\mathbf{u}_k = \overline{\mathbf{u}}_{L+2-k}$ where $\overline{\mathbf{v}}$ indicates the complex conjugate of a vector $\mathbf{v}$, and $\mathbf{u}_k^*\mathbf{X}$ and $\mathbf{u}_{L+2-k}^*\mathbf{X}$ correspond to the same harmonic period, where $\mathbf{v}^*$ denotes the transpose conjugate of $\mathbf{u}_k$.

We compute the elementary matrix by frequency $\mathbf{X}_{B_k}$ as the sum of the two elementary matrices $\mathbf{X}_k$ and $\mathbf{X}_{L+2-k}$, associated with eigenvalues $\widehat{\lambda}_k$ and $\widehat{\lambda}_{L+2-k}$ and frequency $\frac{k-1}{L}$,

$$\begin{aligned}
\mathbf{X}_{B_k} &= \mathbf{X}_k + \mathbf{X}_{L+2-k} \\
&= \mathbf{u}_k\mathbf{u}_k^*\mathbf{X} + \mathbf{u}_{L+2-k}\mathbf{u}_{L+2-k}^*\mathbf{X} \\
&= (\mathbf{u}_k\mathbf{u}_k^* + \overline{\mathbf{u}}_k\overline{\mathbf{u}}_k^*)\mathbf{X} \\
&= 2(R_{\mathbf{u}_k}R_{\mathbf{u}_k}^{'} + I_{\mathbf{u}_k}I_{\mathbf{u}_k}^{'})\mathbf{X}
\end{aligned}$$

where the prime denotes transpose, $R_{\mathbf{u}_k}$ denotes the real part of $\mathbf{u}_k$, and $I_{\mathbf{u}_k}$ its imaginary part. In this way, all the matrices $\mathbf{X}_k, k = 1, \ldots, L$, are real.

3 Separability

As pointed out in [1], another important feature of CSSA is the strong separability of the elementary series as well as those grouped by frequencies, outperforming alternative algorithms. This characteristic is important since many signal extraction procedures assume zero correlation between their underlying components, whereas the estimated signals can be quite correlated. As [5] point out the SSA decomposition can be successful only if the resulting additive components of the series are quite separable from each other. In this section, we deepen in the concept of strong separability of CSSA by developing an example and comparing it to Basic SSA.

For a fixed window length L, given two series $\left\{x_t^{(1)}\right\}$ and $\left\{x_t^{(2)}\right\}$ extracted from the series $\{x_t\}$, we say that they are weakly separable if both their column and row spaces are orthogonal, that is, their trajectory matrices $\mathbf{X}^{(1)}$ and $\mathbf{X}^{(2)}$ are such that $\mathbf{X}^{(1)}\left(\mathbf{X}^{(2)}\right)' = \mathbf{0}_{L\times L}$ and $\left(\mathbf{X}^{(1)}\right)'\mathbf{X}^{(2)} = \mathbf{0}_{N\times N}$. Furthermore, we say that two series $\left\{x_t^{(1)}\right\}$ and $\left\{x_t^{(2)}\right\}$ are strongly separable if they are weakly separable and the two sets of singular values of the trajectory matrices $\mathbf{X}^{(1)}$ and $\mathbf{X}^{(2)}$ associated with $\left\{x_t^{(1)}\right\}$ and $\left\{x_t^{(2)}\right\}$, respectively, are disjoint. When the trajectory matrix of the original time series has not multiple singular values or, equivalently, each elementary reconstructed series belongs to a different harmonic, strong separability is guaranteed according to the previous definition.

We measure separability in terms of **w**-correlation ([5] and [6]), that it is given by

$$\rho_{12}^{w} = \frac{\left\langle \mathbf{x}^{(1)}, \mathbf{x}^{(2)} \right\rangle_w}{\left\| \mathbf{x}^{(1)} \right\|_w \left\| \mathbf{x}^{(1)} \right\|_w},$$

where $\left\langle \mathbf{x}^{(1)}, \mathbf{x}^{(2)} \right\rangle_w = (\mathbf{x}^{(1)})'\mathbf{W}\mathbf{x}^{(2)}$ is the so-called w-inner product and $\left\| \mathbf{x}^{(1)} \right\|_w = \sqrt{\left\langle \mathbf{x}^{(1)}, \mathbf{x}^{(1)} \right\rangle_w}$ and $\mathbf{W} = diag(1, 2, \ldots, \underbrace{L, \ldots, L}_{T-2(L-1)\text{ times}}, \ldots, 2, 1)$. Note that the window length L enters the definition of **w**-correlation. We are interested in producing components with **w**-correlation (ideally) zero because, in this case, we can conclude that the component series are **w**-orthogonal, i.e., $\left\langle \mathbf{x}^{(1)}, \mathbf{x}^{(2)} \right\rangle_w = 0$ and separable (see, [5]). To quickly check how separable are the component series when performing SSA, we will plot the matrix of the absolute values of the **w**-correlations for all the component series, coloring in white the absence of **w**-correlation, in black **w**-correlations in absolute value equal to 1 and in a scale of grey colors, the remaining intermediate values.

3.1 An Illustration of Separability with SSA and CSSA

Circulant SSA produces components that are strongly separable. The main argument is that the real eigenvectors $\sqrt{2}R_{\mathbf{u}_k}$ and $\sqrt{2}I_{\mathbf{u}_k}$ (linked to eigenvalues λ_k and λ_{L+2-k}, respectively, $\lambda_k = \lambda_{L+2-k}$) are orthogonal and have information associated only with frequency $\frac{k-1}{L}$. Those are the only eigenvectors that have information related to this frequency. As eigenvectors can be considered filters [7] these pair of eigenvectors extract elementary series linked to the same frequency without mixing with harmonics of other frequencies. As a result, the two elementary series, when reconstructed in step 4, have spectral correlation close to 1 between them and close to zero with the remaining ones. Taking into account the pairs of reconstructed series per frequency, any grouping of the reconstructed series results in disjoint sets from the point of view of the frequency. Then, Circulant SSA produces components that are approximately strongly separable. In this case, the graph of the **w**-correlation matrix is colored in black in the main diagonal and in white elsewhere as in the ideal case.

To illustrate the separability of Circulant SSA, consider the following example. Let $x_t = x_t^{(1)} + x_t^{(2)}$, where $x_t^{(n)} = A_n \sin(2\pi w_n t)$, $n = 1, 2$ for a sample size $T = 181$. The sine components have different frequency, $w_1 = 1/45$ and $w_2 = 1/10$, but the same amplitude $A_1 = A_2 = 1$. Figure 1 shows the two basic sinusoid components and their sum.

We perform Circulant and Basic SSA with a window length L equal to 90, that is multiple of 45 and 10. Figure 2 shows that the eigenvalues of both, Circulant and Basic SSA are almost identical. In Basic SSA eigenvalues are ordered in a decreasing way, while in CSSA, they are shown related to the frequencies that they identify and that makes easier the spectral identification of the components of the series x_t.

Figures 3 and 4 show the differences between both procedures. Figure 3 shows the w-correlations of the 4 reconstructed elementary series, corresponding to the eigenvalues ordered in a decreasing way. In CSSA we can observe that each pair of elementary reconstructed series corresponding to the same harmonic has w-correlation very close to 1 and are w-orthogonal to the rest.

On the other hand, Fig. 4 shows the scatterplots of the eigenvectors. On the contrary, Basic SSA shows higher w-correlations between the reconstructed series and more mixed patterns in their scatterplots.

In Fig. 5 we plot the component series extracted by Circulant SSA while in Fig. 6 we plot the component series extracted by Basic SSA. Notice that in Basic SSA we cannot separate the two sine series since two elementary components (top left and bottom right in Fig. 6) are quite mixed and do not correspond to a particular frequency (either 1/45 or 1/10). On the contrary, the elementary series extracted from Circulant SSA have only information associated with a particular frequency, either 1/45 or 1/10 (see Fig. 4). This result could have been advanced given the **w**-correlation matrices shown in Fig. 3.

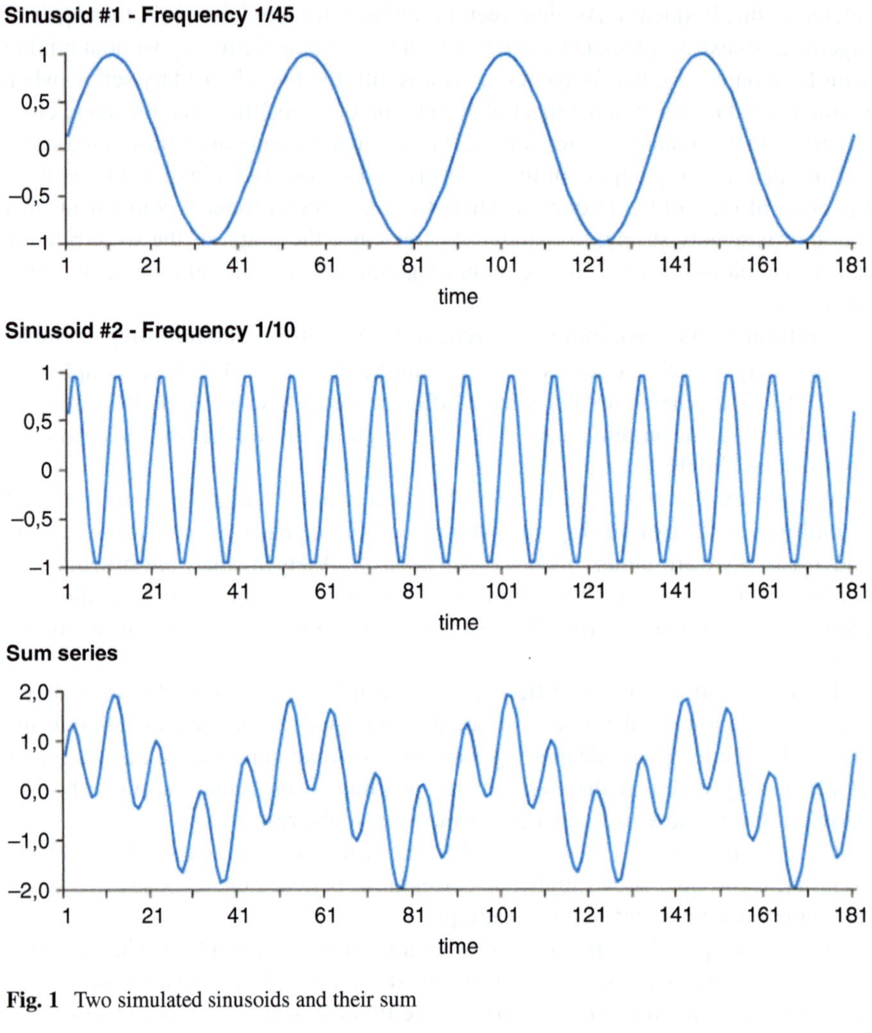

Fig. 1 Two simulated sinusoids and their sum

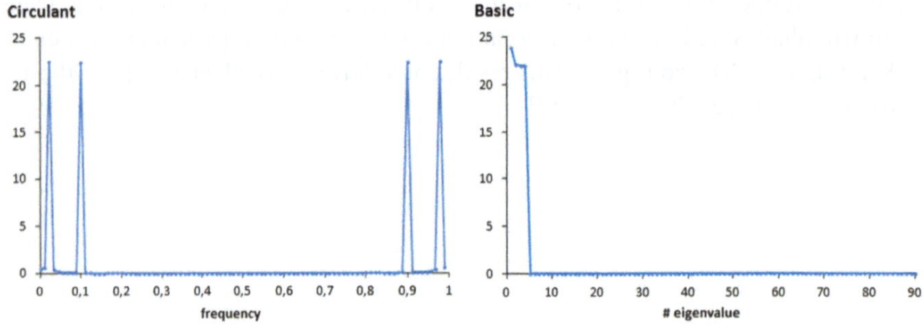

Fig. 2 Eigenvalues obtained through Circulant and Basic SSA

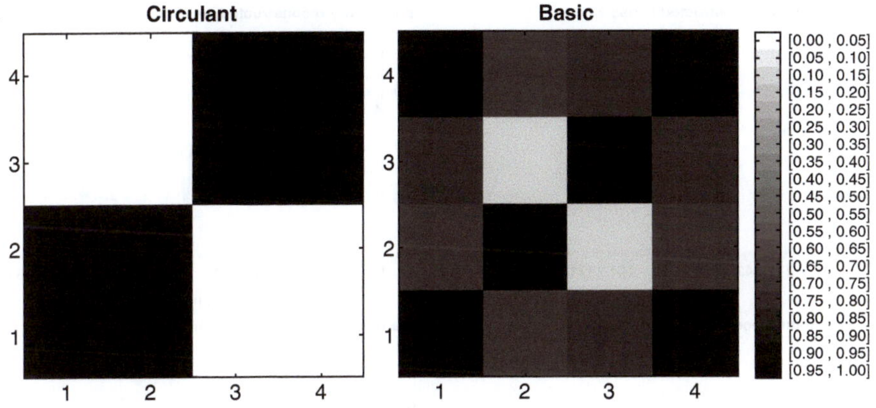

Fig. 3 W-correlations of the reconstructed elementary components obtained through Circulant (left) and Basic (right) SSA

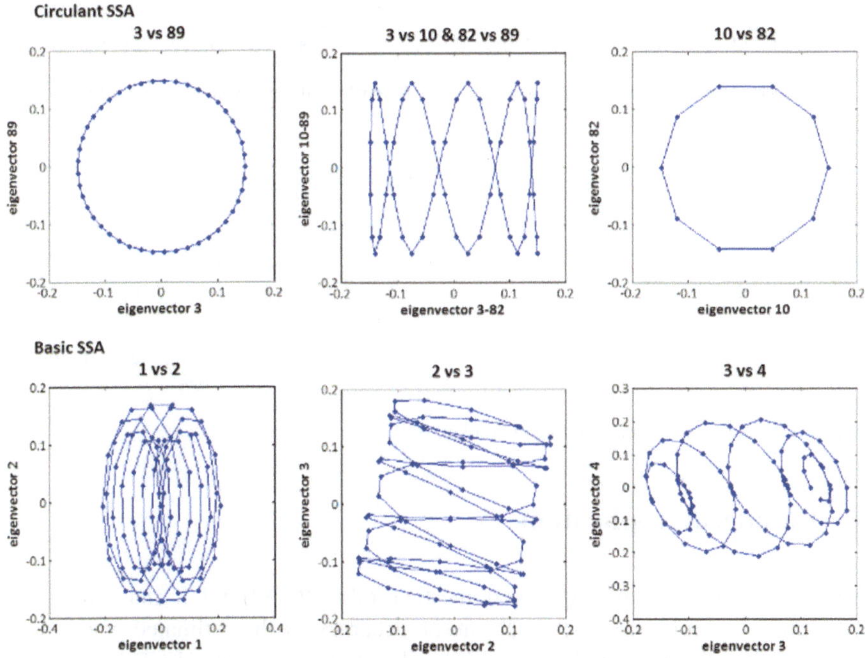

Fig. 4 Plots of pairs of eigenvectors

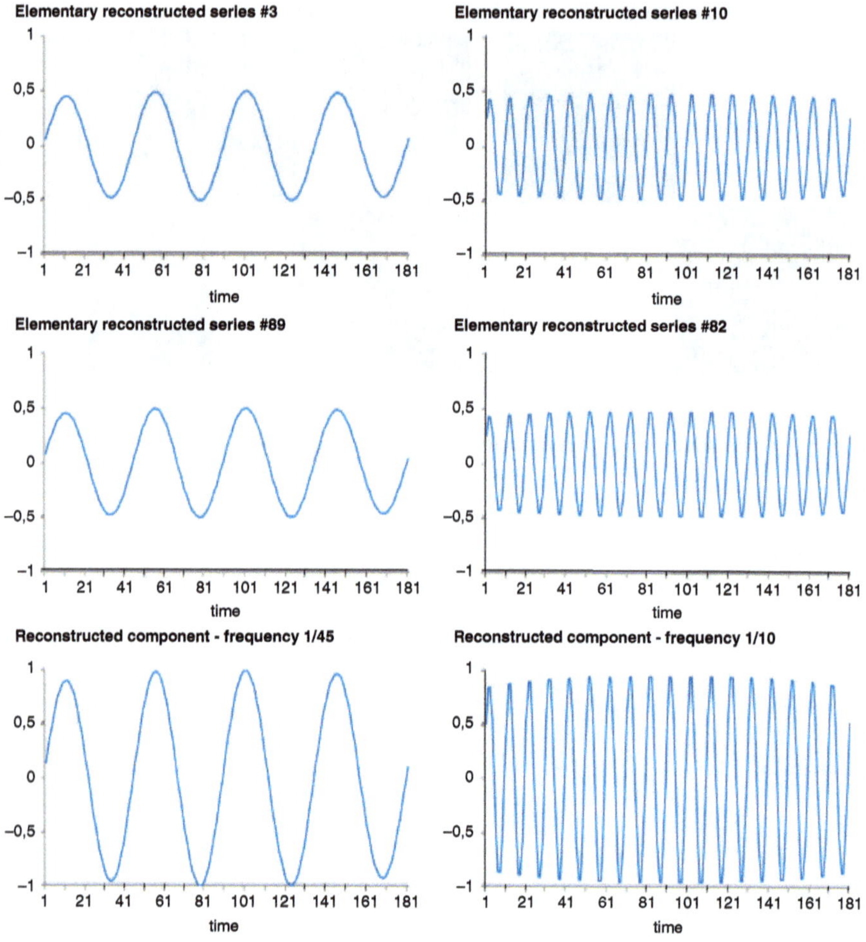

Fig. 5 Components extracted through Circulant SSA

4 Empirical Application

We apply CSSA to quarterly gross domestic product (GDP) in UK and in the USA. Our aim is twofold, first we check on the accuracy of the estimated components by comparing the estimated cycle with the dated recessions by OECD. Second, we check on the strong separability of the components. We pay special attention to the presence of any remaining seasonality in the seasonally adjusted time series, as it is one of the most followed indicators for real time monitoring of economic activity.

Data for UK GDP are taken from the UK Office for National Statistics (www.ons.gov.uk). Data are quarterly original volume chain index numbers where the value of the index for 2010 is 100. Data for US GDP are taken from the US Bureau of

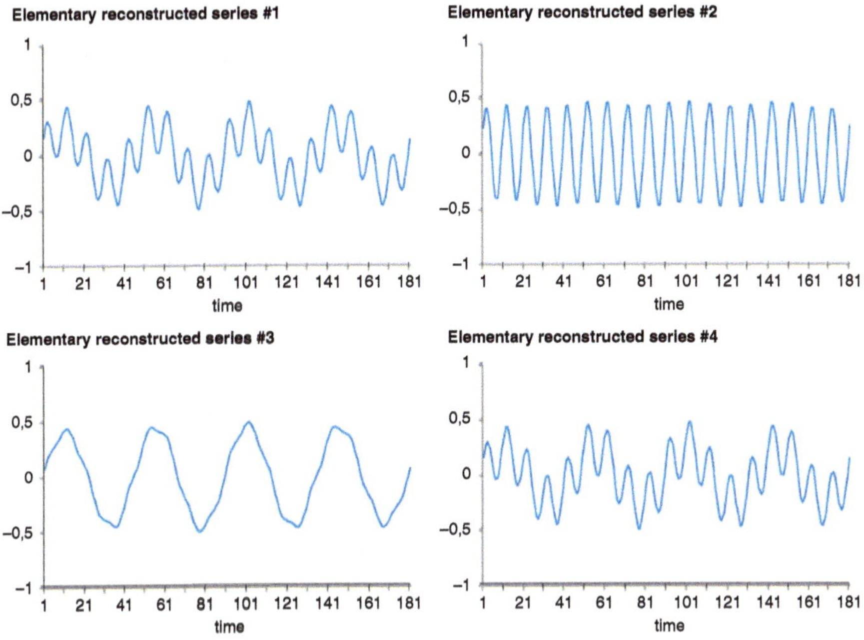

Fig. 6 Components extracted through Basic SSA

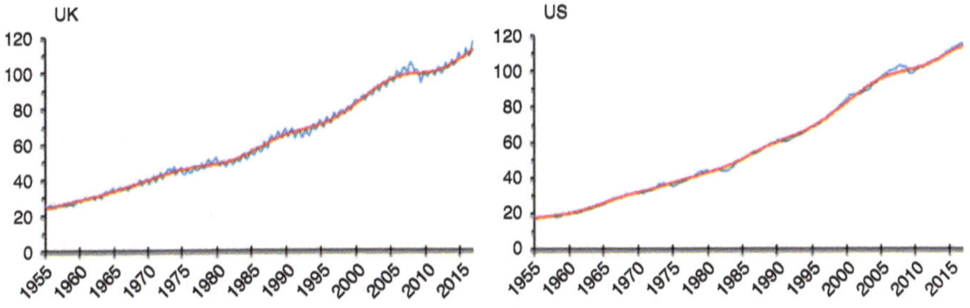

Fig. 7 GDP and estimated trend for US and UK

Economic Analysis (www.bea.gov). US data are only published seasonally adjusted and are index numbers where the value of the index for 2009 is 100. The sample for both indicators goes from the first quarter in 1955 to the last quarter in 2016, therefore we have $T = 248$ observations. Figure 7 shows US and UK GDP.

To select the window length L, we consider a value that is between $T/4$ and $T/3$ and that is multiple of the cycle frequency. In this case, taking into account that business cycle frequencies range between one year and a half and 8 years (32 quarters), we consider $L = 64$. Longer oscillations (16 years, 64 quarters) will be assigned to the trend component.

Table 1 Signal contribution on GDP

	UK	US
Trend	91.79	92.56
Cycle	6.85	6.47
Seasonal	0.35	0.07
Total	98.99	99.09

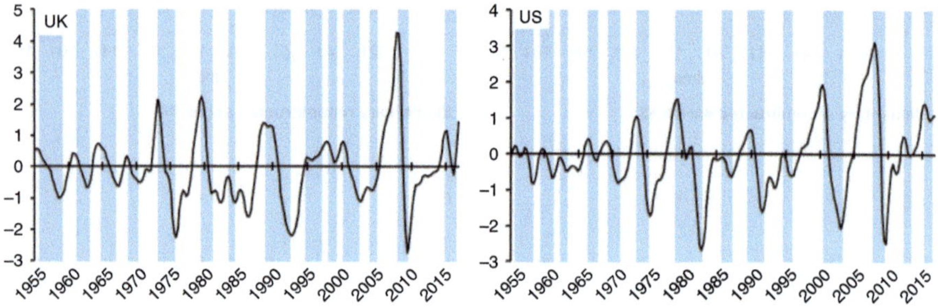

Fig. 8 Estimated GDP cycle for the USA and UK and OECD recession dates (shadowed area)

Let $B_k = \{k, L - k + 2\}$ the elementary pair associated with the frequency $w = \frac{k-1}{L}$. The group associated with the trend will be $I_{trend} = \{B_1, B_2\}$ that corresponds with frequencies 0 and 1/64. The group associated to the cycle will be $I_{cycle} = \{B_3, B_4, B_5, B_6, B_7, B_8, B_9, B_{10}, B_{11}\}$ that corresponds to frequencies 2/63, 3/64, 4/64, 5/64, 6/74,7/64, 8/64, 9/64, and 10/64, while the group associated with the seasonal component is $I_{seasonal} = \{B_{17}, B_{33}\}$, corresponding to the frequencies 1/4 and 1/2.

Figure 7 shows the estimated trend and Table 1 shows the contribution of the different signals to the original data. The adjustment, measured by the contribution of the irregular component in terms of the original corresponding GDP, shows that noise variance is not greater than 1% of that of the original series in both geographical areas. The trend is the signal with greatest contribution, around 92%, the cycle contributes with a little more than 6% and seasonality is low in UK (0.35%) and residual in the USA where the original data were already seasonally adjusted.

Figure 8 shows the business cycle estimated by CSSA and the OECD dated recessions. As it can be seen there is great coincidence except two small differences in UK. The differences occur in 1983, when the OECD dates a shorter recession than CSSA, and 1995 when, on the contrary, the OECD dates a longer recession than CSSA.

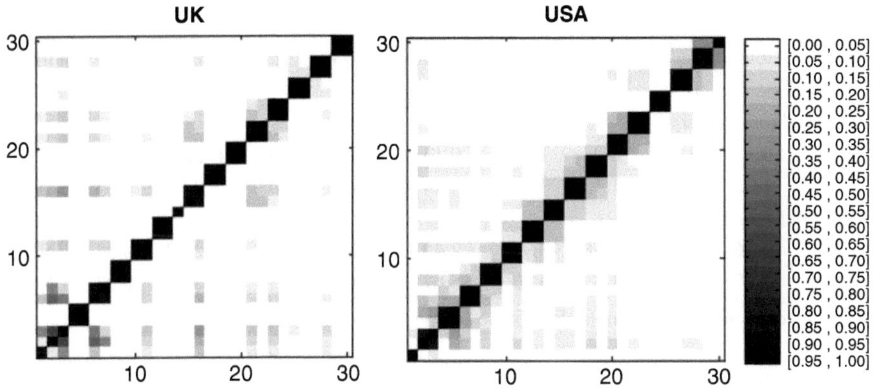

Fig. 9 w-Correlation matrix for the elementary reconstructed series for the 30 greatest eigenvalues

4.1 *Seasonality*

Figure 9 represents w-correlation matrices for the elementary reconstructed series corresponding to the 30 greatest eigenvalues in decreasing order. As it is expected, the plot shows that w-correlations corresponding to elementary reconstructed time series of the group B_k are close to one and generate black diagonal of blocks 2×2. Figure 9 also shows the absence of cross w-correlations with the rest of the reconstructed series. The results also hold for the seasonal frequencies. Notice that in CSSA the eigenvectors of the Circulant matrices do not change for a given window length L so the revisions of the data do not affect so much the results. When seasonally adjusted data are published, they are adjusted for a given time span or published data and, varying the data or the time span can change the conclusions about residual seasonality on the adjusted series.

Findley et al. [4] apply different seasonal residual tests to seasonally adjusted time series. They find that most of the diagnostics must be applied to a subspan (preferably including the most recent data) of the seasonally adjusted series for best residual seasonality detection, but this raises doubts about the idempotency. By that we mean that perhaps the tests do throw different conclusions if applied to the full sample or to some subsamples.

We check the presence of seasonality before and after the adjustment of seasonality to the time series by the combined seasonality test used in X-13 ARIMA-SEATS, [8]. The aim of the combined seasonality test is to determine whether the seasonality of the series can be identifiable. It uses the so-called SI (seasonal-irregular) ratios, calculated as the ratio of the original series to the estimated trend. In other words, SI ratios are estimates of the detrended series. This test comprises four combined tests: stable seasonality (F_S statistic), evolutive seasonality (F_M statistic), identifiable seasonality (statistics $T_1 = \frac{7}{F_S - F_M}$ and $T_2 = \frac{3F_M}{F_S}$), and finally a Kruskal-Wallis

Table 2 Combined seasonal test

	Stable seas.		Evolving seas.		Identifiable seas.			Kruskal-Wallis		I/NI[a]
	F_S	p-val	F_S	p-val	T	T_1	T_2	W	p-val	
Original time series										
UK	189.1	0.00	0.23	1.00	0.14	0.04	0.00	175.8	0.00	I
US	1.02	0.38	1.73	0.00	2.44	6.84	5.08	2.18	0.54	NI
Seasonal adjusted time series										
UK	0.76	0.52	0.58	0.99	2.40	9.24	2.31	2.18	0.54	NI
US	0.20	0.89	1.87	0.00	5.67	35.9	28.3	0.19	0.98	NI

[a]Stands for identifiable and non-identifiable seasonality

test to check the average values of the different seasons. In all the tests, the null hypothesis is the absence of seasonality in the series.

Table 2 shows the combined seasonality tests, and their corresponding p-values, applied both to the original GDP series and to the seasonally adjusted series obtained with Circulant SSA. The top panel is referred to the original series and the bottom panel to the seasonally adjusted series. The consequence is clear: the only series that presents identifiable seasonality is the original UK GDP series, as expected.

5 Concluding Remarks

Circulant SSA is an automated version of SSA that allows the association between desired frequencies and extracted components. This contribution deepens on the property of strong separability between the extracted components. It illustrates the different performance between CSSA and SSA by means of a simulated example finding that the signals extracted by CSSA show smaller w-correlations. One important consequence of the strong separability of the components extracted with CSSA is that the seasonal component should be w-uncorrelated with the remaining components. This means that the adjusted series from seasonality should be clean from seasonal variation. Residual seasonality has been identified in the literature as the most fundamental seasonal adjustment deficiency. In this work, we also test for the absence of residual seasonality with an empirical application to the CSSA seasonally adjusted series obtained from the quarterly GDP series for UK and the USA, showing that the seasonally adjusted series through CSSA pass the tests used in the literature for detecting residual seasonality and, therefore, can be considered clean from seasonality.

Acknowledgements Financial support from the Spanish Ministry of Economy and Competitiveness, project numbers ECO2015-70331-C2-1-R, ECO2015-66593-P, and ECO2016-7618-C3-3-P and Universidad de Alcalá is acknowledged.

The views expressed in this work are those of the authors and should not be attributed to the European Commission.

References

1. Bógalo, J., Poncela, P., & Senra, E. (2017). *Automatic Signal Extraction for Stationary and Non-Stationary Time Series by Circulant SSA*. MPRA Paper 76023. https://mpra.ub.uni-muenchen.de/76023/
2. Burman, J. P. (1980). Seasonal adjustment by signal extraction. *Journal of the Royal Statistical Society. Series A (General), 143*, 321–337.
3. Dagum, E. B. (1978). Modelling, forecasting and seasonally adjusting economic time series with the X-11 ARIMA method. *Journal of the Royal Statistical Society. Series D (The Statistician), 27*(3–4), 203–216.
4. Findley, D. F., Lytras, D. P., & McElroy, T. S. (2017). *Detecting seasonality in seasonally adjusted monthly time series*. Research Report Series Statistics 2017-03. US Census Bureau.
5. Golyandina, N., Nekrutkin, V., & Zhigljavsky, A. (2001). *Analysis of time series structure: SSA and related techniques*. London: Chapman & Hall.
6. Golyandina, N., & Zhigljavsky, A. (2013). *Singular spectrum analysis for time series*. Berlin: Springer.
7. Kume, K. (2012). Interpretation of singular spectrum analysis as complete eigenfilter decomposition. *Advances in Adaptive Data Analysis, 4*(4), 1250023, 1–18.
8. Lothian, J., & Morry, M. (1978). *A Test for the Presence of Identifiable Seasonality When Using the X-11-Arima Program*. Research Paper, Seasonal Adjustment and Time Series Analysis Staff, Statistics Canada.
9. Moulton, B. R., & Cowan, B. D. (2016). Residual seasonality in GDP and GDI: Findings and next steps. *Survey of Current Business, 96*(7), 1–6.

Selection of Window Length in Singular Spectrum Analysis of a Time Series

P. Unnikrishnan and V. Jothiprakash

Abstract Singular Spectrum Analysis (SSA) is a promising non-parametric time series modelling technique that has proved to be successful in data preprocessing in diverse application fields. It is a window length-based method and the appropriate selection of window length plays a crucial role in the accuracy of SSA. However, there are no specific methods depicted in the literature about its selection. In this study, the method of SSA in time series analysis is presented in detail and a sensitivity analysis of window length is carried out based on an observed daily rainfall time series.

1 Introduction

Time series modelling is the study of the temporally arranged data or development of a model for prediction where time is an independent variable [7]. The complexity of the time series model depends on the characteristics of the available time series data [1]. According to the complexity and characteristics of the underlying physical process of the time series, various empirical, conceptual, physically based and data-driven models have been developed in the field of hydrology [9]. Famous time series models in the field of hydrology include conventional stochastic models and data-driven techniques. Recently, models like SSA and wavelet model that works on the internal structure of the time series has found its application in time series modelling. SSA is a model-free and non-parametric time series analysis technique that has proved to be very successful in preprocessing the data by eliminating the unwanted noise [2, 4]. It combines classical time series analysis, multivariate statistics, multivariate geometry, signal processing and dynamical systems [4]. It decomposes the time series into various small subcomponents. The data adaptive nature of the basis functions used in SSA gives the method significant strength

P. Unnikrishnan (✉) · V. Jothiprakash
Indian Institute of Technology, Bombay, India
e-mail: vprakash@iitb.ac.in

P. Bertail et al. (eds.), *Nonparametric Statistics*, Springer Proceedings in Mathematics & Statistics 250, https://doi.org/10.1007/978-3-319-96941-1_21

over classical spectral methods and makes the approach suitable for analysis of some non-linear dynamics [4]. Unlike stochastic models, there is no assumption of stationarity in the method of SSA. SSA can be successfully applied in diverse time series analysis and modelling areas such as extraction of various components of the time series, change point detection, gap filling, image processing, elimination of noise from signal, etc. Window length is the only parameter in the method of SSA. The entire decomposition of the time series will depend on the window length. Thus, it is very important to select window length accurately. In the present study, the importance of window length has been studied and a sensitivity analysis of window length has been carried out on an observed time series. The daily rainfall data of Koyna catchment, Maharashtra, India for 52 years has been utilized in the study. The detailed description of the method of SSA can be seen in next section.

2 Singular Spectrum Analysis (SSA)

Singular Spectrum Analysis involves two major stages: decomposition and reconstruction. Decomposition includes two steps: embedding and Singular Value Decomposition (SVD). Reconstruction phase involves two steps: grouping and diagonal averaging.

2.1 Embedding

It is the stage in SSA where the time series is converted into a matrix upon which multivariate statistics can be carried out. If Y is the time series upon consideration, the embedding stage transfers it into a trajectory matrix X, based on a window length chosen for the time series. A trajectory matrix is a Hankel matrix (anti-diagonals are equal) with dimension L x K, where L is the window length and K is the lag parameter, K =N-L+1, where N is the time series length. If L is the window length and N is the time series length, the embedding stage can be described as below: Let Y be the time series upon consideration

$$Y = y_1, y_2, y_3 \ldots . y_N \tag{1}$$

$$X_i = y_1, y_2, y_3, \ldots y_i + L - 1 \tag{2}$$

Then trajectory matrix is given as:

$$X = [X_1 X_2 X_3 \ldots X_K] \tag{3}$$

2.2 Singular Value Decomposition (SVD)

Singular Value Decomposition (SVD) of the trajectory matrix of time series is the core stage of SSA in which the number and form of decomposed components depend on the window length utilized for decomposition. SVD transforms the matrix into product of three matrices together called Eigen Triple (ET): left singular vector or eigen function (eigen vectors of XX^T), diagonal matrix of eigen values, transpose of right singular vector or principal component (eigen vector of $X^T X$).

$$X_{LxK} = U_{Lxl} D_{LxK} V^*_{KxK} \tag{4}$$

UDV* is called eigen triples, where X is the trajectory matrix, U is the left orthogonal vector system (eigen function), V is the right orthogonal vector system (principal component) and D is the diagonal matrix containing the square roots of eigen values written in descending order. Thus, the trajectory matrix (X) can be interpreted as the sum of d matrices where d is the rank of the trajectory matrix X.

$$X = \sum_{i=1}^{d} X_i \tag{5}$$

$$X_i = U_i \sqrt{\lambda} V_i \tag{6}$$

$\lambda =$ eigen value of$XX^T, \quad i = 1, 2, 3, \cdots d, \quad d = max(i : \sqrt{\lambda_i} > 0).$

2.3 Grouping

Grouping is the stage in SSA where the different components of time series are identified and selected for reconstruction. It is the procedure of arranging matrix terms X_i in Eq. (5). Basically, the method of grouping divides the set of indices 1,. . .d into m disjoint subsets $I_1, I_2, \ldots .I_m$.

$$I = \{i_1, i_2, \ldots ..i_p\} \tag{7}$$

Then, the resultant matrix X_I corresponding to the group I is defined as:

$$X_I = X_{i1} + X_{i2} + \ldots\ldots X_{ip} \tag{8}$$

$$X = X_{I1} + X_{I2} + \ldots\ldots X_{Im} \tag{9}$$

There is no hard and fast rule in the method of grouping. Based on the eigen values, eigen functions and principal components, the different components of the time series need to be identified.

2.4 Diagonal Averaging

In this stage, each matrix X_{Ij} from the grouping stage is transformed to time series of length N. This corresponds to averaging the matrix elements over the anti-diagonals i+j=k+1.

3 Significance and Selection of Window Length

SSA is a window length-based method of time series decomposition that will eliminate the problem of the variation in the frequency behaviour of the time series along the time axis by means of breaking the long-time axis into various time segments. Window length (L) is the only parameter in SSA and it decides the form of the trajectory matrix. Its adequate choice is important in getting better results. The choice of L represents a compromise between information content and statistical confidence. Too small window length can cause mixing of interpretable components and too big window length produce undesirable decomposition of components. There is no universal rule in selecting the window length. However, there are some recommendations in selecting the window length [3]. The general recommendations for its selection are given below:

1. Window length should be less than half of the time series length.
2. If the time series has a periodicity component with known period, it would be better to take a window length proportional to this period.
3. If the window length is relatively big, the results are stable under the small perturbations of the window length.
4. Too small window length would cause mixing of interpretable components.
5. Too big window length may produce undesirable decomposition of components.

4 Sensitivity Analysis of Window Length: A Case Study

In the present study, in order to carry out the sensitivity analysis of window length, a daily rainfall time series has been utilized as a case study. The daily rainfall data pertaining to Koyna catchment in Maharashtra, India from 1st January 1961 to 31st December 2013 has been collected from Koyna Irrigation division office, Government of Maharashtra, India. The details of the study area have been given in Jothiprakash and Magar [6] and Unnikrishnan and Jothiprakash [10]. The time series

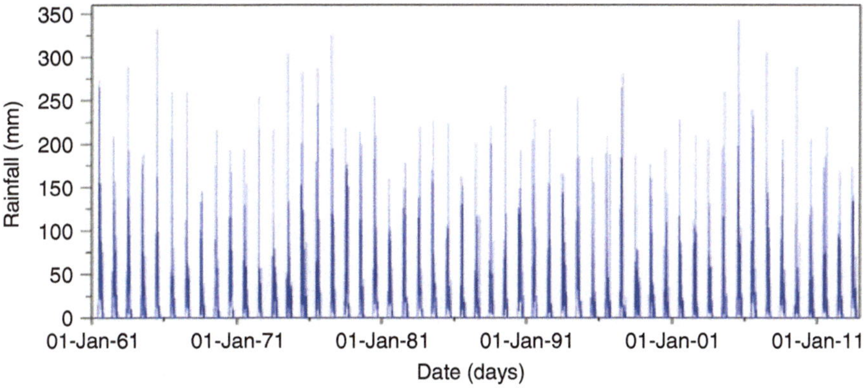

Fig. 1 Daily rainfall of Koyna catchment from 01/01/1961 to 31/12/2013

plot of the Koyna rainfall for the time period of 1st January 1961 to 31st December 2013 is given in Fig. 1. A time series plot is the plot that shows the variation of the given variable with respect to time. In the present case study, the time scale is daily and the variable is rainfall. Figure 1 shows the pattern of the rainfall that repeats every year, with positive rainfall during monsoon season (June–October) and zero rainfall during non-monsoon season (November–May). This repetitive behaviour of the daily rainfall can be accounted for the prevailing periodicity component present in the series. A time series is said to contain a periodic component of period 'T', when the pattern of the series repeats in every 'T' interval of time.

Autocorrelation can be defined as the correlation between the pair of values of the process separated by an interval of length 'k', and the parameter 'k' is generally known as 'lag' [8]. The autocorrelation of a time series 'y_t' for a lag 'k' can be defined as below: As the covariance is independent of time, it can be written as:

$$ACF(k) = \frac{E[(y_t - \mu)(y_{t+k} - \mu)}{\sqrt{Var[y_t]Var[y_{t+k}]}} \tag{10}$$

where ACF(k) is the ACF for a lag 'k' of the time series y_t and y_{t+k} is the time series lagged by 'k' times. ACF can reveal the internal structure of the time series and can be interpreted as the measure of similarity between a realization of the time series (Y(t)) and the same realization shifted by 'k' units [8]. Thus, ACF plot can be used to detect the presence of periodic component in the time series, as it reflects the repetitive nature of the periodic component.

The ACF plot of Koyna daily rainfall series up to a lag of 1000 days is given in Fig. 2. The pattern of ACF plot repeats at every 365 lags. This repetition of pattern of ACF plot indicates the presence of a periodic component with period of 365 days in the time series. Thus, as per recommendation in selection of window length as explained in the previous section, it may be advantageous if we are selecting multiples of 365 as window length. In the present study, in order to carry out

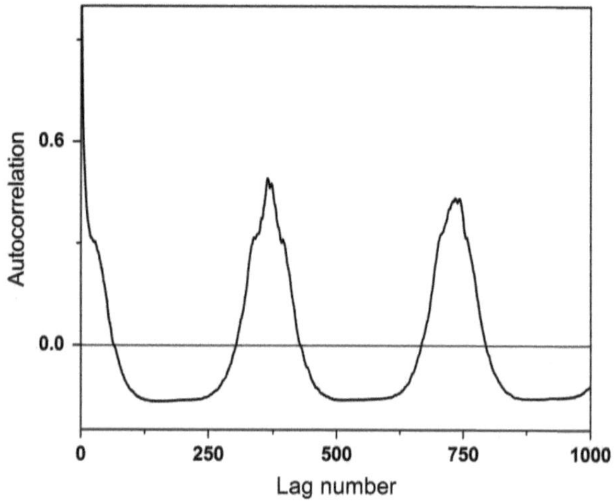

Fig. 2 Autocorrelation function of Koyna daily rainfall data up to a lag of 1000 days

sensitivity analysis in selection of window length, various window lengths have been chosen and the performance of SSA in decomposition is compared. The window length in days chosen are 50, 100, 365, 1000 and 2000 (two values below 365 and two above 365). As discussed in the earlier section, upon SVD of the trajectory matrix, the trajectory matrix is decomposed into product of 'L' number of eigen triples where 'L' is the window length adopted for decomposition. Eigen triples includes eigen function, square root of eigen values and principal components. Each of the eigen function and principal components contains 'L' and 'K' elements, respectively. Hence, each eigen functions and principal components can be treated as a time series. Thus, the choice of window length heavily affects not only the form of trajectory matrix but also that of eigen functions and principal components. Phase space plot of paired eigen functions (one eigen function to the next adjacent eigen function) gives the pictorial representation of the decomposition of the time series. This 2D paired plot between eigen function(t) and eigen function(t+1) will show how the observations can be grouped together for the present decomposition and will give a significant insight to the data analyst about the variation in the data [5]. The 2D paired plots of first 10 eigen functions corresponding to SSA decomposition for various window lengths (50, 100, 365, 1000 and 2000 days) of Koyna daily rainfall data are given in Figs. 3, 4, 5, 6 and 7, respectively.

The 2D paired plots for various window lengths show that for window length 50 as well as for 100, the time series is under-decomposed, which can be interpreted from the open scatter plots of the first few eigen functions shown in Figs. 3 and 4. The 2D paired plots for window lengths 1000 and 2000 (Figs. 6 and 7) show that a number of data points in the same eigen function are sharing the same values which implies a possible over-decomposition of the time series. For more detailed analysis, periodograms of eigen functions for all the decompositions have been used.

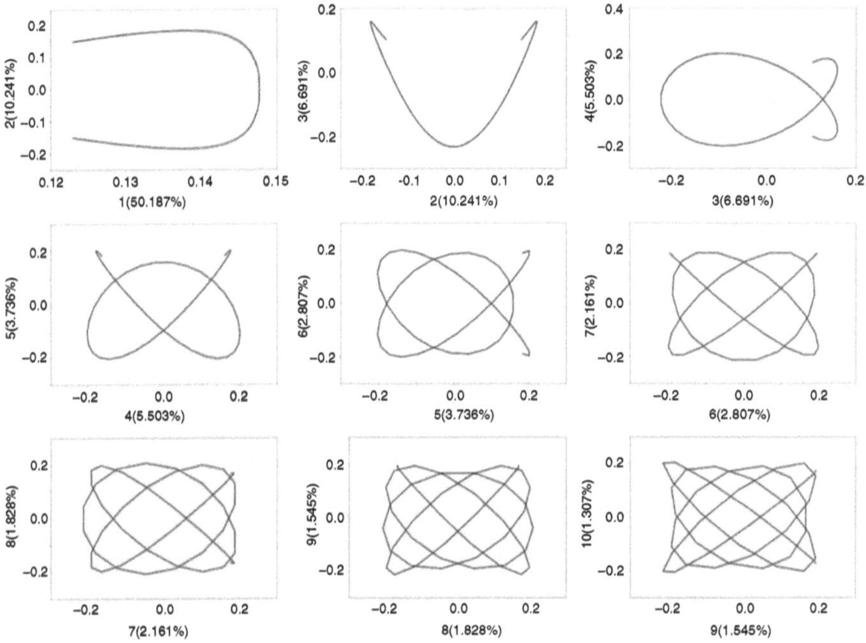

Fig. 3 Paired plot of first 10 eigen functions, SSA decomposition window length=50 days

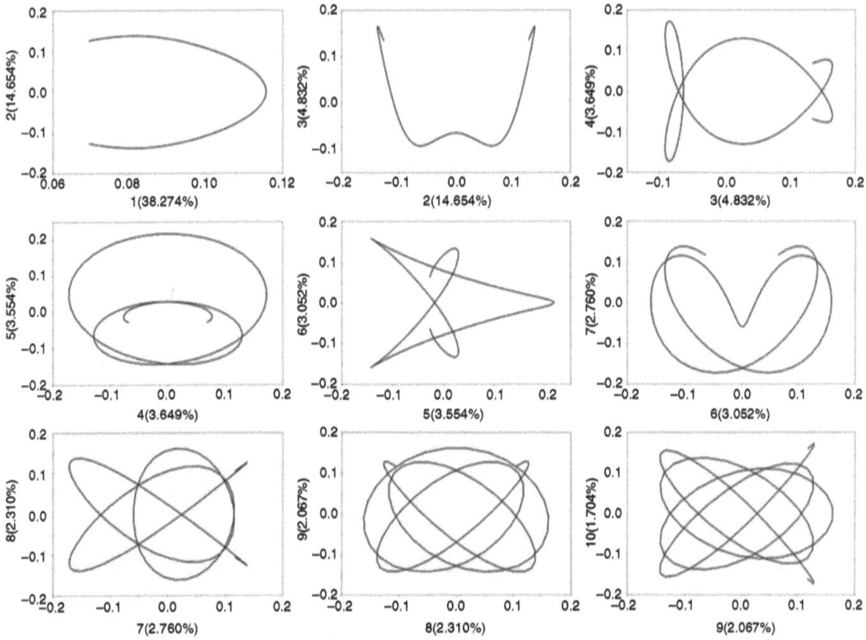

Fig. 4 Paired plot SSA decomposition window length=100 days

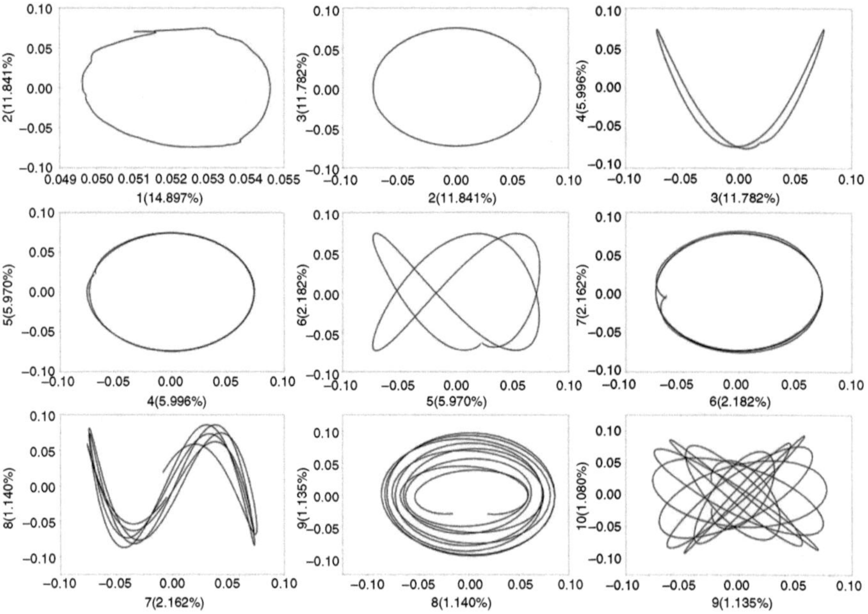

Fig. 5 Paired plot SSA decomposition window length=365 days

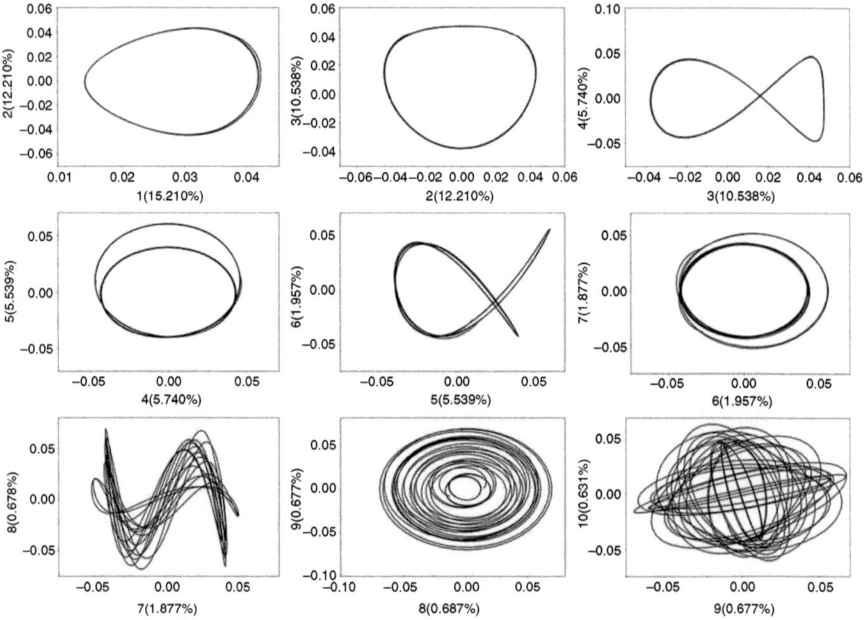

Fig. 6 Paired plot SSA decomposition window length=1000 days

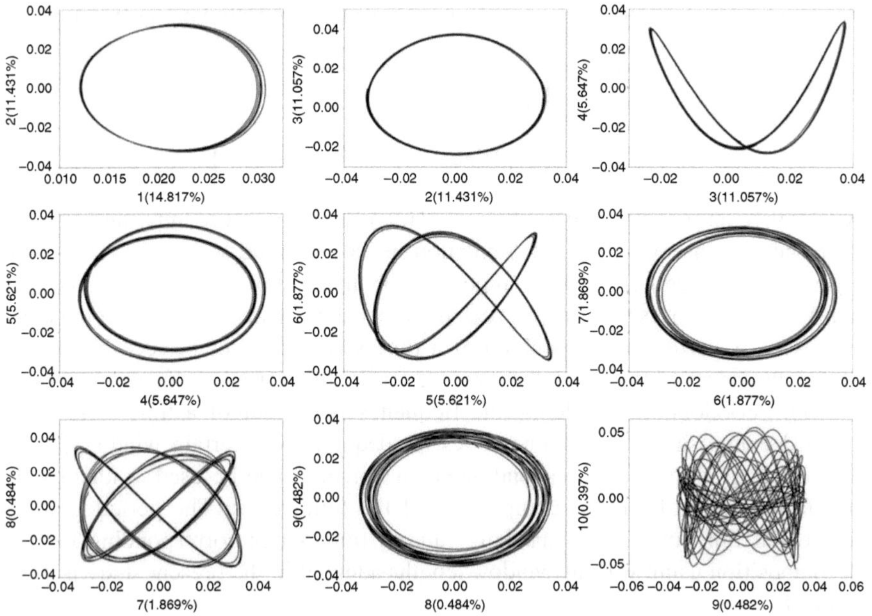

Fig. 7 Paired plot SSA decomposition window length=2000 days

Periodogram can be defined as an estimate of spectral density of a time series and can be used to identify the prevailing periods (or frequencies) of a time series. The periodogram gives a measure of the relative significance of the frequency values present in the time series that might explain the oscillation pattern of the observed data. If the time series under consideration is having 'N' number of observations, then the time series Y_t can be written as the following Fourier representation:

$$Y_t = \sum_{k=0}^{N/2} (a_k \cos \omega_k t + (b_k \sin \omega_k t) \tag{11}$$

where $\omega_k = 2\pi k/N$, k=0,1..N/2, are Fourier frequencies and

$$a_k = \left\{ \begin{array}{ll} \frac{1}{N}\sum_{t=1}^{N} Y_t \cos \omega_k t, & \text{k=0, k=N/2} \\ \frac{2}{N}\sum_{t=1}^{N} Y_t \cos \omega_k t, & \text{k=1,2,..[N-1/2]} \end{array} \right\} \tag{12}$$

$$b_k = \frac{2}{N} \sum_{t=1}^{N} Y_t \sin \omega_k t, k = 1, 2, ..[N - 1/2] \tag{13}$$

The periodogram of the time series $I(\omega_k)$ is given by the following expression:

$$I(\omega_k) = \left\{ \begin{array}{cc} na_0^2, & \text{k=0} \\ \frac{n}{2}(a_k^2 + b_k^2), & \text{k=1,2,..[(n-1)/2]} \\ Na_{N/2}^2, & \text{k=N/2} \end{array} \right\} \tag{14}$$

A relatively large value of periodogram ($I(\omega_k)$) indicates more importance for the frequency ω_k in explaining the oscillation in the observed series. As each of the eigen functions and principal components obtained after SVD of the trajectory matrix are time series themselves, periodogram can be defined for each of the eigen functions. A perfect decomposition of a time series involves perfect distribution of the frequency domain of the time series in the various components (eigen triples). Periodogram can define the frequency behaviour of a time series, thus if the decomposition of the time series is adequate by a certain window length, the periodogram of the corresponding eigen functions can be used to identify the frequency range of various components of the time series. The periodogram of first eigen functions of Koyna daily rainfall time series corresponding to SSA decomposition using various window lengths adopted in the present study is given in Fig. 8.

The periodogram of first eigen function corresponding to window lengths 50 and 100 days (Fig. 8a, b) shows that the frequency range of the non-zero periodogram values is too wide, implying mixing up of components, whereas those corresponding to higher window length (1000 and 2000 days) decomposition (Fig. 8d, e), the frequency range of non-zero periodogram values is too narrow to extract any component implying over-decomposition which result in undesirable components. Also, the periodogram values are also very less in the order of 10^{-6} and 10^{-4} for decomposition using window lengths 2000 and 1000 days, respectively (Fig. 8d, e). It is observed that for window length of 365 days (Fig. 8c), the decomposition is good, decomposing desirable components, and the frequency range of the periodogram is adequate for effective selection of components. The periodogram of other (higher) eigen functions for the various window lengths have also been analysed and also showed a similar implication of wider non-zero frequency range for 50 and 100 and narrow non-zero periodogram frequency range for window lengths 1000 and 2000. In case of SSA decomposition by window lengths of 1000 and 2000 days, the non-zero periodogram values of higher eigen functions are even lesser than that of first eigen function. Thus, bigger window length can result in over-decomposition of the time series and smaller window length can cause under-decomposition and mixing up of components. Selection of appropriate window length is very much essential for proper decomposition and extraction of various components of the time series. A window length equal to the period of periodic component in the time series will be an adequate choice of window length for time series analysis and extraction of various components of the time series.

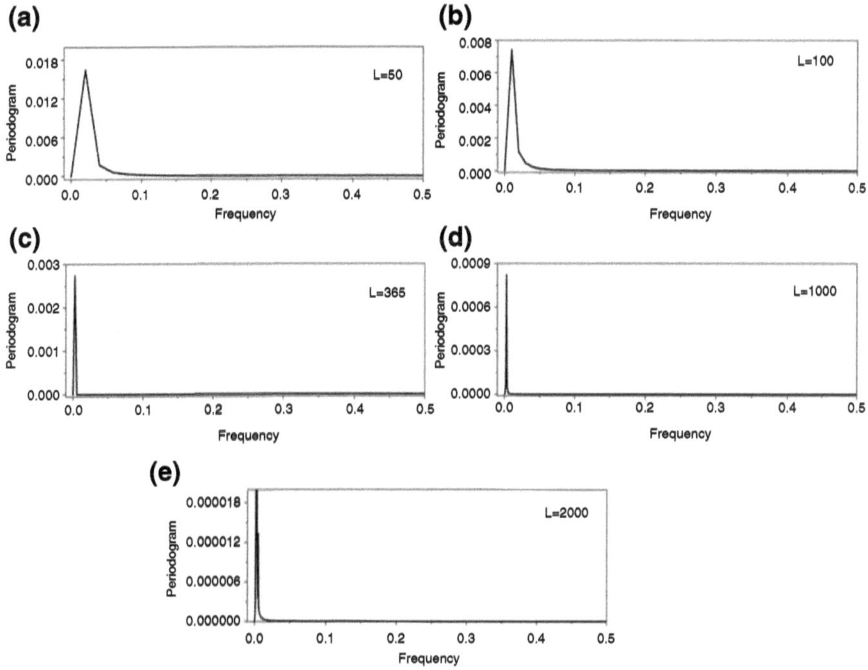

Fig. 8 Periodogram of first eigen function corresponding to the SSA decomposition of the rainfall time series using (**a**) various window length=50 days (**b**) window length=100 days (**c**) window length=365 days (**d**) window length=1000 days (**e**) window length=2000 days

5 Conclusion

SSA is a window length-based method which will break the time series into various small components. Selection of window length is crucial as it defines structure of the trajectory matrix of the time series. In the present study, the importance of window length in the decomposition of an observed hydrologic time series has been studied and a sensitivity analysis of window length has been carried out. 2D paired plots of corresponding eigen functions and periodogram of eigen functions have been used to carry out the sensitivity analysis. The results show that higher window length leads to over-decomposition and wrong interpretation whereas lower window length leads to under-decomposition and mixing up of components. It is observed that a window length proportional to period of the prevailing periodic component of the time series can provide better decomposition and hence can be used for extraction of various components of the time series.

References

1. Box, G. E. P., & Jenkins, G. M. (1976). *Time series analysis-forecasting and control* (2nd ed.). San Francisco: Colorado State University.
2. de Menezes, M. L., & Souza, R. C. (2014). Pessanha JFM combining singular spectrum analysis and PAR(p) structures to model wind speed time series. *Journal of Systems Science and Complexity, 27*, 29–46.
3. Golyandina, N. (2010). On the choice of parameters in singular spectrum analysis and related subspace-based methods. *Statistics and Its Interface, 3*, 259–279.
4. Golyandina, N., & Zhigljavsky, A. (2013). *Singular spectrum analysis for time series*. London: Springer.
5. Helsel, D. R., & Hirsch, R. M. (2002). Statistical methods in water resources. In *Hydrologic analysis and interpretation* (pp. 1–510). U.S .Geological Survey.
6. Jothiprakash, V., & Magar, R. B. (2012). Multi-time-step ahead daily and hourly intermittent reservoir inflow prediction by artificial intelligent techniques using lumped and distributed data. *Journal of Hydrology, 450–451*, 293–307.
7. Machiwal, D., & Jha, M. K. (2012). *Hydrologic time series analysis: Theory and practice*. Berlin: Springer.
8. Priestley, M. (1981). *Spectral analysis and time series*. San Diego: Academic Press.
9. Singh, V. P. (1988). *Hydrologic systems: Vol. 1. Rainfall-runoff modeling*. Eagle Wood Cliffs, NJ: Prentice Hall.
10. Unnikrishnan, P., & Jothiprakash, V. (2015). Extraction of nonlinear rainfall trends using singular spectrum analysis. *Journal of Hydrologic Engineering, 20*, 501–507.

Fourier-Type Monitoring Procedures for Strict Stationarity

S. Lee, S. G. Meintanis, and C. Pretorius

Abstract We consider model-free monitoring procedures for strict stationarity of a given time series. The new criteria are formulated as L2-type statistics incorporating the empirical characteristic function. Monte Carlo results as well as an application to financial data are presented.

1 Introduction

The notion of stationarity plays a very important role in statistical modeling. In its weakest form of first- or second-order stationarity, it implies that the mean or second moment, respectively, are time invariant; see, for instance, [8, 26] and [17]. A more general related notion is that of pth order (weak) stationarity which requires that all joint product moments of order (up to) p are time invariant. Most studies of stationarity are restricted to some form of weak stationarity, which of course is the most suitable concept for linear time series. On the other hand, the property of strict stationarity states that not only moments, but the entire probabilistic structure of a given series is time invariant. This property is of great importance with non-linear time series in which low-order moments are not sufficient for the dynamics of the series, not to mention the case of heavy-tailed time series lacking higher moments [1, 22]. Another divide is between parametric and non-parametric tests

S. Lee
Department of Statistics, Seoul National University, Seoul, South Korea

S. G. Meintanis
National and Kapodistrian University of Athens, Athens, Greece

Unit for Business Mathematics and Informatics, North-West University, Potchefstroom, South Africa
e-mail: simosmei@econ.uoa.gr

C. Pretorius (✉)
Unit for Business Mathematics and Informatics, North-West University, Potchefstroom, South Africa
e-mail: charl.pretorius@nwu.ac.za

P. Bertail et al. (eds.), *Nonparametric Statistics*, Springer Proceedings in Mathematics & Statistics 250, https://doi.org/10.1007/978-3-319-96941-1_22

for stationarity, with the first class containing the majority of procedures in earlier literature. There is also the methodological approach that categorizes the methods which operate either in the time or in the frequency domain. As the existing literature is vast, we provide below only a selected set of references which is by no means exhaustive.

In econometrics, the majority of earlier tests for weak stationarity are either tests for stationarity against unit root or, alternatively, tests of a unit root against stationarity, with the KPSS and the Dickey–Fuller tests being the by far most popular ones and having enjoyed a number of generalizations; see, for instance, [11, 12, 24] and [16]. When it comes to testing for strict stationarity in parametric time series, there exist many tests. These tests typically reduce to testing for a given portion of the parameter space and may often readily be extended to monitoring procedures. We indicatively mention here the work for ARMA, GARCH and DAR models by [4, 9] and [14], respectively. On the other hand, testing for strict stationarity is scarce when performed within a purely nonparametric framework. It appears that [18] was the first to address the issue of testing strict stationarity of the marginal distribution of an arbitrary time series. There is also the method of [15] which is based on the joint characteristic function and the papers of [7] and [21] which test for constancy (of a discretized version) of the marginal quantile process. The interest in testing for stationarity rests on the fact that modelling, predictions and other inferential procedures are invalid if this assumption is violated; see [13] for a good review on this issue. However, although strict stationarity is widely assumed in the literature, it is not truly a realistic assumption when one observes a given time series over a long period of time. On the contrary, it is expected that institutional changes cause structural breaks in the stochastic properties of certain variables, particularly in the macroeconomic and financial world. In this connection, monitoring the stationarity of a stochastic process seems to be of an even greater importance than testing. In this contribution, we propose a sequential procedure for strict stationarity. Our approach uses the characteristic function (CF) as the main tool. The advantage of using this function is that the CF can be estimated purely non-parametrically without the use of smoothing techniques. Moreover, and unlike the case of estimating the joint distribution function, the estimate of the joint CF is easy to obtain and when viewed as a process it is continuous in its argument.

The remainder of the chapter is as follows. In Sect. 2, we introduce the basic idea behind the proposed procedures. Section 3 presents the corresponding detector statistics. A resampling procedure is proposed in Sect. 4 in order to carry out the suggested monitoring method. The results of a Monte Carlo study for the finite-sample properties of the methods are presented in Sect. 5. A short real-world empirical application to market data is presented and discussed in Sect. 6. Finally, we end in Sect. 7 with conclusions and discussion.

2 ECF Statistics

Let $\{X_t\}_{t\in\mathbb{N}}$ be an arbitrary time series, and write $F_t(\cdot)$ for the corresponding distribution function (DF) of the m-dimensional variable $\Upsilon_t = (X_{t-(m-1)}, \ldots, X_t)'$, $m \geq 1$. We are interested in the behaviour of the distribution of Υ_t over time, i.e. to monitoring the joint distribution of the observations X_t of a given dimension m. The null hypothesis is stated as:

$$\mathscr{H}_0 : F_t \equiv F_m \text{ for all } t \geq m+1, \tag{1}$$

against the alternative

$$\mathscr{H}_1 : F_t \equiv F_m,\ t \leq t_0 \text{ and } F_t(y) \neq F_m(y),\ t > t_0, \tag{2}$$

for some $y \in \mathbb{R}^m$, where F_m, F_t, as well as the threshold t_0 are considered unknown. Clearly, $m = 1$ corresponds to monitoring the marginal distribution of X_t, $m = 2$ corresponds to the joint bivariate distribution of $(X_{t-1}, X_t)'$, and so on. As it is typical in monitoring the studies, we assume that there exist a set of observations $X_1, \ldots, X_T$ (often termed *training data*), which involve no change, and that monitoring starts after time T.

To motivate our procedure, let $\psi_Y(u) := E(e^{iu'Y})$, $i = \sqrt{-1}$, be the characteristic function (CF) of an arbitrary random vector Y. We will compare a nonparametric estimator of the joint CF $\psi_{\Upsilon_j}(u)$, of Υ_j, $j = m, m+1, \ldots, T$, with the same estimator obtained from observations beyond time T. Then, the quantity of interest is

$$D_{T,t} = \int_{\mathbb{R}^m} \left|\widehat{\psi}_T(u) - \widehat{\psi}_{T+t}(u)\right|^2 w(u)du, \tag{3}$$

where

$$\widehat{\psi}_J(u) = \frac{1}{J_m}\sum_{j=m}^{J} e^{iu'\Upsilon_j},\ J_m = J - m + 1, \tag{4}$$

is the empirical characteristic function (ECF) computed from the observations Υ_j, $j = m, \ldots, J$, and $w(\cdot)$ is a weight function which will be discussed below.

Our motivation for considering (3) as our main tool is that the null hypothesis (1) may equivalently be stated as:

$$\mathscr{H}_0 : \psi_{\Upsilon_t} \equiv \psi_{\Upsilon_m} \text{ for all } t \geq m+1, \tag{5}$$

and therefore we expect $D_{T,t}$ to be 'small' under the null hypothesis (5). Moreover, and unlike equivalent approaches based on the empirical DF, the ECF approach enjoys the important feature of computational simplicity. Specifically, by straightforward algebra we have, from (3):

$$D_{T,t} = \frac{1}{T_m^2} \sum_{j,k=m}^{T} W_{j,k} + \frac{1}{(T_m+t)^2} \sum_{j,k=m}^{T+t} W_{j,k} - \frac{2}{T_m(T_m+t)} \sum_{j=m}^{T} \sum_{k=m}^{T+t} W_{j,k}, \tag{6}$$

where $W_{j,k} = W(\Upsilon_j - \Upsilon_k)$ with

$$W(x) = \int_{\mathbb{R}^m} \cos(u'x) w(u) du. \tag{7}$$

A standard choice is to set $w(u) = e^{-a\|u\|^2}$, $a > 0$, which leads to

$$W(x) = Ce^{-\|x\|^2/4a}, \tag{8}$$

where $C = (\pi/a)^{m/2}$, and hence renders our statistic in closed form. Another interesting choice results by considering the statistic $\widetilde{D}_{T,t} = -D_{T,t}$ (in which case of course large negative values of $\widetilde{D}_{T,t}$ are significant). Then, we may write

$$\widetilde{D}_{T,t} = \frac{1}{T^2} \sum_{j,k=1}^{T} \widetilde{W}_{j,k} + \frac{1}{(T+t)^2} \sum_{j,k=1}^{T+t} \widetilde{W}_{j,k} - \frac{2}{T(T+t)} \sum_{j=1}^{T} \sum_{k=1}^{T+t} \widetilde{W}_{j,k}, \tag{9}$$

where $\widetilde{W}_{j,k} = \widetilde{W}(\Upsilon_j - \Upsilon_k)$ with

$$\widetilde{W}(x) = \int_{\mathbb{R}^m} (1 - \cos(u'x)) w(u) du. \tag{10}$$

If we let in (10) $w(u) = \|u\|^{-(m+a)}$, $0 < a < 2$, then

$$\widetilde{W}(x) = \widetilde{C}\|x\|^a,$$

where $\widetilde{C}$ is a known constant depending only on m and a, and hence in this case too our statistic comes in a closed-form expression suitable for computer implementation. Note that this weight function was first used by Székely and Rizzo [25], and later employed by Matteson and James [23] in change-point analysis.

The choice for the weight function $w(\cdot)$ is usually based upon computational considerations. In fact, if $w(\cdot)$ integrates to one (even after some scaling) and satisfies $w(-u) = w(u)$, then the function $W(\cdot)$ figuring in (7) can be interpreted as the CF of a symmetric around zero random variable having density $w(\cdot)$. In this connection, $w(\cdot)$ can be chosen as the density of any such distribution. Hence, the choice $e^{-a\|u\|^2}$ corresponds to the multivariate normal density but for computational purposes any density with a simple CF will do.

Another important user-specified parameter of our procedure is the order m that determines the dimension of the joint distribution which is monitored for stationarity. Of course, having a sample size T of training data already imposes the obvious restriction $m \leq T$, but if m is only slightly smaller than T, then we do not expect the ECF to be a reliable estimator of its population counterpart. The situation is similar to the problem of order-choice when estimating correlations from the available data; see [6, p. 221] and [5, p. 33] for some general guidelines. In our Monte Carlo results, we only consider cases where $m \leq 4$ (very small compared to T), but it is reasonable to assume that we could let m grow as T increases.

3 Detector Statistics and Stopping Rule

As already mentioned, we consider online procedures whereby the test is applied sequentially on a dynamic data set which is steadily updated over time with the arrival of new observations. In this context, the null hypothesis is rejected when the value of a suitable detector statistic exceeds an appropriately chosen constant *for the first time*. Otherwise, we continue monitoring. These statistics are commonly defined by a corresponding stopping rule. In order to define this stopping rule, and based on asymptotic considerations, we need to introduce an extra weight function in order to control the large-sample probability of type-I error. In particular, we employ the detector statistic

$$\Delta_{T,t} = \frac{1}{q_\gamma^2\left(\frac{t}{T}\right)} \left(\frac{T+t-m+1}{\sqrt{T-m+1}}\right)^2 D_{T,t}, \tag{11}$$

where $D_{T,t}$ is defined by (3), and

$$q_\gamma(s) = (1+s)\left(\frac{s}{s+1}\right)^\gamma, \quad \gamma \in [0, 1/2). \tag{12}$$

Here, q_γ denotes an extra weight function needed to control (asymptotically) the probability of type-I error for the sequential test procedure. The parameter γ figuring in (12) gives some flexibility to the resulting procedure. Specifically, if early violations are expected, then the value of γ should be close to 1/2, while values closer to zero are appropriate for detecting changes occurring at later stages; see [3].

As already mentioned, it is clear that since the training data $\{X_1, \ldots, X_T\}$ are assumed to involve no change, the monitoring period begins with time $t = T + 1$. Typically, this monitoring continues until time $T(L+1)$, where L denotes a fixed integer, and if $L < \infty$ we call the corresponding procedure closed-end. Otherwise (i.e. if $L = \infty$), we have an open-end procedure. The corresponding stopping rule is specified as:

$$\tau(T; L) = \tau(T) = \begin{cases} \inf\{1 < t \le LT : \Delta_{T,t} > c_\alpha\}, \\ +\infty, \text{ if } \Delta_{T,t} \le c_\alpha \text{for all } 1 < t \le LT, \end{cases} \tag{13}$$

where c_α is a constant that guarantees that the test has size equal to α, asymptotically.

The main problem is then to find an approximation for the critical value c_α and to investigate consistency of the test procedures. Particularly, we require that under $\mathscr{H}_0$ and for a prechosen value of $0 < \alpha < 1$,

$$\lim_{T\to\infty} P_{\mathscr{H}_0}(\tau(T) < \infty) = \alpha, \tag{14}$$

while under alternatives we want

$$\lim_{T\to\infty} P(\tau(T) < \infty) = 1. \tag{15}$$

4 The Resampling Procedure

The asymptotic distribution of the detector statistic in (11) under the null hypothesis $\mathscr{H}_0$ will be reported elsewhere. However, despite its theoretical interest, this limit distribution depends on factors that are unknown in the current, entirely nonparametric, context. For this reason, we suggest a resampling procedure in order to compute critical values and actually implement the new method. Specifically, we employ an adapted version of the moving block bootstrap procedure see, e.g. [19, 20] in order to approximate the critical value c_α of the test. This is done using only the training data, i.e. all data available at time T. Given a block size ℓ, define the overlapping blocks $b_k = (X_k, \ldots, X_{k+\ell-1})$, $k = 1, \ldots, T - \ell + 1$, and proceed as follows:

1. From $\{b_k\}_{k=1}^{T-\ell+1}$, randomly sample $\lceil T/\ell \rceil$ blocks with replacement to obtain the bootstrapped blocks $\{b_k^*\}_{k=1}^{\lceil T/\ell \rceil}$. Throughout, $\lceil x \rceil$ denotes the ceiling of $x \in \mathbb{R}$.
2. Concatenate the b_k^* and select the first T observations as the bootstrap sample $X_1^*, \ldots, X_T^*$.
3. Treat the first $\tilde{T} = \lceil T/(1 + L) \rceil$ bootstrap observations as a *pseudo training sample* and calculate $\Delta_{\tilde{T},t}^* = \Delta_{\tilde{T},t}(X_1^*, \ldots, X_{\tilde{T}+t}^*)$ for each $t = 1, \ldots, T - \tilde{T}$, i.e. run the *monitoring procedure on the remaining data.*
4. Calculate $M^* = \max_{1 \le t \le T-\tilde{T}} \Delta_{\tilde{T},t}^*$.

Repeat steps 1–4 a large number of times, say B, to obtain the (ordered) statistics $M^*_{(1)} \leq \cdots \leq M^*_{(B)}$. An approximate value for c_α is then given by $M^*_{(\lfloor B(1-\alpha)\rfloor)}$. Throughout, $\lfloor x \rfloor$ denotes the floor of $x \in \mathbb{R}$.

In order to choose an appropriate block size, we utilize the correlation between two consecutive observations. To this end, define the following plug-in estimate of lag order due to [2] with an additional upper bound introduced by [26]:

$$p_T = \min\left\{\left\lfloor \left(\frac{3T}{2}\right)^{1/3}\left(\frac{2\widehat{\rho}}{1-\widehat{\rho}^2}\right)^{2/3}\right\rfloor, \left\lfloor 8\left(\frac{T}{100}\right)^{1/3}\right\rfloor\right\},$$

where $\widehat{\rho}$ is an estimator for the first-order autocorrelation of $\{X_t\}_{t=1}^T$. Further, define $\bar{p}_T$ the same as above except with $\widehat{\rho}$ replaced by an estimator for the first-order autocorrelation of the squared observations $\{X_t^2\}_{t=1}^T$. Based on this, a *data-dependent block size* is given by $\ell = \ell_T = \max\{p_T, \bar{p}_T\}$. This choice of ℓ is used throughout the simulations discussed in the following section.

5 Monte Carlo Results

We investigate the performance of the monitoring procedure defined by (11) with criterion given by (6), and weight function $w(u) = e^{-a\|u\|^2}$, $a > 0$. The results corresponding to criterion (9) are similar and therefore are not reported. We only report here Monte Carlo results on the actual level of the procedure. Corresponding power results will be reported elsewhere. We consider the following data generating processes:

$$\text{DGP.S1: } X_t = \varepsilon_t,$$

$$\text{DGP.S2: } X_t = 0.5X_{t-1} + \varepsilon_t,$$

$$\text{DGP.S3: } X_t = \sqrt{h_t}\varepsilon_t, \text{ with } h_t = 0.2 + 0.3X_{t-1}^2,$$

$$\text{DGP.S4: } X_t = \sqrt{h_t}\varepsilon_t, \text{ with } h_t = 0.1 + 0.3X_{t-1}^2 + 0.3h_{t-1},$$

where $X_0 = 0$, and $\{\varepsilon_t\}_{t\in\mathbb{N}}$ is an iid sequence of $N(0, 1)$ random variables. These processes satisfy the null hypothesis of strict stationarity and are introduced to study the size of our monitoring procedure for finite samples. DGP.S1 consists of iid observations, whereas DGP.S2–DGP.S4 introduce some dependence structure without violating the null hypothesis.

The values in Tables 1, 2, 3 and 4 represent the percentage of times that $\mathcal{H}_0$ was rejected out of 2000 independent Monte Carlo repetitions. To estimate the critical

Table 1 Size results for DGP.S1–DGP.S4 with $m = 1$

		DGP.S1					DGP.S2					DGP.S3					DGP.S4				
T	L	$a = 0.1$	0.5	1.0	1.5	5.0	0.1	0.5	1.0	1.5	5.0	0.1	0.5	1.0	1.5	5.0	0.1	0.5	1.0	1.5	5.0
100	1	5	5	5	5	5	8	10	10	10	10	6	6	6	6	6	7	5	5	5	5
	2	5	5	5	5	5	8	10	10	10	11	6	5	5	5	5	6	7	7	6	6
	3	5	5	6	6	5	9	10	10	10	11	5	5	5	5	5	6	5	5	5	6
200	1	5	6	5	5	5	7	8	8	8	8	6	5	5	5	5	5	5	5	5	5
	2	5	5	5	5	5	7	8	8	8	9	5	5	5	5	5	5	5	5	6	6
	3	5	5	5	5	5	7	8	8	8	9	6	5	5	5	5	6	6	6	6	7
300	1	5	5	4	4	5	7	7	8	8	8	5	5	5	5	5	4	4	5	5	5
	2	5	5	5	5	5	7	7	8	8	8	6	5	5	5	5	5	5	5	6	6
	3	5	5	5	5	5	7	8	8	8	8	5	5	5	5	5	4	3	3	4	5

Table 2 Additional size results for DGP.S2 with $m = 1$

		a				
T	L	0.1	0.5	1.0	1.5	5.0
500	1	6	7	7	7	7
	2	6	7	8	8	8
	3	6	7	7	7	7
1000	1	6	7	7	7	7

value c_α, we employed the warp-speed method of [10]. For a significance level of $\alpha = 5\%$, Tables 1 and 2 contain the results for the case of strict stationarity of order $m = 1$, while Tables 3 and 4 supply the corresponding results for the cases $m = 2$ and $m = 4$.

An overall assessment of the figures in Tables 1, 2, 3 and 4 brings forward certain desirable features of the method. A reasonable degree of approximation of the nominal level, and the size of the test seems to converge to the nominal level as the sample size is increased. A further factor influencing the percentage of rejection is the value of the weight parameter a, and in this respect it seems that an intermediate value $0.5 \leq a \leq 1.5$ is preferable in terms of level accuracy, at least in most cases. At the same time though, there is some degree of size distortion with the autoregressive process DGP.S2, even up to sample size $T = 300$. To further account for this phenomenon, we have selectively run the procedure for this process with higher sample size. The resulting figures (reported in Table 2) with sample sizes $T = 500$ and $T = 1000$ suggest that the problem still persists but certainly with decreasing intensity.

6 Real-Data Application

We now demonstrate the potential of the monitoring procedure to detect breaks in strict stationarity. Consider the daily percentage change in the USD/GBP and ZAR/USD exchange rates, which are depicted in Fig. 1. According to the test by Hong et al. [15], these two series exhibit no change in distribution (at $\alpha = 5\%$ for $m = 2, 4, 8$) over the period 1 January to 31 December 2015. Starting on 1 January 2016, the monitoring procedure was run sequentially until the first break in stationarity was detected, or until 31 December 2016, whichever came first. The results of the monitoring procedure are given in Table 5.

Based on the results from the Monte Carlo study, we chose the tuning parameter $a = 1$ for practical application. For each stock, the critical value c_α of the test was approximated by $B = 2000$ bootstrap replications using the resampling procedure described in Sect. 4. The observed run length (in weeks, denoted by $\hat{\tau}$) of the monitoring procedure until a break in stationarity was identified and is reported along with the corresponding calendar date. Additionally, we also supply an approximate p-value, calculated as $\hat{p} = \frac{1}{B}\sum_{b=1}^{B} \mathrm{I}(M^*_{(b)} \geq \max_{1\leq t\leq LT} \Delta_{T,t})$, where $M^*_{(b)}$ is as defined in Sect. 4.

Table 3 Size results for DGP.S1–DGP.S4 with $m = 2$

		DGP.S1					DGP.S2					DGP.S3					DGP.S4				
T	L	$a = 0.1$	0.5	1.0	1.5	5.0	0.1	0.5	1.0	1.5	5.0	0.1	0.5	1.0	1.5	5.0	0.1	0.5	1.0	1.5	5.0
100	1	5	5	5	5	5	6	8	9	9	9	6	5	5	5	5	6	5	5	5	5
	2	4	5	5	5	5	5	7	8	9	9	6	5	5	5	5	6	6	6	6	5
	3	4	5	5	5	5	5	8	9	9	10	6	5	5	5	5	5	5	5	5	5
200	1	5	5	5	5	5	5	7	7	8	8	5	5	5	5	5	4	4	5	5	5
	2	5	5	5	5	5	5	7	7	8	8	5	5	5	5	5	5	5	5	5	5
	3	5	5	5	5	5	6	8	8	8	9	5	5	5	5	5	6	5	5	5	6
300	1	5	5	5	5	5	5	7	7	7	7	5	5	5	5	5	4	4	4	5	5
	2	5	5	5	5	6	6	7	7	8	8	6	5	5	5	5	5	5	5	5	6
	3	5	5	6	6	6	6	7	8	8	8	6	6	5	5	5	4	3	3	3	4

Table 4 Size results for DGP.S1–DGP.S4 with $m = 4$

		DGP.S1					DGP.S2					DGP.S3					DGP.S4				
T	L	$a = 0.1$	0.5	1.0	1.5	5.0	0.1	0.5	1.0	1.5	5.0	0.1	0.5	1.0	1.5	5.0	0.1	0.5	1.0	1.5	5.0
100	1	2	4	5	5	4	1	6	7	7	8	5	4	4	4	3	4	4	4	3	2
	2	2	5	5	5	5	1	5	7	7	8	5	5	4	4	3	6	5	4	4	2
	3	2	4	5	5	4	1	6	8	8	9	5	5	4	4	2	5	4	4	4	2
200	1	3	5	5	5	5	1	6	7	7	8	5	5	5	5	4	4	4	4	4	4
	2	2	4	5	5	5	1	6	7	7	7	5	5	4	4	4	5	4	4	4	4
	3	2	4	5	5	4	1	6	7	8	8	6	5	5	5	4	5	5	5	5	4
300	1	4	5	5	5	5	2	6	6	7	7	5	5	5	5	5	3	3	4	4	4
	2	4	5	5	5	5	2	6	6	7	7	5	5	5	5	5	5	4	5	5	5
	3	4	5	5	5	6	2	6	7	7	8	5	5	5	5	4	4	2	2	2	3

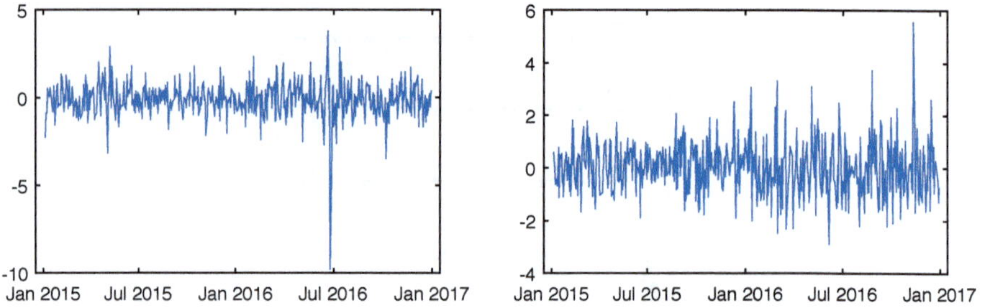

Fig. 1 Standardized daily percentage change in the USD/GBP (left) and ZAR/USD (right) exchange rates. Data source: South African Reserve Bank

Table 5 Results of the monitoring procedure for the two considered series

	USD/GBP			ZAR/USD		
m	$\hat{p}$	$\hat{\tau}$	Date	$\hat{p}$	$\hat{\tau}$	Date
2	0.041	153	2016/08/08	< 0.001	75	2016/04/19
4	0.016	133	2016/07/11	< 0.001	71	2016/04/13
8	0.012	134	2016/07/12	< 0.001	80	2016/04/26

According to our monitoring procedures, both series exhibit breaks in stationarity. For the USD/GBP exchange rate, the first significant break is detected in July/August 2016, shortly after the outcome of the Brexit referendum. For the ZAR/USD exchange rate, a break is detected very early in 2016, amidst a period of political instability and increased volatility in the South African markets.

7 Conclusion

We suggest a procedure for online monitoring of strict stationarity. The criteria involved are entirely model-free and therefore apply to arbitrary time series. In particular, we employ a non-parametric estimate of the joint characteristic function of the underlying process and suggest to monitor an integrated functional of this estimate in order to capture structural breaks in the joint distribution of the observations. Since the limit null distribution depends on the stochastic structure of the series being sampled, a modification of the block bootstrap is used in order to actually implement the procedure. This bootstrap is in turn applied to simulated data, and shows satisfactory performance in terms of level. The same procedure is applied to real data from the financial market. We close by noting that our results, although presented for scalar observations, may readily be extended to multivariate time series.

References

1. Andrews, B., Calder, M., & Davis, R. A. (2009). Maximum likelihood estimation for α-stable autoregressive processes. *The Annals of Statistics, 37*, 1946–1982.
2. Andrews, D. W. K. (1991). Heteroskedasticity and autocorrelation consistent covariance matrix estimation. *Econometrica, 59*, 817–858.
3. Aue, A., Horváth, L., Hŭsková, M., & Kokoszka, P. (2006). Change-point monitoring in linear models with conditionally heteroskedastic errors. *The Econometrics Journal, 9*, 373–403.
4. Bai, J. (1994). Weak convergence of the sequential empirical processes of residuals in ARMA models. *The Annals of Statistics, 22*, 2051–2061.
5. Box, G. E. P., & Jenkins, G. M. (1970). *Time series analysis: Forecasting and control.* San Francisco: Holden-Day.
6. Brockwell, P. J., & Davis, R. A. (1991). *Time series: Theory and methods*. New York: Springer.
7. Busetti, F., & Harvey, A. (2010). Tests of strict stationarity based on quantile indicators. *Journal of Time Series Analysis, 31*, 435–450.
8. Dwivedi, Y., & Subba Rao, S. (2011). A test for second order stationarity of a time series based on the discrete Fourier transform. *Journal of Time Series Analysis, 32*, 68–91.
9. Francq, C., & Zakoïan, J. (2012). Strict stationarity testing and estimation of explosive and stationary generalized autoregressive conditional heteroscedasticity models. *Econometrica, 80*, 821–861.
10. Giacomini, R., Politis, D., & White, H. (2013). A warp-speed method for conducting Monte Carlo experiments involving bootstrap estimators. *Econometric Theory, 29*, 567–589.
11. Gil-Alana, L. (2003). Testing the power of a generalization of the KPSS-tests against fractionally integrated hypotheses. *Computational Economics, 22*, 23–38.
12. Giraitis, L., Kokoszka, P., Leipus, R., & Teyssiére, G. (2003). Rescaled variance and related tests for long memory in volatility and levels. *Journal of Econometrics, 112*, 265–294.
13. Guégan, D. (2010). Non-stationary samples and meta-distribution. In A. Basu, T. Samanta, & A. Sen Gupta (Eds.), *Statistical paradigms. Recent advances and reconciliations*. Singapore: World Scientific.
14. Guo, S., Li, D., & Li, M. (2016). Strict stationarity testing and global robust quasi-maximum likelihood estimation of DAR models, Forthcoming. https://pdfs.semanticscholar.org/3c27/34fd9e878f65749c65aba055c5439521d5f9.pdf
15. Hong, Y., Wang, X., & Wang, S. (2017). Testing strict stationarity with applications to macroeconomic time series. *International Economic Review, 58*, 1227–1277.
16. Horváth, L. P., Kokoszka, P., & Rice, G. (2014). Testing stationarity of functional time series. *Journal of Econometrics, 179*, 66–82.
17. Jentsch, C., & Subba Rao, S. (2015). A test for second order stationarity of a multivariate time series. *Journal of Econometrics, 185*, 124–161.
18. Kapetanios, G. (2009). Testing for strict stationarity in financial variables. *Journal of Banking & Finance, 33*, 2346–2362.
19. Künsch, H. (1989). The jackknife and the bootstrap for general stationary observations. *The Annals of Statistics, 17*, 1217–1241.
20. Lahiri, S. (2003). *Resampling methods for dependent data*. New York: Springer.
21. Lima, L., & Neri, B. (2013). A test for strict stationarity. In V. Huynh, et al. (Eds.), *Uncertainty analysis in econometrics with applications*. Berlin: Springer.
22. Loretan, M., & Phillips, P. (1994). Testing the covariance stationarity of heavy-tailed time series. *Journal of Empirical Finance, 1*, 211–248.
23. Matteson, D. S., & James, N. A. (2014). A nonparametric approach for multiple change point analysis of multivariate data. *Journal of the American Statistical Association, 109*, 334–345.

24. Müller, U. (2005). Size and power of tests of stationarity in highly autocorrelated time series. *Journal of Econometrics, 128*, 195–213
25. Székely, G., & Rizzo, M. (2005). Hierarchical clustering via joint between-within distances: Extending Ward's minimum variance method. *Journal of Classification, 22*, 151–183.
26. Xiao, Z., & Lima, L. (2007). Testing covariance stationarity. *Econometric Reviews, 26*, 643–667.

Nonparametric and Parametric Methods for Change-Point Detection in Parametric Models

G. Ciuperca

Abstract We consider a posteriori and a priori change-point models. The parametric regression functions of the each phase can be nonlinear or linear, and moreover, in the linear case, the number of the explanatory variables could be large. Theoretical results and simulations are presented for each model. For a posteriori change-point nonlinear model, the results obtained by two estimation techniques are given in the case when the change-point number is known. So, the quantile and empirical likelihood nonparametric methods are considered. If the number of the change-points is unknown, a consistent criterion is proposed. When the change-point model is linear with a large number of explanatory variables, it would make the automatic selection of variables. The adaptive LASSO quantile method is then proposed and studied. On the other hand, we propose a nonparametric test based on the empirical likelihood, in order to test if the model changes. For detecting in real time a change in model, we consider two cases, nonlinear and linear models. For a nonlinear model, a hypothesis test based on CUSUM of LS residuals is constructed. For a linear model with large number of explanatory variables, we propose a CUSUM test statistic based on adaptive LASSO residuals.

1 Introduction

A change-point model is a model which changes its form at unknown times, the change location being generally unknown. This problem was introduced in the quality control, for checking whether there is a change in the performance of a machine. Generally, there are two types of change-point problem:

- *a posteriori (retrospective)* when the data are completely known at the end of the experiment. Then, a posteriori, the question if model has changed is considered. In the affirmative case, we must find the number of changes, their location and

G. Ciuperca (✉)
Université Lyon 1, CNRS, UMR 5208, Institut Camille Jordan, Villeurbanne Cedex, France
e-mail: Gabriela.Ciuperca@univ-lyon1.fr

P. Bertail et al. (eds.), *Nonparametric Statistics*, Springer Proceedings
in Mathematics & Statistics 250, https://doi.org/10.1007/978-3-319-96941-1_23

finally we need to estimate each phase. We will present results for a change-point nonlinear model by a parametric method: the quantile method, and a nonparametric estimation method, the empirical likelihood (EL). If the change-point model is linear, but with a large number of explanatory variables, then the automatic selection of variables must be carried out. Methods of LASSO type are then used. On the other hand, in order to test the change of the model, a nonparametric test based on the EL method is proposed.

- *a priori (sequential, on-line)* when the detection is performed in real time. The most used technique is the cumulative sum (CUSUM) method. In this talk, two cases are presented: for a nonlinear model a hypothesis test based on weighed CUSUM of least squares (LS) residuals is constructed and for a linear model with a large number of explanatory variables, we propose a CUSUM test statistic based on adaptive LASSO residuals.

We present here some results obtained by author and its co-authors in the papers: [1–5].

Some general notations are used throughout in this chapter. All vectors are column and $\mathbf{v}^t$ denotes the transpose of $\mathbf{v}$. All vectors and matrices are in bold. For a vector, $\|.\|_2$ is the euclidean norm, $\|.\|_1$ is the L_1 norm and for a matrix $\mathbf{M}$, $\|\mathbf{M}\|_1$ is the subordinate norm to the vector norm $\|.\|_1$. For a vector (matrix), if $\mathscr{A}$ is a index set, then the vector (matrix) with subscript $\mathscr{A}$ is a subvector (submatrix) with the index in the set $\mathscr{A}$.

This contribution is organized as follows. In Sect. 2 are given theoretical results and simulations for a priori change-point model, while in Sect. 3, the CUSUM method is used for sequential detection of a change-point.

2 Off-Line Procedures

In this section we first consider a nonlinear model and then a linear model with a large number of explanatory variables. For change-point nonlinear model, using quantile method, a consistent criterion for finding the number of changes is proposed, the location of the changes and the model in each phase are estimated. The EL method allows to find an asymptotic test statistic for detection of the changes in model.

For a linear model in high-dimension, we use adaptive LASSO quantile method for automatic selection of relevant variables in each phase and for finding the number of change. In order to have more accurate parameter estimators and a better adjustment of the dependent variable, we use the EL method.

Studied models in this section have the following form:

$$Y_i = h(\mathbf{X}_i, \ldots; t_i) + \varepsilon_i, \qquad i = 1, \ldots, n. \tag{1}$$

The function h, for $\mathbf{x} \in \mathbb{R}^d$, $t \in \mathbb{R}$, has the form:

$$h(\mathbf{x}, \ldots; t) = f_1(\mathbf{x}, \ldots)\mathbb{1}_{t \leq l_1} + f_2(\mathbf{x}, \ldots)\mathbb{1}_{l_1 < t \leq l_2} + \ldots \; f_{K+1}(\mathbf{x}, \ldots)\mathbb{1}_{l_K < t}, \qquad (2)$$

We say that model (1) with (2) has K change-points or $K + 1$ phases. The classical regression models are single-phase. In model (1), Y is the response variable, $\mathbf{X}$ is a vector of explicative variables, t is another variable in relation to which the change may occurs. The variables $\mathbf{X}$ and t can be either random or deterministic. Knowing n observations for $(Y, \mathbf{X}, t)$, parametric or nonparametric techniques can be used to study models (1) and (2). In this talk, the regression function depends only on $\mathbf{x} \in \Upsilon \subseteq \mathbb{R}^d$ and on a parameter vector $\boldsymbol{\beta} \in \Gamma \subseteq \mathbb{R}^p$. So $f : \Upsilon \times \Gamma \to \mathbb{R}$ and it is known up to the parameter $\boldsymbol{\beta}$. The variable t is the number of observation.

The main difficulties for studying model (1) are: if K unknown, the model is unidentifiable and then we must first determinate K and afterwards estimate the model; moreover, the objective function is not regular with respect to the parameters $l_1, \cdots, l_K$.

Throughout this contribution, the errors (ε_i) are supposed i.i.d. with φ_ε its density and F_ε its distribution function.

2.1 Nonlinear Model

In this subsection, we suppose that the function f is nonlinear.

2.1.1 Estimation in a Change-Point Nonlinear Quantile Model

The results presented in this subsubsection are published in [4].

Let us consider the following nonlinear model, where the changes $l_1, \cdots, l_K$ of relation (2) occur with respect to the observations:

$$Y_i = \sum_{r=0}^{K} f(\mathbf{X}_i, \boldsymbol{\beta}_{r+1})\mathbb{1}_{l_r \leq i < l_{r+1}} + \varepsilon_i, \qquad i = 1, \cdots, n,$$

with $l_0 = 1, l_{K+1} = n$. We suppose that the set Γ is compact. The parameters of the model are: *the regression parameters* $\mathfrak{d}_1 \equiv (\boldsymbol{\beta}_1, \cdots, \boldsymbol{\beta}_{K+1})$ and *the change-points* $\mathfrak{d}_2 \equiv (l_1, \cdots, l_K) \in \mathbb{N}^K$. The true values are $\mathfrak{d}_1^0$ and $\mathfrak{d}_2^0$.

Classically, the model errors ε_i are supposed with mean zero and bounded variance. If these conditions are not satisfied or if model contains outliers, then the LS estimators of the model parameters can have large errors. A very interesting and robust alternative method was proposed by Koenker and Bassett [9] by introduction of quantile method. For a fixed quantile index $\tau \in (0, 1)$, the check function $\rho_\tau : \mathbb{R} \to \mathbb{R}$ is defined by $\rho_\tau(u) = u[\tau - \mathbb{1}_{u \leq 0}]$. We define the quantile estimators of parameters $\eth_1$ and $\eth_2$ by

$$(\hat{\eth}^{(\tau)}_{1n}, \hat{\eth}^{(\tau)}_{2n}) \equiv \underset{(\eth_1,\eth_2)}{\text{argmin}} \sum_{r=1}^{K+1} \sum_{i=l_{r-1}+1}^{l_r} \rho_\tau(Y_i - f(\mathbf{X}_i, \boldsymbol{\beta}_r)).$$

The components of $\hat{\eth}^{(\tau)}_{1n}$ are $(\hat{\boldsymbol{\beta}}^{(\tau)}_1, \cdots, \hat{\boldsymbol{\beta}}^{(\tau)}_{K+1})$ and of $\hat{\eth}^{(\tau)}_{2n}$ are $(\hat{l}^{(\tau)}_1, \cdots, \hat{l}^{(\tau)}_K)$. Note that for the particular case $\tau = 1/2$ we obtain the least absolute deviation estimator.

The following assumptions are considered for the errors, design and the regression function. The design X_i is deterministic. For the errors $(\varepsilon_i)_{1\le i\le n}$ we suppose:
(NLQ1) There exist two constants $c_0, \delta_0 > 0$ such that for all $|x| \le \delta_0$, we have: $\min(F_\varepsilon(|x|) - F_\varepsilon(0), F_\varepsilon(0) - F_\varepsilon(-|x|)) \ge c_0|x|$.

The regression function $f(\mathbf{x}, \boldsymbol{\beta})$ is supposed twice differentiable in $\boldsymbol{\beta}$ on Γ and continuous on Υ. In the following, for $\mathbf{x} \in \Upsilon$ and $\boldsymbol{\beta} \in \Gamma$ we use notation $\dot{\mathbf{f}}(\mathbf{x}, \boldsymbol{\beta}) \equiv \partial f(\mathbf{x}, \boldsymbol{\beta})/\partial \boldsymbol{\beta}$ and $\ddot{\mathbf{f}}(\mathbf{x}, \boldsymbol{\beta}) \equiv \partial^2 f(\mathbf{x}, \boldsymbol{\beta})/\partial \boldsymbol{\beta}^2$ and $\boldsymbol{\beta} o$ is the true value of $\boldsymbol{\beta}$. Moreover, for the function f, the following assumptions are considered:
(NLQ2) There exist two constants $c_2, c_3 > 0$ and $n_0 \in \mathbb{N}$ such that for all $\boldsymbol{\beta}_1, \boldsymbol{\beta}_2 \in \Gamma$ and $n \ge n_0$: $c_2\|\boldsymbol{\beta}_1 - \boldsymbol{\beta}_2\|_2 \le \left(n^{-1}\sum_{i=1}^n[f(\mathbf{X}_i, \boldsymbol{\beta}_1) - f(\mathbf{X}_i, \boldsymbol{\beta}_2)]^2\right)^{1/2} \le c_3\|\boldsymbol{\beta}_1 - \boldsymbol{\beta}_2\|_2$. Moreover, we have that, $n^{-1}\sum_{i=1}^n \dot{\mathbf{f}}(\mathbf{X}_i, \boldsymbol{\beta}^{\mathbf{0}})\dot{\mathbf{f}}^t(\mathbf{X}_i, \boldsymbol{\beta}^{\mathbf{0}})$ converges, as $n \to \infty$, to a positive definite matrix. Furthermore, $\max_{1\le i\le n} n^{-1/2}\|\dot{\mathbf{f}}(\mathbf{X}_i, \boldsymbol{\beta}^{\mathbf{0}})\|_2 \to 0$, as $n \to \infty$.
(NLQ3) For all $\boldsymbol{\beta} \in \Gamma, \mathbf{x} \in \Upsilon$, we have that $\|\dot{\mathbf{f}}(\mathbf{x}, \boldsymbol{\beta})\|_2$ is bounded.
(NLQ4) For all $\boldsymbol{\beta} \in \Gamma, \mathbf{x} \in \Upsilon$, we have that $\|\ddot{\mathbf{f}}(\mathbf{x}, \boldsymbol{\beta})\|_1$ is bounded.

Concerning the change-point location, we suppose that each phase contains a large number of observations:
(NLQ5) $l_{r+1} - l_r \ge n^a, a > 1/2$, for all $r = 0, \cdots, K$, with $l_0 = 1$ and $l_{K+1} = n$.

For fixed (known) K, under these assumptions we have the following theorem which obtains that the distance between the change-point quantile estimator and the true value is finished. The asymptotic distribution of the quantile estimators is also found. The asymptotic distribution of the regression parameter estimator is Gaussian and the change-point estimator has an asymptotic distribution depending on the quantile index τ and on the difference of the values of f for two consecutive phases.

Theorem 1 *Under assumptions (NLQ1)–(NLQ5), if density function φ_ε of ε is differentiable in a neighbourhood of 0 and $\varphi'_\varepsilon(x)$ is bounded in this neighbourhood, then:*

(i) $\|\hat{\eth}^{(\tau)}_{2n} - \eth^0_2\|_2 = O_{\mathbb{P}}(1)$.

(ii) for each $r = 1, \cdots, K$, *we have* $(\hat{l}^{(\tau)}_r - l^0_r) \xrightarrow[n\to\infty]{\mathcal{L}} \text{argmin}_{j\in\mathbb{Z}} Z^{(\tau)}_{r,j}$, *with:*

$$Z^{(\tau)}_{r,j} \equiv \begin{cases} \sum_{i=l^0_r+1}^{l^0_r+j} \left[\rho_\tau(\varepsilon_i - f(\mathbf{X}_i, \boldsymbol{\beta}^0_r) + f(\mathbf{X}_i, \boldsymbol{\beta}^0_{r+1})) - \rho_\tau(\varepsilon_i)\right], & \text{for } j = 1, 2, \cdots \\ \sum_{i=l^0_r+j}^{l^0_r} \left[\rho_\tau(\varepsilon_i - f(\mathbf{X}_i, \boldsymbol{\beta}^0_{r+1}) + f(\mathbf{X}_i, \boldsymbol{\beta}^0_r)) - \rho_\tau(\varepsilon_i)\right], & \text{for } j = -1, -2, \cdots \end{cases}$$

(iii) for each $r = 1, \cdots, K+1$, *we have*

$$(\hat{l}_r^{(\tau)} - \hat{l}_{r-1}^{(\tau)})^{1/2}(\hat{\boldsymbol{\beta}}_r^{(\tau)} - \boldsymbol{\beta}_r^0)\boldsymbol{\Sigma}_r^{1/2} \xrightarrow[n\to\infty]{\mathscr{L}} \mathscr{N}(\boldsymbol{0}, \frac{\tau(1-\tau)}{\varphi_\varepsilon^2(0)}\boldsymbol{I}_p),$$

with $\boldsymbol{\Sigma}_r \equiv (l_r^0 - l_{r-1}^0)^{-1}\sum_{i=l_{r-1}^0+1}^{l_r^0} \dot{f}(\mathbf{X}_i, \boldsymbol{\beta}_r^0)\dot{f}^t(\mathbf{X}_i, \boldsymbol{\beta}_r^0)$ *and* $\boldsymbol{I}_p$ *the identity matrix of order p.*

The asymptotic results of Theorem 1 are obtained when the number of changes is known, this is rarely the case in applications. Thus, based on the quantile method and on Schwarz criterion, we propose a general criterion for determining the phase number in a model:

$$\hat{K}_n^{(\tau)} \equiv \operatorname*{argmin}_K \left(n \log\left(n^{-1}S_n(\tau, \hat{\mathfrak{d}}_{1n}^{(\tau)}(K), \hat{\mathfrak{d}}_{2n}^{(\tau)}(K))\right) + P(K,p)B_n\right), \qquad (3)$$

where:

$S_n(\tau, \mathfrak{d}_1, \mathfrak{d}_2) \equiv \sum_{j=0}^K \sum_{i=l_j+1}^{l_{j+1}} \rho_\tau(Y_i - f(\mathbf{X}_i, \boldsymbol{\beta}_{j+1}))$;

$(\hat{\mathfrak{d}}_{1n}^{(\tau)}(K), \hat{\mathfrak{d}}_{2n}^{(\tau)}(K))$ is the quantile estimator of $(\mathfrak{d}_1, \mathfrak{d}_2)$ for a fixed K;

$(B_n)_n$ a deterministic sequence converging to ∞ such that $B_n n^{-a} \to 0$, $B_n n^{-1/2} \to \infty$, as $n \to \infty$, with $a > 1/2$;

penalty function $P(K,p)$ is such that $P(K_1,p) \le P(K_2,p)$, for all number change-points $K_1 \le K_2$.

Let K_0 be the true value of K. The following theorem shows that proposed criterion (3) is consistent.

Theorem 2 *Under the same conditions of Theorem 1, if moreover* $\mathbb{E}[\rho_\tau(\varepsilon)] > 0$ *and* $\mathbb{E}[\rho_\tau^2(\varepsilon)] < \infty$, *then* $\mathbb{P}[\hat{K}_n^{(\tau)} = K_0] \to 1$, *as* $n \to \infty$.

Simulations To evaluate the performance of the quantile method in a change-point nonlinear model, simulations are realized using Monte Carlo replications for $n = 100$ observations. Moreover, in order to demonstrate the usefulness of the proposed estimators, we compare the performances of the LS and of the quantile methods for a growth model with two change-points: $f(x, \boldsymbol{\beta}) = b_1 - \exp(-b_2 x)$, with $\boldsymbol{\beta} = (b_1, b_2)$. We denote by $\hat{l}_1, \hat{l}_2, \hat{\boldsymbol{\beta}}_1, \hat{\boldsymbol{\beta}}_2, \hat{\boldsymbol{\beta}}_3$ the estimations by LS or by quantile method for the two change-points l_1, l_2 and for three vectors of the three phases. For the errors ε, three distributions were considered: standard normal, Laplace and Cauchy (Table 1, see also [4]). We obtain in all situations that the median of the change-point estimations is very close to the true values. When the errors are Gaussian, very good results are obtained by the two methods. For Laplace errors, the results deteriorate slightly. For Cauchy errors, the quantile method gives very satisfactory results, while by LS method the obtained estimations are biased and with a large variation.

Table 1 Median of the change-point estimations, mean of the regression parameter estimations by LS and quantile methods for a growth model with two change-points in $l_1^0 = 20$, $l_2^0 = 85$ and the true parameters $\boldsymbol{\beta}_1^0 = (0.5, 1)$, $\boldsymbol{\beta}_2^0 = (1, -0.5)$, $\boldsymbol{\beta}_3^0 = (2.5, 1)$

Estimation method	ε law	Median($\hat{l}_1$)	Median($\hat{l}_2$)	Mean($\hat{\boldsymbol{\beta}}_1$) *sd*($\hat{\boldsymbol{\beta}}_1$)	Mean($\hat{\boldsymbol{\beta}}_2$) *sd*($\hat{\boldsymbol{\beta}}_2$)	Mean($\hat{\boldsymbol{\beta}}_3$) *sd*($\hat{\boldsymbol{\beta}}_3$)
LS	$\varepsilon \sim \mathcal{N}$	19	84	(0.52, 1.06)	(0.98, -0.5)	(2.52, 1.07)
				(0.17, 0.64)	*(0.09, 0.02)*	*(0.15, 0.5)*
	$\varepsilon \sim \mathcal{L}$	19	84	(0.58, 1.28)	(0.98, -0.5)	(2.65, 1.13)
				(0.42, 1.56)	*(0.22, 0.05)*	*(0.46, 1.1)*
	$\varepsilon \sim \mathcal{C}$	22	85	(2.51, 1.26)	(2.34, -0.24)	(7.7, 1.75)
				(18.7, 2.2)	*(12.7, 0.96)*	*(42, 3.4)*
Quantile	$\varepsilon \sim \mathcal{N}$	19	84	(0.52, 1.1)	(0.99, -0.5)	(2.53, 1.1)
				(0.16, 0.78)	*(0.09, 0.02)*	*(0.18, 0.8)*
	$\varepsilon \sim \mathcal{L}$	19	84	(0.57, 1.17)	(1, -0.5)	(2.6, 1.45)
				(0.37, 1.28)	*(0.13, 0.04)*	*(0.32, 3.2)*
	$\varepsilon \sim \mathcal{C}$	20	84	(0.58, 1.2)	(0.98, -0.48)	(2.7, 1.75)
				(0.55, 1.1)	*(0.29, 0.23)*	*(0.6, 3.1)*

2.1.2 Empirical Likelihood Test by Off-Line Procedure for a Nonlinear Model

Above we proposed a criterion for finding the number of changes. Now, we propose a statistical test to detect a change in regression model.

Under null hypothesis H_0, we test that there is no change in the regression parameters of model:

$H_0: \quad Y_i = f(\mathbf{X}_i; \boldsymbol{\beta}) + \varepsilon_i, \qquad i = 1, \cdots, n,$

against the hypothesis that the parameters change from $\boldsymbol{\beta}_1$ to $\boldsymbol{\beta}_2$ at an unknown observation k:

$$H_1: \quad Y_i = \begin{cases} f(\mathbf{X}_i; \boldsymbol{\beta}_1) + \varepsilon_i & i = 1, \cdots, k, \\ f(\mathbf{X}_i; \boldsymbol{\beta}_2) + \varepsilon_i & i = k+1, \cdots, n. \end{cases}$$

We test then $H_0: \boldsymbol{\beta}_1 = \boldsymbol{\beta}_2 = \boldsymbol{\beta}$ against $H_1: \boldsymbol{\beta}_1 \neq \boldsymbol{\beta}_2$.

In order to introduce the EL method, let $y_1, \cdots, y_k, y_{k+1}, \cdots, y_n$ be observations for the random variables $Y_1, \cdots, Y_k, Y_{k+1}, \cdots, Y_n$. Consider the following sets $I \equiv \{1, \ldots, k\}$ and $J \equiv \{k+1, \ldots, n\}$ which contain the observation index of the two segments for the model under hypothesis H_1. Corresponding to these sets, we consider the probabilities for observing the value y_i (respectively y_j) of the dependent variable Y_i (respectively Y_j) : $p_i \equiv \mathbb{P}[Y_i = y_i]$, for $i \in I$ and $q_j \equiv \mathbb{P}[Y_j = y_j]$, for $j \in J$.

The regression function f is supposed thrice differentiable in $\boldsymbol{\beta}$ on Γ and continuous on Υ.

Consider the following random vector $\mathbf{g}_i(\boldsymbol{\beta}) = \dot{\mathbf{f}}(\mathbf{X}_i, \boldsymbol{\beta})[Y_i - f(\mathbf{X}_i, \boldsymbol{\beta})]$. Under H_0, the profile EL ratio for $\boldsymbol{\beta}$ has the form

$$\mathscr{R}'_{0,nk}(\boldsymbol{\beta}) = \sup_{(p_1,\cdots,p_k)} \sup_{(q_{k+1},\cdots,q_n)} \Big\{ \prod_{i\in I} kp_i \prod_{j\in J}(n-k)q_j; \sum_{i\in I} p_i = 1, \sum_{j\in J} q_j = 1, \sum_{i\in I} p_i \mathbf{g}_i(\boldsymbol{\beta}) = \sum_{j\in J} q_j \mathbf{g}_j(\boldsymbol{\beta}) = \mathbf{0}_d \Big\}.$$

and under H_1, the profile EL ratio for $\boldsymbol{\beta}_1, \boldsymbol{\beta}_2$ has the form

$$\mathscr{R}'_{1,nk}(\boldsymbol{\beta}_1, \boldsymbol{\beta}_2) = \sup_{(p_1,\cdots,p_k)} \sup_{(q_{k+1},\cdots,q_n)} \Big\{ \prod_{i\in I} kp_i \prod_{j\in J}(n-k)q_j; \sum_{i\in I} p_i = 1, \sum_{j\in J} q_j = 1, \sum_{i\in I} p_i \mathbf{g}_i(\boldsymbol{\beta}_1) = \mathbf{0}_d, \sum_{j\in J} q_j \mathbf{g}_j(\boldsymbol{\beta}_2) = \mathbf{0}_d \Big\}.$$

For testing H_0 against H_1 we consider the profile EL ratio:

$$\frac{\mathscr{R}'_{0,nk}(\boldsymbol{\beta})}{\mathscr{R}'_{1,nk}(\boldsymbol{\beta}_1, \boldsymbol{\beta}_2)}. \tag{4}$$

Under hypothesis H_1, we have a Wilks theorem on each segment (see [10]). Then, since the observations are independent we have:
$-2\log \mathscr{R}'_{1,nk}(\boldsymbol{\beta}_1, \boldsymbol{\beta}_2) \xrightarrow[n\to\infty]{\mathscr{L}} \chi^2(2p)$. Consequently, since the denominator of (4) does not asymptotically depend on $\boldsymbol{\beta}_1$ and $\boldsymbol{\beta}_2$, we are going to consider the test statistic $-2\log \mathscr{R}'_{0,nk}(\boldsymbol{\beta})$. On the other hand, since in expression of $\mathscr{R}'_{0,nk}(\boldsymbol{\beta})$ they are constraints, using the Lagrange multiplier method, we have that maximizing this statistic is equivalent to maximizing the following statistic with respect to $\boldsymbol{\beta}$, η_1, η_2, $\boldsymbol{\lambda}_1, \boldsymbol{\lambda}_2$,

$$\sum_{i\in I}[\log p_i - n\boldsymbol{\lambda}_1^t p_i \mathbf{g}_i(\boldsymbol{\beta})] + \sum_{j\in J}[\log q_j + n\boldsymbol{\lambda}_2^t q_j \mathbf{g}_j(\boldsymbol{\beta})] + \eta_1(\sum_{i\in I} p_i - 1) + \eta_2(\sum_{j\in J} q_j - 1),$$

with $\eta_1, \eta_2, \boldsymbol{\lambda}_1, \boldsymbol{\lambda}_2$ the Lagrange multipliers. The statistic $-2\log \mathscr{R}'_{nk,0}(\boldsymbol{\beta})$ becomes

$$2\sum_{i\in I} \log\big[1 + \frac{n}{k}\boldsymbol{\lambda}_1^t \mathbf{g}_i(\boldsymbol{\beta})\big] + 2\sum_{j\in J} \log\big[1 - \frac{n}{n-k}\boldsymbol{\lambda}_2^t \mathbf{g}_j(\boldsymbol{\beta})\big]. \tag{5}$$

The maximizer of (5) in $\boldsymbol{\beta}$ satisfies $\sum_{i\in I} p_i \boldsymbol{\lambda}_1^t \dot{\mathbf{g}}_i(\boldsymbol{\beta}) - \sum_{j\in J} q_j \boldsymbol{\lambda}_2^t \dot{\mathbf{g}}_j(\boldsymbol{\beta}) = 0$. In order to have a single parameter $\boldsymbol{\lambda}$, we restrict the study to a particular case:

$$\mathbf{V}_{1n}(\boldsymbol{\beta})\boldsymbol{\lambda}_1 = \mathbf{V}_{2n}(\boldsymbol{\beta})\boldsymbol{\lambda}_2, \text{ with } \mathbf{V}_{1n}(\boldsymbol{\beta}) \equiv k^{-1}\sum_{i\in I} \dot{\mathbf{g}}_i(\boldsymbol{\beta}),$$

$$\mathbf{V}_{2n}(\boldsymbol{\beta}) \equiv (n-k)^{-1}\sum_{j\in J} \dot{\mathbf{g}}_j(\boldsymbol{\beta}).$$

Then, statistic (5) becomes:

$$2\sum_{i\in I}\log[1+\frac{n}{k}\boldsymbol{\lambda}^t\mathbf{g}_i(\boldsymbol{\beta})] + 2\sum_{j\in J}\log[1-\frac{n}{n-k}\boldsymbol{\lambda}^t\mathbf{V}_{1n}(\boldsymbol{\beta})\mathbf{V}_{2n}^{-1}(\boldsymbol{\beta})\mathbf{g}_j(\boldsymbol{\beta})]. \quad (6)$$

On the other hand, in order that parameters belong to a bounded set, in the place of k we consider $\theta_{nk} \equiv k/n$. Let be $\hat{\boldsymbol{\lambda}}(\theta_{nk})$ and $\hat{\boldsymbol{\beta}}(\theta_{nk})$ the maximizers of (6). Then $\hat{\boldsymbol{\lambda}}(\theta_{nk})$ and $\hat{\boldsymbol{\beta}}(\theta_{nk})$ are the solution of the score equations. By a computational proof enough (see [5]), we show that, under H_0, $\|\hat{\boldsymbol{\lambda}}(\theta_{nk})\|_2 \overset{a.s.}{\underset{n\to\infty}{\longrightarrow}} 0$ and $\hat{\boldsymbol{\beta}}(\theta_{nk})$ is a consistent estimator of $\boldsymbol{\beta}$. Then, we can consider the following test statistic

$$T_{nk}(\theta_{nk},\boldsymbol{\lambda},\boldsymbol{\beta}) = 2\sum_{i\in I}\log(1+\frac{1}{\theta_{nk}}\boldsymbol{\lambda}^t\mathbf{g}_i(\boldsymbol{\beta})) + 2\sum_{j\in J}\log(1-\frac{1}{1-\theta_{nk}}\boldsymbol{\lambda}^t\mathbf{g}_j(\boldsymbol{\beta})).$$

Because the regression function is nonlinear, in order to that the maximum EL always exists, we consider that the parameter θ_{nk} belongs to closed interval $[\Theta_{1n}, \Theta_{2n}]$, which is included in open interval $(0, 1)$. Finally, the test statistic for testing H_0 against H_1 is:

$$\tilde{T}_n \equiv \max_{\theta_{nk}\in[\Theta_{1n},\Theta_{2n}]} T_{nk}(\theta_{nk}, \hat{\boldsymbol{\lambda}}(\theta_{nk}), \hat{\boldsymbol{\beta}}(\theta_{nk})),$$

with $T_{nk}(\theta_{nk},\boldsymbol{\lambda},\boldsymbol{\beta}) = 2\sum_{i\in I}\log(1+\frac{1}{\theta_{nk}}\boldsymbol{\lambda}^t\mathbf{g}_i(\boldsymbol{\beta}))+2\sum_{j\in J}\log(1-\frac{1}{1-\theta_{nk}}\boldsymbol{\lambda}^t\mathbf{g}_j(\boldsymbol{\beta}))$.

In order to present the asymptotic distribution of $\tilde{T}_n$ we introduce the following functions $A(x) \equiv (2\log x)^{1/2}$, $D(x) \equiv 2\log x + \log\log x$, for $x > 0$, respectively $x > 1$ and the sequence $u(n) \equiv \frac{1+\Theta_{1n}\Theta_{2n}}{\Theta_{1n}(1-\Theta_{2n})}$.

Theorem 3 *Under hypothesis H_0, for classical assumptions on f, $\mathbf{X}$ and $\boldsymbol{\beta}$ (see Theorem 3 of [5]) we have, for all $t \in \mathbb{R}$, that*

$$\lim_{n\to\infty} \mathbb{P}\{A(\log u(n))(\tilde{T}_n)^{\frac{1}{2}} \le t + D(\log u(n))\} = exp(-e^{-t}).$$

The following theorem gives that test statistic $\tilde{T}_n$ has the asymptotic power equal to one (see Theorem 5 of [5]).

Theorem 4 *The power of the EL ratio test $\tilde{T}_n$ converges to 1, as $n \to \infty$.*

As a consequence of these two theorems, for a fixed size α, we can deduct the critical test region:

$$(\tilde{T}_n)^{1/2} \geq \frac{-\log(-\log\alpha) + D(\log u(n))}{A(\log u(n))}.$$

Theorems 3 and 4 allow us to provide an estimator of the change-point location:

$$\hat{k}_n \equiv n \min\{\tilde{\theta}_{nk};\ \tilde{\theta}_{nk} = \underset{\theta_{nk}\in[\Theta_{1n},\Theta_{2n}]}{\operatorname{argmax}} \ T_{nk}(\theta_{nk}, \hat{\boldsymbol{\lambda}}(\theta_{nk}), \hat{\boldsymbol{\beta}}(\theta_{nk}))\}. \tag{7}$$

Simulations We report a Monte Carlo simulation study in order to evaluate the performance of the proposed test statistic $\tilde{T}_n$. We consider the following nonlinear function: $f(x, \boldsymbol{\beta}) = a/b(1 - x^b)$, with $\boldsymbol{\beta} = (a, b)$. Under null hypothesis H_0, the model is $Y_i = a/b(1 - X_i^b) + \varepsilon_i$, for $i = 1, \cdots, n$, with the true values of the parameters $a^0 = 10$, $b^0 = 2$. Under H_1, the model is: $Y_i = a_1/b_1(1 - X_i^{b_1})1\!\!1_{i\leq k_0} + a_2/b_2(1 - X_i^{b_2})1\!\!1_{i>k_0} + \varepsilon_i$, for $i = 1, \cdots, n$, with the true values of the parameters $a_1^0 = 10$, $b_1^0 = 2$, $a_2^0 = 7$, $b_2^0 = 1.75$. In order to obtain more precise false probabilities, we calculate by Monte Carlo replications a new critical value. Accordingly, the empirical test size is 0.05. For the new critical value, we obtain in all situations the empirical power equal to 1. In Table 2 (see [5]), we give the summarized results (mean, standard-deviation, median) for the estimations $\hat{k}_n$ of (7), corresponding to 5000 Monte Carlo replications, for errors of Gaussian or Cauchy distribution. The change occurs at observation 200 or 400. In view of results presented in Table 2, we deduce that the proposed estimation method approaches very well the true value k^0. Note that in all situations, the median and the mean of the change-point estimations coincide or are very close to the true value of k^0.

2.2 *Linear Models in High-Dimension*

We consider now the case of a phenomenon (dependent variable) as a function of one very large variable number, with unknown change-point number. A very significant advancement in variable selection for a model without change-point was

Table 2 Descriptive statistics for the estimations of the change-point by EL method, for $n = 1000$

		$\hat{k}_n$		
Error distribution	k^0	Mean($\hat{k}_n$)	sd($\hat{k}_n$)	Median($\hat{k}_n$)
$\varepsilon_i \sim \mathcal{N}(0, 1)$	200	204	12	200
	400	400	5	400
$\varepsilon_i \sim 1/\sqrt{6}(\chi^2(3) - 3)$	200	204	12	200
	400	400	5	400

realized by Tibshirani [11] proposing the LASSO method. Then, the estimation and model selection are simultaneously treated as a single minimization problem. If the model has change-points, the LASSO method would allow at the same time to estimate the parameters on every segment and eliminates the irrelevant predictive regressors without using sequential hypothesis test.

2.2.1 Adaptive LASSO Quantile for Multiphase Model

When the dimension of $\boldsymbol{\beta}$ is large, it is interesting that a regression parameter estimator to satisfy the *oracle properties*: the nonzero parameter estimator is asymptotically normal and zero parameters are shrunk directly to 0 with a probability converging to one (*sparsity property*).

For a linear model without change-point, [12] proposed an adaptive LASSO estimator which satisfies the oracle properties, while for weaker error conditions, [1] proposes the adaptive LASSO quantile estimator method which we will consider here in a model with change-points. As for the nonlinear case, we first suppose that the number K of change-points is fixed:

$$Y_i = \mathbf{X}_i^t \boldsymbol{\beta}_1 1\!\!1_{1 \leq i < l_1} + \mathbf{X}_i^t \boldsymbol{\beta}_2 1\!\!1_{l_1 \leq i < l_2} + \cdots + \mathbf{X}_i^t \boldsymbol{\beta}_{K+1} 1\!\!1_{l_K \leq i \leq n} + \varepsilon_i, \quad i = 1, \cdots, n.$$

In order to study the properties of LASSO quantile estimator in a change-point model, we suppose for the deterministic design $(\mathbf{X}_i)$ that: $n^{-1} \max_{1 \leq i \leq n} \mathbf{X}_i' \mathbf{X}_i \underset{n \to \infty}{\longrightarrow} 0$ and for any $r = 1, \cdots, K+1$, the matrix $(l_r - l_{r-1})^{-1} \sum_{i=l_{r-1}+1}^{l_r} \mathbf{X}_i \mathbf{X}_i' \underset{n \to \infty}{\longrightarrow} \mathbf{C}_r$.

We define the adaptive LASSO quantile estimator of the change-point location that the minimizer of the quantile process penalized with a weighted L_1 norm:

$$(\hat{l}_1^*, \cdots, \hat{l}_K^*) \equiv \underset{(l_1, \cdots, l_K)}{\operatorname{argmin}} \inf_{(\boldsymbol{\beta}_j, b_j)} \sum_{r=1}^{K+1} [\sum_{i=l_{r-1}+1}^{l_r} \rho_\tau (Y_i - b_r - \mathbf{X}_i^t \boldsymbol{\beta}_r) + \lambda_{(l_{r-1}; l_r)} \hat{\boldsymbol{\omega}}_{(l_{r-1}; l_r)}^t |\boldsymbol{\beta}_r|],$$

with $(b_1, \cdots, b_{K+1})$ the τth quantile of ε on each phase, the weight $\hat{\boldsymbol{\omega}}_{(l_{r-1}; l_r)} \equiv |\hat{\boldsymbol{\beta}}_{(l_{r-1}; l_r)}|^{-g}$, and $\hat{\boldsymbol{\beta}}_{(l_{r-1}; l_r)}$ the quantile estimator of $\boldsymbol{\beta}_r$, calculated on the observations $l_{r-1}+1, \cdots, l_r$. The tuning parameter $\lambda_{n;(l_{r-1}, l_r)}$ depends on the sample size in every segment.

Between l_{r-1} and l_r, we define the regression parameter estimators by adaptive LASSO quantile method as:

$$\hat{\boldsymbol{\beta}}^*_{(l_{r-1}, l_r)} = \underset{\boldsymbol{\beta}}{\operatorname{argmin}} [\sum_{i=l_{r-1}+1}^{l_r} \rho_\tau (Y_i - b_r - \mathbf{X}_i^t \boldsymbol{\beta}_r) + \lambda_{(l_{r-1}; l_r)} \hat{\boldsymbol{\omega}}_{(l_{r-1}; l_r)}^t |\boldsymbol{\beta}_r|].$$

In order to study the oracle properties of the regression parameter adaptive LASSO quantile estimators, we consider for each two consecutive true change-points l_{r-1}^0 and l_r^0, for $r = 1, \cdots, K+1$, the set with the index of nonzero components of the true regression parameters:

$$\mathscr{A}_r^0 \equiv \{k \in \{1, \cdots, p\}; \quad \beta^0_{r,k} \neq 0\},$$

and similarly, the index set of nonzero components of adaptive LASSO quantile estimator of the regression parameters:

$$\hat{\mathscr{A}}^*_{n,r} \equiv \{j \in \{1, \cdots, p\}; \quad \hat{\boldsymbol{\beta}}^*_{(\hat{l}^*_{r-1};\hat{l}^*_r),j} \neq 0\}.$$

The presence of the change-points in model makes that the important sparsity property is not obvious. By Theorem 3.2 of [3] we obtain the convergence rate of the change-point estimators, which implies using with Karush-Kuhn-Tucker conditions that on every segment the sparsity property is true.

Theorem 5 (Theorem 3.3 of [3]) *If the tuning parameter sequence* $(\lambda_{(l_{r-1};l_r)})_{1 \leq r \leq K+1}$ *is a sequence, depending on n, converging to zero, such that* $(l_r - l_{r-1})^{1/2}\lambda_{(l_{r-1};l_r)} \to \infty$ *and if* $(c_{(l_{r-1},l_r)})$ *is another deterministic sequence, such that* $c_{(l_{r-1},l_r)} \to 0$, $(l_r - l_{r-1})c^2_{(l_{r-1},l_r)}/\log(l_r - l_{r-1}) \to \infty$, $\lambda_{(l_{r-1};l_r)}c^{-2}_{(l_{r-1},l_r)} \to 0$, *and* $(l_r - l_{r-1})^{(g-1)/2}\lambda_{(l_{r-1};l_r)} \to \infty$, *as* $n \to \infty$, *we have:*

(i) $(\hat{l}^*_r - \hat{l}^*_{r-1})^{1/2}(\hat{\boldsymbol{\beta}}^*_{(\hat{l}^*_{r-1};\hat{l}^*_r)} - \boldsymbol{\beta}^0_r)_{\mathscr{A}^0_r} \xrightarrow[n\to\infty]{\mathscr{L}} \mathscr{N}(\mathbf{0}, (\tau(1-\tau))\varphi_\varepsilon^{-2}(b^0_r)(\mathbb{V}_{\mathscr{A}^0_r})^{-1})$

(ii) For every $r = 1, \cdots, K$, *we have* $\lim_{n\to\infty} \mathbb{P}\left[\hat{\mathscr{A}}^*_{n,r} = \mathscr{A}^0_r\right] = 1$

Remark 1 To estimate the true change-points number we can propose a criterion, similarly to Schwarz criterion.

Simulations For a model with three phases, we calculate by Monte Carlo simulations the percentage of zero coefficients estimated to zero (true 0) and the percentage of nonzero coefficients estimated to zero (false 0) by two methods: adaptive LASSO and adaptive LASSO quantile (see Table 3). In Table 3 (see also [3]), ε_j is generic error in the jth phase with $j = 1, 2, 3$. Their distributions are Gaussian($\mathscr{N}$) or Exponential($\mathscr{E}$): $\mathscr{E}_1$ for $\mathscr{E}xp(-4.5, 1)$, $\mathscr{E}_2$ for $\mathscr{E}xp(-6.5, 1)$. For Gaussian errors, the two methods give very satisfactory results when the number of observations in

Table 3 Model with three phases

	Interval $(1, \hat{l}_1)$				Interval $(\hat{l}_1, \hat{l}_2)$				Interval $(\hat{l}_2, n)$			
	% true 0		% false 0		% true 0		% false 0		% true 0		% false 0	
Error distribution	aQ	aLS	aQ	aLS	aQ	aLS	aQ	aLS	aQ	aLS	aQ	aLS
$\varepsilon_1, \varepsilon_2, \varepsilon_3 \sim \mathscr{E}_1$	96	82	1	11	99	92	0	2	100	100	0	2
$\varepsilon_1, \varepsilon_2, \varepsilon_3 \sim \mathscr{N}$	97	88	1	12	98	96	0	3	99	100	0	1
$\varepsilon_1 \sim \mathscr{E}_2, \varepsilon_2, \varepsilon_3 \sim \mathscr{E}_1$	97	74	8	10	99	93	0	2	99	100	0	4

$\boldsymbol{\beta}^0_1 \neq \boldsymbol{\beta}^0_2 \neq \boldsymbol{\beta}^0_3$. Percentage (%) of true 0 and of false 0 by adaptive LASSO quantile (aQ) and adaptive LASSO for LS model (aLS)

each phase is large enough. If the errors are of exponential distribution, the adaptive LASSO quantile methods give the best results in terms of true zeros or false zeros percentage.

Then, the adaptive LASSO quantile method is more precise than the adaptive LASSO for LS model.

2.2.2 Empirical Likelihood Test for High-Dimensional Two-Sample Model

In order to have more accurate parameter estimators and a better adjustment for the dependent variable, when p converges to infinity or when the model has outliers, it is more appropriate to use the EL method. The presented results are proved in [6].

We test the following linear model:

$$H_0 : Y_i = \mathbf{X}_i^t \boldsymbol{\beta} + \varepsilon_i, \qquad i = 1, \cdots, n,$$

against

$$H_1 : Y_i = \begin{cases} \mathbf{X}_i^t \boldsymbol{\beta} + \varepsilon_i, & 1 \leq i \leq k, \\ \mathbf{X}_i^t \boldsymbol{\beta}_2 + \varepsilon_i, & k < i \leq n, \end{cases}$$

with $\boldsymbol{\beta} \in \mathbb{R}^p$, $p \to \infty$ as $n \to \infty$ and $\lim_{n\to\infty} k/n \in (0, 1)$. Remark that with respect to the adaptive LASSO quantile model, we consider that the number of explanatory variables is divergent. The two hypotheses can be also written, $H_0 : \boldsymbol{\beta}_2 = \boldsymbol{\beta}$ and $H_1 : \boldsymbol{\beta}_2 \neq \boldsymbol{\beta}$.

If we denote: $\mathbf{z}_i(\boldsymbol{\beta}) \equiv \mathbf{X}_i(Y_i - \mathbf{X}_i^t \boldsymbol{\beta})$, similarly to nonlinear model, we obtain that the EL ratio statistic is:

$$\mathbb{EL}_{nk}(\boldsymbol{\beta}) \equiv 2\sum_{i=1}^{k} \log\left(1 + \frac{n}{k}\boldsymbol{\lambda}^t \mathbf{z}_i(\boldsymbol{\beta})\right) + 2\sum_{j=k+1}^{n} \log\left(1 - \frac{n}{n-k}\boldsymbol{\lambda}^t \mathbf{z}_j(\boldsymbol{\beta})\right),$$

with $\boldsymbol{\lambda} \in \mathbb{R}^p$ the Lagrange multiplier.

The errors ε_i are considered i.i.d. with $\mathbb{E}[\varepsilon_i] = 0$ and $Var(\varepsilon_i) = \sigma^2 < \infty$. Let us consider the following square matrix $\mathbf{V}_n^0 \equiv nk^{-2}\sigma^2 \sum_{i=1}^{k} \mathbf{X}_i \mathbf{X}_i^t + n(n-k)^{-2}\sigma^2 \sum_{j=k+1}^{n} \mathbf{X}_j \mathbf{X}_j^t$.

The following assumptions are considered for ε_i and $\mathbf{X}_i$:

(ELHD1) There exist two constants $C_0, C_1 > 0$, such that $0 < C_0 < \inf_n \gamma_1(\mathbf{V}_n^0) \leq \sup_n \gamma_1(\mathbf{V}_n^0) < C_1 < \infty$.

(ELHD2) $p^{-1} \sum_{s=1}^{p} |X_{i,s}|^q < C_3$, $1 \leq i \leq n$, for some $C_3 > 0$, and $q \geq 4$;

(ELHD3) $\mathbb{E}|\epsilon_1|^{2q} < C_4$, for some $C_4 > 0$.

(ELHD4) $p\, k^{(2-q)/(2q)} \to 0$ and $p\, (n-k)^{(2-q)/(2q)} \to 0$, as $n, k \to \infty$.

(ELHD5) $\sum_{r,s=1}^{p} \alpha^{rrss} = O(p^2)$.

(ELHD6) $\sum_{r,s,u=1}^{p} \alpha^{rsu}\alpha^{rsu} = O(p^{5/2})$ and $\sum_{r,s,u=1}^{p} \alpha^{rss}\alpha^{suu} = O(p^{5/2})$.

(ELHD7) For all $i = 1, \ldots, n$, for $l \in \mathbb{N}^*$, $j_1, \ldots, j_l \in \{1, \ldots, p\}$, and whenever $\sum_{i=1}^{l} d_i \leq 6$, there exists a constant $0 < C_5 < \infty$, such than $\mathbb{E}[w_{i,j_1}^{d_1} \cdots w_{i,j_l}^{d_l}] \leq C_5$.

For assumptions (ELHD5)–(ELHD7) we have used the notations:
$\mathbf{w}_i^0 = (w_{i,1}^0, \cdots, w_{i,p}^0) \equiv (\mathbf{V}_n^0)^{-1/2}\mathbf{z}_i(\boldsymbol{\beta}^0)$,
$\alpha^{t_1 \cdots t_r} \equiv n^{r-1}k^{-r}\sum_{i=1}^{k}\mathbb{E}[w_{i,t_1}^0 \cdots w_{i,t_r}^0] + n^{r-1}(n-k)^{-r}\sum_{i=k+1}^{n}\mathbb{E}[w_{i,t_1}^0 \cdots w_{i,t_r}^0]$.

We first establish an asymptotic approximation of the EL ratio statistic under the null hypothesis.

Proposition 1 *If hypothesis H_0 is true, under assumptions (ELHD1)–(ELHD5), we have $\mathbb{EL}_n(\boldsymbol{\beta}^0) = n\boldsymbol{\psi}_n^{0t}(\mathbf{V}_n^0)^{-1}\boldsymbol{\psi}_n^0 + o_{\mathbb{P}}(p^{1/2})$, with $\boldsymbol{\psi}_n^0 \equiv k^{-1}\sum_{i=1}^{k}\mathbf{X}_i\varepsilon_i - (n-k)^{-1}\sum_{j=k+1}^{n}\mathbf{X}_j\varepsilon_j$.*

By constructing a martingale and applying the martingale limit theorem, we obtain by the following theorem that under hypothesis H_0, the statistic $\mathbb{EL}_{nk}(\boldsymbol{\beta}^0) - p$ converges in law to a standard Gaussian distribution, where Δ_n is a variance of standardization.

Theorem 6 *Under hypothesis H_0, if assumptions (ELHD1)–(ELHD7) hold and if $p = o(n^{1/3})$, then*

$$\frac{\mathbb{EL}_{nk}(\boldsymbol{\beta}^0) - p}{\Delta_n/n} \overset{\mathbb{P}}{=} \frac{n\boldsymbol{\psi}_n^{0t}(\mathbf{V}_n^0)^{-1}\boldsymbol{\psi}_n^0 - p}{\Delta_n/n} \xrightarrow[n\to\infty]{\mathscr{L}} \mathscr{N}(0,1),$$

We remark that when the dimension p converges to infinity, the asymptotic distribution of the EL statistic is Gaussian, which is different from the obtained law for a model with a fixed variable number (see Theorem 3).

Simulations We conducted a Monte Carlo simulation study to evaluate the proposed method in terms of empirical sizes (Table 4, see [6]) for nominal size $\alpha = 0.05$. The model errors are either Exponential or Gaussian and the p-vector $\boldsymbol{\beta}^0 = (1, \cdots, p)$. We obtain that empirical powers are equal to 1, that is, if there are changes in the coefficients of the second phase of the model, the test statistic given by Theorem 6 detects always this change. In order to ameliorate the empirical

Table 4 Empirical size ($\hat{\alpha}$) for Exponential and Gaussian errors by EL test for high-dimensional two-sample model

			Exponential errors	Gaussian errors
n	k	p	$\hat{\alpha}$	$\hat{\alpha}$
600	350	2	0.08	0.07
		10	0.10	0.09
		20	0.11	0.11

Table 5 Empirical critical value $\hat{c}_{1-\alpha/2}$ and corresponding $\hat{\alpha}$, $\hat{\pi}$, by EL test for high-dimensional two-sample model

		Exponential errors			Gaussian errors		
n	k	$\hat{c}_{1-\alpha/2}$	$\hat{\alpha}$	$\hat{\pi}$	$\hat{c}_{1-\alpha}$	$\hat{\alpha}$	$\hat{\pi}$
600	225	3.40	0.03	1	2.85	0.03	1
	375		0.03	1		0.02	1

sizes $\hat{\alpha}$, we recalculate the critical values and then $\hat{\alpha}$ are less than 0.05, for empirical powers always equal to 1 (Table 5, see [6]), for $p = 50$ and $\boldsymbol{\beta}_2^0 = 1 - \boldsymbol{\beta}^\mathbf{0}$.

3 On-Line Procedures

Let us now consider that there is only one model on the first m observations, called also historical data. Then, we will test in real time if the model changes at each observation starting with observation $m + 1$. We will first test the change in real time in a nonlinear model using the statistic of the CUSUM of LS residuals and afterwards in a linear model with large number of variables using the CUSUM statistic of the adaptive LASSO residuals.

3.1 *In Nonlinear Model*

Let us consider the following nonlinear model with independent observations:

$$Y_i = f(\mathbf{X}_i; \boldsymbol{\beta}_i) + \varepsilon_i, \qquad i = 1, \cdots, m, \cdots, m + T_m,$$

with $f : \Upsilon \times \Gamma \to \mathbb{R}$ known up to the parameters $\boldsymbol{\beta}_i$, $\Gamma \subseteq \mathbb{R}^p$, $\Upsilon \subseteq \mathbb{R}^d$ and $\mathbf{X}_i$ the random vector of regressors. We suppose that on the first m observations, no change in the regression parameter has occurred

$$\boldsymbol{\beta}_i = \boldsymbol{\beta}^\mathbf{0}, \qquad \text{for } i = 1, \cdots, m,$$

with $\boldsymbol{\beta}^\mathbf{0}$ (unknown) the true value. For observations after m, we test $H_0 : \boldsymbol{\beta}_i = \boldsymbol{\beta}^\mathbf{0}$, for all $m + 1 \leq i \leq m + T_m$ against

$$H_1 : \text{there exists } k_m^0 \geq 1, \text{ such that } \begin{cases} \boldsymbol{\beta}_{i,m} = \boldsymbol{\beta}^\mathbf{0}, & \text{for } \ m + 1 \leq i \leq m + k_m^0, \\ \boldsymbol{\beta}_{i,m} = \boldsymbol{\beta}_m^1 \neq \boldsymbol{\beta}^\mathbf{0}, & \text{for } \ m + k_m^0 + 1 \leq i \leq m + T_m. \end{cases}$$

The value of $\boldsymbol{\beta}_m^1$ is also unknown. This problem has been addressed in literature when function f is linear $f(\mathbf{x}; \boldsymbol{\beta}) = \mathbf{x}^t \boldsymbol{\beta}$ (see [7, 8]). The following notations are considered in this subsection: $\mathbf{A} \equiv \mathbb{E}[\dot{\mathbf{f}}(\mathbf{X}; \boldsymbol{\beta}^0)]$, $\mathbf{B} \equiv \mathbb{E}[\dot{\mathbf{f}}(\mathbf{X}; \boldsymbol{\beta}^0)\dot{\mathbf{f}}^t(\mathbf{X}; \boldsymbol{\beta}^0)]$ and $\mathscr{D} \equiv \left[\mathbf{A}^t \mathbf{B}^{-1} \mathbf{A}\right]^{1/2}$. Matrix $\mathbf{B}$ is supposed positive definite.

The regression function, the random vector $\mathbf{X}_i$, and the error ε_i satisfy the following assumptions:

(NLS1) $(\varepsilon_i)_{1 \leq i \leq n}$ are i.i.d. $\mathbb{E}[\varepsilon_i] = 0$, $Var[\varepsilon_i] = \sigma^2$ and $\mathbb{E}[|\varepsilon_i|^\nu] < \infty$ for some $\nu > 2$.

(NLS2) $\ddot{\mathbf{f}}(\mathbf{x}, \boldsymbol{\beta})$ is bounded for all $\boldsymbol{\beta}$ in a neighbourhood of $\boldsymbol{\beta}^0$, for all $\mathbf{x} \in \mathbb{R}^d$.

(NLS3) For $i = 1, \cdots, T_m$, the errors ε_i are independent of the random vectors $\mathbf{X}_j$, for $j = 1, \cdots, m + T_m$.

(NLS4) $(m+l)^{-1} \sum_{i=1}^{m+l} f(\mathbf{X}_i; \boldsymbol{\beta}^0) \xrightarrow[m\to\infty]{a.s.} \mathbb{E}[f(\mathbf{X}; \boldsymbol{\beta}^0)]$,

$(m+l)^{-1} \sum_{i=1}^{m+l} \dot{\mathbf{f}}(\mathbf{X}_i; \boldsymbol{\beta}^0) \xrightarrow[m\to\infty]{a.s.} \mathbb{E}[\dot{\mathbf{f}}(\mathbf{X}; \boldsymbol{\beta}^0)]$,

$(m+l)^{-1} \sum_{i=1}^{m+l} \dot{\mathbf{f}}(\mathbf{X}_i; \boldsymbol{\beta}^0)\dot{\mathbf{f}}^t(\mathbf{X}_i; \boldsymbol{\beta}^0) \xrightarrow[m\to\infty]{a.s.} \mathbf{B}$, for all $l = 0, 1, \cdots, T_m$.

(NLS5) $(m + k_m^0 + l)^{-1} \sum_{i=1}^{m+k_m^0+l} f(\mathbf{X}_i; \boldsymbol{\beta}_m^0) \xrightarrow[m\to\infty]{a.s.} \mathbb{E}[f(\mathbf{X}; \boldsymbol{\beta}_m^0)]$,

$(m + k_m^0 + l)^{-1} \sum_{i=1}^{m+k_m^0+l} \dot{\mathbf{f}}(\mathbf{X}_i; \boldsymbol{\beta}_m^0) \xrightarrow[m\to\infty]{a.s.} \mathbb{E}[\dot{\mathbf{f}}(\mathbf{X}; \boldsymbol{\beta}_m^0)]$,

$(m + k_m^0 + l)^{-1} \sum_{i=1}^{m+k_m^0+l} \dot{\mathbf{f}}(\mathbf{X}_i; \boldsymbol{\beta}_m^0)\dot{\mathbf{f}}^t(\mathbf{X}_i; \boldsymbol{\beta}_m^0) \xrightarrow[m\to\infty]{a.s.} \mathbf{B}$, $\forall l = 0, 1, \cdots, T_m - k_m^0$.

Let us consider the following statistic, built as the weighted cumulative sum of the residuals, $0 \leq \gamma < 1/2$, $k = 1, \cdots, T_m$,

$$G(m, k, \gamma) \equiv \sum_{m+1 \leq i \leq m+k} [Y_i - f(\mathbf{X}_i, \hat{\boldsymbol{\beta}}_m)]/g(m, k, \gamma),$$

with, $g(m, k, \gamma) \equiv m^{1/2}(1 + \frac{k}{m})(\frac{k}{k+m})^\gamma$, where $\hat{\boldsymbol{\beta}}_m \equiv \operatorname{argmin}_\beta \sum_{j=1}^m [Y_j - f(\mathbf{X}_j; \boldsymbol{\beta})]^2$ is the LS estimator of $\boldsymbol{\beta}$. Let us also consider the corresponding residuals $\hat{\varepsilon}_i \equiv Y_i - f(\mathbf{X}_i; \hat{\boldsymbol{\beta}}_m)$. We build a test statistic based on the residuals $\hat{\varepsilon}_i$ after the observation m by estimating the parameter $\boldsymbol{\beta}$ on the historical data. The CUSUM of the residuals is $\sum_{i=m+1}^{m+k} \hat{\varepsilon}_i$.

The following theorem (Theorem 3.1 of [1]) gives the asymptotic distribution of the test statistic under the null hypothesis.

Theorem 7 *Under hypothesis H_0, if assumptions (NLS1)–(NLS4) are true, we have for all real $c > 0$ that:*

(i) If $T_m = \infty$ or ($T_m < \infty$ and $\lim_{m\to\infty} T_m/m = \infty$), then

$$\lim_{m\to\infty} \mathbb{P}[\frac{1}{\hat{\sigma}_m} \sup_k | \sum_{i=m+1}^{m+k} \frac{\hat{\varepsilon}_i}{g(m,k,\gamma)} | \le c] = \mathbb{P}[\sup_{0\le t\le \mathscr{D}^{-2}} \frac{(1+t-\mathscr{D}^2 t)|W(t)|}{t^\gamma} \le c]. \tag{8}$$

(ii) If $T_m < \infty$ and $\lim_{m\to\infty} T_m/m = T < \infty$, then the left-hand side of (8) is equal to $\mathbb{P}\left[\sup_{0\le t\le \frac{T}{1+\mathscr{D}^2 T}} \frac{(1+t-\mathscr{D}^2 t)|W(t)|}{t^\gamma} \le c\right]$.

The covariance function of the Wiener process $\{W(t), 0 \le t < \infty\}$ is $Cov(W(s), W(t)) = min(s,t)$, with $s, t \in [0, \mathscr{D}^{-2}]$ for (i) and $s, t \in [0, \frac{T}{1+\mathscr{D}^2 T}]$ for (ii).

We remark that, unlike to the linear case considered by Horváth et al. [7], the asymptotic distribution of the test statistic, under H_0, depends on the regression function f (by $\mathscr{D}$) and on the true parameter $\boldsymbol{\beta}^0$.

In order to have a test statistic for testing H_0 against H_1, it is necessary also to study the behaviour of the statistic under alternative hypothesis H_1. By the following theorem, the statistic converges in probability to infinity as m converges to infinity. For this, we suppose that the change-point k_m^0 is not very far from the last observation of historical data. This supposition poses no problem for practical applications since if H_0 was not rejected until an observation k_m of order m, we reconsider as historical data the observations of one up to k_m. Another assumption is that, before and after the break, on an average, the model is different.

Theorem 8 *Under H_1, if $k_m^0 = O(m)$, assumptions (NLS1)–(NLS4) are true and if $\mathbb{E}[f(\mathbf{X}; \boldsymbol{\beta}^0)] \ne \mathbb{E}[f(\mathbf{X}; \boldsymbol{\beta}_m^1)]$ hold, then*

$$\frac{1}{\hat{\sigma}_m} \sup_{1\le k\le T_m} \left[\left| \sum_{i=m+1}^{m+k} \hat{\varepsilon}_i \right| / g(m,k,\gamma) \right] \xrightarrow[m\to\infty]{\mathbb{P}} \infty.$$

Therefore, we can deduce from Theorems 7 and 8 that the null hypothesis H_0 is rejected in the change-point

$$\hat{k}_m \equiv \begin{cases} \inf\left\{1 \le k \le T_m, \hat{\sigma}_m^{-1}|G(m,k,\gamma)| \ge c_\alpha(\gamma)\right\} \\ \infty, \quad \text{if } \hat{\sigma}_m^{-1}|G(m,k,\gamma)| < c_\alpha(\gamma), \quad \text{for every } 1 \le k \le T_m, \end{cases}$$

where $c_\alpha(\gamma)$ is the $(1-\alpha)$ quantile of the asymptotic distribution obtained by Theorem 7. The proofs of these last two theorems are found in [1].

3.2 In Linear Model with Large Number of Explanatory Variables

We consider now that on the first m observations, we have a classical model of linear regression:

$$Y_i = \mathbf{X}_i^t \boldsymbol{\beta} + \varepsilon_i, \qquad i = 1, \cdots, m, \tag{9}$$

with $\boldsymbol{\beta} \in \mathbb{R}^p$, p fixed but with the possibility that p is very close to m.

For automatic selection of the variables, [12] proposes the adaptive LASSO estimator as:

$$\hat{\boldsymbol{\beta}}_m^* \equiv \underset{\boldsymbol{\beta} \in \mathbb{R}^p}{\operatorname{argmin}} [\sum_{i=1}^m (Y_i - \mathbf{X}_i^t \boldsymbol{\beta})^2 + \lambda_m \sum_{j=1}^p \hat{\omega}_j |\beta_{,j}|],$$

with $\hat{\omega}_j \equiv |\hat{\beta}_{m,j}|^{-g}$, $\hat{\beta}_{m,j}$ the j-th component of the LS estimator $\hat{\boldsymbol{\beta}}_m$, $\beta_{,j}$ the j-th component of $\boldsymbol{\beta}$ and $g > 0$. The tuning parameter sequence (λ_m) is such that $\lambda_m \to \infty$ as $m \to \infty$.

For errors ε_i we have the following assumptions: ε_i i.i.d. $\mathbb{E}[\varepsilon_1] = 0$, $\sigma^2 = Var(\varepsilon_1) < \infty$, $\mathbb{E}[|\varepsilon_1|^\nu] < \infty$, $\nu > 2$.

Let be the set of index of nonzero components of the adaptive LASSO estimator $\hat{\mathscr{A}}_m^* \equiv \{j \in \{1, \cdots, p\};\ \ \hat{\beta}_{m,j}^* \neq 0\}$ and $\mathscr{A} \equiv \{j \in \{1, \cdots, p\};\ \ \beta_{,j}^0 \neq 0\}$ the set of index of nonzero components of the true value $\boldsymbol{\beta}^0$.

For σ^2, we consider the following estimator:

$$\hat{\sigma}_m^{*2} \equiv \frac{1}{m - Card(\hat{\mathscr{A}}_m^*)} \sum_{i=1}^m (Y_i - \mathbf{X}_i^t \hat{\boldsymbol{\beta}}_m^*)^2.$$

As for nonlinear model, we test $H_0 : \boldsymbol{\beta}_i = \boldsymbol{\beta}^0$ for all $i = m+1, m+2, \cdots$ against

$$H_1 : \text{ there exists } k^0 \geq 1 \text{ such that } \begin{cases} \boldsymbol{\beta}_i = \boldsymbol{\beta}^0, & i = m+1, \cdots, m+k^0 \\ \boldsymbol{\beta}_i = \boldsymbol{\beta}^1, & i = m+k^0+1, \cdots \end{cases}$$

with $\boldsymbol{\beta}^0 \neq \boldsymbol{\beta}^1$.

The residuals corresponding to $\hat{\boldsymbol{\beta}}_m^*$ are $\hat{\varepsilon}_i^* = Y_i - \mathbf{X}_i \hat{\boldsymbol{\beta}}_m^*$, for $i = 1, \cdots, m, m+1, \cdots, m+k$.

The following results come from [2]. The tuning parameter (λ_m) satisfies the additional conditions $m^{-1/2}\lambda_m \to 0$, $m^{(g-1)/2}\lambda_m \to \infty$, as $m \to \infty$. Moreover, we assume that the model is significant, i.e. at least one of the regressors affects significantly to the response variable: $\exists j \in \{1, \cdots, p\}$, such that $\beta_{,j}^0 \neq 0$.

Theorem 9

1) If hypothesis H_0 holds, then we have for any real $c > 0$:

$$\lim_{m\to\infty} \mathbb{P}\left[\frac{1}{\hat{\sigma}^*_m} \sup_{1\leq k<\infty} |\sum_{i=m+1}^{m+k} \hat{\varepsilon}^*_i|/g(m,k,\gamma) \leq c\right] = \mathbb{P}\left[\sup_{0\leq t\leq 1} \frac{|W(t)|}{t^\gamma} \leq c\right],$$

with $\{W(t); 0 \leq t < \infty\}$ a Wiener process.

2) If hypothesis H_1 holds, then

$$\frac{1}{\hat{\sigma}^*_m} \sup_{1\leq k<\infty} |\sum_{i=m+1}^{m+k} \hat{\varepsilon}^*_i|/g(m,k,\gamma) \xrightarrow[m\to\infty]{\mathbb{P}} \infty.$$

The proof of this theorem is completely different from that of the classical linear model for LS residuals. For adaptive LASSO estimator, we don't know its explicit form and further, we must take into account the automatic selection of the nonzero parameters. Then, we mainly use the KKT optimality conditions for the CUSUM statistic study.

Simulations The empirical sizes and powers of the CUSUM test statistic corresponding to LS and to adaptive LASSO residuals are given in Table 6 (see [2]), for $p = 400$, $\mathscr{A} = \{3, 30, 90\}$, $k^0 \in \{5, 25\}$, $\gamma \in \{0.25, 0.49\}$, sizes $\alpha \in \{0.025, 0.05\}$. The nonzero components of $\boldsymbol{\beta}^0$ have the values $\beta^0_{,3} = 5$, $\beta^0_{,30} = 2$, $\beta^0_{,90} = -1$. Under hypothesis H_1 only components 90 and 91 change in 0 and -1, respectively. In all cases, the empirical test size corresponding to the adaptive LASSO method is smaller than the theoretical size α. On the other hand, the CUSUM test with LS residuals gives many false alarms. This shows that when the number of variables is very large, the test doesn't work well. In Fig. 1 (see [2]) we represented the empirical density of the stopping time $\hat{k}_m$: with solid line for the density corresponding to LS framework and with doted line for the density corresponding to adaptive LASSO

Table 6 Empirical test size $\hat{\alpha}$ and power $\hat{\pi}$ by CUSUM test statistic for LS and adaptive LASSO (*aLASSO*) residuals, for $\gamma \in \{0.25, 0.49\}$

				$\gamma = 0.25$		$\gamma = 0.49$	
k^0	(m, T)	Method	$\hat{\alpha}, \hat{\pi}$	$\alpha = 0.025$	$\alpha = 0.05$	$\alpha = 0.025$	$\alpha = 0.05$
5	(410,80)	LS	$\hat{\alpha}$	0.94	0.96	0.99	0.99
			$\hat{\pi}$	0.98	0.99	1	1
		aLASSO	$\hat{\alpha}$	0.002	0.002	0.004	0.01
			$\hat{\pi}$	0.97	0.98	0.99	0.99
25	(410,100)	LS	$\hat{\alpha}$	0.97	0.98	1	1
			$\hat{\pi}$	0.98	0.99	1	1
		aLASSO	$\hat{\alpha}$	0	0	0.01	0.02
			$\hat{\pi}$	0.96	0.98	0.96	0.97

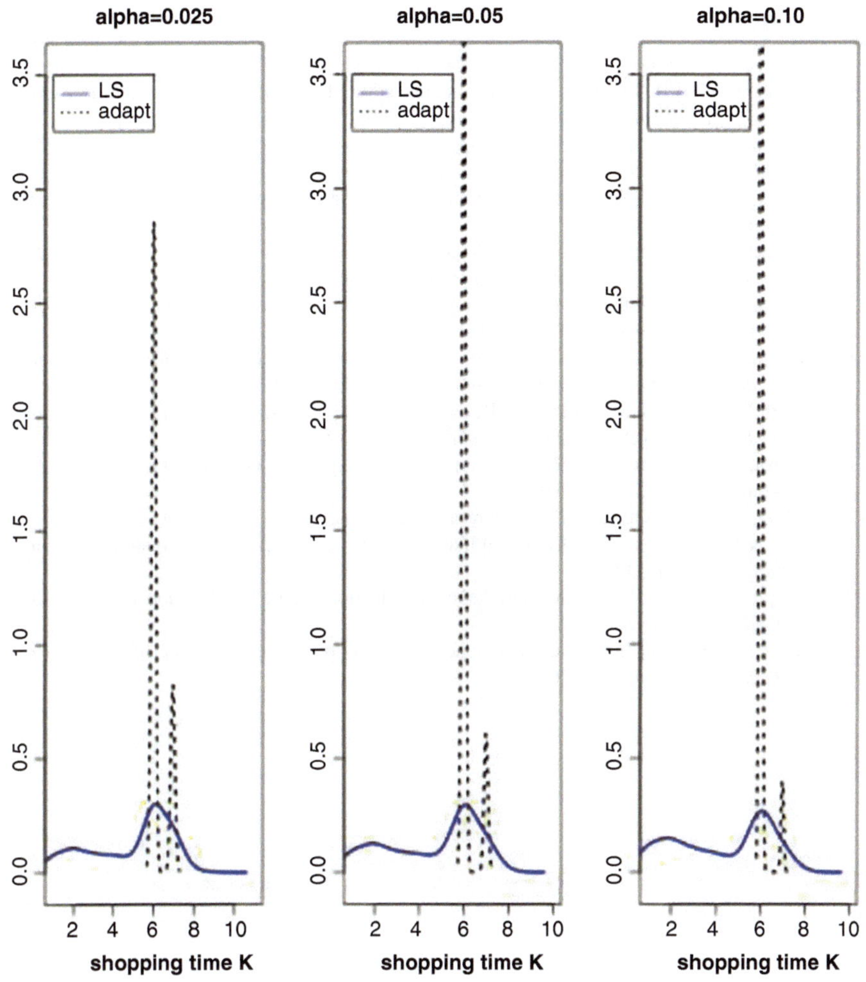

Fig. 1 Estimated density of the stopping time, corresponding to LS and adaptive LASSO residuals, for $\alpha \in \{0.025, 0.05, 0.10\}$

framework. We deduce that the estimation of k^0 by adaptive LASSO method is very accurate and unbiased, when $k^0 = 5$, $\gamma = 0.25$ and for $\alpha \in \{0.025, 0.05, 0.10\}$. The estimation of k^0 by LS method is biased and detects the change before it occurs (false change). These results are consistent with those previously found (in Table 6) for the empirical sizes. The empirical density shape of the stopping time $\hat{k}_m$ using the LS residuals also indicates that the variability of $\hat{k}_m$ is very large.

References

1. Ciuperca, G. (2013). Two tests for sequential detection of a change-point in a nonlinear model. *Journal of Statistical Planning and Inference, 143*(10), 1621–1834.
2. Ciuperca, G. (2015). Real time change-point detection in a model by adaptive LASSO and CUSUM. *Journal de la Société Francaise de Statistique, 156*, 113–132.
3. Ciuperca, G. (2016). Adaptive LASSO model selection in a multiphase quantile regression. *Statistics, 50*(5), 1100–1131.
4. Ciuperca, G. (2017). Estimation in a change-point nonlinear quantile model. *Communications in Statistics – Theory and Methods, 46*(12), 6017–6034.
5. Ciuperca, G., & Salloum, Z. (2015). Empirical likelihood test in a posteriori change-point nonlinear model. *Metrika, 78*(8), 919–952.
6. Ciuperca, G., & Salloum, Z. (2016). Empirical likelihood test for high-dimensional two-sample model. *Journal of Statistical Planning and Inference, 178*, 37–60.
7. Horváth, L., Hušková, M., Kokoszka, P., & Steinebach, J. (2004). Monitoring changes in linear models. *Journal of Statistical Planning and Inference, 126*, 225–251.
8. Hušková, M., & Kirch, C. (2012). Bootstrapping sequential change-point tests for linear regression, *Metrika, 75*, 673–708.
9. Koenker, R., & Bassett, G. (1978). Regression quantiles. *Econometrica, 46*(1), 33–50.
10. Qin, J., & Lawless, J. (2004). Empirical likelihood and general estimating equations. *Annals of Statistics, 22*(4), 300–325.
11. Tibshirani, R. (1996). Regression shrinkage and selection via the LASSO. *Journal of the Royal Statistical Society, Series B, 58*, 267–288.
12. Zou, H. (2006). The adaptive Lasso and its oracle properties. *Journal of the American Statistical Association, 101*(476), 1418–1428.

Variance Estimation Free Tests for Structural Changes in Regression

Barbora Peštová and Michal Pešta

Abstract A sequence of time-ordered observations possesses a trend, which is possibly subject to change at most once at some unknown time point. The aim is to test whether such an unknown change has occurred or not. The change point methods presented here rely on ratio type test statistics based on maxima of the cumulative sums. These detection procedures for the change in regression are also robustified by considering a general score function. The main advantage of the proposed approach is that the variance of the observations neither has to be known nor estimated. The asymptotic distribution of the test statistic under the no change null hypothesis is derived. Moreover, we prove the consistency of the test under alternatives. The results are illustrated through a simulation study, which demonstrates computational efficiency of the procedures. A practical application to real data is presented as well.

1 Introduction and Main Goals

The problem of an unknown change in linear regression models, particularly the *trending regression models*, is studied and procedures for detection of such change within the observed time-ordered sequence are presented. The considered underlying stochastic model allows *at most one change*. The ratio type test statistic elaborated here is derived from the non-ratio type test statistics based on partial sums of the residuals that are commonly used in the change point analysis. They do not need to be standardized by any variance estimate, which makes them a suitable

B. Peštová
Department of Medical Informatics and Biostatistics, Institute of Computer Science, The Czech Academy of Sciences, Prague, Czech Republic
e-mail: pestova@cs.cas.cz

M. Pešta (✉)
Faculty of Mathematics and Physics, Department of Probability and Mathematical Statistics, Charles University, Prague, Czech Republic
e-mail: michal.pesta@mff.cuni.cz

P. Bertail et al. (eds.), *Nonparametric Statistics*, Springer Proceedings in Mathematics & Statistics 250, https://doi.org/10.1007/978-3-319-96941-1_24

alternative for the non-ratio type test statistics, most of all in situations, when it is *difficult to find a variance estimate* with satisfactory properties. Such difficulty can occur especially under alternatives.

A procedure for testing the change in linear regression with equidistant design was considered by Jarušková [10]. Limit distribution for over-all maximum type test statistics under the assumption of no change was given. Antoch and Hušková [1] described the detection of structural changes in a general regression setup. Nonlinear polynomial regression model from the change point perspective was studied by Aue et al. [2]. M-tests for the detection of changes in the linear models were presented in [6]. Furthermore, [7] performed permutation type tests in the linear models. Bootstrapping with and without replacement in the change point analysis for the linear regression models was discussed in [8]. Bai and Perron [3] gave an extension into multiple structural changes, occurring at unknown time points, in the linear regression model estimated by least squares. In [5], the ratio type test statistics for the change in mean were introduced. Lately, [15] considered procedures for detecting the change of regression parameters in the linear model when both the regressors and the errors are weakly dependent in the sense of L_p-m approximability. Peštová and Pešta [13] applied the ratio type test statistics for detection of the structural changes in panel data.

This chapter is structured as follows: Sect. 2 introduces a change point model in regression together with stochastic assumptions. A ratio type test statistic for the change point detection is proposed in Sect. 3. Consequently, the asymptotic behavior of the considered test statistic is derived, which covers the main theoretical contribution. Section 4 contains a simulation study that illustrates performance of the asymptotic test. It numerically emphasizes the advantages and disadvantages of the proposed procedure. A practical application of the developed approach is presented in Sect. 5. Proofs are given in the Appendix.

2 Change Point Problem in Regression

We assume to have a set of observations $Y_{1,n}, \ldots, Y_{n,n}$ obtained at time-ordered points and that these data follow a linear regression model. Particularly, we are interested in studying a situation, where a *change in regression parameters* may occur at some unknown time point τ. Such situation can be formally described as

$$Y_{k,n} = \boldsymbol{h}^\top(k/n)\boldsymbol{\beta} + \boldsymbol{h}^\top(k/n)\boldsymbol{\delta}_n I\{k > \tau_n\} + \varepsilon_k, \quad k = 1, \ldots, n, \tag{1}$$

where $\boldsymbol{\beta} = (\beta_1, \ldots, \beta_p)^\top$, $\boldsymbol{\delta}_n = (\delta_1, \ldots, \delta_p)^\top$, and τ_n are unknown parameters. Functions $\boldsymbol{h}(t) = (h_1(t), \ldots, h_p(t))^\top$ are such that $h_1(t) = 1$ for $t \in [0, 1]$ and $h_j(t),\ j = 2, \ldots, p$ are continuously differentiable functions on [0, 1]. We are going to assume that the error terms $\varepsilon_1, \ldots, \varepsilon_n$ are independent and identically distributed (iid) random variables, satisfying $\mathbb{E}\varepsilon_k = 0$ and $\mathbb{V}\varepsilon_k = \sigma^2 > 0$ for $k = 1, \ldots, n$.

Despite the fact that the observed data $\{Y_{k,n}\}_{k=1,n=1}^{n,\infty}$ form a stochastic *triangular array*, the random disturbances $\{\varepsilon_n\}_{n=1}^{\infty}$ are just a single sequence of random variables. So, the errors remain the same for each row of the triangular array of the observed variables. For the sake of convenience, we suppress the index n in the observations $Y_{k,n}$ as well as in the parameters $\boldsymbol{\delta}_n$ and τ_n (and in the variables depending on the latter) whenever possible. However, we have to keep in mind that in the asymptotic results both $\boldsymbol{\delta}_n \equiv \boldsymbol{\delta}$ and $\tau \equiv \tau_n$ may change, as n increases over all bounds. Then, model (1) can be rewritten as the *regression change point* model

$$Y_k = \boldsymbol{h}^\top(k/n)\boldsymbol{\beta} + \boldsymbol{h}^\top(k/n)\boldsymbol{\delta} I\{k > \tau\} + \varepsilon_k, \quad k = 1, \ldots, n. \tag{2}$$

Model (2) corresponds to the situation where the first τ observations follow the linear model with the regression parameter $\boldsymbol{\beta}$ and the remaining $n - \tau$ observations follow the linear regression model with the changed regression parameter $\boldsymbol{\beta} + \boldsymbol{\delta}$. The parameter τ is called the *change point*.

The basic question is whether a change in the regression parameters occurred at some unknown time point τ or not. Using the above introduced notation, the null hypothesis of no change can be expressed as

$$H_0 : \tau = n. \tag{3}$$

We are going to test this null hypothesis against the alternative that the change occurred at some time point τ prior to the latest observed time n, i.e.,

$$H_1 : \tau < n, \ \boldsymbol{\delta} \neq \boldsymbol{0}. \tag{4}$$

A graphical illustration of the change point model (2) for regression parameters under the alternative can be seen in Fig. 1.

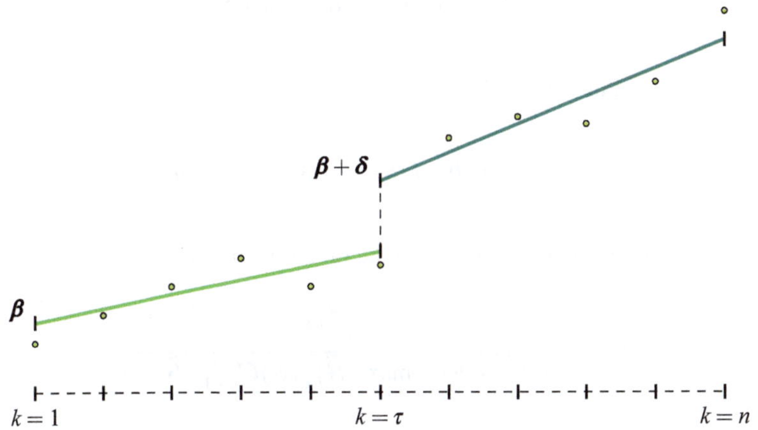

Fig. 1 Illustration of the change point problem in regression

3 Ratio Type Test Statistic for Detection of a Change

For the situation described above, test statistics based on the *weighted partial sums of residuals* are often used, i.e., statistics of the form

$$\boldsymbol{S}_{j,k}(\psi) = \sum_{i=1}^{j} \boldsymbol{h}(i/n)\psi\left(Y_i - \boldsymbol{h}^\top(i/n)\boldsymbol{b}_k(\psi)\right), \quad j,k = p+1,\ldots,n,\ j \leq k,$$

which can also be rewritten elementwise (for the lth element of $\boldsymbol{S}_k$)

$$S_{j,k}^{(l)}(\psi) = \sum_{i=1}^{j} h_l(i/n)\psi\left(Y_i - \boldsymbol{h}^\top(i/n)\boldsymbol{b}_k(\psi)\right),$$

for $l = 1,\ldots,p;\ j,k = p+1,\ldots,n,\ j \leq k$. Here, ψ is a score function and $\boldsymbol{b}_k(\psi)$ is an M-estimate of the regression parameter $\boldsymbol{\beta}$ based on observations $Y_1,\ldots,Y_k$ from model (2) with $\tau = n$ (under the null), i.e., it is a solution of the equation

$$\sum_{i=1}^{k} \boldsymbol{h}(i/n)\psi\left(Y_i - \boldsymbol{h}^\top(i/n)\boldsymbol{b}\right) = \boldsymbol{0}$$

with respect to $\boldsymbol{b}$. Let us similarly denote

$$\widetilde{\boldsymbol{S}}_{j,k}(\psi) = \sum_{i=j+1}^{n} \boldsymbol{h}(i/n)\psi\left(Y_i - \boldsymbol{h}^\top(i/n)\widetilde{\boldsymbol{b}}_k(\psi)\right), \quad j,k = 1,\ldots,n-p-1,\ k \leq j,$$

where $\widetilde{\boldsymbol{b}}_k(\psi)$ is an M-estimate of the parameter $\boldsymbol{\beta}$ based on observations $Y_{k+1},\ldots,Y_n$. That means, it is a solution of the equation

$$\sum_{i=k+1}^{n} \boldsymbol{h}(i/n)\psi\left(Y_i - \boldsymbol{h}^\top(i/n)\boldsymbol{b}\right) = \boldsymbol{0}$$

with respect to $\boldsymbol{b}$. Further, we denote

$$\boldsymbol{C}_{j,k} = \sum_{i=j}^{k} \boldsymbol{h}(i/n)\boldsymbol{h}^\top(i/n), \quad j,k = 1,\ldots,n,\ j \leq k. \tag{5}$$

Using this notation, we may now define the ratio type test statistic

$$\mathscr{R}_n(\psi) = \max_{n\gamma \leq k \leq n-n\gamma} \frac{\max_{1 \leq j \leq k} \boldsymbol{S}_{j,k}^\top(\psi)\boldsymbol{C}_{1,k}^{-1}\boldsymbol{S}_{j,k}(\psi)}{\max_{k \leq j \leq n-1} \widetilde{\boldsymbol{S}}_{j,k}^\top(\psi)\boldsymbol{C}_{k+1,n}^{-1}\widetilde{\boldsymbol{S}}_{j,k}(\psi)}, \tag{6}$$

where $0 < \gamma < 1/2$ is a given constant.

Let us note that the matrices $\boldsymbol{C}_{1,k}$ and $\boldsymbol{C}_{k+1,n}$ become regular after adding Assumption M2 (see below) and considering k and $n-k$ sufficiently large. Being particular, $k-1$ and $n-k-1$ have to be at least as large as p, i.e., the dimension of $\boldsymbol{h}(\cdot)$. Inverses of these matrices in (6) exist, since γ is a fixed constant known in advance and the test statistic $\mathscr{R}_n(\psi)$ is mainly studied from the asymptotic point of view. This ensures that the number of summands in (5) is larger than fixed p.

The idea behind the construction of the test statistic $\mathscr{R}_n(\psi)$ in (6) lies in comparing two total distances of weighted residuals from their center of gravity (by evaluating the ratio of the numerator and the denominator). This view comes from the fact that $\boldsymbol{S}_{j,k}(\psi)$ is a sum of weighted residuals and $\boldsymbol{C}_{1,k}$ acts as a distance measure in the Mahalanobis sense. Similarly for the denominator of (6).

3.1 Asymptotic Properties of the Robust Test Statistic

We proceed with deriving asymptotic properties of the robust ratio type test statistic $\mathscr{R}_n(\psi)$, under the null hypothesis as well as under the alternatives. Before stating the main asymptotic results, we introduce several model assumptions. The following four assumptions apply to the model's errors $\varepsilon_1, \ldots, \varepsilon_n$ and the score function ψ.

Assumption R1 *The random error terms* $\{\varepsilon_i, i \in \mathbb{N}\}$ *are iid random variables with a distribution function* F*, that is symmetric around zero.*

Assumption R2 *The score function* ψ *is a non-decreasing and antisymmetric function.*

Assumption R3

$$0 < \int \psi^2(x)dF(x) < \infty$$

and

$$\int |\psi(x+t_2) - \psi(x+t_1)|^2 dF(x) \leq C_1|t_2 - t_1|^{\eta}, \quad |t_j| \leq C_2,\ j = 1, 2$$

for some constants $\eta > 0$ *and* $C_1, C_2 > 0$.

Assumption R4 *Let us denote* $\lambda(t) = -\int \psi(e-t)dF(e)$ *for* $t \in \mathbb{R}$*. We assume that* $\lambda(0) = 0$ *and that there exists a first derivative* $\lambda'(\cdot)$ *that is Lipschitz in the neighborhood of* 0 *and satisfies* $\lambda'(0) > 0$.

The conditions regarding ψ reduce to the moment restrictions for $\psi_{L_2}(x) = x$ (L_2 method) taking $\eta = 2$. For $\psi_{L_1}(x) = \text{sgn}(x)$ (L_1 method), the conditions reduce to F being a symmetric distribution, having continuous density f in a neighborhood of 0 with $f(0) > 0$, and $\eta = 1$. Similarly, we may consider the derivative of the Huber loss function, i.e.,

$$\psi_H(x) = x\, I\{|x| \le C\} + C\,\mathrm{sgn}(x)\, I\{|x| > C\} \tag{7}$$

for some $C > 0$. In that case to satisfy Assumptions R2–R4, we need to assume F being a symmetric distribution function with the continuous density f in a neighborhood of C and $-C$ satisfying $f(C) > 0$ and $f(-C) > 0$ with $\eta = 2$. Furthermore, the use of score function

$$\psi_\beta(x) = \beta - I\{x < 0\}, \quad x \in \mathbb{R},\ \beta \in (0, 1)$$

results in test procedures related to the β-regression quantiles.

Although we consider only a symmetric distribution function F in our approach, the results can be generalized to include an asymmetric random error distribution. Such distribution is a common source of outlying observations that occur in practical applications. To generalize our approach for the asymmetric distribution function F, one needs to assure that there exists some unique $t_0 \in \mathbb{R}$ such that $\lambda(t_0) = 0$ by modified Assumption R4. Correspondingly, all other assumptions required to hold in the neighborhood of $t = 0$ need to be satisfied in the neighborhood of $t = t_0$.

The next pair of assumptions refer to the system of covariate functions $\boldsymbol{h} = (h_1, \ldots, h_p)^\top$, which represent the model design.

Assumption M1 $h_1(t) = 1,\ t \in [0, 1]$.

Assumption M2 $h_2(\cdot), \ldots, h_p(\cdot)$ *are continuously differentiable functions on* $[0, 1]$ *such that*

$$\int_0^1 h_j(t)dt = 0, \quad j = 2, \ldots, p.$$

The $p \times p$ *matrix functions*

$$\boldsymbol{C}(t) = \left(\int_0^t h_j(x) h_l(x) dx \right)_{j,l=1,\ldots,p}, \quad t \in [0, 1]$$

and $\widetilde{\boldsymbol{C}}(t) = \boldsymbol{C}(1) - \boldsymbol{C}(t)$ *are regular for each* $t \in (0, 1]$ *and* $t \in [0, 1)$, *respectively.*

Let us remark that Assumption M2 covers important situations like polynomial and harmonic polynomial regression.

Now, we may characterize the limit behavior of the test statistic under the null hypothesis.

Theorem 1 (Under null) *Suppose that* $Y_1, \dots, Y_n$ *follow model* (2) *and assume that Assumptions R1–R4 and M1–M2 hold. Then, under null hypothesis* (3),

$$\mathscr{R}_n(\psi) \xrightarrow[n\to\infty]{\text{dist}} \sup_{\gamma\le t\le 1-\gamma} \frac{\sup_{0\le s\le t} \boldsymbol{S}^\top(s,t)\boldsymbol{C}^{-1}(t)\boldsymbol{S}(s,t)}{\sup_{t\le s\le 1} \widetilde{\boldsymbol{S}}^\top(s,t)\widetilde{\boldsymbol{C}}^{-1}(t)\widetilde{\boldsymbol{S}}(s,t)}, \tag{8}$$

such that

$$\boldsymbol{S}(s,t) = \int_s^t \boldsymbol{h}(x)d\mathscr{W}(x) - \boldsymbol{C}(s)\boldsymbol{C}^{-1}(t)\int_0^t \boldsymbol{h}(x)d\mathscr{W}(x), \quad 0\le s\le t\le 1,\ t\ne 0$$

and

$$\widetilde{\boldsymbol{S}}(s,t) = \int_t^s \boldsymbol{h}(x)d\widetilde{\mathscr{W}}(x) - \widetilde{\boldsymbol{C}}(s)\widetilde{\boldsymbol{C}}^{-1}(t)\int_t^1 \boldsymbol{h}(x)d\widetilde{\mathscr{W}}(x), \quad 0\le t\le s\le 1,\ t\ne 1,$$

where $\{\mathscr{W}(x), 0\le x\le 1\}$ *is a standard Wiener process and* $\widetilde{\mathscr{W}}(x) = \mathscr{W}(1) - \mathscr{W}(x)$.

Realizing the property of a standard Wiener process, the definition of a Brownian bridge $\mathscr{B}(x) = \mathscr{W}(x) - x\mathscr{W}(1)$, $x\in[0,1]$, and using stochastic calculus together with Assumption M2, we end up with

$$\int_0^s \boldsymbol{h}(x)\mathrm{d}\mathscr{W}(x) - \boldsymbol{C}(s)\boldsymbol{C}^{-1}(t)\int_0^t \boldsymbol{h}(x)\mathrm{d}\mathscr{W}(x) = \int_0^s \boldsymbol{h}(x)\mathrm{d}\mathscr{B}(x) - \boldsymbol{C}(s)\boldsymbol{C}^{-1}(t)\int_0^t \boldsymbol{h}(x)\mathrm{d}\mathscr{B}(x).$$

Therefore, one can still have the same limit distribution when $\mathrm{d}\mathscr{W}(x)$ is replaced by $\mathrm{d}\mathscr{B}(x)$ and $\mathrm{d}\widetilde{\mathscr{W}}(x)$ is replaced by $\mathrm{d}\widetilde{\mathscr{B}}(x)$, where $\{\mathscr{B}(x), 0\le x\le 1\}$ and $\{\widetilde{\mathscr{B}}(x), 0\le x\le 1\}$ are independent Brownian bridges.

The next theorem describes a situation under some local alternatives.

Theorem 2 (Under Local Alternatives) *Suppose that* $Y_1, \dots, Y_n$ *follow model* (2), *assume that*

$$\|\boldsymbol{\delta}_n\| \to 0 \quad and \quad \sqrt{n}\|\boldsymbol{\delta}_n\| \to \infty, \quad as \quad n\to\infty, \tag{9}$$

and $\tau = [\zeta n]$ *for some* $\gamma < \zeta < 1-\gamma$ *(alternative* (4) *holds). Then, under Assumptions R1–R4 and M1–M2,*

$$\mathscr{R}_n(\psi) \xrightarrow[n\to\infty]{\P} \infty.$$

The previous theorem provides asymptotic consistency of the studied test statistic. The null hypothesis is rejected for large values of the ratio type test statistic.

Table 1 Simulated critical values corresponding to the asymptotic distribution of the test statistic $\mathscr{R}_n(\psi)$ under the null hypothesis and to the covariate functions $h_1(t) = 1$ and $h_2(t) = t - 1/2$

	90%	95%	97.5%	99%	99.5%
$\gamma = 0.1$	7.629223	9.923813	13.114384	17.339711	20.891231
$\gamma = 0.2$	4.351720	5.981026	7.583841	11.048126	14.371703

Table 2 Simulated critical values corresponding to the asymptotic distribution of the test statistic $\mathscr{R}_n(\psi)$ under the null hypothesis and to the covariate functions $h_1(t) = 1$, $h_2(t) = t - 1/2$, and $h_3(t) = 4t^2 - 4t + 2/3$

	90%	95%	97.5%	99%	99.5%
$\gamma = 0.1$	5.638486	7.062320	8.633249	12.225500	13.631059

Being more formal, we reject H_0 at significance level α if $\mathscr{R}_n(\psi) > r_{1-\alpha,\gamma}$, where $r_{1-\alpha,\gamma}$ is the $(1-\alpha)$-quantile of the asymptotic distribution from (8).

3.2 Asymptotic Critical Values

The explicit form of the limit distribution (8) is not known. The critical values may be determined by *simulation of the limit distribution* from Theorem 1. Theorem 2 ensures that we reject the null hypothesis for large values of the test statistic. We simulated the asymptotic distribution from (8) by discretizing the stochastic integrals present in $\boldsymbol{S}(s,t)$ and $\widetilde{\boldsymbol{S}}(s,t)$ and using the relationship of a random walk to a Wiener process. We considered 1000 as the number of discretization points within [0, 1] interval and the number of simulations equal to 1000. We also tried to use higher numbers of discretization points, but only small differences in the critical values were acquired. In Table 1, we present several critical values for covariate functions $h_1(t) = 1$ and $h_2(t) = t - 1/2$.

Moreover, Table 2 shows critical values for covariate functions $h_1(t) = 1$, $h_2(t) = t - 1/2$, and $h_3(t) = 4t^2 - 4t + 2/3$ with $\gamma = 0.1$. The system of covariate functions was chosen in the way that those functions are orthogonal in the $L_2([0,1])$ sense.

A possible extension of the proposed methods, which will be part of the future research, is *bootstrapping*. Using the bootstrap techniques implemented similarly as by Peštová and Pešta [14] for the change in means, one can obtain critical values in an alternative way compared to the presented asymptotic approaches.

4 Simulation Study

A simulation experiment was conducted to study the *finite sample properties* of the asymptotic test for an unknown change in the regression parameters. Performance of the tests based on the ratio type test statistic $\mathscr{R}_n(\psi)$ with $\psi_{L_2}(x) = x$ and

Table 3 Empirical size of the test for the change in regression under H_0 using the asymptotic critical values from $\mathscr{R}_n(\psi)$, considering various significance levels α and $n = 100$

Score	Error distribution	$\alpha = 0.01$	$\alpha = 0.05$	$\alpha = 0.10$
L_2	N(0, 1)	0.023	0.111	0.214
	t_5	0.050	0.191	0.294
L_1	N(0, 1)	0.003	0.045	0.136
	t_5	0.003	0.039	0.112

$\psi_{L_1}(x) = \text{sgn}(x)$ is studied from a numerical point of view. In particular, the interest lies in the *empirical size* of the proposed tests under the null hypothesis and in the empirical *rejection rate* (power) under the alternatives. Random samples of data (1000 repetitions) are generated from the linear regression change point model (2) with $h_1(t) = 1$ and $h_2(t) = t - 1/2$. The number of observations considered is $n = 100$. Higher sample sizes were also tried and the effect of number of observations will be discussed at the end of this section. Parameter γ is set to 0.1.

The innovations are obtained as iid random variables from a standard normal N(0, 1) or Student t_5 distribution. The regression parameters $\boldsymbol{\beta}$ is chosen as $(2, 3)^\top$. Simulation scenarios are produced by varying combinations of these settings. Table 3 provides the empirical size for the asymptotic version of the regression change point test, where the theoretical significance level is $\alpha = 0.01, 0.05$, and 0.10.

Generally, the empirical sizes in case of the asymptotic tests are higher than they should be, especially for the L_2 case. That means the test rejects the null hypothesis more often than one would expect. Possible explanation of this difficulty can be that the test statistics converge only very slowly to the theoretical asymptotic distribution under the null hypothesis. Better performance of the asymptotic test under the null hypothesis is achieved, when the L_1 score function is chosen for $\mathscr{R}_n(\psi)$ compared to the L_2 method. The L_2 method appears to be too liberal in rejecting the null hypothesis. There seems to be no significant effect of the errors' distribution on the empirical rejection rates based on this simulation study.

The performance of the testing procedure under H_1 in terms of the empirical rejection rates is shown in Table 4, where the change point is set to $\tau = n/2$ or $\tau = n/4$. The values of $\boldsymbol{\delta}$ are chosen as $\boldsymbol{\delta} = (1, 1)^\top$ and $\boldsymbol{\delta} = (2, 3)^\top$.

The test power drops when switching from a change point located in the middle of the time series to a change point closer to the beginning or the end of the time series. The errors with heavier tails (represented by the Student t_5 distribution) yield slightly smaller power than the errors with lighter tails (standard normal distribution). When using the L_1 method, power of the test decreases compared to the L_2 method. On the other hand, it keeps the theoretical significance level under the null hypothesis better.

We have shown the simulated powers of the tests only for two choices of $\boldsymbol{\beta}$, however, several other values were tried in the simulation scenarios. There was no visible effect of the value of regression parameter on the test's power. It brought the results that one would naturally expect: the higher the value of change in the

Table 4 Empirical power of the test for the change in regression under H_1 using the asymptotic critical values from $\mathscr{R}_n(\psi)$, considering various significance levels α and $n = 100$

Score	Error distribution	δ	τ	$\alpha = 0.01$	$\alpha = 0.05$	$\alpha = 0.10$
L_2	N(0, 1)	$(1, 1)^\top$	$n/2$	0.110	0.332	0.476
			$n/4$	0.065	0.197	0.330
		$(2, 3)^\top$	$n/2$	0.600	0.855	0.922
			$n/4$	0.096	0.296	0.465
	t_5	$(1, 1)^\top$	$n/2$	0.108	0.319	0.469
			$n/4$	0.087	0.247	0.348
		$(2, 3)^\top$	$n/2$	0.429	0.690	0.801
			$n/4$	0.106	0.281	0.426
L_1	N(0, 1)	$(1, 1)^\top$	$n/2$	0.017	0.145	0.291
			$n/4$	0.002	0.086	0.211
		$(2, 3)^\top$	$n/2$	0.105	0.508	0.715
			$n/4$	0.005	0.115	0.241
	t_5	$(1, 1)^\top$	$n/2$	0.009	0.112	0.237
			$n/4$	0.005	0.072	0.190
		$(2, 3)^\top$	$n/2$	0.091	0.436	0.651
			$n/4$	0.003	0.102	0.233

regression parameter, the higher the power is achieved. Better results in terms of power may be obtained by considering larger sample size or an alternative more far away from the null hypothesis. The choice of $\boldsymbol{\delta} = (5, 5)^\top$ rapidly increases the power of the tests. By considering 250 and 500 observations, one can conclude that the power of the tests increases as the number of observations grows, which is expected.

5 Application to Surface Temperature Data

The analyzed data come from a large data set based on long-term surface temperature measurements at several meteorological stations around the world (for more details see [12], data set HadCRUT3). In Fig. 2, we see the data together with already estimated regression curves. The data represent *temperature anomalies*, i.e., differences from what is expected to be measured in some particular area at some particular time of the year. Each observation corresponds to monthly measurements at the chosen area located in the South Pacific Ocean, close to New Zealand (the center of the 5×5 degree area is located at 177.5 W and 32.5 S). The data covers the period of years 1947–1987 including 485 months.

We took L_2 score function with $p = 3$, $h_1(x) = 1$, $h_2(x) = x - 1/2$, $h_3(x) = 4x^2 - 4x + 2/3$, and $\gamma = 0.1$. The null hypothesis of no change in the parameters of the quadratic regression model based on the asymptotic test (95%

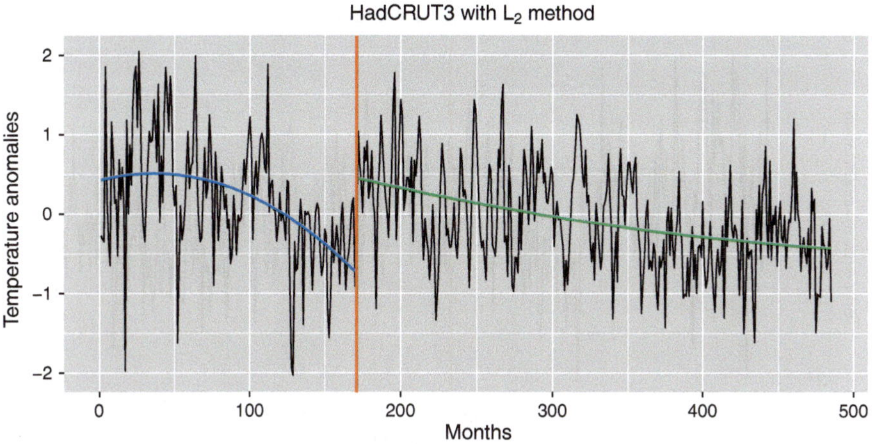

Fig. 2 The surface temperature data analyzed by the L_2 method. Estimated change point is depicted by the orange vertical line and estimated regression curves are drawn by the blue and green lines

critical value equals 7.06232) is rejected both for the L_2 method and the L_1 method, since $\mathscr{R}_{485}(\psi_{L_2}) = 30.59436$ and $\mathscr{R}_{485}(\psi_{L_1}) = 8.477089$.

We estimate the time of change τ by maximizing the numerator in (6) when using all the observations for the statistic in the numerator, i.e.,

$$\widehat{\tau} = \arg\max_k \boldsymbol{S}_{k,n}^\top(\psi)\boldsymbol{C}_{1,n}^{-1}\boldsymbol{S}_{k,n}(\psi). \tag{10}$$

For the L_2 score function, we get $\widehat{\tau} = 171$. Using L_1 approach, we obtain $\widehat{\tau} = 212$.

The estimates of the regression parameters can then be obtained as

$$\boldsymbol{b}_{\widehat{\tau}} = \boldsymbol{C}_{1,\widehat{\tau}}^{-1} \sum_{i=1}^{\widehat{\tau}} \boldsymbol{h}(i/n)\, Y_i \quad \text{and} \quad \widetilde{\boldsymbol{b}}_{\widehat{\tau}} = \boldsymbol{C}_{\widehat{\tau}+1,n}^{-1} \sum_{i=\widehat{\tau}+1}^{n} \boldsymbol{h}(i/n)\, Y_i. \tag{11}$$

The fitted quadratic curves for the surface temperature data before and after the estimated change point are shown in Fig. 2 for the L_2 method and in Fig. 3 for the L_1 method.

Note that the estimated change points using the L_2 and L_1 method are not very close to each other. As a consequence, the estimated quadratic regression parameter corresponding to the fitted curve before the estimated change point using the L_2 method possesses the opposite sign compared to the estimated quadratic regression parameter corresponding to the fitted curve before the estimated change point using the L_1 method. Similarly for the estimated quadratic regression parameter corresponding to the fitted curve after the estimated change point. One of the possible reasons is that there exist more change points in such a long observation history.

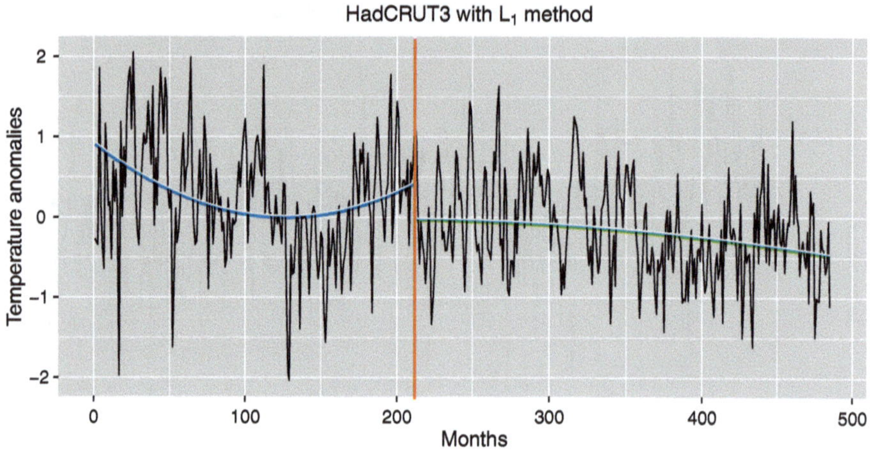

Fig. 3 The surface temperature data analyzed by the L_1 method. Estimated change point is depicted by the orange vertical line and estimated regression curves are drawn by the blue and green lines

6 Conclusion

The ratio type test statistics provide an alternative to the non-ratio type test statistics in situations, in which variance estimation is problematic. The change point detection of at most one *change in the regression parameters* of the regression model is discussed. Asymptotic behavior of the ratio type test statistics is studied under the null hypothesis of no change and under the local alternatives of a change occurring at some unknown time point. We robustify testing procedures by assuming a *general score function* in the test statistics. To obtain the critical values, *approximations of the limit distribution* are used. The simulation study illustrates that even for a relatively small length of the time series, the presented approaches work fine, while various simulation scenarios are considered. The simulations reveal that the methods keep the significance level under the null and provide reasonable powers under the alternatives. Finally, the proposed methods are applied to the real data.

Acknowledgements Institutional support to Barbora Peštová was provided by RVO:67985807. The research of Michal Pešta was supported by the Czech Science Foundation project "DYME—Dynamic Models in Economics" No. P402/12/G097.

Appendix: Proofs

Proof (of Theorem 1) Asymptotic representation for the M-estimate of regression parameter $\boldsymbol{\beta}$ can be obtained by Jurečková et al. [11], Section 5.5:

$$\boldsymbol{b}_k(\psi) - \boldsymbol{\beta} = \boldsymbol{C}_{1,k}^{-1}\frac{1}{\lambda'(0)}\sum_{i=1}^{k}\boldsymbol{h}(i/n)\psi(\varepsilon_i) + O_{\mathsf{P}}(k^{-1}) \tag{12}$$

as $k \to \infty$ and $n\gamma \le k \le n(1-\gamma)$. Moreover, by the Hájek-Rényi-Chow inequality [4] for each $A > 0$, $\varphi \in (0, 1/2]$, and $\boldsymbol{t} \in \mathbb{R}^p$,

$$\P\left[\max_{1\le l\le k/2} k^{-1/2+\varphi}l^{-\varphi}\left|\sum_{i=1}^{l} h_j(i/n)\Big(\psi(\varepsilon_i - \boldsymbol{h}^\top(i/n)\boldsymbol{t}k^{-1/2}) - \psi(\varepsilon_i) + \lambda(\boldsymbol{h}^\top(i/n)\boldsymbol{t}k^{-1/2})\Big)\right| \ge A\right]$$

$$\le D_1A^{-2}k^{-1+2\varphi}\sum_{i=1}^{[k/2]} l^{-2\varphi}\int\left(\psi(\varepsilon - \boldsymbol{h}^\top(i/n)\boldsymbol{t}k^{-1/2}) - \psi(\varepsilon)\right)^2 \mathrm{d}F(\varepsilon)$$

$$\le D_2A^{-2}\left(k^{-1/2}\|\boldsymbol{t}\|\right)^{\eta}, \quad j = 1, \ldots, p, \tag{13}$$

with some constants $D_1, D_2 > 0$, where η is the constant from Assumption R3. Similarly,

$$\P\left[\max_{k/2\le l\le k-1} k^{-1/2+\varphi}(k-l)^{-\gamma}\left|\sum_{i=l+1}^{k} h_j(i/n)\Big(\psi(\varepsilon_i - \boldsymbol{h}^\top(i/n)\boldsymbol{t}k^{-1/2}) - \psi(\varepsilon_i)\right.\right.$$

$$\left.\left. + \lambda(\boldsymbol{h}^\top(i/n)\boldsymbol{t}k^{-1/2})\Big)\right| \ge A\right] \le D_3A^{-2}\left(k^{-1/2}\|\boldsymbol{t}\|\right)^{\eta}, \quad j = 1, \ldots, p, \tag{14}$$

with some constant $D_3 > 0$. Combining (12)–(14), we get

$$\max_{1\le j\le k-1}\frac{1}{\sqrt{k}}\left(\frac{j(k-j)}{k^2}\right)^{-\varphi}\left\|\sum_{i=1}^{j}\boldsymbol{h}(i/n)\psi\left(Y_i - \boldsymbol{h}^\top(i/n)\boldsymbol{b}_k(\psi)\right)\right.$$

$$\left. - \left(\sum_{i=1}^{j}\boldsymbol{h}(i/n)\psi(\varepsilon_i) - \boldsymbol{C}_{1,j}\boldsymbol{C}_{1,k}^{-1}\sum_{i=1}^{k}\boldsymbol{h}(i/n)\psi(\varepsilon_i)\right)\right\| = o_{\mathsf{P}}(1), \tag{15}$$

as $k \to \infty$. Using again the same arguments, we also have as $(n-k) \to \infty$

$$\max_{k+1\le j\le n-1}\frac{1}{\sqrt{n-k}}\left(\frac{(n-j)(j-k)}{(n-k)^2}\right)^{-\varphi}\left\|\sum_{i=j+1}^{n}\boldsymbol{h}(i/n)\psi\left(Y_i - \boldsymbol{h}^\top(i/n)\widetilde{\boldsymbol{b}}_k(\psi)\right)\right.$$

$$\left. - \left(\sum_{i=j+1}^{n}\boldsymbol{h}(i/n)\psi(\varepsilon_i) - \boldsymbol{C}_{j+1,n}\boldsymbol{C}_{k+1,n}^{-1}\sum_{i=k+1}^{n}\boldsymbol{h}(i/n)\psi(\varepsilon_i)\right)\right\| = o_{\mathsf{P}}(1). \tag{16}$$

Hence with respect to (15) and (16), the limit distribution of

$$\left(\max_{1\le j\le k} \boldsymbol{S}_{j,k}^{\top}(\psi)\boldsymbol{C}_{1,k}^{-1}\boldsymbol{S}_{j,k}(\psi),\ \max_{k\le j\le n-1} \widetilde{\boldsymbol{S}}_{j,k}^{\top}(\psi)\boldsymbol{C}_{k+1,n}^{-1}\widetilde{\boldsymbol{S}}_{j,k}(\psi)\right)$$

is the same as that of

$$\left(\max_{1\le j\le k}\left\{\frac{1}{\sqrt{k}}\left(\sum_{i=1}^{j}\boldsymbol{h}(i/n)\psi(\varepsilon_i)-\boldsymbol{C}_{1,j}\boldsymbol{C}_{1,k}^{-1}\sum_{i=1}^{k}\boldsymbol{h}(i/n)\psi(\varepsilon_i)\right)^{\top}\left(\frac{1}{k}\boldsymbol{C}_{1,k}\right)^{-1}\right.\right.$$
$$\left.\frac{1}{\sqrt{k}}\left(\sum_{i=1}^{j}\boldsymbol{h}(i/n)\psi(\varepsilon_i)-\boldsymbol{C}_{1,j}\boldsymbol{C}_{1,k}^{-1}\sum_{i=1}^{k}\boldsymbol{h}(i/n)\psi(\varepsilon_i)\right)\right\},$$
$$\max_{k\le j\le n-1}\left\{\left(\sum_{i=j+1}^{n}\boldsymbol{h}(i/n)\psi(\varepsilon_i)-\boldsymbol{C}_{j+1,n}\boldsymbol{C}_{k+1,n}^{-1}\sum_{i=k+1}^{n}\boldsymbol{h}(i/n)\psi(\varepsilon_i)\right)^{\top}\boldsymbol{C}_{k+1,n}^{-1}\right.$$
$$\left.\left.\left(\sum_{i=j+1}^{n}\boldsymbol{h}(i/n)\psi(\varepsilon_i)-\boldsymbol{C}_{j+1,n}\boldsymbol{C}_{k+1,n}^{-1}\sum_{i=k+1}^{n}\boldsymbol{h}(i/n)\psi(\varepsilon_i)\right)\right\}\right),$$

which by denoting $k=[nt]$ for $t\in(0,1)$ weakly converges in $\mathrm{dist}^2[\gamma,1-\gamma]$ to

$$\left(\sup_{0\le s\le t}\left\{\left(\int_0^s\boldsymbol{h}(x)\mathrm{d}\mathscr{W}(x)-\boldsymbol{C}(s)\boldsymbol{C}^{-1}(t)\int_0^t\boldsymbol{h}(x)\mathrm{d}\mathscr{W}(x)\right)^{\top}\boldsymbol{C}^{-1}(t)\right.\right.$$
$$\left.\left(\int_0^s\boldsymbol{h}(x)\mathrm{d}\mathscr{W}(x)-\boldsymbol{C}(s)\boldsymbol{C}^{-1}(t)\int_0^t\boldsymbol{h}(x)\mathrm{d}\mathscr{W}(x)\right)\right\},$$
$$\sup_{t\le s\le 1}\left\{\left(\int_s^1\boldsymbol{h}(x)\mathrm{d}\widetilde{\mathscr{W}}(x)-\widetilde{\boldsymbol{C}}(s)\widetilde{\boldsymbol{C}}^{-1}(t)\int_t^1\boldsymbol{h}(x)\mathrm{d}\widetilde{\mathscr{W}}(x)\right)^{\top}\widetilde{\boldsymbol{C}}^{-1}(t)\right.$$
$$\left.\left.\left(\int_s^1\boldsymbol{h}(x)\mathrm{d}\widetilde{\mathscr{W}}(x)-\widetilde{\boldsymbol{C}}(s)\widetilde{\boldsymbol{C}}^{-1}(t)\int_t^1\boldsymbol{h}(x)\mathrm{d}\widetilde{\mathscr{W}}(x)\right)\right\}\right),$$

as $n\to\infty$. The weak distributional convergence holds due to [9] (Theorem 1) and Assumption M2, together with the facts that

$$\sup_{0\le s\le t}\left\{\left(\int_0^s\boldsymbol{h}(x)\mathrm{d}\mathscr{W}(x)-\int_0^s\left[\boldsymbol{h}(x)\int_0^t\boldsymbol{h}^{\top}(x)\boldsymbol{C}^{-1}(t)\boldsymbol{h}(y)\mathrm{d}\mathscr{W}(y)\right]\mathrm{d}x\right)^{\top}\boldsymbol{C}^{-1}(t)\right.$$
$$\left.\left(\int_0^s\boldsymbol{h}(x)\mathrm{d}\mathscr{W}(x)-\int_0^s\left[\boldsymbol{h}(x)\int_0^t\boldsymbol{h}^{\top}(x)\boldsymbol{C}^{-1}(t)\boldsymbol{h}(y)\mathrm{d}\mathscr{W}(y)\right]\mathrm{d}x\right)\right\}$$
$$=\sup_{0\le s\le t}\left\{\left(\int_0^s\boldsymbol{h}(x)\mathrm{d}\mathscr{W}(x)-\boldsymbol{C}(s)\boldsymbol{C}^{-1}(t)\int_0^t\boldsymbol{h}(x)\mathrm{d}\mathscr{W}(x)\right)^{\top}\boldsymbol{C}^{-1}(t)\right.$$
$$\left.\left(\int_0^s\boldsymbol{h}(x)\mathrm{d}\mathscr{W}(x)-\boldsymbol{C}(s)\boldsymbol{C}^{-1}(t)\int_0^t\boldsymbol{h}(x)\mathrm{d}\mathscr{W}(x)\right)\right\},$$

and that

$$\sup_{t\leq s\leq 1}\Bigg\{\Bigg(\int_s^1 \boldsymbol{h}(x)\mathrm{d}\widetilde{\mathscr{W}}(x)-\int_s^1\Big[\boldsymbol{h}(x)\int_t^1\boldsymbol{h}^\top(x)\widetilde{\boldsymbol{C}}^{-1}(t)\boldsymbol{h}(y)\mathrm{d}\widetilde{\mathscr{W}}(y)\Big]\mathrm{d}x\Bigg)^\top\widetilde{\boldsymbol{C}}^{-1}(t)$$
$$\Bigg(\int_s^1 \boldsymbol{h}(x)\mathrm{d}\widetilde{\mathscr{W}}(x)-\int_s^1\Big[\boldsymbol{h}(x)\int_t^1\boldsymbol{h}^\top(x)\widetilde{\boldsymbol{C}}^{-1}(t)\boldsymbol{h}(y)\mathrm{d}\widetilde{\mathscr{W}}(y)\Big]\mathrm{d}x\Bigg)\Bigg\}$$
$$=\sup_{t\leq s\leq 1}\Bigg\{\Bigg(\int_s^1 \boldsymbol{h}(x)\mathrm{d}\widetilde{\mathscr{W}}(x)-\widetilde{\boldsymbol{C}}(s)\widetilde{\boldsymbol{C}}^{-1}(t)\int_t^1\boldsymbol{h}(x)\mathrm{d}\widetilde{\mathscr{W}}(x)\Bigg)^\top\widetilde{\boldsymbol{C}}^{-1}(t)$$
$$\Bigg(\int_s^1 \boldsymbol{h}(x)\mathrm{d}\widetilde{\mathscr{W}}(x)-\widetilde{\boldsymbol{C}}(s)\widetilde{\boldsymbol{C}}^{-1}(t)\int_t^1\boldsymbol{h}(x)\mathrm{d}\widetilde{\mathscr{W}}(x)\Bigg)\Bigg\}.$$

Then, the assertion of the theorem directly follows by the continuous mapping theorem. □

Proof (of Theorem 2) Let us choose $k>\tau+1$ and $k=[\xi n]$ for some $\zeta<\xi<1-\gamma$. Moreover, let us take into account assumption (9). Using the same arguments as in (15) and due to the fact that the local alternatives hold, we have, as $n\to\infty$,

$$\max_{1\leq j\leq k}\boldsymbol{S}_{j,k}^\top(\psi)\boldsymbol{C}_{1,k}^{-1}\boldsymbol{S}_{j,k}(\psi)\geq\boldsymbol{S}_{\tau,k}^\top(\psi)\boldsymbol{C}_{1,k}^{-1}\boldsymbol{S}_{\tau,k}(\psi)$$
$$=\left(\sum_{i=1}^{\tau}\boldsymbol{h}(i/n)\psi\left(Y_i-\boldsymbol{h}^\top(i/n)\boldsymbol{b}_k(\psi)\right)\right)^\top\boldsymbol{C}_{1,k}^{-1}\left(\sum_{i=1}^{\tau}\boldsymbol{h}(i/n)\psi\left(Y_i-\boldsymbol{h}^\top(i/n)\boldsymbol{b}_k(\psi)\right)\right)$$
$$=A_{k1}+2A_{k2}+A_{k3}+o_{\mathsf{P}}(1),$$

where

$$A_{k1}=\left(\sum_{i=1}^{\tau}\boldsymbol{h}(i/n)\psi(\varepsilon_i)-\boldsymbol{C}_{1,\tau}\boldsymbol{C}_{1,k}^{-1}\sum_{i=1}^{k}\boldsymbol{h}(i/n)\psi(\varepsilon_i)\right)^\top\boldsymbol{C}_{1,k}^{-1}$$
$$\left(\sum_{i=1}^{\tau}\boldsymbol{h}(i/n)\psi(\varepsilon_i)-\boldsymbol{C}_{1,\tau}\boldsymbol{C}_{1,k}^{-1}\sum_{i=1}^{k}\boldsymbol{h}(i/n)\psi(\varepsilon_i)\right),$$

$$A_{k2}=\left(\sum_{i=1}^{\tau}\boldsymbol{h}(i/n)\psi(\varepsilon_i)-\boldsymbol{C}_{1,\tau}\boldsymbol{C}_{1,k}^{-1}\sum_{i=1}^{k}\boldsymbol{h}(i/n)\psi(\varepsilon_i)\right)^\top\boldsymbol{C}_{1,k}^{-1}$$
$$\left(\sum_{i=1}^{\tau}\boldsymbol{h}(i/n)\boldsymbol{h}^\top(i/n)-\boldsymbol{C}_{1,\tau}\boldsymbol{C}_{1,k}^{-1}\boldsymbol{C}_{\tau+1,k}\right)\boldsymbol{\delta},$$

$$A_{k3} = \boldsymbol{\delta}^\top \left(\sum_{i=1}^{\tau} \boldsymbol{h}(i/n)\boldsymbol{h}^\top(i/n) - \boldsymbol{C}_{\tau+1,k}\boldsymbol{C}_{1,k}^{-1}\boldsymbol{C}_{1,\tau} \right) \boldsymbol{C}_{1,k}^{-1} \left(\sum_{i=1}^{\tau} \boldsymbol{h}(i/n)\boldsymbol{h}^\top(i/n) - \boldsymbol{C}_{1,\tau}\boldsymbol{C}_{1,k}^{-1}\boldsymbol{C}_{\tau+1,k} \right) \boldsymbol{\delta}.$$

Then from the proof of Theorem 1 we get $A_{k1} = O_{\mathsf{P}}(1)$ as $n \to \infty$. Furthermore, with respect to assumption (9),

$$A_{k3} = \boldsymbol{\delta}^\top \left(\boldsymbol{C}_{1,\tau} - \boldsymbol{C}_{\tau+1,k}\boldsymbol{C}_{1,k}^{-1}\boldsymbol{C}_{1,\tau} \right) \boldsymbol{C}_{1,k}^{-1} \left(\boldsymbol{C}_{1,\tau} - \boldsymbol{C}_{1,\tau}\boldsymbol{C}_{1,k}^{-1}\boldsymbol{C}_{\tau+1,k} \right) \boldsymbol{\delta}$$
$$= \boldsymbol{\delta}^\top \boldsymbol{C}_{1,\tau}\boldsymbol{C}_{1,k}^{-1}\boldsymbol{C}_{1,\tau}\boldsymbol{C}_{1,k}^{-1}\boldsymbol{C}_{1,\tau}\boldsymbol{C}_{1,k}^{-1}\boldsymbol{C}_{1,\tau}\boldsymbol{\delta} \xrightarrow[n\to\infty]{\P} \infty.$$

Finally, $|A_{k2}| \leq \sqrt{A_{k1}A_{k3}}$. Therefore, under the considered assumptions, the term A_{k3} is asymptotically dominant over the remaining terms. It follows that

$$\max_{1\leq j\leq k} \boldsymbol{S}_{j,k}^\top(\psi)\boldsymbol{C}_{1,k}^{-1}\boldsymbol{S}_{j,k}(\psi) \xrightarrow[n\to\infty]{\P} \infty.$$

For $\tau + 1 < k = [\xi n]$, the denominator in (6) has the same distribution as under the null hypothesis and it is, therefore, bounded in probability. It follows that the maximum of the ratio has to tend to infinity as well, as $n \to \infty$. □

References

1. Antoch, J., & Hušková, M. (2003). Detection of structural changes in regression. *Tatra Mountains Mathematical Publications, 26*, 201–215.
2. Aue, A., Horváth, L., Hušková, M., & Kokoszka, P. (2008). Testing for changes in polynomial regression. *Bernoulli, 14*(3), 637–660
3. Bai, J., & Perron, P. (1998). Estimating and testing linear models with multiple structural changes. *Econometrica, 66*(1), 47–78.
4. Chow, Y.S., & Teicher, H. (2003). *Probability theory: Independence, interchangeability, martingales* (3rd edn.). New York: Springer.
5. Horváth, L., Horváth, Z., & Hušková, M. (2008). Ratio tests for change point detection. In N. Balakrishnan, E. A. Peña, & M. J. Silvapulle (Eds.), *Beyond parametrics in interdisciplinary research: Festschrift in honor of Professor Pranab K. Sen* (Vol. 1, pp. 293–304). Beachwood, OH: IMS Collections.
6. Hušková, M., & Picek, J. (2002). *M*-tests for detection of structural changes in regression. In *Statistical data analysis based on the L_1-norm and related methods* (pp. 213–227). New York: Springer.
7. Hušková, M., & Picek, J. (2004). Some remarks on permutation type tests in linear models. *Discussiones Mathematicae Probability and Statistics, 24*, 151–182.
8. Hušková, M., & Picek, J. (2005). Bootstrap in detection of changes in linear regression. *Sankhyā: The Indian Journal of Statistics, 67*(2), 200–226.

9. Jandhyala, V., & MacNeill, I. (1997). Iterated partial sum sequences of regression residuals and tests for changepoints with continuity constraints. *Journal of the Royal Statistical Society: Series B (Statistical Methodology), 59*(1), 147–156.
10. Jarušková, D. (2003). Asymptotic distribution of a statistic testing a change in simple linear regression with equidistant design. *Statistics and Probability Letters, 64*(1), 89–95.
11. Jurečková, J., Sen, P.K., & Picek, J. (2012). *Methodology in robust and nonparametric statistics*. Boca Raton, FL: CRC Press.
12. Met Office Hadley Centre. (2008). Met Office Hadley Centre observations datasets. HadCRUT3. [Online; Available from http://www.metoffice.gov.uk/hadobs/; Updated February 2, 2008; Accessed October 02, 2017].
13. Peštová, B., & Pešta, M. (2015). Testing structural changes in panel data with small fixed panel size and bootstrap. *Metrika, 78*(6), 665–689.
14. Peštová, B., & Pešta, M. (2017). Change point estimation in panel data without boundary issue. *Risks, 5*(1), 7.
15. Prášková, Z., & Chochola, O. (2014). M-procedures for detection of a change under weak dependence. *Journal of Statistical Planning and Inference, 149*, 60–76.

Bootstrapping Harris Recurrent Markov Chains

Gabriela Ciołek

Abstract The main objective of this paper is to present bootstrap uniform functional central limit theorem for Harris recurrent Markov chains over uniformly bounded classes of functions. We show that the result can be generalized also to the unbounded case. To avoid some complicated mixing conditions, we make use of the well-known regeneration properties of Markov chains. Regenerative properties of Markov chains can be applied in order to extend some concepts in robust statistics from i.i.d. to a Markovian setting. It is possible to define an influence function and Fréchet differentiability on the torus which allows to extend the notion of robustness from single observations to the blocks of data instead. We present bootstrap uniform central limit theorems for Fréchet differentiable functionals in a Markovian case.

1 Introduction

The naive bootstrap for independent and identically distributed random variables was introduced by Efron [10]. Since then, many resampling methods have been established for dependent data. Hall [12], Carlstein [7], Liu and Singh [17] were among the first to propose block bootstrap methods designed for dependent processes. Block bootstrap methods constitute an active field of research (see [15] for extensive survey on many theoretical and methodological aspects of those procedures). However, they all struggle with problem of the choice of length of the blocks which can abuse the dependence structure of the data. Another approach is to resample residuals or to use wild bootstrap which is tailor-made for heteroscedastic models (see [13] for a huge survey on bootstrap for dependent data).

G. Ciołek (✉)
LTCI, Télécom ParisTech, Université Paris-Saclay, Paris, France

P. Bertail et al. (eds.), *Nonparametric Statistics*, Springer Proceedings in Mathematics & Statistics 250, https://doi.org/10.1007/978-3-319-96941-1_25

In this framework we are interested in bootstrap methods when the data are Markovian. Rajarshi [23] proposed the Markovian bootstrap with estimated transition densities and [20, 21] used transition distribution functions. Kulperger and Prakasa Rao [14] investigated bootstrap procedures for Markov chains with finite state space. Athreya and Fuh [1] extended the approach to countable state spaces. Athreya and Fuh [1, 2] proposed methods which rely on the renewal properties of Markov chains when a (recurrent) state is visited infinitely often. Main idea when working with regenerative processes is to cut the trajectory into data segments which are i.i.d. The regenerative processes are especially appealing to us, since they allow in a natural way to extend the results from the i.d.d. case into Markovian one. Datta and McCormick [9] proposed procedure for bootstrapping additive functionals $1/n \sum_{i=1}^{n} f(X_i)$ when the chain possesses an atom; however, their method is not second-order correct. Bertail and Clémençon [6] modified this procedure, i.e., it is second-order correct in the stationary case. Bertail and Clémençon [4] formulated the regenerative block bootstrap (RBB) method for atomic chains and approximate block bootstrap method (ARBB) for general Harris recurrent Markov chains and proved their consistency. Both methods are a natural generalization of standard non-parametric bootstrap procedure for the i.d.d. data since in a Markovian case we draw regeneration data blocks (instead of single observations) from the empirical distribution function based on blocks. Thus, both procedures are data-driven and do not require a choice of the length of the blocks which is problematic when one uses block bootstrap methods. In parallel to the results of [4], [22] has proved bootstrap CLT for the mean under minimal moment assumptions on renewal times for atomic Markov chains.

This work is aimed to extend the results of [4, 22]. Radulović [22] has proved uniform bootstrap CLT for atomic Markov chains when the class of functions $\mathcal{F}$ is uniformly bounded. We show that the bootstrap uniform CLT also holds when the chain is Harris recurrent. Moreover, we relax the uniform boundedness conditions and impose moment conditions on the envelope. The main difficulty when deriving the asymptotic results for Markov chains is random number of (pseudo-) regeneration blocks. We show that it is feasible to replace this number by its deterministic equivalent which enables to revert to standard theory of empirical processes indexed by classes of functions in the i.i.d. case. It is noteworthy that this chapter is just a short survey based on the paper of [8] and we refer to this paper for more details, comments, and explanations.

Let $X = (X_n)_{n \in \mathbb{N}}$ be a homogeneous Markov chain on a countably generated state space $(E, \mathcal{E})$ with transition probability Π and initial probability ν. Denote by $\mathbb{P}_x$ (resp. $\mathbb{P}_\nu$) the probability measure such that $X_0 = x$ and $X_0 \in E$ (resp. $X_0 \sim \nu$), and $\mathbb{E}_x(\cdot)$ is the $\mathbb{P}_x$-expectation (resp. $\mathbb{E}_\nu(\cdot)$ the $\mathbb{P}_\nu$-expectation). Assume that X is *ψ-irreducible* and aperiodic. Suppose further that X is positive recurrent Harris Markov chain. By the results of [19], we know that any Harris recurrent Markov chain can be extended to a chain that possesses an atom A.

We define the sequence of regeneration times $(\tau_A(j))_{j \geq 1}$. Let

$$\tau_A = \tau_A(1) = \inf\{n \geq 1 : X_n \in A\}$$

be the first time when the chain hits regeneration set A and

$$\tau_A(j) = \inf\{n > \tau_A(j-1), X_n \in A\} \text{ for } j \geq 2$$

are next consecutive visits of X to regeneration set A. By the strong Markov property, given any initial law ν, the sample paths of X can be divided into i.i.d. blocks corresponding to the consecutive visits of the chain to the regeneration set A. The segments of data are of the form:

$$\mathscr{B}_j = (X_{1+\tau_A(j)}, \cdots, X_{\tau_A(j+1)}),\ j \geq 1$$

and take values in the torus $\cup_{k=1}^{\infty} E^k$.

Throughout the paper, we write $l_n = \sum_{i=1}^{n} \mathbb{I}\{X_i \in A\}$ for the total number of consecutive visits of the chain to the atom A. We make the convention that $B_{l_n}^{(n)} = \emptyset$ when $\tau_A(l_n) = n$. Denote by $l(B_j) = \tau_A(j+1) - \tau_A(j),\ j \geq 1$, the length of regeneration blocks. Let $f : E \to \mathbb{R}$ be μ-integrable function. By $u_n(f) = \frac{1}{\tau_A(l_n)-\tau_A(1)} \sum_{i=1}^{n} f(X_i)$ we denote the estimator of the unknown asymptotic mean $\mathbb{E}_\mu(f(X_1))$. Refer to [5] and [18] for a detailed review of the theory of atomic regenerative Markov chains.

Remark 1 We discard the first and the last non-regenerative blocks in order to avoid large bias of estimators build from blocks.

We apply the so-called *splitting technique* (see [19]) in order to artificially construct a regeneration set for chain X. The splitting technique requires existence of so-called small sets, which have the following property: there exists a parameter $\delta > 0$, a positive probability measure Φ supported by S and an integer $m \in N^*$ such that

$$\forall x \in S,\ A \in \mathscr{E}\ \ \Pi^m(x, A) \geq \delta\ \Phi(A), \tag{1}$$

where Π^m denotes the m-th iterate of the transition probability Π. We call (1) the minorization condition and denote by $\mathscr{M}$. We assume that the family of the conditional distributions $\{\Pi(x, dy)\}_{x\in E}$ and the initial distribution ν are dominated by a σ-finite measure λ of reference, so that $\nu(dy) = f(y)\lambda(dy)$ and $\Pi(x, dy) = p(x, y)\lambda(dy)$, for all $x \in E$. The minorization condition requests that Φ is absolutely continuous with respect to λ and that $p(x, y) \geq \delta\phi(y)$, $\lambda(dy)$ a.s. for any $x \in S$, with $\Phi(dy) = \phi(y)dy$. We apply the Nummelin's splitting technique and obtain a split chain $X^{\mathscr{M}} = (X_n, Y_n)_{n\in\mathbb{N}}$, where $(Y_n)_{n\in\mathbb{N}}$ is the sequence of independent Bernoulli r.v.'s with parameter δ. It is known that $X^{\mathscr{M}}$ possesses an atom $A_{\mathscr{M}}$ and inherits all the stability and communication properties of the chain X. Throughout the paper unless specified otherwise, we assume that $\mathbb{E}_{A_{\mathscr{M}}}(\tau_{A_{\mathscr{M}}})^2 < \infty$. The regenerative blocks of the split chain are i.i.d. In practice, one wants to approximate the Nummelin's construction. The main idea is to approximate successive hitting times of $A_{\mathscr{M}} = S \times \{1\}$ by the sequence $\hat{\tau}_{A_{\mathscr{M}}}(i),\ i = 1, \cdots, \hat{l}_n$, where $\hat{l}_n = \sum_{i=1}^{n} \mathbb{I}\{X_i \in S, \hat{Y}_i = 1\}$ is the total number of visits of the split chain to $A_{\mathscr{M}}$ up to time n and the random vector $\hat{\mathbb{Y}}_n = (\hat{Y}_1, \cdots, \hat{Y}_n)$ is the approximation of the vector $(Y_1, \cdots, Y_n)$ via

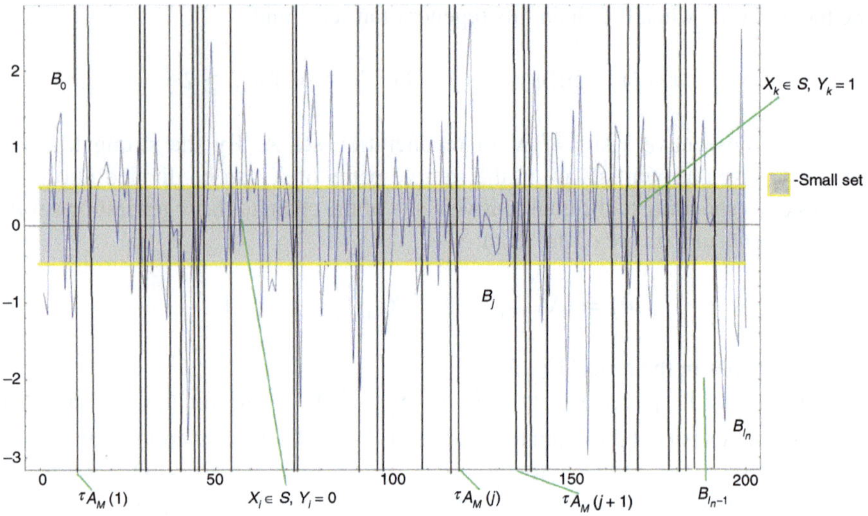

Fig. 1 Regeneration block construction for AR(1) model

approximating procedure described in [4]. The approximated blocks are of the form:

$$\hat{\mathscr{B}}_0 = (X_1, \cdots, X_{\hat{\tau}_{A_{\mathscr{M}}}(1)}), \cdots, \hat{\mathscr{B}}_j = (X_{\hat{\tau}_{A_{\mathscr{M}}}(j)+1}, \cdots, X_{\hat{\tau}_{A_{\mathscr{M}}}(j+1)}), \cdots,$$

$$\hat{\mathscr{B}}_{\hat{l}_n-1} = (X_{\hat{\tau}_{A_{\mathscr{M}}}(\hat{l}_n-1)+1}, \cdots, X_{\hat{\tau}_{A_{\mathscr{M}}}(\hat{l}_n)}), \hat{\mathscr{B}}^{(n)}_{\hat{l}_n} = (X_{\hat{\tau}_{A_{\mathscr{M}}}(\hat{l}_n)+1}, \cdots, X_{n+1}).$$

Figure 1 illustrates a regeneration block construction for an autoregressive model of order one. Moreover, we denote by $\hat{n}_{A_{\mathscr{M}}} = \hat{\tau}_{A_{\mathscr{M}}}(\hat{l}_n) - \hat{\tau}_{A_{\mathscr{M}}}(1) = \sum_{i=1}^{\hat{l}_n-1} l(\hat{B}_j)$ the total number of observations after the first and before the last pseudo-regeneration times. Let $\sigma_f^2 = \frac{1}{\mathbb{E}_{A_{\mathscr{M}}}(\tau_{A_{\mathscr{M}}})}\mathbb{E}_{A_{\mathscr{M}}}\left(\sum_{i=1}^{\tau_{A_{\mathscr{M}}}}\{f(X_i) - \mu(f)\}^2\right)$ be the asymptotic variance. Furthermore, we set that $\hat{\mu}_n(f) = \frac{1}{\hat{n}_{A_{\mathscr{M}}}}\sum_{i=1}^{\hat{l}_n-1} f(\hat{B}_j)$, where

$$f(\hat{B}_j) = \sum_{i=1+\hat{\tau}_{A_{\mathscr{M}}}(j)}^{\hat{\tau}_{A_{\mathscr{M}}}(j+1)} f(X_i) \text{ and } \hat{\sigma}_n^2(f) = \frac{1}{\hat{n}_{A_{\mathscr{M}}}}\sum_{i=1}^{\hat{l}_n-1}\left\{f(\hat{B}_i) - \hat{\mu}_n(f)l(\hat{B}_i)\right\}^2.$$

2 Bootstrapping Harris Recurrent Markov Chains

To establish our uniform bootstrap results we need to generate bootstrap blocks $B_1^*, \cdots, B_k^*$ which are obtained from approximate regenerative block bootstrap algorithm (ARBB) introduced by [4]. For completeness of exposition we recall the ARBB procedure below.

Algorithm 1 (ARBB Procedure)

1. *Draw sequentially bootstrap data blocks* $B_1^*, \cdots, B_k^*$ *(we denote the length of the blocks by* $l(B_j^*)$, $j = 1, \cdots, k$*) independently from the empirical distribution function*

$$\hat{\mathscr{L}}_n = \frac{1}{\hat{l}_n - 1} \sum_{i=1}^{\hat{l}_n - 1} \delta_{\hat{B}_i},$$

where $\hat{B}_i$, $i = 1, \cdots, \hat{l}_n - 1$ *are initial pseudo-regeneration blocks. We draw the bootstrap blocks until* $l^*(k) = \sum_{i=1}^{k} l(B_i^*)$ *exceeds n. We denote* $l_n^* = \inf\{k : l^*(k) > n\}$.
2. *Bind the bootstrap blocks obtained in step 1 in order to construct the ARBB bootstrap sample* $X^{*(n)} = (X_1^*, \cdots, X_{l_n^*-1}^*)$.
3. *Compute the ARBB statistic and its ARBB distribution, namely* $T_n^* = T(X^{*(n)}) = T(B_1^*, \cdots, B_{l_n^*-1}^*)$ *and its standardization* $S_n^* = S(X^{*(n)}) = S(B_1^*, \cdots, B_{l_n^*-1}^*)$.
4. *Compute the ARBB distribution*

$$H_{ARBB}(x) = \mathbb{P}^*(S_n^{*-1}(T_n^* - T_n) \leq x),$$

where $\mathbb{P}^*$ *is the conditional probability given the data.*

Few more pieces of notation: we denote by $n_{A_{\mathcal{M}}}^* = \sum_{i=1}^{l_n^*-1} l(B_j^*)$ the length of the bootstrap sample, where $l_n^* = \inf\left\{k : l^*(k) = \sum_{i=1}^{k} l(B_j^*) > n\right\}$. Moreover, we write $\mu_n^*(f) = \frac{1}{n_{A_{\mathcal{M}}}^*} \sum_{i=1}^{l_n^*-1} f(B_i^*)$ and $\sigma_n^{*2}(f) = \frac{1}{n_{A_{\mathcal{M}}}^*} \sum_{i=1}^{l_n^*-1} \{f(B_i^*) - \mu_n^*(f) l(B_j^*)\}^2$ for bootstrap estimates of mean and variance.

We derive our results under assumptions on Harris chain formulated in [4]. We recall them below for the reader's convenience (see [4], page 700, for details and discussion on the conditions).

Let $(\alpha_n)_{n \in \mathbb{N}}$ be a sequence of nonnegative numbers that converges to zero. We impose that the positive recurrent Harris Markov chain X fulfills the following conditions (compare with Theorems 3.2 and 3.3 in [4]):

A1. S is chosen so that $\inf_{x \in S} \phi(x) > 0$ and the transition density p is estimated by p_n at the rate α_n (usually we consider $\alpha_n = \frac{\log(n)}{n}$) for the mean squared error (MSE) when error is measured by the L^∞ loss over S^2.

Let $k \geq 2$ be a real number.

$\mathscr{H}_1(f, k, \nu)$. The small set S is such that

$$\sup_{x \in S} \mathbb{E}_x \left[\left(\sum_{i=1}^{\tau_S} |f(X_i)| \right)^k \right] < \infty \ \textit{and} \ \mathbb{E}_\nu \left[\left(\sum_{i=1}^{\tau_S} |f(X_i)| \right)^k \right] < \infty.$$

$\mathscr{H}_2(k, \nu)$. The set S is such that $\sup_{x \in S} \mathbb{E}_x(\tau_S^k) < \infty$ and $\mathbb{E}_\nu(\tau_S^k) < \infty$.

$\mathscr{H}_3$. Density $p(x, y)$ is estimated by $p_n(x, y)$ at the rate α_n for the MSE when error is measured by the L^∞ loss over $S \times S$:

$$\mathbb{E}_\nu \left(\sup_{(x,y) \in S \times S} |p_n(x, y) - p(x, y)|^2 \right) = O(\alpha_n), \quad \text{as } n \to \infty.$$

$\mathscr{H}_4$. The density ϕ is such that $\inf_{x \in S} \phi(x) > 0$.

$\mathscr{H}_5$. The transition density $p(x, y)$ and its estimate $p_n(x, y)$ are bounded by a constant $R < \infty$ over S^2.

Denote by $BL_1(\mathscr{F})$ the set of all 1-Lipschitz bounded functions on $l^\infty(\mathscr{F})$. We define the bounded Lipschitz metric on $l^\infty(\mathscr{F})$ as

$$d_{BL_1}(X, Y) = \sup_{b \in BL_1(l^\infty(\mathscr{F}))} |\mathbb{E}b(X) - \mathbb{E}b(Y)|; \quad X, Y \in l^\infty(\mathscr{F}).$$

We choose to work with bounded Lipschitz metric since it metrizes the weak convergence of empirical processes.

When one formulates bootstrap procedure in order to obtain bootstrap estimate of sampling distribution, it is crucial to know if the two distributions are sufficiently close.

Definition 1 We say that $\mathbb{Z}_n^*$ is weakly consistent of $\mathbb{Z}_n$ if $d_{BL_1}(\mathbb{Z}_n^*, \mathbb{Z}_n) \xrightarrow{P} 0$.

2.1 *Bootstrap Uniform Central Limit Theorems for Harris Recurrent Markov Chains*

To establish uniform bootstrap CLT over permissible, uniformly bounded classes of functions $\mathscr{F}$, we need to control the size of $\mathscr{F}$. We require the finiteness of its *covering number* $N_p(\epsilon, Q, \mathscr{F})$ which is interpreted as the minimal number of balls with radius ϵ needed to cover $\mathscr{F}$ in the norm $L_p(Q)$ and Q is a measure on E with finite support. Moreover, we impose the finiteness of the *uniform entropy integral* of $\mathscr{F}$, namely $\int_0^\infty \sqrt{\log N_2(\epsilon, \mathscr{F})} d\epsilon < \infty$, where $N_2(\epsilon, \mathscr{F}) = \sup_Q N_2(\epsilon, Q, \mathscr{F})$. In the following we consider process $\mathbb{Z}_n$ defined as

$$\mathbb{Z}_n = \hat{n}_{A_{\mathscr{M}}}^{1/2} \left[\frac{1}{\hat{n}_{A_{\mathscr{M}}}} \sum_{i=1}^{\hat{l}_n - 1} \left\{ f(\hat{B}_i) - l(\hat{B}_i)\mu(f) \right\} \right]. \tag{2}$$

Theorem 1 *Suppose that* (X_n) *is a positive recurrent Harris Markov chain and the assumptions A1,* $\mathscr{H}_1(f,\rho,\nu)$, $\mathscr{H}_2(\rho,\nu)$ *with* $\rho \geq 4$, $\mathscr{H}_3$, $\mathscr{H}_4$, *and* $\mathscr{H}_5$ *are satisfied by* (X_n). *Assume further, that* $\mathscr{F}$ *is a permissible, uniformly bounded class of functions and* $\int_0^\infty \sqrt{\log N_2(\epsilon,\mathscr{F})}d\epsilon < \infty$. *Then, the process*

$$\mathbb{Z}_n^* = n_{A_{\mathscr{M}}}^{*1/2}\left[\frac{1}{n_{A_{\mathscr{M}}}^*}\sum_{i=1}^{l_n^*-1} f(B_i^*) - \frac{1}{\hat{n}_{A_{\mathscr{M}}}}\sum_{i=1}^{\hat{l}_n-1} f(\hat{B}_i)\right] \tag{3}$$

converges in probability under $\mathbb{P}_\nu$ *to a gaussian process G indexed by* $\mathscr{F}$ *whose sample paths are bounded and uniformly continuous with respect to the metric* $L_2(\mu)$.

Proof The proof is based on the bootstrap central limit theorem introduced by [11]. Finite dimensional convergence follows directly from Theorem 5.9 from [16] coupled with Theorems 3.2 and 3.3 in [4] (see [8] for details). Next, we need to verify if for every $\epsilon > 0$

$$\lim_{\delta\to 0}\limsup_{n\to\infty}\mathbb{P}^*(\|\mathbb{Z}_n^*\|_{\mathscr{F}_\delta} > \epsilon) = 0 \text{ in probability under } \mathbb{P}_\nu, \tag{4}$$

where $\|R\|_{\mathscr{F}_\delta} := \sup\{|R(f) - R(g)| : \rho(f,g) < \delta\}$ and $R \in l^\infty(\mathscr{F})$. Moreover, $\mathscr{F}$ must be totally bounded in $L_2(\mu)$. The total boundedness of $\mathscr{F}$ was shown by [16]. In order to show (4), we replace the random numbers $n_{A_{\mathscr{M}}}^*$ and l_n^* by their deterministic equivalents. By the same arguments as in the proof of Theorem 3.3 in [4], we have that

$$\mathbb{Z}_n^*(f) = \frac{1}{\sqrt{n}}\left[\sum_{i=1}^{1+\left\lfloor \frac{n}{\mathbb{E}_{A_{\mathscr{M}}}(\tau_A)}\right\rfloor}\{f(B_i^*) - \hat{\mu}_n(f)l(B_i^*)\}\right] + o_{\mathbb{P}^*}(1),$$

where $\lfloor\cdot\rfloor$ is an integer part. Thus, we can switch to the analysis of the process

$$\mathbb{U}_n^*(f) = \frac{1}{\sqrt{n}}\left[\sum_{i=1}^{1+\left\lfloor \frac{n}{\mathbb{E}_{A_{\mathscr{M}}}(\tau_A)}\right\rfloor}\{f(B_i^*) - \hat{\mu}_n(f)l(B_i^*)\}\right].$$

Note, that $\{f(B_i^*) - \hat{\mu}_n(f)l(B_i^*)\},\ i \geq 1$ are i.i.d.

Next, take $h = f - g$. Denote by $w_n = 1 + \left\lfloor \frac{n}{\mathbb{E}_{A_{\mathcal{M}}}(\tau_A)} \right\rfloor$ and $Y_i = l(B_i^*) - \hat{\mu}_n(h)l(B_i^*)$. We introduce one more piece of notation: $Y_i = h(B_i^* - \hat{\mu}_n(f)l(B_i^*)$. Then, note that

$$\mathbb{P}^*(\|Y_1 + \cdots + Y_{w_n}\|_{\mathcal{F}_\delta} > \sqrt{n}\epsilon) \leq \mathbb{P}^*\left(\|h(B_1^*) + \cdots + h(B_{w_n}^*)\|_{\mathcal{F}_\delta} > \frac{\sqrt{n}\epsilon}{2}\right)$$
$$+ \mathbb{P}^*\left(\|l(B_1^*)\hat{\mu}_{n,h} + \cdots + l(B_{w_n}^*)\hat{\mu}_{n,h}\|_{\mathcal{F}_\delta} > \frac{\sqrt{n}\epsilon}{2}\right) = I + II. \quad (5)$$

We treat terms I and II separately. Observe that by Markov's inequality we obtain for the first term

$$\mathbb{P}^*\left(\|h(B_1^*) + \cdots + h(B_{w_n}^*)\|_{\mathcal{F}_\delta} > \frac{\sqrt{n}\epsilon}{2}\right) \leq \frac{4\mathbb{E}^*(\|h(B_1^*) + \cdots + h(B_{w_n}^*)\|_{\mathcal{F}_\delta})^2}{n}$$
$$= \frac{4w_n\mathbb{E}^*(\|h(B_1)\|_{\mathcal{F}_\delta})^2}{n}$$

by the fact that $h(B_i^*),\ i \geq 1$ are i.i.d. Next, we deduce that

$$\mathbb{E}^*(\|h(B_1^*)\|_{\mathcal{F}_\delta})^2 = \frac{1}{w_n}\sum_{i=1}^{w_n}\|h(B_1)\|_{\mathcal{F}_\delta}^2 \to \mathbb{E}_{A_{\mathcal{M}}}(\|h(B_1)\|_{\mathcal{F}_\delta})^2 \ \ a.s.$$

Moreover,

$$\mathbb{E}_{A_{\mathcal{M}}}(\|h(B_1)\|_{\mathcal{F}_\delta})^2 = \mathbb{E}_{A_{\mathcal{M}}}\left(\|\sum_{i=1}^{\tau_{A_{\mathcal{M}}}} h(X_i)\|_{\mathcal{F}_\delta}\right)^2$$
$$= \mathbb{E}_{A_{\mathcal{M}}}\left(\|\sum_{i=1}^{\tau_{A_{\mathcal{M}}}} h^2(X_i)\|_{\mathcal{F}_\delta}\right) + 2\mathbb{E}_{A_{\mathcal{M}}}\left(\|\sum_{i=1}^{\tau_{A_{\mathcal{M}}}}\sum_{i\neq j} h(X_i)h(X_j)\|_{\mathcal{F}_\delta}\right)$$
$$\leq \delta^2\mathbb{E}_{A_{\mathcal{M}}}(\tau_{A_{\mathcal{M}}}) + 2\delta^2\mathbb{E}_{A_{\mathcal{M}}}(\tau_{A_{\mathcal{M}}})^2 \to 0 \quad (6)$$

in $\mathbb{P}_\nu$-probability as $\delta \to 0$. Thus, we conclude that

$$\mathbb{P}^*\left(\|h(B_1^*) + \cdots + h(B_{w_n}^*)\|_{\mathcal{F}_\delta} > \frac{\sqrt{n}\epsilon}{2}\right) \to 0 \ \ in \ \ \mathbb{P}_\nu - probability \quad (7)$$

by (6) and $w_n/n \leq 1$.

In order to control the term II, we apply Markov's inequality and get

$$\begin{aligned} &\mathbb{P}^*\left(\|l(B_1^*)\hat{\mu}_{n,h}+\cdots+l(B_{w_n}^*)\hat{\mu}_{n,h}\|_{\mathscr{F}_\delta}>\frac{\sqrt{n}\epsilon}{2}\right)\\ &\le \frac{4\mathbb{E}^*\left(\|l(B_1^*)\hat{\mu}_{n,h}+\cdots+l(B_{w_n}^*)^2\hat{\mu}_{n,h}\|_{\mathscr{F}_\delta}\right)^2}{n}\\ &=\frac{4w_n\mathbb{E}^*(l(B_1^*))^2\|\hat{\mu}_{n,h}\|^2_{\mathscr{F}_\delta}}{n}. \end{aligned} \tag{8}$$

Obviously, $w_n/n \le 1$ and $\|\hat{\mu}_{n,h}\|_{\mathscr{F}_\delta} \to 0$ in $\mathbb{P}_\nu$-probability since $\mathbb{Z}_n$ is stochastically equicontinuous. Bertail and Clémençon [4] have showed that

$$\mathbb{E}^*\left(l(B_1^*)^2|X^{(n+1)}\right) \to \mathbb{E}_{A_{\mathscr{M}}}(\tau_{A_{\mathscr{M}}})^2<\infty \tag{9}$$

in $\mathbb{P}_\nu$-probability along the sample as $n \to \infty$. Thus, by (8) and (9) we have that

$$\mathbb{P}^*\left(\|l(B_1^*)\hat{\mu}_{n,h}+\cdots+l(B_{w_n}^*)\hat{\mu}_{n,h}\|_{\mathscr{F}_\delta}>\frac{\sqrt{n}\epsilon}{2}\right)\to 0 \tag{10}$$

in $\mathbb{P}_\nu$-probability along the sample as $n \to \infty$.

The stochastic equicontinuity of $\mathbb{Z}_n^*$ is implied by (7) and (10). Thus, we can apply bootstrap central limit theorem introduced by [11] which yields the result. □

Remark 2 Note that the reasoning from the proof of the above theorem can be directly applied to the proof of Theorem 2.2 in [22]. In order to show the asymptotic stochastic equicontinuity of the bootstrap version of the empirical process indexed by uniformly bounded class of functions $\mathscr{F}$, we switch from the process $\mathbb{Z}_n^*(f)_{f\in\mathscr{F}} := \sqrt{n^*}\{\mu_{n^*}(f)-\mu_{n_A}(f)\}$, where $n_A = \tau_A(l_n)-\tau_A$ to the process $\mathbb{U}_n^*(f) = \frac{1}{\sqrt{n}}\left[\sum_{i=1}^{1+\left\lfloor\frac{n}{\mathbb{E}_A(\tau_A)}\right\rfloor}\{f(B_i^*)-\mu_{n_A}(f)l(B_i^*)\}\right]$ and the standard probability inequalities applied to the i.i.d. blocks of data yield the result.

In the following, we show that we can weaken the assumption of uniform boundedness imposed on the class $\mathscr{F}$.

Theorem 2 *Suppose that (X_n) is a positive recurrent Harris Markov chain and the assumptions A1, $\mathscr{H}_1(f,\rho,\nu)$, $\mathscr{H}_2(\rho,\nu)$ with $\rho \ge 4$, $\mathscr{H}_3$, $\mathscr{H}_4$, and $\mathscr{H}_5$ are satisfied by (X_n). Assume further, that $\mathscr{F}$ is a permissible class of functions and such that the envelope F satisfies $\mathbb{E}_A\left[\sum_{\tau_A<j\le\tau_A(2)}F(X_j)\right]^2<\infty$. Suppose further, that*

$\int_0^\infty \sqrt{\log N_2(\epsilon, \mathcal{F})}d\epsilon < \infty$. Then, the process

$$\mathbb{Z}_n^* = n_{A_\mathcal{M}}^{*1/2}\left[\frac{1}{n_{A_\mathcal{M}}^*}\sum_{i=1}^{l_n^*-1} f(B_i^*) - \frac{1}{\hat{n}_{A_\mathcal{M}}}\sum_{i=1}^{\hat{l}_n-1} f(\hat{B}_i)\right] \tag{11}$$

converges in probability under $\mathbb{P}_\nu$ to a gaussian process G indexed by $\mathcal{F}$ whose sample paths are bounded and uniformly continuous with respect to the metric $L_2(\mu)$.

Proof The proof of the theorem goes analogously to the proof of Theorem 1 with few natural modifications. Theorem 4.3 from [24] combined with Theorems 3.2 and 3.3 establishes finite dimensional convergence of $\mathbb{Z}_n^*$ to G. It is shown in [24] that $\mathcal{F}$ is totally bounded in $L_2(\mu)$ when $\mathcal{F}$ fulfills only the condition that the envelope F is in $L_2(\mu)$. The proof of stochastic equicontinuity of $\mathbb{Z}_n^*$ follows the same lines as in the proof of Theorem 1.

3 Bootstrapping Fréchet Differentiable Functionals

Robust statistics provides tools to deal with data when we suspect that they include a small proportion of outliers. We denote by $\mathcal{P}$ the set of all probability measures on E and $(\vartheta, \|\cdot\|)$ a separable Banach space. Let $T : \mathcal{P} \to \vartheta$ be a functional on $\mathcal{P}$. If the limit

$$\frac{T((1-t)\mu + t\delta_x) - T(\mu)}{t}, \quad \text{as } t \to 0$$

is finite for all $\mu \in \mathcal{P}$ and for any $x \in E$, then we say that the influence function $T^{(1)} : \mathcal{P} \to \vartheta$ of the functional T is well defined and for all

$$x \in E \ \ T^{(1)}(x, \mu) = \lim_{t\to 0} \frac{T((1-t)\mu + t\delta_x) - T(\mu)}{t}.$$

Let d be some metric on $\mathcal{P}$.

Definition 2 We say that the functional $T : \mathcal{P} \to \mathbb{R}$ is Fréchet differentiable at $\mu_0 \in \mathcal{P}$ for a metric d, if there exists a continuous linear operator DT_{μ_0} and a function $\epsilon^{(1)}(\cdot, \mu_0) : \mathbb{R} \to (\vartheta, \|\cdot\|)$, which is continuous at 0 and $\epsilon^{(1)}(0, \mu_0) = 0$ such that

$$\forall\, \mu \in \mathcal{P}, \ \ T(\mu) - T(\mu_0) = DT_{\mu_0}(\mu - \mu_0) + R^{(1)}(\mu, \mu_0),$$

where $R^{(1)}(\mu, \mu_0) = d(\mu, \mu_0)\epsilon^{(1)}(d(\mu, \mu_0), \mu_0)$. Furthermore, we say that T has an influence function $T^{(1)}(\cdot, \mu_0)$ if for DT_{μ_0}:

$$\forall \mu_0 \in \mathcal{P}, \quad DT_{\mu_0}(\mu - \mu_0) = \int_E T^{(1)}(x, \mu_0)\mu(dx).$$

In the following, we will work with metric $d_{\mathcal{F}}$ defined as

$$d_{\mathcal{F}}(P, Q) := \sup_{h\in\mathcal{F}} \left| \int hd(P - Q) \right|$$

for any $P,\ Q \in \mathcal{P}$. It is noteworthy that the choice of the metric must be done carefully, in this framework we decided to work with the Kolmogorov's distance since, roughly speaking, it ensures a precise control of the remainder term $d(\mu_n, \mu)$.

Theorem 3 *Let $\mathcal{F}$ be a permissible, uniformly bounded class of functions, such that $\int_0^\infty \sqrt{\log N_2(\epsilon, \mathcal{F})}d\epsilon < \infty$. Suppose that $sup_{x\in A_{\mathcal{M}}}\mathbb{E}_x(\tau_{A_{\mathcal{M}}})^{2+\gamma} < \infty,\ \gamma > 0$ (fixed). Assume further, that the conditions of Theorem 1 hold and $T : \mathcal{P} \to \mathbb{R}$ is Fréchet differentiable functional at μ. Then, in general Harris positive recurrent case, we have that $n^{1/2}(T(\mu_n^*) - T(\hat{\mu}_n))$ converges weakly in $l^\infty(\mathcal{F})$ to a gaussian process G_μ indexed by $\mathcal{F}$, whose sample paths are bounded and uniformly continuous with respect to the metric $L_2(\mu)$.*

Remark 3 The above theorem works also in the regenerative case.

Proof Assume that $\mathbb{E}_\mu T^{(1)}(x, \mu) = 0$. It follows directly from the definition of Fréchet differentiability that

$$\sqrt{n}(T(\mu_n^*) - T(\hat{\mu}_n)) = \sqrt{n}\left(DT_{\hat{\mu}_n}(\mu_n^* - \hat{\mu}_n)\right) + \sqrt{n}\left(d_{\mathcal{F}}(\hat{\mu}_n, \mu)\epsilon^{(1)}(d_{\mathcal{F}}(\hat{\mu}_n, \mu), \mu)\right)$$
$$+ \sqrt{n}\left(d_{\mathcal{F}}(\mu_n^*, \mu)\epsilon^{(1)}(d_{\mathcal{F}}(\mu_n^*, \mu), \mu)\right).$$

The distance $d_{\mathcal{F}}(\hat{\mu}_n, \mu) = O_{\mathbb{P}_\nu}(n^{-1/2})$ by Theorem 5.9 from [16]. Next, observe that

$$d_{\mathcal{F}}(\mu_n^*, \mu) \le d_{\mathcal{F}}(\mu_n^*, \hat{\mu}_n) + d_{\mathcal{F}}(\hat{\mu}_n, \mu).$$

From Theorem 1 it is easy to conclude that $d_{\mathcal{F}}(\mu_n^*, \hat{\mu}_n) = O_{\mathbb{P}_\nu}\left(n^{-1/2}\right)$ and thus, $d_{\mathcal{F}}(\mu_n^*, \mu) = O_{\mathbb{P}_\nu}(n^{-1/2})$. Next, we apply Theorem 1 to $\sqrt{n}(T(\hat{\mu}_n) - T(\mu)) = \sqrt{n}(DT_\mu(\hat{\mu}_n - \mu)) + o_{\mathbb{P}_\nu}(1)$. The linear part in the above equation is gaussian as long as

$$0 < \mathbb{E}_\mu T^{(1)}(X_i, \mu)^2 \le C_1(\mu)^2\mathbb{E}_\mu F^2(X) < \infty$$

(see [3]), but that assumption is of course fulfilled since $\mathscr{F}$ is uniformly bounded. Thus, it is easy to deduce that

$$\sqrt{n}(T(\hat{\mu}_n) - T(\mu)) \xrightarrow{L} DT_\mu G_\mu$$

and

$$\sqrt{n}(T(\mu_n^* - T(\mu))) = \sqrt{n}\left[\frac{1}{n^*_{\mathscr{A}\mathscr{M}}}\sum_{i=1}^{n^*_{\mathscr{A}\mathscr{M}}} T^{(1)}(X_i^*, \mu)\right] + o_{\mathbb{P}_\nu}(1)$$

in $l^\infty(\mathscr{F})$. Thus, we have that

$$\sqrt{n}[T(\mu_n^*) - T(\hat{\mu}_n)] \xrightarrow{L} DT_\mu G_\mu$$

and this completes the proof.

Theorem 3 can be easily generalized to the case when $\mathscr{F}$ is unbounded and has an envelope in $L_2(\mu)$.

Theorem 4 *Let $\mathscr{F}$ be a permissible class of functions such that the envelope F satisfies*

$$\mathbb{E}_A\left[\sum_{\tau_A < j \leq \tau_A(2)} F(X_j)\right]^2 < \infty. \tag{12}$$

Suppose that $\int_0^\infty \sqrt{\log N_2(\epsilon, \mathscr{F})}d\epsilon < \infty$. Assume further, that the conditions of Theorem 2 hold and that $T : \mathscr{P} \to \mathbb{R}$ is Fréchet differentiable functional at μ. Then, in general Harris positive recurrent case, we have that $n^{1/2}(T(\mu_n^*) - T(\hat{\mu}_n))$ converges weakly in $l^\infty(\mathscr{F})$ to a gaussian process G_μ indexed by $\mathscr{F}$, whose sample paths are bounded and uniformly continuous with respect to the metric $L_2(\mu)$.

The proof of Theorem 4 follows analogously to the proof of Theorem 3. Apply the results of [24] and Theorem 2 instead one of [16] and Theorem 1 to control the remainder terms. Then, the reasoning goes line by line as in the proof of Theorem 3.

Remark 4 In particular, Theorem 4 is also true in the regenerative case.

Acknowledgements This work was supported by a public grant as part of the Investissement d'avenir, project reference ANR-11-LABX-0056-LMH.

References

1. Athreya, K. B., & Fuh, C. D. (1992). Bootstrapping Markov chains: Countable case. *Journal of Statistical Planning and Inference, 33*, 311–331.
2. Athreya, K. B., & Fuh, C. D. (1993). Central limit theorem for a double array of Harris chains. *Sankhyä: The Indian Journal of Statistics, Series A, 55*, 1–11.
3. Barbe, Ph., & Bertail, P. (1995). *The weighted bootstrap. Lecture notes in statistics* (Vol. 98). New-York: Springer.
4. Bertail, P., & Clémençon, S. (2006a). Regenerative block bootstrap for Markov chains. *Bernoulli, 12*, 689–712.
5. Bertail, P., & Clémençon , S. (2006b). Regeneration-based statistics for Harris recurrent Markov chains. In *Dependence in probability and statistics. Lecture notes in statistics* (Vol. 187). New York: Springer.
6. Bertail, P., & Clémençon, S. (2007). Second-order properties of regeneration-based bootstrap for atomic Markov chains. *Test, 16*, 109–122.
7. Carl Carlstein, E. (1986). The use of subseries values for estimating the variance of a general statistics from a stationary time series. *The Annals of Statistics, 14*, 1171–1179.
8. Ciołek, G. (2016). Bootstrap uniform central limit theorems for Harris recurrent Markov chains. *Electronic Journal of Statistics, 10*(2), 2157–217.
9. Datta, S., & McCormick, W. (1995). Some continuous Edgeworth expansions for Markov chains with applications to bootstrap. *Journal of Multivariate Analysis, 52*, 83–106.
10. Efron, B. (1979). Bootstrap methods: another look at the jackknife. *The Annals of Statistics, 7*, 1–26.
11. Giné, E., & Zinn, J. (1990). Bootstrapping general empirical measures. *Annals of Probability, 18*, 851–869.
12. Hall, P. (1985). Resampling a coverage pattern. *Stochastic Processes and their Applications, 20*, 231–246.
13. Kreissa, J. P., & Paparoditis, E. (2011). Bootstrap methods for dependent data: A review. *Journal of the Korean Statistical Society, 40*(4), 357–378.
14. Kulperger, R. J., & Prakasa Rao, B. L. S. (1989). Bootstrapping a finite state Markov chain. *Sankhyä Series A, 51*, 178–191.
15. Lahiri, S. (2003). *Resampling methods for dependent data*. New York: Springer.
16. Levental, S. (1988). Uniform limit theorems for Harris recurrent Markov chains. *Probability Theory and Related Fields, 80*, 101–118.
17. Liu, R. Y., & Singh, K. (1992). Moving blocks jackknife and bootstrap capture weak dependence. In R. LePage & L. Billard (Eds.), *Exploring the limits of bootstrap* (pp. 225–248). New York: Wiley.
18. Meyn, S., & Tweedie, R. (1996). *Markov chains and stochastic stability*. New York: Springer.
19. Nummelin, E. (1978). A splitting technique for Harris recurrent Markov chains. *Zeitschrift für Wahrscheinlichkeitstheorie und Verwandte Gebiete, 43*, 309–318.
20. Paparoditis, E., & Politis, D. N. (2000). The local bootstrap for kernel estimators under general dependence conditions. *Annals of the Institute of Statistical Mathematics, 52*, 139–159.
21. Paparoditis, E., & Politis, D. N. (2001). A Markovian local resampling scheme for nonparametric estimators in time series analysis. *Econometric Theory, 17*, 540–566.
22. Radulović, D. (2004). Renewal type bootstrap for Markov chains. *Test, 13*(1), 147–192.
23. Rajarshi, M. B. (1990). Bootstrap in Markov sequences based on estimates of transition density. *Annals of the Institute of Statistical Mathematics, 42*, 253–268.
24. Tsai, T. H. (1998). *The Uniform CLT and LIL for Markov Chains*. Ph.D. thesis, University of Wisconsin.

Index

P. Bertail et al. (eds.), *Nonparametric Statistics*, Springer Proceedings in Mathematics & Statistics 250, https://doi.org/10.1007/978-3-319-96941-1

MIX
Papier aus verantwortungsvollen Quellen
Paper from responsible sources
FSC® C105338

If you have any concerns about our products,
you can contact us on
ProductSafety@springernature.com

In case Publisher is established outside the EU,
the EU authorized representative is:
Springer Nature Customer Service Center GmbH
Europaplatz 3, 69115 Heidelberg, Germany

Printed by Libri Plureos GmbH
in Hamburg, Germany